T0139319

MOLECULAR BIOLOGY
INTELLIGENCE
UNIT

Bacterial Integrative Mobile Genetic Elements

Adam P. Roberts, PhD
Peter Mullany, PhD
Department of Microbial Diseases
UCL Eastman Dental Institute
University College London
London, UK

LANDES BIOSCIENCE
AUSTIN, TEXAS
USA

Bacterial Integrative Mobile Genetic Elements

Molecular Biology Intelligence Unit

Landes Bioscience

Printed in the USA.

Please address all inquiries to the publisher:
Landes Bioscience, 1806 Rio Grande, Austin, Texas 78701, USA
Phone: 512/ 637 6050; Fax: 512/ 637 6079
www.landesbioscience.com

The chapters in this book are available in the Madame Curie Bioscience Database.
http://www.landesbioscience.com/curie

ISBN: 978-1-58706-660-3

Library of Congress Cataloging-in-Publication Data

Bacterial integrative mobile genetic elements / [edited by] Adam P. Roberts, Peter Mullany.
    p. ; cm. -- (Molecular biology intelligence unit)
  Includes bibliographical references and index.
  ISBN 978-1-58706-660-3 (alk. paper)
  I. Roberts, Adam P. II. Mullany, Peter, III. Series: Molecular biology intelligence unit (Unnumbered : 2003)
  [DNLM: 1. Interspersed Repetitive Sequences. 2. Conjugation, Genetic. 3. DNA Transposable Elements. 4. DNA, Bacterial--genetics. 5. Genome, Bacterial. QU 58]

  616.9'11042--dc23
            2012039893

Dedication

For little Emma —APR

About the Editors...

ADAM P. ROBERTS is a Senior Lecturer in molecular microbiology at University College London and runs the International Transposon Registry for newly identified bacterial transposons and is the founding Editor-in-Chief of the journal *Mobile Genetic Elements*. His main research interests are the biology of mobile genetic elements, including the antibiotic resistance that is often carried by them and the transcriptional and translational regulation which controls their movement.

About the Editors...

PETER MULLANY is a Professor in molecular microbiology at University College London and has investigated mobile genetic elements and the nosocomial pathogen *Clostridium difficile* for over 30 years. He has pioneered research into using these mobile elements as tools for investigating the molecular biology of the clostridia.

CONTENTS

EDITORS

Adam P. Roberts
Peter Mullany
Department of Microbial Diseases
UCL Eastman Dental Institute
University College London
London, UK
Chapter 9

CONTRIBUTORS

Vicki Adams
Department of Microbiology
School of Biomedical Sciences
Monash University
Victoria, Australia
Chapter 7

Catherine C. Adley
Microbiology Laboratory
Department of Chemical
    and Environmental Sciences
University of Limerick
Limerick, Ireland
Chapter 11

Yvonne Agersø
Department of Microbiology
    and Risk Assessment
National Food Institute
Technical University of Denmark
Lyngby, Denmark
Chapter 9

Melanie B. Berkmen
Department of Chemistry
    and Biochemistry
Suffolk University
Boston, Massachusetts, USA
Chapter 12

E. Fidelma Boyd
Department of Biological Sciences
University of Delaware
Newark, Delaware, USA
Chapter 17

Mathieu Brochet
The Welcome Trust Sanger Institute
Cambridge, UK
Chapter 15

Vincent Burrus
Département de Biologie
Faculté des Sciences
Université de Sherbrooke
Sherbrooke, Québec, Canada
Chapter 13

Piklu Roy Chowdhury
The i3 Institute
University of Technology
Sydney, Australia
Chapter 3

Lena Ciric
Department of Microbial Diseases
UCL Eastman Dental Institute
University College London
London, UK
Chapter 9

Nancy L. Craig
Department of Molecular Biology
    and Genetics
Howard Hughes Medical Institute
Johns Hopkins University
    School of Medicine
Baltimore, Maryland, USA
Chapter 1

Violette Da Cunha
Laboratoire Evolution et Génomique
  Bactériennes
CNRS URA2171
Institut Pasteur
Paris, France
Chapter 15

Emilie Esnault
CNRS UMR8621
Université Paris-Sud
Institut de Génétique et Microbiologie
Orsay, France
Chapter 8

Yoshikazu Furuta
Department of Medical Genome Sciences
Graduate School of Frontier Sciences
and
Institute of Medical Science
University of Tokyo
Tokyo, Japan
Chapter 5

Jeffrey F. Gardner
Department of Microbiology
University of Illinois
Urbana, Illinois, USA
Chapter 14

Geneviève Garriss
Département de Biologie
Faculté des Sciences
Université de Sherbrooke
Sherbrooke, Québec, Canada
Chapter 13

Bridget K. Giarusso
Department of Chemistry
  and Biochemistry
Suffolk University
Boston, Massachusetts, USA
Chapter 12

Philippe Glaser
Laboratoire Evolution et Génomique
  Bactériennes
CNRS URA2171
Institut Pasteur
Paris, France
Chapter 15

J. Peter Gogarten
Department of Molecular
  and Cell Biology
University of Connecticut
Storrs, Connecticut, USA
Chapter 4

Romain Guérillot
Laboratoire Evolution et Génomique
  Bactériennes
CNRS URA2171
Institut Pasteur
Paris, France
Chapter 15

Xiao Han
Department of Infection Control Science
Graduate School of Medicine
Juntendo University
Tokyo, Japan
Chapter 18

Keiichi Hiramatsu
Department of Infection Control Science
Graduate School of Medicine
Juntendo University
Tokyo, Japan
Chapter 18

Teruyo Ito
Department of Infection Control Science
Graduate School of Medicine
Juntendo University
Tokyo, Japan
Chapter 18

Azmiza Jasni
Department of Microbial Diseases
UCL Eastman Dental Institute
University College London
London, UK
Chapter 9

Priscilla A. Johanesen
Department of Microbiology
School of Biomedical Sciences
Monash University
Victoria, Australia
Chapter 7

Ichizo Kobayashi
Department of Medical Genome Sciences
Graduate School of Frontier Sciences
and
Department of Biophysics
and Biochemistry
Graduate School of Science
and
Institute of Medical Science
University of Tokyo
Tokyo, Japan
Chapter 5

Kyoko Kuwahara-Arai
Department of Infection Control Science
Graduate School of Medicine
Juntendo University
Tokyo, Japan
Chapter 18

Maurizio Labbate
The i3 Institute
University of Technology
Sydney, Australia
Chapter 3

Thierry Lambert
Département de Microbiologie
UFR de Pharmacie
Université Paris
Châtenay-Malabry, France
Chapter 10

Stephanie J. Laurer
Department of Chemistry
and Biochemistry
Suffolk University
Boston, Massachusetts, USA
Chapter 12

Zaoping Li
Department of Microbiology
Cornell University
Ithaca, New York, USA
Chapter 1

Dena Lyras
Department of Microbiology
School of Biomedical Sciences
Monash University
Victoria, Australia
Chapter 7

Max Mergeay
Unit of Microbiology
Belgian Nuclear Research Centre
(SCK•CEN)
Mol, Belgium
Chapter 11

Sofia Mindlin
Institute of Molecular Genetics
Russian Academy of Sciences
Moscow, Russia
Chapter 2

Marco Minoia
Department of Fundamental
Microbiology
University of Lausanne
Lausanne, Switzerland
Chapter 16

Ryo Miyazaki
Department of Fundamental
Microbiology
University of Lausanne
Lausanne, Switzerland
Chapter 16

Michael G. Napolitano
Department of Biological Sciences
University of Delaware
Newark, Delaware, USA
Chapter 17

Mai Nguyen
Section of Digestive Diseases
  and Nutrition
University of Illinois
Chicago, Illinois, USA
Chapter 6

J. Tony Pembroke
Molecular Biochemistry Laboratory
Department of Chemical
  and Environmental Sciences
University of Limerick
Limerick, Ireland
Chapter 11

Jean-Luc Pernodet
CNRS UMR8621
Université Paris-Sud
Institut de Génétique et Microbiologie
Orsay, France
Chapter 8

Joseph E. Peters
Department of Microbiology
Cornell University
Ithaca, New York, USA
Chapter 1

Mayya Petrova
Institute of Molecular Genetics
Russian Academy of Sciences
Moscow, Russia
Chapter 2

Nicolas Pradervand
Department of Fundamental
  Microbiology
University of Lausanne
Lausanne, Switzerland
Chapter 16

Alain Raynal
CNRS UMR8621
Université Paris-Sud
Institut de Génétique et Microbiologie
Orsay, France
Chapter 8

Friedrich Reinhard
Department of Fundamental
  Microbiology
University of Lausanne
Lausanne, Switzerland
Chapter 16

Rodrigo Romero
Department of Biology
Suffolk University
Boston, Massachusetts, USA
Chapter 12

Julian I. Rood
Department of Microbiology
School of Biomedical Sciences
Monash University
Victoria, Australia
Chapter 7

Michael P. Ryan
Microbiology Laboratory
Department of Chemical
  and Environmental Sciences
University of Limerick
Limerick, Ireland
Chapter 11

Abigail A. Salyers
Department of Microbiology
University of Illinois
Urbana, Illinois, USA
Chapter 14

Vladimir Sentchilo
Department of Fundamental
  Microbiology
University of Lausanne
Lausanne, Switzerland
Chapter 16

Nadja B. Shoemaker
Department of Microbiology
University of Illinois
Urbana, Illinois, USA
Chapter 14

H.W. Stokes
The i3 Institute
University of Technology
Sydney, Australia
Chapter 3

Sandra Sulser
Department of Fundamental
    Microbiology
University of Lausanne
Lausanne, Switzerland
Chapter 16

Kristen S. Swithers
Department of Molecular
    and Cell Biology
University of Connecticut
Storrs, Connecticut, USA
Chapter 4

Ariane Toussaint
Laboratoire de Bioinformatique
    des Génomes et des Réseaux (BiGRe)
Université Libre de Bruxelles
Brussels, Belgium
Chapter 11

Sae Tsubakishita
Department of Infection Control Science
Graduate School of Medicine
Juntendo University
Tokyo, Japan
Chapter 18

Jan Roelof van der Meer
Department of Fundamental
    Microbiology
University of Lausanne
Lausanne, Switzerland
Chapter 16

Rob Van Houdt
Unit of Microbiology
Belgian Nuclear Research Centre
    (SCK•CEN)
Mol, Belgium
Chapter 11

Gayatri Vedantam
Deptartment of Veterinary Science
    and Microbiology
University of Arizona
and
Southern Arizona VA Healthcare System
Tucson, Arizona, USA
Chapter 6

Lisbeth Elvira de Vries
Department of Microbiology
    and Risk Assessment
National Food Institute
Technical University of Denmark
Lyngby, Denmark
and
Department of Veterinary Disease Biology
University of Copenhagen
Frederiksberg, Denmark
Chapter 9

PREFACE

A s our understanding of mobile genetic elements continues to grow we are gaining a deeper appreciation of their importance in shaping the bacterial genome and in the properties they confer to their bacterial hosts. These include, but are by no means limited to, resistance to antibiotics, and heavy metals, toxin production and increased virulence, production of antibiotics and the ability to utilise a diverse range of metabolic substrates. We are also gaining an understanding of diversity of these elements and their interactions with each other; a property which continually complicates any attempt to classify them. We are learning more about the molecular mechanisms by which they translocate to new genomic sites both within genomes and between different bacteria. This book provides a timely, state of the art update on the properties of an important selection of different bacterial integrative mobile genetic elements and the myriad of different ways in which they move and influence the biology of the host bacterium.

The book begins with a comprehensive chapter we consider a tour de force on the biology of Tn7; a very well-studied transposon capable of transposition by one of two different mechanisms. Next is a chapter describing the transposons which carry resistance genes to mercury illustrating how diverse the elements are which have acquired this resistance. Chapters 3 and 4 deal with integrons, and inteins and introns respectively. The detailed molecular biology described in these two chapters illustrates different translocation mechanisms which have evolved highlighting the way different solutions for the same problem can evolve. Next is an insightful chapter on restriction modification systems as mobile genetic elements. We then move onto the elements capable of not only transposition but also intercellular transfer. Chapters 6 and 7 review the current knowledge of the mobilisable transposons of *Bacteroides* and *Clostrdium* spp. respectively and then there are nine chapters on different families of integrative and conjugative elements (ICEs), also known as conjugative transposons (CTns). The differences between the various ICE families are explored in detail in these chapters. The penultimate chapter is a comprehensive and thought provoking review of genomic islands, and the final chapter describes the SCCmec element form *Staphylococcus aureus* which shows that these integrative elements can have devastating consequences for humans.

Obviously there are many different types of transposable elements that we have not covered in this book since an up to date review exists elsewhere and therefore was not warranted at this time. The chapters are all written by authors who have undertaken pioneering work in their respective fields and we think this book is vital reading for all who are interested in the biology of bacteria and the mobile elements they carry.

Adam P. Roberts, PhD
Peter Mullany, PhD
Department of Microbial Diseases, UCL Eastman Dental Institute
University College London, London, UK

Acknowledgments

On behalf of all the authors of the chapters the editors would like to thank the myriad funding agencies throughout the world that have made this research possible. Without these funds our knowledge of the fundamental processes involved in genomic mobility would be far from what it is today.

CHAPTER 1

Transposon *Tn7*

Zaoping Li,[1] Nancy L. Craig*,[2] and Joseph E. Peters[1]

Abstract

*T*n7 is a bacterial DNA cut and paste transposon that is distinguished by its ability to use several different pathways for target selection. It may insert into a single specific chromosomal site called *attTn7* that provides a "safe haven", insert preferentially into a conjugating plasmid that can facilitate its dispersal through bacterial populations or not insert into these targets because of cis-acting target immunity. This sophisticated choice of target sites is mediated by a complex transposition machine including a novel two-polypeptide transposase, pathway-specific targeting proteins and an ATP-dependent regulator that bridges the donor DNA and the target DNA interactions. *Tn7*-like transposons are widespread in nature and are found in bacterial chromosomes, in some cases forming genomic islands, and also on plasmids.

Introduction

The bacterial transposon *Tn7* is a sophisticated mobile genetic element that is widespread in the environment in highly divergent bacteria. In contrast to most other transposable elements that only recognize a single type of target, *Tn7* mobilizes through several alternate targeting mechanisms that probably account for its wide distribution. In one targeting pathway, *Tn7* transposes at a high frequency into a specific site on the chromosome, the *attTn7* site or attachment site of *Tn7*, which is highly conserved in all bacteria examined to date. Insertion into this site does not have any obvious negative effect on the host and thus this pathway serves to direct *Tn7* to a "safe haven" in the host and likely facilitates the vertical transmission of *Tn7*. In another pathway, *Tn7* insertions occur preferentially into plasmids capable of moving between bacteria called conjugal plasmids. Given that conjugal plasmids can often have a broad host-range, this pathway promotes the dissemination of *Tn7* via horizontal gene transfer. There is also evidence that *Tn7* elements in the environment may have mutations that favor a broader array of insertion sites to help meet the selection environment (see below).

Transposon *Tn7* encodes five proteins for transposition, TnsA, TnsB, TnsC, TnsD, and TnsE (Fig. 1). Of these five proteins, TnsABC+D are required for transposition into the *attTn7* site in the chromosome, while TnsABC+E are required for targeting conjugal plasmids for transposition.[33,110,152] TnsABC is shared by the two pathways and is therefore sometimes called the core transposition machinery. TnsD and TnsE are alternative target site selectors that activate TnsABC and promote insertion into specific preferred target DNAs. No transposition occurs with wild-type TnsABC alone. *Tn7* has another type of regulation that impacts both the TnsD-mediated and TnsE-mediated pathways of transposition; *Tn7* transposition is strongly inhibited from inserting into a target DNA that already has a copy of *Tn7*,[6,33,59] a phenomenon called "target immunity" that is also found with other transposons such as bacteriophage Mu[119] and Tn3.[123]

[1]Department of Microbiology, Cornell University, Ithaca, New York, USA; [2]Howard Hughes Medical Institute, Department of Molecular Biology & Genetics, Johns Hopkins University School of Medicine, Baltimore, Maryland, USA.
*Corresponding Author: Nancy L. Craig—Email: ncraig@jhmi.edu

Bacterial Integrative Mobile Genetic Elements, edited by Adam P. Roberts and Peter Mullany.
©2013 Landes Bioscience.

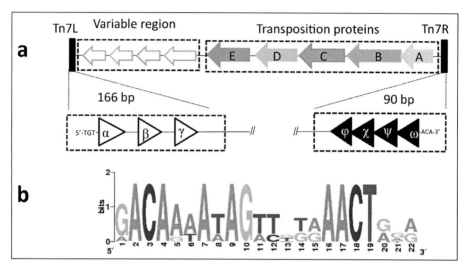

Figure 1. Map of Transposon *Tn7*. a) A schematic representation of *Tn7* and its end structure. The transposition genes are carried in the right side of *Tn7* as an array that is in synteny in *Tn7*-like elements as indicated as A, B, C, D, E for *tnsA, tnsB, tnsC, tnsD,* and *tnsE* in the figure. The variable region contains genes not related to *Tn7* transposition and the composition of this region varies in different *Tn7*-like elements. In the original *Tn7*, antibiotic resistant genes, *dhfr* (trimethoprim resistence), *sat* (streptothricin resistance), and *aadA* (streptomycin and spectinomycin resistance) are found in a defective class 2 integron cassette system. The left (Tn7L) and right (Tn7R) ends are the cis-acting factor for transposition. Both ends contain multiple TnsB homologous binding sites represented as triangles: three discrete sites in Tn7L (α, β, γ from exterior to interior) and four overlapping sites (ω, Ψ, χ, φ from exterior to interior) in Tn7R. b) Sequence conservation of the seven TnsB binding sites of *Tn7*. The sequence logo was generated with the online tool WebLogo[37] after aligning the sequences of the TnsB binding sites by MUSCLE algorithm in UGene software http://ugene.unipro.ru.

At the center of *Tn7* transposition regulation is the ATPase protein TnsC that conveys signals between the target selecting protein and the transposase TnsAB.[33,142] The use of an ATPase as a molecular switch in the control of transposition is also found with other transposons[142] including bacteriophage Mu,[131,155] Tn5090,[111] IS21,[118] to name a few.

Tn7 transposes by a cut-and-paste mechanism where the element is excised from the donor DNA by double-stranded breaks at the ends of the element while being directly joined into the target DNA via the 3'-ends of the element (Fig. 2).[9] Recombination requires the formation of a nucleoprotein complex with all the required transposition proteins, the *Tn7* ends, and the preferred target DNA substrate. One of the hallmarks of *Tn7* transposition is that no DNA breakage and joining is initiated in the absence of any of these components.[9,138] Because the joining of the transposon ends into the target DNA occurs with a stagger of 5 base pairs, subsequent repair of the staggered break in the target DNA by cellular processes generates the 5 base pair (bp) target site duplication flanking the element at the new insertion site, a signature of *Tn7* transposition[33] (Fig. 2).

Tn7 Transposition Functions

The Tn7L and Tn7R Ends are Structurally and Functionally Different

The cis-acting functions for transposition are the left (Tn7L) and right (Tn7R) ends of *Tn7*. These ends contain multiple recognition sites for the TnsB subunit of the TnsAB transposase. Because the sequences recognized by the transposase are in opposite orientation in the left and right ends, these ends are sometimes referred to as the inverted repeat sequences. Like the cis-acting

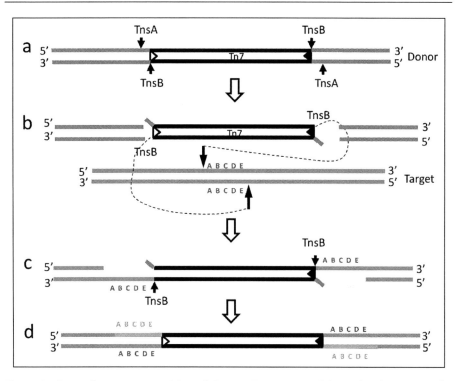

Figure 2. Cut-and-paste transposition of *Tn7*. a) *Tn7* is excised from the donor DNA by double-strand breaks, where TnsB cuts at the 3'-ends of the transposon, but TnsA cleaves 3 bp outside the element in the flanking DNA at the 5'-ends. b-c)TnsB joins the free 3'-OH to 5 bp-staggered positions on the target DNA, generating single insertion products with gaps flanking the 5'-end. d) Repair of the small gaps by cellular processes result in the 5-bp target site duplication characteristic for *Tn7*.

end sequences found in other transposons[38] there are two functional aspects to the ends of the elements: one for transposase binding and the other for signaling the point of DNA cleavage. The very terminal trinucleotide 5'-TGT...ACA-3', which is a highly conserved as the terminal sequence in many transposable elements, is essential for the breakage and joining reaction.[50,149] DNA breakage exposes the 3'-end of the element and the exposed 3'-A-OH is subsequently joined to the 5' end of target DNA[9,50] (Fig. 2). Multiple transposase binding sites are found within each of the transposon ends that are required for *Tn7* transposition. The natural and fully functional *Tn7* ends are rather large[5,7,80,93] (Fig. 1). The Tn7L end is 166 bp with four sites for TnsB binding. The Tn7R end is 90 bp with three sites for TnsB binding.[6] The TnsB binding sites are similar, but not identical, with the spacing of the sequences relative to one another also being important.

Tn7L and Tn7R are also functionally distinct. *Tn7* transposition has a strong orientation bias, in both TnsD-mediated transposition into the *attTn7* site,[9,33,80] and TnsE-mediated transposition into conjugal plasmids and other targets[13-15,80,108,154] (see below). An interesting, but unanswered question is what molecular mechanism accounts for the orientation specificity. The different arrangement of the TnsB binding sites probably dictates the different roles of Tn7L and Tn7R in organizing and assembling the transposition complex. It is important to note that this orientation must be set in the binding of the transposase to the ends and then this signal must also be coordinated with the binding of TnsC to appropriate target complexes. An interesting observation of unknown significance is that miniTn7 elements with two Tn7R ends are competent for transposition in vivo and in vitro, but elements with two Tn7L ends are not.[6,50]

The Core TnsABC Machinery

As noted above transposon *Tn7* encodes five proteins for transposition.[104] This is in contrast to what is found with most other transposons, which more commonly encode one or two proteins for catalyzing transposition. In *Tn7* elements, genes encoding the five transposition proteins are located next to the right end as an array, where all genes are oriented in the same direction with their 5′-end toward the right end of *Tn7* (Fig. 1). Except for the fact that TnsB has homology to proteins of the transposase/retroviral integrase superfamily, the other four proteins each belongs to a unique family that only consists of homologous proteins from other *Tn7*-like elements. The novelty of the Tns proteins and the functional nucleoprotein complex that forms between these proteins and the ends of the element for transposition to occur make *Tn7* a fascinating model for studying protein-DNA interaction, protein-protein interaction, and assembly of multi-subunit molecular machinery.

TnsAB the Transposase

In other transposition systems, the transposase consists of a single type of polypeptide, which binds specifically to the ends of transposon and does all the chemistry required for transposition. However, from early work it was clear that no strand breakage and joining activity was detectable with TnsB alone in the *Tn7* system. Instead, multiple lines of evidence indicated that TnsB+TnsA together form a heteromeric transposase.[18,88,128] Another feature that sets the *Tn7* mechanism apart from other transposable elements is the way the element is cut away from the donor DNA during cut-and-paste transposons. With many DNA transposons the element excises from the flanking donor DNA by double-strand breaks through formation of hairpin structures.[63] In the case of Tn5, Tn10, and *piggyBac*, there is initially a cut at the 3′-end and the exposed 3′-OH then attacks the opposite strand to form a hairpin at each end of the element.[17,33,72,95] Although the excision of the transposon *Hermes*, a member of the *h*AT transposon family, also involves formation of hairpin, the initial DNA breakage is at the 5′-ends of the transposon and the hairpins are formed on the donor DNA. This process is similar to the process performed by the RAG recombinase during the V(D)J recombination process.[158] The Tc1/*mariner* transposons are excised by sequential cleavage of both the 5′- and 3′-ends by the transposase.[44,150] For *Tn7* transposition, the 3′-ends of the element are cleaved by TnsB, but TnsA is responsible for nicking at the 5′-ends.[88,128] The TnsB-mediated process of cleavage at the 3′-ends of the element occurs as a direct joining event to the target DNA as is found with bacteriophage Mu, but in the case of *Tn7* this process only occurs in the presence of TnsA (Fig. 2).[128]

TnsA: a Restriction Enzyme in Recombination

TnsA is responsible for cleavage at the 5′-ends of *Tn7* element. Inactivating the strand cleavage activity of TnsA results in a switch from cut-and-paste to replicative transposition both in vivo and in vitro.[88,128] The N-terminal domain of TnsA is structurally homologous to catalytic domain of Type II restriction endonuclease FokI, which cuts DNA nonspecifically at a fixed distance from its recognition sequence (Fig. 3).[64] Structural comparisons and mutational studies unequivocally identified the active site as a catalytic triad composed of E63, D114, and K132,[64,88,128] which coordinates two Mg^{2+} ions as a cofactor even in the absence of substrate DNA. Residue D114 directly contacts the metal ion and replacing residue D114 with cysteine alters the metal specificity from Mg^{2+} to Mn^{2+}.[64,88,128] The crystal structure of TnsA also revealed a surface in the C-terminus of the protein that likely provides the interface for the interaction with TnsB.[64] Due to the fact that TnsA alone does not possess detectable DNA binding ability,[10] it is likely that positioning of TnsA for catalysis occurs though an interaction with the TnsB proteins that are bound to the ends of the element. In support of this idea, TnsA gain-of-function mutants containing mutations in this region can promote transposition with gain-of-function TnsB mutants in the absence of other transposition proteins (Fig. 3).[83] In addition to its enzymatic activity, TnsA also plays a role in controlling the activity of TnsB, as TnsB-mediated 3′-end joining to the target DNA requires the presence of TnsA suggesting the TnsA and TnsB proteins have evolved together to form a true heteromeric transposase capable of catalysis on both strands of DNA.[128]

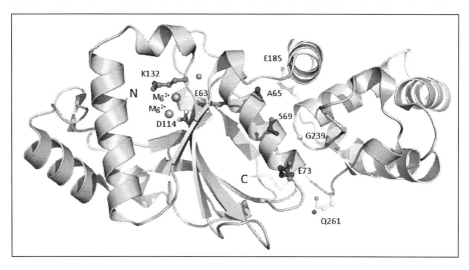

Figure 3. TnsA and its gain-of-function mutations. In the ribbon diagram, the N-terminal catalytic domain of TnsA is shown in blue and the C-terminal domain is in purple. The active site of TnsA consists of E63, D114, and K132. Class I mutations of TnsA (A65, S69, and E73) that allow transposition with wild type TnsB and TnsC are on the same α-helix with E63. Class II mutations of TnsA (E185 and Q261) can promote transposition together with TnsB gain-of-function mutants (TnsB[M366I] and TnsB[A325T]) independent of TnsC. (Image was a kind gift from Alison Hickman, NIH, NIDDK.) A color version of this image is available online at www.landesbioscience.com/curie.

TnsB is a Member of the Large Transposes/Retroviral Integrase Superfamily

TnsB is a member of the transposase/retroviral integrase superfamily that includes transposases of other bacterial transposons and retrotransposons, and retroviral integrases, despite the early finding that only limited sequence similarity can be found between TnsB and other members of this family.[43,58,63,120,121] Three functional domains may be present in TnsB: the N-terminal DNA binding domain, the catalytic domain, and a C-terminal domain for protein-protein interaction with TnsC, like the other recombinases.[121]

The central core region is defined as the part of TnsB that has limited, but significant, sequence similarity with members of the transposase/retroviral integrase superfamily.[33,64,128] The catalytic domain of the transposase/retroviral integrase superfamily has a conserved RNaseH fold, featuring a DD(35)E (DDE) signature motif that coordinates the two divalent metal ions at the active site assisting the various nucleophilic attack reactions involved in DNA cleavage and strand transfer.[63,121] TnsB mutants bearing site-specific mutations of the DDE motif residues are defective in *Tn7* end cleavage and joining activities in vitro and inactive for TnsABC+D and TnsABC+E transposition in vivo.[128] Furthermore, when the TnsB DDE mutants are used for in vitro TnsABC+D transposition specific intermediates accumulate, indicating that multiple active sites are involved in the transposition complex.[128] Other work supports the idea that multiple TnsB proteins are present in the transposition complex (see below).[68]

TnsB is the only Tns protein shown to specifically bind the *Tn7* ends (Fig. 1). TnsB occupies its multiple binding sites on each end in a progressive and sequential manner, indicating that the various binding sites have different apparent affinity to TnsB.[5] The most interior site of Tn7L (γ site) and χ site of Tn7R are occupied first (Fig. 1). However, the sites immediately adjacent to the transposon termini, where DNA breakage and joining occur, are occupied only after the interior sites are bound, likely ensuring no DNA cleavage occurs until the full transposition complex is assembled. The fact that the Φ site of Tn7R is not bound until all the other three sites on Tn7R are filled, together with its dispensable role for transposition in vivo,[6] indicates that it is not essential for assembly of the transposition complex[5](Fig. 1). TnsB binding to its recognition site results in asymmetric DNA bending in regions where it binds weakly, which may play a role in the assembly of the intricate transposition complex.[5] The sequence-specific DNA binding function of TnsB is located at the N-terminus of the protein,[33] which may consist of two subdomains, akin to the composition of the consensus sequence binding domain of MuA by domain Iβγ.[26,29,30,130] However, no canonical DNA binding motif is readily identifiable in this domain of TnsB and exactly how TnsB binds to the TnsB-binding sites at the ends of *Tn7* is yet to be determined.

TnsB alone can bring the transposon ends together to form the so-called "synapsis" structure,[138] which is a key step in transposition initiation. Transposases of Mu, Tn5/Tn10, Mos1 of the Tc1/mariner family and the P element also form synapsed structures for transposition.[4,25,27,40,122,148] Given that these transposase are all members of the transposase/retroviral integrase superfamily, but not close relatives, it may be that there is a common DNA binding and synapsis strategy for transposases in this superfamily. In the well-characterized Mu and Tn5 systems, it has been shown that catalysis occurs in trans, where proteins bound to the left end actually do the chemistry on the right end of the element and vice versa.[4,40,129] It will be interesting to investigate if the TnsB-mediated strand breakage is also catalyzed in trans with *Tn7*. Another interesting question is how many TnsB proteins are in the core machinery. Although seven TnsB proteins can be detected in the post-transposition complex (see below),[68] it is possible that just two TnsB proteins are required for the 3′-end breakage and joining activity, but others are important for transpososome assembly, similar to the roles of multiple MuA proteins bound to the ends of bacteriophage Mu.[26,97,99]

TnsC, an AAA+ Regulator

TnsC is the regulator of *Tn7* transposition that communicates between the target selecting protein TnsD or TnsE and the transposase (Fig. 4).[10,142] Whether or not a potential target is used for transposition is determined by the "state" of TnsC bound on the target DNA. In a productive transposition process, TnsC is likely recruited to the target DNA through interactions with the target DNA-protein (TnsD or TnsE) complex. TnsC then likely recruits the transposase-donor complex to form the transpososome. TnsA may also be recruited via interaction with TnsC to form a ACD-att*Tn7* target complex[68] before the assembly of the transpososome with the TnsB-donor complex. In establishment of target site immunity, TnsC is actively displaced by TnsB from a potential target (see below).

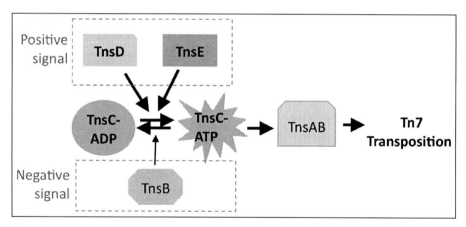

Figure 4. TnsC plays the central role in *Tn7* transposition regulation. TnsC in the ATP-bound state is the active form. TnsD and TnsE recruit TnsC to the corresponding targets allowing active TnsC. By contrast, TnsB inactivates TnsC and displaces it from targets with a preexisting copy of *Tn7*. Active TnsC can interact with the heteromeric transposase TnsAB and activate its DNA-breakage and DNA-joining activities.

TnsC is a Member of the AAA+ Superfamily

The AAA+ (ATPases Associated with a wide variety of cellular Activities) superfamily is characterized by a conserved nucleotide phosphate-binding motif, where a Walker A motif binds the β-gamma phosphate moieties of the bound nucleotide and a Walker B motif coordinates a Mg^{2+} cation at the active site.[57] An AAA domain can be confidently identified in region 128–292 of TnsC by searching major protein domain databases including Interpro,[70] Pfam,[46] SMART,[79] and CDD (Fig. 5a)[86] with the protein sequence. The Walker A motif of TnsC is located at position 136–144 and the Walker B motif at position 228–233, both of these regions are highly conserved throughout TnsC homologs (Fig. 5b). Purified TnsC protein was shown to specifically bind adenine nucleotides (ATP, ADP, AMP, and non-hydrolyzable analogs) and to hydrolyze ATP at a moderate rate with Mg^{2+}.[49,139] ATP or Mg^{2+} seem to have profound effect on the conformation of TnsC: TnsC readily forms insoluble aggregates in low-salt solutions, but both ATP and Mg^{2+} were found to improve the solubility of TnsC and stabilize TnsC in solution.[49]

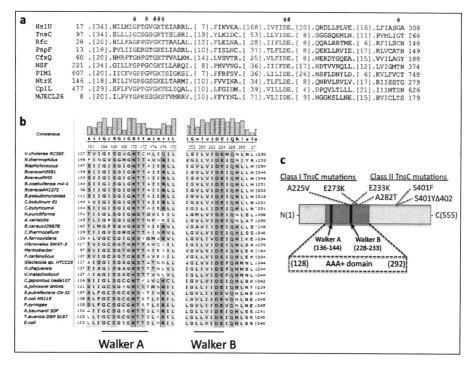

Figure 5. The nucleotide-binding motif in TnsC and TnsC gain-of-function mutations. a) Alignment of TnsC with other members of the AAA+ superfamily. Alignment was generated by CDD and catalytically important residues of Walker A motif (GxxxxGK[S/T], where x is any residue) and Walker B motif (hhhh[D/E], where h is a hydrophobic residue) are marked by #. b) Nucleotide binding motifs are highly conserved in TnsC homologs. Protein sequences of TnsC homologs are named by the hosts of *Tn7*-like elements. Sequences are aligned by Muscle program in UGENE software with Clustal coloring scheme http://ugene.unipro.ru. c) TnsC gain-of-function mutations that allow transposition independent of TnsD or TnsE. Class I mutations (A225V and E273K) are still responsive to signals presented by TnsD or TnsE but class II mutations (E233K, A282T, S401F, and S401YΔD402) are not. Walker A and Walker B motifs and the AAA+ domain are indicated with the positions relative to TnsC from *Tn7*. A color version of this image is available online at www.landesbioscience.com/curie.

ATP-Bound TnsC is the Active Form

The ability of TnsC to bind DNA is a prerequisite for its ability to select a target site. TnsC is an ATP-dependent nonspecific DNA binding protein.[49] No DNA binding is detected in the presence of ADP or AMP or in the absence of adenine nucleotide. However, non-hydrolyzable ATP analogs (AMP-PNP and ATP-γ-S) enhance TnsC DNA binding ability. Under standard in vitro transposition conditions, TnsABC+D transposition requires ATP.[10] In addition, while no transposition is found with TnsABC alone, TnsABC-mediated transposition will occur constitutively in the presence of AMP-PNP in vitro (i.e., without an specific targeting protein or structure).[10] Of further note, gain-of-function TnsC mutants that allow transposition in the absence of TnsD or TnsE have either a reduced ATPase activity or are able to bind DNA with an increased stability (Fig. 5c).[139,141] These findings indicate that the active form of TnsC is in an ATP-bound state and the target complex is able to help switch TnsC to its active form.[10,33,83,141]

TnsC is Implicated in Multiple Interactions

TnsC is believed to play a central regulatory role in transposition as the "matchmaker" or communicator protein conveying signals between the target selecting protein and the transposase. As expected, TnsC is involved in multiple protein-protein interactions in addition to its ability to bind DNA. The C-terminal portion of TnsC (TnsC 504–555) was found to directly interact with TnsA.[83,124,139] The N-terminal domain (TnsC1–293) interacts with TnsD in a yeast two-hybrid assay.[96] TnsC has also been shown to interact with TnsB,[138] but the region responsible for this interaction is yet to be determined. Additionally, TnsC forms a homodimer,[124,139] but the dimerization interface in TnsC again has not been identified.

In theory there could be two regions in TnsC that bind DNA, one for the target DNA and another for the transposon ends. A run of positive residues between 495 and 504 in TnsC have been shown to have an effect on TnsAB-mediated donor DNA cleavage likely through binding the transposon ends.[124] The C-terminus of TnsC also appears to have a known DNA binding motif. 388–407 region of TnsC may contain a Helix-Turn-Helix DNA binding motif (UniProtKB/Swiss-Prot accession P05846)[31] which could be responsible for target DNA binding.

It is interesting to note that the C-terminus of TnsC is poorly conserved but the N-terminus is almost identical among TnsC homologs (Z. Li and J.E. Peters, unpublished observation). This could indicate that interactions in the C-terminus are more structural in nature, or evolving more quickly.

Clues to the molecular mechanism of transposition activation may come from a deletion mutant of TnsC, TnsC$^{\Delta1–293}$, which is active in promoting transposition with TnsAB, but is not responsive to positive signals presented by TnsD or TnsE.[141] This together with other findings discussed above indicates that the C-terminal domain of TnsC is for activating TnsAB, but that the N-terminal domain is functionally interacting with the target complex. One model that accommodates the data are that the N-terminal domain keeps TnsC in an "off" state by masking the C-terminal domain of TnsC; conformational changes induced by ATP binding (ATP binding site in the N-terminal domain) could then reveal the C-terminal domain and allow it to activate the transposase. Crystal structures of this protein alone and/or in combination with its interacting proteins will be important for a detailed understanding of all these protein-protein interactions.

Toggling between the ATP-bound and ADP-bound state (via hydrolysis) would be a mechanism for TnsC to switch between an active form that can stimulate recombination and an inactive form for transposition to produce a target searching behavior for TnsC.[142] In theory stimulating the hydrolysis activity of TnsC would drive TnsC molecules into the "off state" while inhibiting hydrolysis or stimulating nucleotide exchange could drive molecules into the "on state." Such a mechanism is used in bacteriophage Mu transposition where the ATP-bound MuB protein can bind the target DNA and deliver it to the MuA transposase but ADP-bound MuB cannot.[3,131,155] An appealing idea is that target complex may hold TnsC in the ATP-bound state. This idea is supported by the increased ATP binding observed when TnsC is incubated with TnsD and *attTn7*. By contrast, TnsB may stimulate the ATPase activity of TnsC resulting in the ADP-bound conformation of TnsC that cannot bind the target.[33,142]

It is interesting to note that the two classes of TnsC mutations that allow TnsABC core machinery to work with little target specificity appear to be present in naturally occurring TnsC homologs.[106] It would be of interest to test if these mutations could account for the existence of *Tn7*-like elements in locations other than the *attTn7* sites in the chromosome.

Target Selection in the TnsABC+D Transposition Pathway

TnsD is a sequence-specific DNA binding protein that mediates *Tn7* transposition into a specific site, called *attTn7*, on the chromosome at high frequency (Fig. 6). The binding sequence recognized by TnsD is within the 3'-end of the glucosamine-6-phosphate synthase (*glmS*) gene, which is about 25 bp upstream of the actual insertion point in the transcription termination region.[10,151] Insertion into the *attTn7* site has no detectable negative effect on the host.[51] Therefore, the *attTn7* site is considered as the "safe haven" for *Tn7* propagation with host DNA replication.[34,110] While probably not relevant in the environment, when the *attTn7* site is not available, in the laboratory TnsD can direct *Tn7* transposition at a very low frequency into so-called pseudo-*attTn7* sites that show homology to *attTn7*.[74]

Sequence Requirements of attTn7

Within the *attTn7* site region, the region actually bound by TnsD is the only critical determinant for target activity; the sequence of the actual insertion point can be varied without significant effect.[55] Within *attTn7*, TnsD-binding sequence is located at +22 to +55 (relative to the point of insertion) (Fig. 6),[10,92,151] which actually encodes the active site of the GlmS protein (Glucosamine-fructose-6-phosphate aminotransferase) involved in cell wall synthesis.[33,96] GlmS analogs have been found in organisms from bacteria, archae, and eukaryotes. The TnsD-binding sequence is highly conserved in *glmS* genes where variation is limited to the "wobble" positions of each codon.[75,96] Changing any of the conserved sites to the opposite type, i.e purine to pyrimidine or vice versa, affected the binding affinity of TnsD and consequently the frequency of insertion adjacent to the TnsD binding site.[96] Modeling of the target DNA sequence as a B-form DNA revealed that all of the sites that had significant effects on TnsD binding and target activity (+31, +33, +42, +43, +45, +51 and +54) are on one face of the DNA, consistent with the model that TnsD binds the major groove of the target DNA.[33,76] Several nucleotides on the top strand that

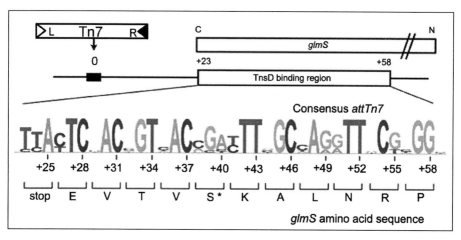

Figure 6. Organization of *attTn7* and its sequence conservation. Within *attTn7*, the central base of the 5 bp sequence duplicated upon *Tn7* insertion is designated as position "0" and base pairs toward the *glmS* gene is in positive numbers. TnsD binding sequence (+23-+55) is in the open reading frame of *glmS* gene and consists of the region that encodes the active site of GlmS. Conservation of the TnsD binding sequence is shown in the sequence logo. Figure is modified from a previous publication.[96]

caused more than 100-fold decrease in transposition when mutated, were found to be in close contact with TnsD protein,[96] indicating an important role of these sites for TnsD binding.

Consistent with the sequence requirement for TnsD-mediated transposition, *Tn7* elements are almost exclusively found at the *attTn7* sites in the chromosomes of divergent bacterial hosts.[34,106,107] TnsD has also been shown to bind the human *glmS* homologs *gfpt-1* and *gfpt-2*,[75] the *Drosophila gfat-1* and *gfat-2*, and the zebrafish *gfpt-1*.[96] In vitro transposition into human *attTn7* homologs was found to be correlated with the binding affinity of TnsD to these sequences, but transposition decreased when this DNA was assembled into nucleosomes in vitro.[75] How the dynamic structure of eukaryotic chromatin might affect *Tn7* targeting is yet to be tested. When studied in *E. coli*, *Tn7* is able to recognize human *gfpt* sequence as an active target for transposition in a site-and-orientation specific manner.[28]

Target DNA Binding by TnsD

Sequence alignment of TnsD homologs revealed a highly conserved N-terminal region (N1–170) containing a $C^{124}C^{127}C^{152}H^{155}$ motif characteristic of a zinc finger.[96] Evidence that this is a zinc-finger like DNA binding motif came from the finding that TnsD binding to *attTn7* is diminished when Zn^{2+} in the buffer is specifically chelated away by 1,10-phenanthroline. In addition, changing any residue of the CCCH motif into a serine abolishes TnsD binding in vitro and also results in about 1,000-fold decrease in TnsABC+D transposition into *attTn7* in vivo. Interestingly, however, the TnsD N-terminus alone does not bind *attTn7*; rather almost the entire protein appears to be required for DNA binding. A screen for missense mutations with a dominant-negative phenotype (i.e., null mutants that prevented transposition activity when in combination with the wild-type protein) in TnsD revealed many mutations over the entire length of TnsD that all affected DNA binding. These results indicate that there are more elements important for DNA binding spread across the protein, in addition to the CCCH zinc binding motif.[96]

TnsD binding to *attTn7* target sequence shows a core region of protection, extending from +30 to +55 and the interaction is primarily within the major groove of the *attTn7* DNA. A second, albeit weaker, protected region is located +22 to +30 where a DNA distortion is imposed upon TnsD-binding. In the TnsC-TnsD-*attTn7* complex the distorted region is covered and the protection extends past the insertion site to position -15.[76] More importantly, in the TnsC-TnsD-*attTn7* complex TnsC occupies the minor groove of DNA at the insertion site, which leaves the major groove accessible for the transposase TnsAB to act,[76] consistent with the 5bp staggered joining of the transposon ends to the target DNA (which would predict that the transposase acts on the major groove). As discussed below, the TnsD-induced distortion is actually the signal that attracts TnsC, whereas interaction between TnsC and TnsD may be important for activation of TnsC.

TnsD Interaction with TnsC

TnsD interacts with TnsC in a yeast two-hybrid analysis where, presumably, a solution interaction is occurring as opposed to an interaction on the yeast chromosome at the site homologous to *attTn7* (to which TnsD does not bind[75]). C-terminal truncated TnsD (TnsD 1–309) binds TnsC better than full-length TnsD, suggesting some level of regulation in the reaction between TnsD and TnsC. The investigators were unable to define a smaller interaction domain; a greater truncation of TnsD (TnsD 1–293) abolishes its ability to bind TnsC and deleting several residues from the N-terminus of TnsD also results in a polypeptide unable to bind TnsC. Together these results suggest that the domain for TnsD to interact with TnsC is located in the N-terminus.

Host Factors in TnsD-Mediated Transposition

It is also important to note that host factors also participate in TnsD-mediated transposition.[133] Two host factors, L29, a component of the 50S ribosomal subunit, and ACP, acyl carrier protein essential for fatty-acid biosynthesis, were found to stimulate TnsD binding to the target DNA in vitro. The two proteins function together to enhance the apparent affinity of TnsD for *attTn7* by more than 20-fold. ACP and L29 also stimulate TnsABCD transposition more than 3-fold in vitro, despite that the TnsABC+D in vitro transposition system has been highly optimized. More importantly, one of these host factors seems to be critical for TnsABC+D transposition in vivo; mutating L29 specifically decreased TnsABC+D transposition by more than 100-fold. ACP could not be tested as easily becasue of its essential role in the cell. Given the important role of L29 and ACP in cellular metabolism, this may provide a way of regulating TnsD-mediated transposition according to the cellular conditions.

Target Selection in the TnsABC+E Transposition Pathway

While TnsABC+D pathway is exceptional at choosing a specific site for efficient transposition via a DNA sequence, TnsE-mediated transposition targets sites without sequence similarity.[33,110,152] Instead, TnsE recognizes a complex that is frequently available in discontinuous DNA replication (Fig. 7).[105] TnsABC+E pathway preferentially inserts into actively conjugating plasmids,[154] and filamentous phage M13,[45] thereby facilitating horizontal transfer of Tn7 among bacteria.

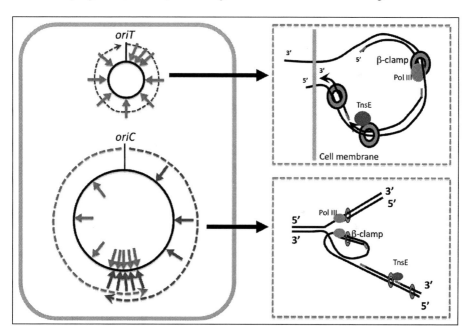

Figure 7. TnsE-mediated transposition targets lagging strand DNA replication for transposition. *Tn7* transposition events mediated by TnsE into both conjugal plasmids (upper left) and the chromosome (bottom left) are all oriented in one left-to-right orientation with respect to the lagging strand DNA replication of the relevant replicon (right panels). Within each DNA, TnsE-mediated transposition has a preference for insertion into the leading region during conjugation and into the region where DNA replication terminates in the chromosome. Two factors important for this targeting are a 3′-recessed ends and the β clamp processivity factor. On the left panel, each arrow represents an idealized distribution of TnsE-mediated transposition event and blue and green arrows indicate opposite orientations. Position of *oriT* is labeled on the circular plasmid DNA showing the direction of DNA replication of the element as a purple dashed line. On the circular chromosome, bidirectional replication from *oriC* origin is indicated as blue and green dashed lines for each repichore (see text). Schematic representation of the DNA replication of the conjugal plasmid in the recipient cell (upper) and the chromosome (lower) is shown in the right, with all of the important proteins labeled (see text). A color version of this figure is available online at www.landesbioscience.com/curie.

TnsE-Mediated Transposition Targets Lagging Strand DNA Replication

TnsE-mediated transposition is stimulated many fold in the presence of a conjugating plasmid. In addition, these insertions preferentially target the conjugal plasmid; greater than 90% of transposition events occur into the plasmid, even though the plasmid only represents less than 5% of the cellular DNA.[154] Early work indicated that transposition events occurred in one orientation in conjugal plasmids and it was later shown that these transposition events occurred as the element entered a new bacterial host[14-16,154] (Fig. 7).

At a very low frequency, TnsE-mediated transposition also targets the chromosome, and *Tn7* insertions in the chromosome occur in opposite orientations in the two replichores of the circular chromosome (Fig. 7).[108,109] The *E. coli* chromosome is replicated using a single origin (*oriC*), which initiates replication in both directions. Replication forks meet at a position generally equidistant from the origin. The two regions of the chromosome that are replicated by separate replisomes are called replichores.[134] Interestingly when TnsE-mediated insertions in the chromosome were mapped, a striking pattern emerged. While there was a clear bias for the region where DNA replication terminates, it was also found that all of the insertions occurred in the same orientation within each replichore (An idealized distribution is shown in Fig. 7). Together with the orientation bias found with insertions into conjugal plasmids in the recipient cells, it could be concluded that TnsE-mediated transposition occurs in one orientation with the DNA strand undergoing lagging-strand replication (Fig. 7).[108,110] Interestingly, transposons unrelated to Tn7, Tn917 and IS903, also have been shown to preferentially target where DNA replication terminates.[135,147] However, because an orientation bias cannot be established with Tn917 and IS903 it is unclear if these elements also share an attraction for lagging strand DNA replication.

Factors Important for TnsE Targeting of Lagging-Strand DNA Replication

What feature of lagging strand DNA synthesis attracts TnsE-mediated transposition events? TnsE is a DNA-binding protein which has a specific preference for DNA structures with 3'-recessed ends. TnsE mutants that promote higher transposition frequency (up to a 1000-fold increase) in vivo bind structures with 3'-recessed ends much better than the wild type protein,[108] indicating that the ability of TnsE to bind 3'-recessed DNA is involved in target site selection. A specific affinity for 3'-recessed DNA ends was intriguing because DNA ends would be expected to be overrepresented during discontinuous DNA synthesis, whereas recessed ends would not generally be expected during leading strand synthesis. Moreover, when progress of the lagging strand DNA synthesis is blocked by various impediments, the DNA polymerase would simply abandon the ongoing DNA synthesis, but take on primers downstream to resume DNA replication.[65,90,94] DNA structures with 3'-recessed end or ssDNA gap would be generated in these processes. The possibility of DNA structures with 3'-recessed ends as a bio-active target for TnsE-mediated transposition was directly tested in vitro.[105] The in vitro system was set up with purified proteins, a donor plasmid (containing a miniTn7 element but with a conditional replication origin) and a gapped DNA structure on a circular plasmid to simulate structures with 3'-recessed ends found in vivo. This system allowed monitoring of the in vitro transposition events by in vivo output from transformation of the deproteinized in vitro reaction mix. *Tn7* transposition was found to occur only with a gapped DNA structure and transposition depended on TnsE. The fact that no transposition was detectable with nicked DNA substrate is consistent with the model that TnsE targeting requires 3'-recessed ends. However, transposition events occurred randomly all over the gapped plasmid in both orientations, indicating that something else was involved in determining the polarity of a potential target in addition to the 5'-3' polarity of DNA.

TnsE interaction with the β clamp processivity factor of DNA polymerase III holoenzyme proved to be the other important factor for TnsE targeting. β clamp confers DNA replication processivity by tethering DNA polymerases to the template DNA. Like the 3'-recessed DNA structures or gaps on DNA, β clamp is another factor that accumulates on the DNA strand that undergoes lagging strand DNA synthesis.[143,157] For leading strand DNA synthesis, in theory, the β clamp only has to be loaded once and the DNA replicase will be able to finish replicating the full length of the replichore with the same processivity factor. By contrast, a new β clamp is loaded for every Okazaki fragment during lagging strand DNA synthesis. While β clamps are eventually recycled in the process, there is likely a period of time when these clamps are available to interact with other proteins. In fact, many proteins involved in processing the newly-synthesized Okazaki fragments have been shown to use the clamps left behind as a mobile platform for their various functions.[71,82] Proteins suggested or shown to interact with the β clamp include DNA polymerase I (which is involved in removing the 5'-primer of a previous Okazaki fragment while filling the gap between two Okazaki fragments), DNA ligase (which seals two adjacent Okazaki fragments), and MutS and MutL (which may use the clamp to differentiate nascent DNA strand for mismatch repair). These and other β-binding proteins all interact with the asymmetric ring-shaped β clamp primarily via a β-binding motif that binds specifically to a hydrophobic pocket at the C-terminal face of the β ring.[39] TnsE also contains such a β-binding motif at the N-terminus, which is found in a highly conserved region in the TnsE homologs (Fig. 8). Substitution any of the conserved residues

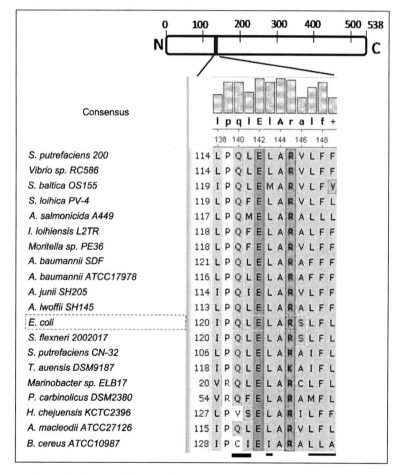

Figure 8. β clamp binding motif in TnsE homologs. TnsE protein found in *E. coli* is represented as a bar on the top with the amino (N) and carboxy (C) termini indicated. TnsE homologs from hosts indicated are aligned with MUSCLE program in UGENE software (http://ugene. unipro.ru) and the region between residues 120 and 131 that contains the β clamp binding motif is shown. Sequence conservation and the consensus sequences are shown above the alignment. Underlined residues are changed individually or in combination to study their importance in β clamp binding.

with alanine residues reduces or abolishes TnsE-mediated transposition in vivo. A weak interaction between TnsE and the β clamp was detected by yeast two-hybrid assay, protein mobility-shift assay, and far-western assay, and quantified by Surface Plasmon Resonance (SPR) analysis. The β-binding motif is important for this interaction, as judged from the observation that when TnsE proteins with the alanine replacements in the β clamp-binding motif were used in the assays the interaction sharply decreased. It is interesting to note that the TnsE-β clamp interaction does not appear to interrupt the normal traffic on the β clamp, as only when extremely high levels of TnsE are induced do you find an SOS response in the cell.[105] Under these conditions of very high TnsE overexpression it could be shown that SOS was induced because of the interaction with β clamp. Proof of an important role for the β clamp came from the TnsE-reconstituted transposition system. While gapped DNA substrates proved to be productive targets for in vitro TnsE-mediated transposition, unexpectedly these insertions occurred in two orientations. However, when the gapped substrate was populated with β clamps a striking change in the transposition profile was observed. In the substrates populated with β clamps, transposition events were found to occur in predominantly only one orientation, with the right end of *Tn7* close to the 3'-end of the single-strand gap as occurs in vivo. In addition, a hotspot of insertion was observed on the gapped plasmid, recapitulating what was observed in vivo with insertions on the chromosome.[105] The successful reconstitution of an in vitro transposition system indicates the primary target of TnsE-mediated transposition is the complex of the β clamp at the primer/template junction. Exactly how TnsE gains access to the clamp in vivo and where it binds on the clamp are intriguing questions to be answered with further research. TnsE may need more signals from the replisome to target DNA replication given the finding that in vitro TnsE-mediated transposition was only found when strong gain-of-activity mutants of TnsE were used in the assay.

The fact that lagging-strand DNA synthesis during conjugation is the most preferred target for TnsE-mediated transposition may indicate that the uncoupled DNA replication fork is more accessible to TnsE (Fig. 7). For chromosomal DNA replication, the replicases for leading-strand and lagging-strand DNA synthesis are coupled by the τ subunit of the DNA polymerase III holoenzyme.[89,103] However, such an elaborate coordination cannot be expected for leading strand and lagging strand DNA synthesis of the conjugating plasmid, since they occur in different cells, the donor cell and the recipient cell, respectively. In an uncoupled DNA replication fork, the DNA replicase may dissociate more frequently from the β-clamp; thereby allowing the gain of access of TnsE.

TnsE-mediated transposition into both conjugal plasmids and the *E. coli* chromosome also display a preference for specific regions, the leading region and the terminus region, respectively (Fig. 7). This could be explained by the persistence of gap structures in these regions during the final stage of replication (Fig. 9). Chromosomal DNA replication terminates upon the convergence of the replication forks in the terminus region, a process facilitated by the many *ter* (termination) sites in this region.[42,66] Circular duplex DNA with a short gap in the terminus region of the nascent strand has been observed after resolution of the converging forks[2,23,62,145] and DNA polymerase I-mediated repair-like processes are implicated in closing the gaps.[2,23,87] Such a gap structure may also be generated by re-circularizing the transferred strand upon the completion of conjugation.[153] TnsE could efficiently use such structures by directly competing with DNA polymerase I.

Other DNA Replication Processes Targeted by TnsE

TnsE is able to target replicating filamentous bacteriophages for transposition.[45] Insertion into the bacteriophage M13 also occurred almost exclusively in a single left-to-right orientation with respect to the replication of the minus strand of the phage genome, which would presumably occur in a leading-strand-like manner. Its targeting by TnsE-mediated transposition suggests that it is somewhat different from the replication of the conjugal plasmid in the donor cell.

TnsE-mediated transposition is stimulated by induction of double strand breaks (DSBs) in the chromosome.[109] This is true for the direct induction of DSBs in the chromosome or with DSBs associated with various treatments, like exposure to UV light, mitomycin C, or phleomycin.[136]

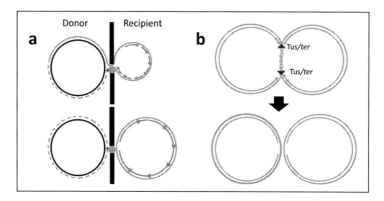

Figure 9. Persistence of gap structures may lead to regional preference of TnsE-mediated transposition into both plasmid and chromosome. a) Plasmid conjugation and concurrent DNA synthesis in the donor and recipient cells. The direction of DNA replication is represented by the arrow heads. The transferred strand is indicated by the blue circle and the nascent strand is indicated by the dashed lines in blue in the donor and green recipient cells, respectively. The relaxasome that mediates plasmid conjugation is shown as the blue oval. Upon the completion of conjugation, the transferred strand is re-circularized by the relaxasome and a gap structure would be generated as indicated in the figure. b) Resolution of the converging forks results in short gaps in the nascent strands of daughter chromosomes. The template strands are indicated by the blue circles and the nascent strands are indicated by the green circles, with arrow heads indicating the direction of replication. The terminus region is indicated by the Tus/*ter* complexes. A color version of this image is available online at www. landesbioscience.com/curie.

Interestingly, following the induction of a single DSB in the *E. coli* chromosome, *Tn7* insertions do not occur immediately at the site of the DSB, but instead across a region proximal to the point of DNA break.[109] Of further interest is that transposition occurs at a series of hotspots hundreds of kilobases away from the original break site.[136] The orientation of transposition events in this system suggests that TnsE targets the repair-associated replication. The absence of such hotspots in cells without DSB induction suggests that the repair-associated replication machinery is different from that of chromosomal DNA replication and may be particularly vulnerable for DNA replication roadblocks. Further studies, which would have implications for the mechanism of replication-mediated repair, are needed to test this model.

Why Target Lagging-Strand DNA Replication?

Lagging-strand DNA synthesis has subsequently been found to be sensitive to multiple forms of recombination. Transposons of the IS200/IS605 family and Group II introns,[101] which mobilize by very different mechanisms, and bacteriophage lambda RED recombination,[32] all seem to prefer DNA undergoing lagging strand DNA replication as their target for recombination. These observations suggest that recombination into the DNA strand undergoing lagging strand DNA replication may be advantageous or that DNA undergoing lagging strand replication may naturally be more vulnerable to recombination. Support for the latter idea comes from the finding that in many bacterial species small oligonucleotides can be integrated into the lagging stand even in the absence of recombination proteins.[146]

Recognition of a Positive Target Signal by TnsC

How does TnsC recognize a potential target for transposition? Why does insertion occur into some sites but not into others? The fact that virtually no transposition occurs with only the core TnsABC machinery in vivo indicates an important role of TnsD or TnsE in activating this core machinery and probably recruiting the core machinery to the potential target site. However, isolation of TnsC gain-of-function mutants that allow transposition in the absence of TnsD and TnsE indicates that the role of the target site selecting proteins is linked through the transposition regulator protein TnsC.[141] As discussed before, gain-of-function mutants of TnsC have either slower ATPase activity or altered DNA binding ability,[139] both of which may keep TnsC at the active state.

Important insight into how TnsC locates a potential target came from work with the mutant core machinery, TnsABC^{A225V}, where the mutant TnsC^{A225V} is used in combination with wild type transposase TnsAB. In in vitro reactions using duplex DNA targets, transposition events promoted by the mutant core machinery occurred in both orientations with very low target site specificity; transposition events dispersed throughout the target plasmid and no apparent sequence similarity was found among the insertion sites or sequences flanking the target sites.[19,116,117] This picture was completely changed when transposition reaction were performed with a plasmid to which was attached a pyrimidine triplex-forming DNA oligonucleotide.[85,116,117] More than 70% of the transposition events examined located at a specific small region near the triplex region. Of additional interest, 98% of the insertions that occurred in the hotspot immediately adjacent to the triplex occurred in one orientation with the right end of *Tn7* proximal to the triplex. This site- and orientation-specificity of TnsABC^{A225V}-promoted transposition into a triplex-containing DNA substrate is comparable to that observed with the TnsABC+D pathway (Fig. 10). Interestingly, TnsC was also found to bind preferentially to triplex containing DNAs,[116,117] similar to the specific

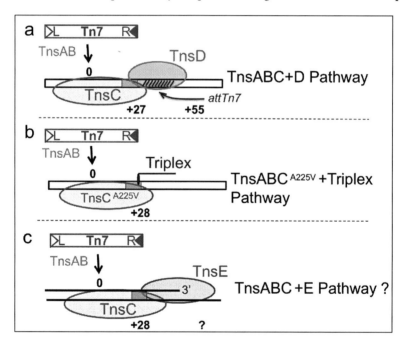

Figure 10. Activation of TnsC by various target complexes. Comparison of TnsD-mediated transposition with those of TnsABC mutant machinery indicates that TnsC recognizes a distortion on the target DNA. Both TnsD binding to *attTn7* and formation of triplex DNA induce conformational change in the duplex DNA, which is indicated by the shaded box. The relative position and orientation of *Tn7* insertions to the DNA distortion found in both TnsABC+D and TnsABC^{A225V} transposition system emphasize the critical role of target DNA structure in *Tn7* transposition. Also, it appears that protein-protein interaction of TnsC N-terminus with TnsD is important for activating TnsC. Since TnsABC+E pathway shares the core machinery with the TnsABC+D pathway, it is likely that the same mechanism is involved in TnsABC+E transposition. Interaction of TnsC with the target selecting protein(s) is indicated by the overlapping of the representing circles.

binding of TnsC to the TnsD-*attTn7* complex. As discussed above, binding of TnsD imposes a distortion in a region of the *attTn7* target DNA. Annealing of a triplex forming oligomer to its duplex target DNA also results in conformational changes at the triplex-duplex junction. The ability of TnsC to bind both triplex-containing DNA and TnsD-*attTn7* target complex and the relative position of *Tn7* insertions to the DNA distortions on both kinds of DNA substrates indicate that a DNA-based signal may play a central role in target site recognition by TnsC (Fig. 10).

It is important to note that DNA distortion alone is not sufficient to activate wild type TnsC. A triplex DNA substrate could not stimulate wild type TnsABC-mediated transposition. This, together with the presence of TnsD in the post-transposition complex (see below), indicates that the role of TnsD is not simply an "assembly" factor that helps recruit TnsC by inducing a conformational change on DNA. Direct protein-protein interaction between TnsC and TnsD may be important for activating TnsC. The two signals, TnsC-TnsD and TnsC-distortions interactions may help make sure that activation occurs only in the targeting complex (Fig. 10).

How TnsC is attracted to the target complex in TnsE-mediated transposition pathway is not yet clear. Given that TnsABC+D and TnsABC+E pathways of *Tn7* transposition share the same core machinery, one would expect that TnsC would recognize the same kind of signals to locate potential targets in TnsE-mediated transposition (Fig. 10). It remains to be determined if TnsE binding to target DNA induces conformational change. However, there seems to be evidence for TnsE and TnsC functional interaction in controlling the target selection in TnsE-mediated transposition (Q. Shi and J. E. Peters, unpublished data).

Target Immunity

Not only does *Tn7* transposition respond to positive signals presented by the target site selecting proteins TnsD and TnsE, *Tn7* transposition is also sensitive to a process called target immunity in both the TnsD- and TnsE-mediated pathways, in which the presence of a transposon copy in a target DNA inhibits further insertion of the element into that target DNA.[6,59] A similar process occurs with other transposable elements, such as Tn3-like elements[53,54,78,100,123] and bacteriophage Mu.[119] Target immunity with *Tn7* was found in vivo both on a plasmid or chromosome and has also been established in vitro.[6,9,41,137,140] The degree of inhibition seems inversely related to the distance between a potential target site and the preexisting copy of *Tn7*,[6,41] indicating that immunity is a local and not a global effect. A small (60 kb) derivative of the *E. coli* F plasmid was rendered inactive for transposition when the plasmid contains *Tn7* end sequences.[6]

The basis of the immunity signal is the TnsB binding sites found in the ends of *Tn7*[140]: both the right end and left end of *Tn7* can impose immunity on a target DNA and a single TnsB binding site is effective at imposing immunity when studied in vitro, indicating that the other features of *Tn7* ends are not important for target immunity. It is also worth noting that not all TnsB-binding sites are equally effective in imposing target immunity, which is probably due to the different binding affinities of TnsB to these sites, as Tn7R41 containing the weakest TnsB-binding site ω does not confer target site immunity in vivo[6] or in vitro.[9]

The establishment of target immunity appears to depend on a TnsB-TnsC interaction that triggers displacement of TnsC from *Tn7*-containing targets, a process accompanied by the increased ATPase activity of TnsC. Without hydrolysis, immunity is bypassed and *Tn7* readily inserts into target DNA with a preexisting copy of *Tn7*.[140] In TnsABC+D transposition in vitro, target immunity was completely abolished when non-hydrolyzable ATP analogs were used.[10] Conceivably, the loss of target immunity observed with the TnsC[E233K] mutant is also because of the loss of ATPase activity, as the mutation is at a critical residue of the Walker B motif and mutating this residue has been shown to abolish the ATPase activity in other ATPases.[57]

The ability of *Tn7* to evaluate potential targets based on their distance from a pre-existing copy likely provides another mechanism that facilitates *Tn7* survival and dispersion. This is because target immunity will prevent potentially hazardous events, such as insertion of *Tn7* into itself or at close vicinity, which will disrupt the element directly or as a result of homologous recombination, but favor the dissemination of elements to distant sites within a DNA molecule or to different DNA molecules. While discouraging transposition within or near the element is a benefit of target site immunity for all transposons, there is yet another benefit for target site immunity for *Tn7*. Because very high efficiency transposition is found with insertions into the *attTn7* site in the TnsD-pathway, multiple insertions would continue to accumulate very quickly without the immunity conferring process operating. Therefore, target site immunity can be thought of as having a second and more important benefit beyond that found with other transposons in essentially turning off the TnsD-mediated insertion pathway once an insertion resides at the *attTn7* site.

Expansion of the *Tn7* Family by Virtue of Vertical and Horizontal Transposition

It is perhaps not surprising that *Tn7*-like elements are widespread in the environment given its exceptional level of control. *Tn7* elements with transposition proteins identical to the original *Tn7* isolate from *E. coli* have been found in different bacterial species isolated from clinical settings. Some of these elements have different antibiotic resistant genes in the class II integron embedded in the variable region of *Tn7*.[20,112-115] More importantly, *Tn7*-like elements, all of which have the *tnsABCDE* genes in synteny, have been found in a wide variety of bacterial hosts from diverse environments. So far, more than 50 *Tn7*-like elements have been found in genomes of members of γ-proteobacteria, δ-proteobacteria, low-G+C gram-positive firmicutes, and cyanobacteria.[106] Most of the elements reside in the equivalent *attTn7* sites in these hosts, but elements in plasmids and non-*attTn7* chromosomal locations are also found, indicating the occurrence of transpositions mediated by both pathways in nature.[106,107] A trend of decreased presence of *Tn7* in plasmids, but increased presence in the chromosome has been reported in trimethoprim-resistant *E. coli*,[73] suggesting a continuous transposition of *Tn7*-like elements into its "safe haven" in the nature. Transposition of *Tn7*-like elements into the *attTn7* site has been shown to lead to the formation of genomic islands (Fig. 11), as clearly demonstrated in *Shewanella* genus[107] and *Acinetobacter baumannii*,[125] due to the vast capacity of the variable region in *Tn7* to carry genes of different functions and potential recombination sequences for other mobile genetic elements. While at first it might seem surprising to find examples in the environment where multiple *Tn7*-like elements reside in the same location because of target immunity (see above), divergence of the end sequences and TnsB likely accounts for this phenomenon. The presence of non-identical *Tn7*-like elements in tandem in the same locus is probably because the TnsB homolog of one *Tn7*-like element is not able to recognize the transposon end sequences of another, thus obscuring the potent target site immunity found with *Tn7*.[107] Comparison of the transposition protein homologs from *Tn7*-like elements may provide clues for studying the structure-and-function of these proteins, as has been demonstrated in study of TnsB,[137] TnsD,[96] and TnsE.[105] It is interesting to note that many gain-of-function mutations isolated in the laboratory have been found in naturally occurring transposition protein homologs,[106] which may suggest that *Tn7*-like elements might be more active. While one would expect the broad distribution of a *Tn7*-like element in various hosts (given the ability of TnsE-mediated transposition to target mobile plasmids), phylogenetic analysis of the TnsABCD proteins actually suggests that *Tn7*-like elements have limited distribution outside a limited phylogenetic distribution of bacteria.[106] Why *Tn7*-like elements might show some level of species or genus specific adaptation will require further research, but may be related to a need for transposon-encoded proteins to adapt to certain host factors.

Internal Networking of the Core Machinery

Interaction between TnsA and TnsB

Functional studies have indicated that activities of TnsA and TnsB are inter-dependent. Although TnsA and TnsB have distinct roles in the chemical steps in transposition and their cleavage activities can be separated using inactivated TnsA or TnsB, no transposon end breaking or joining of the free transposon end to the target DNA is detectable in the absence of either

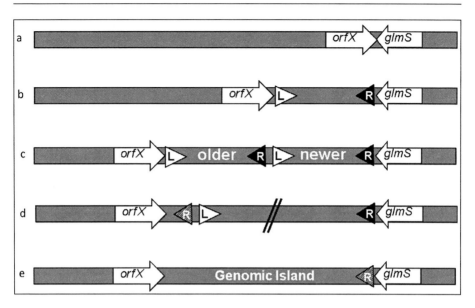

Figure 11. *Tn7*-like elements mediate formation of genomic island. A *Tn7*-like element is bounded by the left and right ends (open and filled triangles). A sequence of events may lead to the formation of genomic island(s) in the attachment site, which may eventually lack of any recognizable feature of Tn7. a) Empty *attTn7* sites can be identified adjacent to the *glmS* gene. b) Insertion of one *Tn7*-like element into the *attTn7* site of would bring its variable region possibly with other mobile genetic elements and potential recombination sequences for integrons, bacteriophages, or other transposons. c) Accumulation of multiple *Tn7*-like elements (C) in the attachment site would promote homologous recombination between these elements. d) Over time elements are inactivated by deletions, insertions and point mutations, which would potentially inactivate the elements. e) Inactivated elements are always subject to further reductive evolution and eventually only the highly selected components may remain.

one of the other proteins.[128] However, under modified in vitro reaction conditions, i.e., high glycerol and Mn^{2+}, wild-type TnsAB can promote double-stranded breakage at *Tn7* ends and intra-molecular joining where one end of transposon is joined to the other end instead of other DNA molecules.[18] TnsA mutants with mutations in the C-terminal domain, in combination with TnsB mutants containing mutations in positions close to the active-site residues, can promote intermolecular transposition independently.[83] As TnsA alone cannot bind the ends of *Tn7*,[10] an interaction between TnsA and TnsB provides a mechanism for TnsB to recruit TnsA to *Tn7* ends and for TnsA to modulate activity of TnsB.

Interaction between TnsA and TnsC

Direct interaction between TnsA and TnsC has been demonstrated by pull-down assays,[139] gel filtration,[124] and protein protease footprinting analysis and crystallographic.[83,124] TnsA and TnsC forms 2:2 complex in solution, probably via the dimerization of TnsC since TnsA appears to be a monomer while TnsC is dimer in solution. The last 52 residues of TnsC(504–555) are necessary and sufficient for a TnsA interaction.[124] In the co-crystal structure of TnsA/TnsC(504–555) (Fig. 12),[124] the TnsC 504–555 fragment is in an extended conformation covering a large, exposed hydrophobic surface on TnsA (1400Å), consistent with the increased stability of TnsA/C complex compared with TnsA alone. The N-terminal domain of TnsA interacts with the TnsC fragment.[83,124] All of the class I gain-of-function mutations of TnsA ((A65V, S69N, and E73K)),[83] which allow activation of TnsAB transposase by TnsC in the absence of TnsD (or TnsE), are located in the TnsA/C interface (Figs. 3 and 12).

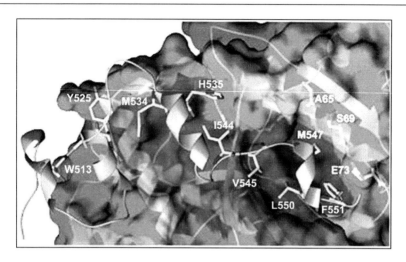

Figure 12. TnsC C-terminal peptide on the hydrophobic surface of TnsA. Surface-exposed hydrophobic residues of TnsA is shown in green and polar or charged residues are in gray. The main chain of TnsC C-terminal peptide (504–555) is shown in gold. TnsC residues involved in hydrophobic interaction with TnsA are labeled. Peach, red, blue, and yellow represent carbon, oxygen, nitrogen, and sulfur, respectively. Gain-of-function TnsA mutations (A65, S69, and E73) are shown with white carbon atoms. Reprinted from reference 124. A color version of this image is available online at www.landesbioscience.com/curie.

TnsC(495–555) appears to be able to stimulate the donor DNA cleavage activity of TnsAB in conditions with high concentration of glycerol and Mn^{2+},[124] which, as we discussed above, requires formation of the complete transposition complex under normal conditions.[9] However, when important residues in the region of TnsA for TnsAC interaction are substituted with alanines, the resulting TnsA mutants (TnsA[L70A/E71A/W72A] and TnsA[D78A/E81A]) are not active in transposition or forming TnsACD-*attTn7* complex.[83] The TnsAC interaction may be also critical for formation of pre-transposition complex and channeling a TnsC-mediated transposition signal (see below).

Interaction between TnsB and TnsC

Interaction between TnsB and TnsC determines whether a target complex is formed with TnsC that is productive or abortive, which can be further influenced by TnsA (see below). A productive interaction between TnsB and TnsC is required for assembly of a pre-transposition complex (see below) that is a prerequisite in *Tn7* transposition.[138] A negative interaction between the proteins is required for target site immunity in that interaction of TnsB with TnsC can promote TnsC dissociation, as described above.[140] Interestingly, the same TnsBC interaction is involved in transposition and target immunity and the region of TnsB that mediates this interaction is located at the C-terminal end (Fig. 13).[137,138] C-terminal truncated TnsB is not active in transposition even though it is able to bind *Tn7* ends and mutating important residues in the C-terminal domain of TnsB abolishes the formation of pre-transposition complex.[138] TnsB mutants that allow compromised target immunity bear mutations in this C-terminal region of TnsB and high concentrations of TnsB C-terminal peptide can impose target immunity in vitro.[137]

The apparent paradox involving productive and inhibitory interactions between the same region of TnsB and TnsC is resolved by the role of TnsA. TnsA stabilizes TnsBCD-*attTn7* complex[138] and also inhibit the TnsB-provoked dissociation of TnsC from a potential target,[137] suggesting a role of TnsA in channeling TnsBC interaction toward productive transposition. The TnsAC interaction, and probably the TnsAB interaction, described above must be important for this function of TnsA. Interplay among protein-protein interactions in the TnsABC core machinery is therefore essential in the regulation of transposition.

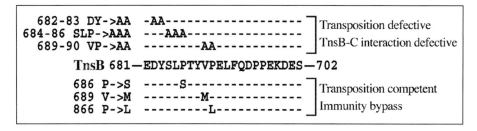

Figure 13. TnsB C-terminus is involved in interaction with TnsC and mutations that affect target immunity and transposition overlap. The TnsB immunity bypass mutations (top) were localized to the C-terminus of TnsB (middle). Substitution of several residues with alanine in the same region blocks TnsB–TnsC interaction and transposition (bottom).

The relationship between TnsB and TnsC is somewhat similar to the relationship between the MuA and MuB of bacteriophage Mu.[26] For Mu transposition, the MuA transposase binds multiple sites at the ends of the Mu element and mediates 3′-end cleavage and joining for transposition, while MuB is the ATP-dependent protein that binds to the target DNA and activates MuA for transposition. Like the effect of TnsB on TnsC in the *Tn7* system, MuA is able to impose target immunity by provoking the ATPase activity of MuB, thereby dissociating MuB from immune targets that already have a copy of Mu. Like TnsB, the same C-terminal domain of MuA is involved in interaction with MuB that leads to either transposition or target immunity. However, the decision whether or not an insertion will occur into a potential target for phage Mu seems to depend on the different interaction between MuA and MuB without input from an addition targeting protein like TnsD or TnsE.

Assembly of the Transpososome for TnsABC+D Transposition

At the heart of Tn7 transposition regulation is the assembly of the transposition complex or transpososome with all of the transposition proteins, an appropriate target DNA, and the transposon ends in the donor DNA. Multiple protein-DNA and protein-protein interactions are required to form such an active complex. For TnsABC+D transposition, assembly of the transposition complex is a prerequisite; no recombination intermediates or products can be produced in the absence of any components of the full reconstitution.[10,88,128] This is different from other transposons like Tn5 and Tn10, where excision of the transposon occurs before the engagement of the appropriate target site.[52,127] For bacteriophage Mu transposition, excision can occur either before or after the engagement of target DNA (although this flexibility may be an in vitro "artifact").[35]

Pre-Transposition Complex

Transposition complexes can be formed in vitro which do not carry out transposition using Ca^{2+} in the reaction; transposition complexes can be further stabilized through the use of a cross-linking reagent, which allow the identification of pre-transposition complexes in TnsABC+D transposition as slow-migration complexes in the native gel electrophoresis (Fig. 14). Like many other transposases,[121,126,144] TnsB alone can bring together the left- and right- ends of *Tn7* to form a complex.[138] Higher order nucleoprotein complex was formed with TnsABCD and this complex was active in performing cut-and-paste transposition when isolated and incubated with Mg^{2+}, indicating that this complex (TnsABCD-DNA complex) is a relevant transposition intermediate. Formation of the pre-transposition complex requires that both ends of *Tn7* on the same molecule is mediated through the interaction between TnsB and TnsC.[138]

TnsA may play a critical role in efficient assembly of the pre-transposition complex, even though it is not necessary for the formation of donor-target complex, as a nucleoprotein complex containing both the donor and target DNA can form with just TnsBCD (TnsBCD-DNA complex).

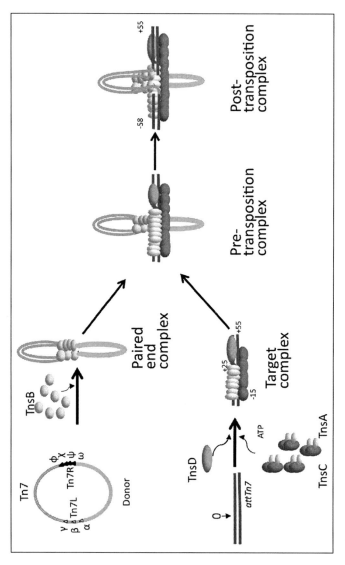

Figure 14. Illustration of TnsABC+D transposition pathway at the transpososome level. Tn7 is represented by green lines with open or filled triangles representing the TnsB-binding sites at the ends of the Tn7. TnsB (cyan) binds and brings together Tn7L and Tn7R to form a paired end complex. TnsD (blue) binds the target *attTn7* (purple lines) and induces a conformational change on the DNA duplex (oblique lines), which recruits the TnsA (yellow)-TnsC (red) complexes to form a target complex. TnsA+TnsB+TnsC interactions bring the paired-end complex together with the target complex to form the pre-transposition complex. The nucleoprotein complex is then actived for the cleavage reactions that remove flanking donor DNA (thick grey lines) and join the transposon ends to the target DNA. In the stable post-transposition complex, all TnsABC+D proteins are present, as well as the TnsD-induced distortion on the target DNA. Protein complex in the post-transposition complex makes extensive interactions with the Tn7::*attTn7* transposition product, especially on the target DNA where a large region left to the insertion point is strikingly protected. The boundaries of protein or protein complexes on the target DNA are indicated in the figure (see text). A color version of this figure is available online at www.landesbioscience.com/curie.

First of all, TnsA is able to bind to the TnsCD-*attTn7* complex and form a higher order complex, the TnsACD-*attTn7*.[83] Furthermore, the amount of TnsABCD-DNA complex is much more than TnsBCD-DNA complex and the former withstands temperature challenge while the latter dissociates rapidly.[138] These experiments suggest the biologically relevant pre-transpositional complex includes TnsA, TnsB, TnsC, TnsD and the DNAs.

The Post-Transposition Complex

Unlike the pre-transposition complexes that are not stable without DNA cross-linking, post-transposition complex, the nucleoprotein complex present after in vitro *Tn7* transposition (Fig. 14) can be isolated following multiple cycles of washing and gel-filtration in the absence of a cross-linking reagent,[68] suggesting that the nucleoprotein complexes involved in transposition become increasingly stable as the transposition process proceeds. This is probably a result of progressive conformational changes of the complex that drive the reactions forward, as has been found with bacteriophage Mu[56,98,156] and also Tn10.[36] Analysis of the post-transposition complex by atomic force microscopy revealed that the *Tn7* element is disconnected from the flanking donor DNA and the *Tn7* ends are held together as a single protein complex. In the post-transposition complex, the DNA product of transposition in which *Tn7* is covalently linked to target *attTn7* is bound all the time by the transposition proteins and only upon deproteinization could the unlinked Tn7 ends be present separately.

The post-transposition complex and products were also analyzed in a non-denaturing gel[68] by exploiting TnsA, TnsB, TnsC, and TnsD proteins individually labeled with a short fluorescent peptide and fluorescently or radiochemically labeled DNA substrates. TnsA, TnsB, TnsC, TnsD are all present in the post-transposition complex. The presence of TnsD, together with the preservation of TnsD-induced distortion on the target DNA in the post-transposition complex, indicates that TnsD does not function only as an assembly factor to recruit TnsC.

Labeling both the transposition proteins and DNA substrates allowed quantitative investigation of the composition of the complex.[68] Since it is very likely that TnsD binds *attTn7* target as a monomer given the asymmetric nature of *attTn7*, the ratio of TnsD to target *attTn7* DNA was set to 1:1, and the abundance of the other transposition proteins was estimated accordingly. Notably TnsB/*attTn7* ratio was 6.3 fold of TnsD/*attTn7*, interestingly close to the TnsB site/ *attTn7* ratio of 7/1, indicating that TnsB likely binds as monomer. Much higher amounts of TnsA and TnsC were present, TnsA/*attTn7* was about 18-fold and TnsC/*attTn7* was about 25. Given that TnsA and TnsC form $TnsA_2$:$TnsC_2$ complexes,[124] it seems likely that some of TnsA and TnsC exist in this form, although some TnsA has dissociated from the post-transposition complex in the course of analysis. It is also possible that a portion of TnsC proteins interact with TnsB, perhaps as 1:1 TnsBC complexes that are distinct from the $TnsA_2$:$TnsC_2$ complexes. Although these observations about Tns protein ratios strongly support the model that proteins of the core transposition machinery function as oligomers, future higher-resolution analysis will be required to unambiguously to determine the oligomeric state of the Tns proteins in the core machinery.

The Tns proteins make extensive interactions with the DNA in the post-transposition complex as revealed by footprinting studies (Fig. 14).[68] All seven TnsB binding sites in the complex are occupied, which is consistent with the observed ratio of 6.3 TnsB/*attTn7*. In comparison to the protection pattern of *Tn7* ends bound by TnsB alone, more protection is found in the post-transposition complex for the terminal α site of Tn7L and Ψ and ω sites of Tn7R; the extreme ends of *Tn7* are also protected, consistent with the presence of TnsA for 5'-end cleavage in the complex and the engagement of *Tn7* ends in the transposition reaction. On the *attTn7* target DNA, TnsD-induced distortion is still evident. Perhaps most intriguingly, a region of 114 bp (-58 - +55) of what was the target DNA is protected, extending well beyond the regions protected by TnsD (+25 - +55)[10] or TnsCD (-15 - + 55).[76] It is not known whether the same number of TnsC molecules is present in the TnsCD-*attTn7* complex and the post-transposition complex, or binding with TnsA and/or TnsB increases the oligomerization of TnsC on target DNA. Since the sequence of DNA left to the insertion point is not important for TnsD targeting,[76] protection of ~60 bp by the protein complex in this region is striking. It may be that this region is bound by TnsA$_2$C$_2$ complex and this asymmetry in Tns-target interaction might determine the orientation specificity of TnsD-mediated transposition.[68]

The highly stabile post-transposition complex imposes an obstacle for cellular repair of the gaps flanking the 5'-ends of *Tn7* in the final transposition product.[68] Repair of the gaps is readily observed with DNA polymerase I and DNA ligase but removal of the protein complex is a prerequisite, indicating that the proteins in the complex block the access of repair enzymes to the gaps. Crude cell lysate can actively disassemble the protein complex and allow repair. This is similar to the disassembly of phage Mu transpososome: Strand transfer in the transposition process creates a stable complex tightly bound by MuA; Disassembly of the transpososome by a series of yet-to-be-defined host proteins, including chaperone ClpX, is required before initiation of phage DNA replication.[1,22,102]

Tn7 as a Genetic Tool

Transposable elements have been an important tool for modifying genes and genomes. The multiple well-defined transposition systems described above make *Tn7* an excellent genetic tool for these kinds of applications.

First, the orientation-specific transposition of *Tn7* by TnsABC+D pathway to a highly conserved site in the chromosome provides a reliable method for targeted DNA delivery. The high frequency of this pathway, about 1–10% of the target inserted without any selection,[41] guarantees the successful isolation of positive clones. TnsABC+D transposition-based cloning has proved to be an efficient and straightforward method for site-specific insertion of desired DNA fragments into the chromosome.[12,69,91] The ability to isolate positive clones and stable maintenance of the cloned genes in the absence of antibiotic selection makes this method particularly useful for efforts to make strains for vaccine production or environmental release.[67,91] In experiments where physiological expression levels of proteins are desired, introducing a single copy of the gene into the chromosome would be advantageous over cloning the same gene on a multi-copy plasmid.[11,91] This site-specific transposition pathway of *Tn7* has also been used for transfer of large DNA fragments into different replicons,[33,47,84] which has great potential in synthetic genomes. It is not unreasonable to suggest that the ability of TnsD-mediated transposition to target eukaryotic *attTn7* sites may eventually be exploited as a tool for gene therapy, avoiding the common mutagenic effects of most current gene therapy vectors.[75]

The development of a highly efficient in vitro transposition system with mutant core machinery, TnsABC^{A225V}, which has little target site specificity, makes *Tn7* transposition an ideal tool for random mutagenesis.[19] Together with the availability of multiple miniTn7 derivatives that allow transcriptional and translational fusions, TnsABC^{A225V} in vitro transposition have been widely used in making mutation libraries of bacteria,[12,21] yeast,[8,24,77,132] and fungi.[81] Furthermore, transposon mutagenesis in vitro is more effective than in vivo mutagenesis as it avoids potential bias imposed by in vivo specific features. in vitro transposition mutagenesis is also an appealing choice for organisms that lack readily tractable genetic systems.

Dissection of TnsE-mediated transposition can provide unique insights into many aspects of DNA replication and repair. TnsE-mediated transposition targets β clamp at the primer/template junction for transposition.[105] Since the β clamp is known to bind numerous proteins of important cellular functions, traffic on the clamp should be tightly controlled.[71] Therefore, it is of particular interest to understand when and where such a complex would be available for *Tn7* transposition. Given that the leading strand DNA replication in fact is also discontinuous,[48,60,61] it is intriguing to ask why TnsE-mediated only targets the lagging-strand DNA synthesis for transposition or under which condition we can isolate transposition events targeting the leading strand DNA synthesis. Another yet to be answered question involves asking how repair-associated DNA replication attracts TnsE-mediated transposition and why regional hotspots are found with the induction of double-stranded breaks. Understanding the exceptional ability of *Tn7* to sense the differences in leading strand vs. lagging strand DNA synthesis and repair-associated replication vs. normal DNA replication will help elucidate the replication fork structure involved in these processes.

Conclusion

Tn7 is a sophisticated genetic element that has evolved to carefully select among multiple targets for transposition. As such, *Tn7* is found globally as an important vector for the horizontal transfer of genetic information playing a role in bacterial evolution through the formation of genomic islands. The use of multiple proteins to execute transposition in each of two transposition pathways distinguishes it from other transposable elements. The study of *Tn7* transposition will also continue to provide valuable insight into how protein-protein and protein-DNA interactions are involved in the assembly of elaborate nucleoprotein complexes. Such interactions are critical in other processes like DNA replication, repair, and other forms of recombination. Since most of the *Tn7*-encoded proteins form their own unique families, the study of *Tn7* will also help understand a large number of proteins present in nature. Furthermore, the targeting of a fundamental cellular process in the TnsABC+E pathway suggests it should be useful in gaining an understanding of basic replication and repair processes in the cell.

Acknowledgments

Work in the lab of JEP is funded by the National Institutes of Health (RO1GM069508). NLC is an Investigator of the Howard Hughes Medical Institute. Work in the NLC lab is funded in part by the National Institutes of Health (RO1GM076425).

References

1. Abdelhakim AH, Sauer RT, Baker TA. The AAA+ ClpX machine unfolds a keystone subunit to remodel the Mu transpososome. Proc Natl Acad Sci USA 2010; 107:2437-42; PMID:20133746; http://dx.doi.org/10.1073/pnas.0910905106.
2. Adams DE, Shekhtman EM, Zechiedrich EL, et al. The role of topoisomerase IV in partitioning bacterial replicons and the structure of catenated intermediates in DNA replication. Cell 1992; 71:277-88; PMID:1330320; http://dx.doi.org/10.1016/0092-8674(92)90356-H.
3. Adzuma K, Mizuuchi K. Target immunity of Mu transposition reflects a differential distribution of Mu B protein. Cell 1988; 53:257-66; PMID:2965985; http://dx.doi.org/10.1016/0092-8674(88)90387-X.
4. Aldaz H, Schuster E, Baker TA. The interwoven architecture of the Mu transposase couples DNA synapsis to catalysis. Cell 1996; 85:257-69; PMID:8612278; http://dx.doi.org/10.1016/S0092-8674(00)81102-2.
5. Arciszewska LK, Craig NL. Interaction of the Tn7-encoded transposition protein TnsB with the ends of the transposon. Nucleic Acids Res 1991; 19:5021-9; PMID:1656385; http://dx.doi.org/10.1093/nar/19.18.5021.
6. Arciszewska LK, Drake D, Craig NL. Transposon Tn7. cis-Acting sequences in transposition and transposition immunity. J Mol Biol 1989; 207:35-52; PMID:2544738; http://dx.doi.org/10.1016/0022-2836(89)90439-7.
7. Arciszewska LK, McKown RL, Craig NL. Purification of TnsB, a transposition protein that binds to the ends of Tn7. J Biol Chem 1991; 266:21736-44; PMID:1657979.
8. Bachman N, Biery MC, Boeke JD, Craig NL. Tn7-mediated mutagenesis of Saccharomyces cerevisiae genomic DNA. Methods Enzymol 2002; 350:230-47; PMID:12073315; http://dx.doi.org/10.1016/S0076-6879(02)50966-6.
9. Bainton R, Gamas P, Craig NL. Tn7 transposition in vitro proceeds through an excised transposon intermediate generated by staggered breaks in DNA. Cell 1991; 65:805-16; PMID:1645619; http://dx.doi.org/10.1016/0092-8674(91)90388-F.
10. Bainton RJ, Kubo KM, Feng JN, Craig NL. Tn7 transposition: target DNA recognition is mediated by multiple Tn7-encoded proteins in a purified in vitro system. Cell 1993; 72:931-43; PMID:8384534; http://dx.doi.org/10.1016/0092-8674(93)90581-A.
11. Bao Y, Lies DP, Fu H, Roberts GP. An improved Tn7-based system for the single-copy insertion of cloned genes into chromosomes of Gram-negative bacteria. Gene 1991; 109:167-8; PMID:1661697; http://dx.doi.org/10.1016/0378-1119(91)90604-A.
12. Barnes RJ, Leung KT, Schraft H, Ulanova M. Chromosomal gfp labelling of Pseudomonas aeruginosa using a mini-Tn7 transposon: application for studies of bacteria-host interactions. Can J Microbiol 2008; 54:48-57; PMID:18388971; http://dx.doi.org/10.1139/W07-118.
13. Barth PT, Datta N. Two naturally occurring transposons indistinguishable from Tn7. J Gen Microbiol 1977; 102:129-34; PMID:915473.

14. Barth PT, Datta N, Hedges RW, Grinter NJ. Transposition of a deoxyribonucleic acid sequence encoding trimethoprim and streptomycin resistances from R483 to other replicons. J Bacteriol 1976; 125:800-10; PMID:767328.
15. Barth PT, Grinter NJ. Map of plasmid RP4 derived by insertion of transposon C. J Mol Biol 1977; 113:455-74; PMID:328901; http://dx.doi.org/10.1016/0022-2836(77)90233-9.
16. Barth PT, Grinter NJ, Bradley DE. Conjugal transfer system of plasmid RP4: analysis by transposon 7 insertion. J Bacteriol 1978; 133:43-52; PMID:338595.
17. Bhasin A, Goryshin IY, Reznikoff WS. Hairpin formation in Tn5 transposition. J Biol Chem 1999; 274:37021-9; PMID:10601258; http://dx.doi.org/10.1074/jbc.274.52.37021.
18. Biery MC, Lopata M, Craig NL. A minimal system for Tn7 transposition: the transposon-encoded proteins TnsA and TnsB can execute DNA breakage and joining reactions that generate circularized Tn7 species. J Mol Biol 2000; 297:25-37; PMID:10704304; http://dx.doi.org/10.1006/jmbi.2000.3558.
19. Biery MC, Stewart FJ, Stellwagen AE, et al. A simple in vitro Tn7-based transposition system with low target site selectivity for genome and gene analysis. Nucleic Acids Res 2000; 28:1067-77; PMID:10666445; http://dx.doi.org/10.1093/nar/28.5.1067.
20. Biskri L, Mazel D. Erythromycin esterase gene ere(A) is located in a functional gene cassette in an unusual class 2 integron. Antimicrob Agents Chemother 2003; 47:3326-31; PMID:14506050; http://dx.doi.org/10.1128/AAC.47.10.3326-3331.2003.
21. Bordi C, Butcher BG, Shi Q, et al. In vitro mutagenesis of Bacillus subtilis by using a modified Tn7 transposon with an outward-facing inducible promoter. Appl Environ Microbiol 2008; 74:3419-25; PMID:18408063; http://dx.doi.org/10.1128/AEM.00476-08.
22. Burton BM, Baker TA. Mu transpososome architecture ensures that unfolding by ClpX or proteolysis by ClpXP remodels but does not destroy the complex. Chem Biol 2003; 10:463-72; PMID:12770828; http://dx.doi.org/10.1016/S1074-5521(03)00102-9.
23. Bussiere DE, Bastia D. Termination of DNA replication of bacterial and plasmid chromosomes. Mol Microbiol 1999; 31:1611-8; PMID:10209736; http://dx.doi.org/10.1046/j.1365-2958.1999.01287.x.
24. Castano I, Kaur R, Pan S, et al. Tn7-based genome-wide random insertional mutagenesis of Candida glabrata. Genome Res 2003; 13:905-15; PMID:12695329; http://dx.doi.org/10.1101/gr.848203.
25. Chaconas G, Lavoie BD, Watson MA. DNA transposition: jumping gene machine, some assembly required. Curr Biol 1996; 6:817-20; PMID:8805293; http://dx.doi.org/10.1016/S0960-9822(02)00603-6.
26. Chaconas GH. Harshey RM. Transposition of phage Mu DNA. In: Craig NL, Craigie R, Gellert M, Lambowitz AM, eds. Mobile DNA II. Washington, DC: ASM Press, 2002:384-402.
27. Chalmers RM, Kleckner N. IS10/Tn10 transposition efficiently accommodates diverse transposon end configurations. EMBO J 1996; 15:5112-22; PMID:8890185.
28. Cleaver SH, Wickstrom E. Transposon Tn7 gene insertion into an evolutionarily conserved human homolog of Escherichia coli attTn7. Gene 2000; 254:37-44; PMID:10974534; http://dx.doi.org/10.1016/S0378-1119(00)00283-3.
29. Clubb RT, Omichinski JG, Savilahti H, et al. A novel class of winged helix-turn-helix protein: the DNA-binding domain of Mu transposase. Structure 1994; 2:1041-8; PMID:7881904; http://dx.doi.org/10.1016/S0969-2126(94)00107-3.
30. Clubb RT, Schumacher S, Mizuuchi K, et al. Solution structure of the I gamma subdomain of the Mu end DNA-binding domain of phage Mu transposase. J Mol Biol 1997; 273:19-25; PMID:9367742; http://dx.doi.org/10.1006/jmbi.1997.1312.
31. Consortium. The Universal Protein Resource (UniProt) in 2010. Nucleic Acids Res 2010; 38:D142-8; PMID:19843607; http://dx.doi.org/10.1093/nar/gkp846.
32. Court DL, Sawitzke JA, Thomason LC. Genetic engineering using homologous recombination. Annu Rev Genet 2002; 36:361-88; PMID:12429697; http://dx.doi.org/10.1146/annurev.genet.36.061102.093104.
33. Craig NL. Tn7. In: Craig NL, Craigie R, Gellert M, Lambowitz AM, eds. Mobile DNA II. Washington, DC: ASM Press, 2002:423-456.
34. Craig NL. Transposon Tn7. Curr Top Microbiol Immunol 1996; 204:27-48; PMID:8556868.
35. Craigie R, Mizuuchi K. Transposition of Mu DNA: joining of Mu to target DNA can be uncoupled from cleavage at the ends of Mu. Cell 1987; 51:493-501; PMID:2822259; http://dx.doi.org/10.1016/0092-8674(87)90645-3.
36. Crellin P, Chalmers R. Protein-DNA contacts and conformational changes in the Tn10 transpososome during assembly and activation for cleavage. EMBO J 2001; 20:3882-91; PMID:11447129; http://dx.doi.org/10.1093/emboj/20.14.3882.
37. Crooks GE, Hon G, Chandonia JM, Brenner SE. WebLogo: a sequence logo generator. Genome Res 2004; 14:1188-90; PMID:15173120; http://dx.doi.org/10.1101/gr.849004.
38. Curcio MJ, Derbyshire KM. The outs and ins of transposition: from mu to kangaroo. Nat Rev Mol Cell Biol 2003; 4:865-77; PMID:14682279; http://dx.doi.org/10.1038/nrm1241.

39. Dalrymple BP, Kongsuwan K, Wijffels G, et al. A universal protein-protein interaction motif in the eubacterial DNA replication and repair systems. Proc Natl Acad Sci USA 2001; 98:11627-32; PMID:11573000; http://dx.doi.org/10.1073/pnas.191384398.

40. Davies DR, Goryshin IY, Reznikoff WS, Rayment I. Three-dimensional structure of the Tn5 synaptic complex transposition intermediate. Science 2000; 289:77-85; PMID:10884228; http://dx.doi.org/10.1126/science.289.5476.77.

41. DeBoy RT, Craig NL. Tn7 transposition as a probe of cis interactions between widely separated (190 kilobases apart) DNA sites in the Escherichia coli chromosome. J Bacteriol 1996; 178:6184-91; PMID:8892817.

42. Duggin IG, Wake RG, Bell SD, Hill TM. The replication fork trap and termination of chromosome replication. Mol Microbiol 2008; 70:1323-33; PMID:19019156; http://dx.doi.org/10.1111/j.1365-2958.2008.06500.x.

43. Dyda F, Hickman AB, Jenkins TM, et al. Crystal structure of the catalytic domain of HIV-1 integrase: similarity to other polynucleotidyl transferases. Science 1994; 266:1981-6; PMID:7801124; http://dx.doi.org/10.1126/science.7801124.

44. Feng X, Colloms SD. In vitro transposition of ISY100, a bacterial insertion sequence belonging to the Tc1/mariner family. Mol Microbiol 2007; 65:1432-43; PMID:17680987; http://dx.doi.org/10.1111/j.1365-2958.2007.05842.x.

45. Finn JA, Parks AR, Peters JE. Transposon Tn7 directs transposition into the genome of filamentous bacteriophage M13 using the element-encoded TnsE protein. J Bacteriol 2007; 189:9122-5; PMID:17921297; http://dx.doi.org/10.1128/JB.01451-07.

46. Finn RD, Mistry J, Tate J, et al. The Pfam protein families database. Nucleic Acids Res 2010; 38:D211-22; PMID:19920124; http://dx.doi.org/10.1093/nar/gkp985.

47. Frengen E, Weichenhan D, Zhao B, et al. A modular, positive selection bacterial artificial chromosome vector with multiple cloning sites. Genomics 1999; 58:250-3; PMID:10373322; http://dx.doi.org/10.1006/geno.1998.5693.

48. Gabbai CB, Marians KJ. Recruitment to stalled replication forks of the PriA DNA helicase and replisome-loading activities is essential for survival. DNA Repair (Amst) 2010; 9:202-9; PMID:20097140; http://dx.doi.org/10.1016/j.dnarep.2009.12.009.

49. Gamas P, Craig NL. Purification and characterization of TnsC, a Tn7 transposition protein that binds ATP and DNA. Nucleic Acids Res 1992; 20:2525-32; PMID:1317955; http://dx.doi.org/10.1093/nar/20.10.2525.

50. Gary PA, Biery MC, Bainton RJ, et al. Multiple DNA processing reactions underlie Tn7 transposition. J Mol Biol 1996; 257:301-16; PMID:8609625; http://dx.doi.org/10.1006/jmbi.1996.0164.

51. Gay NJ, Tybulewicz VL, Walker JE. Insertion of transposon Tn7 into the Escherichia coli glmS transcriptional terminator. Biochem J 1986; 234:111-7; PMID:3010949.

52. Goryshin IY, Reznikoff WS. Tn5 in vitro transposition. J Biol Chem 1998; 273:7367-74; PMID:9516433; http://dx.doi.org/10.1074/jbc.273.13.7367.

53. Goto N, Shoji A, Horiuchi S, Nakaya R. Conduction of nonconjugative plasmids by F' lac is not necessarily associated with transposition of the gamma delta sequence. J Bacteriol 1984; 159:590-6; PMID:6086578.

54. Grindley NDF. The movement of Tn3-like elements: transposition and cointegrate resolution. In: Craig NL, Craigie R, Gellert M, Lambowitz AM, eds. Mobile DNA II. Washington, DC: ASM Press, 2002:272-302.

55. Gringauz E, Orle KA, Waddell CS, Craig NL. Recognition of Escherichia coli attTn7 by transposon Tn7: lack of specific sequence requirements at the point of Tn7 insertion. J Bacteriol 1988; 170:2832-40; PMID:2836374.

56. Gueguen E, Rousseau P, Duval-Valentin G, Chandler M. The transpososome: control of transposition at the level of catalysis. Trends Microbiol 2005; 13:543-9; PMID:16181782; http://dx.doi.org/10.1016/j.tim.2005.09.002.

57. Hanson PI, Whiteheart SW. AAA+ proteins: have engine, will work. Nat Rev Mol Cell Biol 2005; 6:519-29; PMID:16072036; http://dx.doi.org/10.1038/nrm1684.

58. Haren L, Ton-Hoang B, Chandler M. Integrating DNA: transposases and retroviral integrases. Annu Rev Microbiol 1999; 53:245-81; PMID:10547692; http://dx.doi.org/10.1146/annurev.micro.53.1.245.

59. Hauer B, Shapiro JA. Control of Tn7 transposition. Mol Gen Genet 1984; 194:149-58; PMID:6328211; http://dx.doi.org/10.1007/BF00383510.

60. Heller RC, Marians KJ. Replication fork reactivation downstream of a blocked nascent leading strand. Nature 2006; 439:557-62; PMID:16452972; http://dx.doi.org/10.1038/nature04329.

61. Heller RC, Marians KJ. Replisome assembly and the direct restart of stalled replication forks. Nat Rev Mol Cell Biol 2006; 7:932-43; PMID:17139333; http://dx.doi.org/10.1038/nrm2058.

62. Hiasa H, Marians KJ. Tus prevents overreplication of oriC plasmid DNA. J Biol Chem 1994; 269:26959-68; PMID:7929435.
63. Hickman AB, Chandler M, Dyda F. Integrating prokaryotes and eukaryotes: DNA transposases in light of structure. Crit Rev Biochem Mol Biol 2010; 45:50-69; PMID:20067338; http://dx.doi.org/10.3109/10409230903505596.
64. Hickman AB, Li Y, Mathew SV, et al. Unexpected structural diversity in DNA recombination: the restriction endonuclease connection. Mol Cell 2000; 5:1025-34; PMID:10911996; http://dx.doi.org/10.1016/S1097-2765(00)80267-1.
65. Higuchi K, Katayama T, Iwai S, et al. Fate of DNA replication fork encountering a single DNA lesion during oriC plasmid DNA replication in vitro. Genes Cells 2003; 8:437-49; PMID:12694533; http://dx.doi.org/10.1046/j.1365-2443.2003.00646.x.
66. Hill TM, Marians KJ. Escherichia coli Tus protein acts to arrest the progression of DNA replication forks in vitro. Proc Natl Acad Sci USA 1990; 87:2481-5; PMID:2181438; http://dx.doi.org/10.1073/pnas.87.7.2481.
67. Højberg O, Schnider U, Winteler HV, et al. Oxygen-sensing reporter strain of Pseudomonas fluorescens for monitoring the distribution of low-oxygen habitats in soil. Appl Environ Microbiol 1999; 65:4085-93; PMID:10473420.
68. Holder JW, Craig NL. Architecture of the Tn7 posttransposition complex: an elaborate nucleoprotein structure. J Mol Biol 2010; 401:167-81; PMID:20538004; http://dx.doi.org/10.1016/j.jmb.2010.06.003.
69. Howe K, Karsi A, Germon P, et al. Development of stable reporter system cloning luxCDABE genes into chromosome of Salmonella enterica serotypes using Tn7 transposon. BMC Microbiol 2010; 10:197; PMID:20653968; http://dx.doi.org/10.1186/1471-2180-10-197.
70. Hunter S, Apweiler R, Attwood TK, et al. InterPro: the integrative protein signature database. Nucleic Acids Res 2009; 37:D211-5; PMID:18940856; http://dx.doi.org/10.1093/nar/gkn785.
71. Johnson A, O'Donnell M. Cellular DNA replicases: components and dynamics at the replication fork. Annu Rev Biochem 2005; 74:283-315; PMID:15952889; http://dx.doi.org/10.1146/annurev.biochem.73.011303.073859.
72. Kennedy AK, Guhathakurta A, Kleckner N, Haniford DB. Tn10 transposition via a DNA hairpin intermediate. Cell 1998; 95:125-34; PMID:9778253; http://dx.doi.org/10.1016/S0092-8674(00)81788-2.
73. Kraft CA, Timbury MC, Platt DJ. Distribution and genetic location of Tn7 in trimethoprim-resistant Escherichia coli. J Med Microbiol 1986; 22:125-31; PMID:3018250; http://dx.doi.org/10.1099/00222615-22-2-125.
74. Kubo KM, Craig NL. Bacterial transposon Tn7 utilizes two different classes of target sites. J Bacteriol 1990; 172:2774-8; PMID:2158980.
75. Kuduvalli PN, Mitra R, Craig NL. Site-specific Tn7 transposition into the human genome. Nucleic Acids Res 2005; 33:857-63; PMID:15701757; http://dx.doi.org/10.1093/nar/gki227.
76. Kuduvalli PN, Rao JE, Craig NL. Target DNA structure plays a critical role in Tn7 transposition. EMBO J 2001; 20:924-32; PMID:11179236; http://dx.doi.org/10.1093/emboj/20.4.924.
77. Kumar A, Seringhaus M, Biery MC, Sarnovsky RJ, Umansky L, Piccirillo S, et al. Large-scale mutagenesis of the yeast genome using a Tn7-derived multipurpose transposon. Genome Res 2004; 14:1975-86; PMID:15466296; http://dx.doi.org/10.1101/gr.2875304.
78. Lee CH, Bhagwat A, Heffron F. Identification of a transposon Tn3 sequence required for transposition immunity. Proc Natl Acad Sci USA 1983; 80:6765-9; PMID:6316324; http://dx.doi.org/10.1073/pnas.80.22.6765.
79. Letunic I, Doerks T, Bork P. SMART 6: recent updates and new developments. Nucleic Acids Res 2009; 37:D229-32; PMID:18978020; http://dx.doi.org/10.1093/nar/gkn808.
80. Lichtenstein C, Brenner S. Unique insertion site of Tn7 in the E. coli chromosome. Nature 1982; 297:601-3; PMID:6283361; http://dx.doi.org/10.1038/297601a0.
81. Lo C, Adachi K, Shuster JR, et al. The bacterial transposon Tn7 causes premature polyadenylation of mRNA in eukaryotic organisms: TAGKO mutagenesis in filamentous fungi. Nucleic Acids Res 2003; 31:4822-7; PMID:12907724; http://dx.doi.org/10.1093/nar/gkg676.
82. López de Saro FJ, O'Donnell M. Interaction of the beta sliding clamp with MutS, ligase, and DNA polymerase I. Proc Natl Acad Sci USA 2001; 98:8376-80; PMID:11459978; http://dx.doi.org/10.1073/pnas.121009498.
83. Lu F, Craig NL. Isolation and characterization of Tn7 transposase gain-of-function mutants: a model for transposase activation. EMBO J 2000; 19:3446-57; PMID:10880457; http://dx.doi.org/10.1093/emboj/19.13.3446.
84. Luckow VA, Lee SC, Barry GF, Olins PO. Efficient generation of infectious recombinant baculoviruses by site-specific transposon-mediated insertion of foreign genes into a baculovirus genome propagated in Escherichia coli. J Virol 1993; 67:4566-79; PMID:8392598.

85. Mancuso M, Sammarco MC, Grabczyk E. Transposon Tn7 preferentially inserts into GAA*TTC triplet repeats under conditions conducive to Y*R*Y triplex formation. PLoS ONE 2010; 5:e11121; PMID:20559546; http://dx.doi.org/10.1371/journal.pone.0011121.
86. Marchler-Bauer A, Lu S, Anderson JB, et al. CDD: a Conserved Domain Database for the functional annotation of proteins. Nucleic Acids Res 2011; 39:D225-9; PMID:21109532; http://dx.doi.org/10.1093/nar/gkq1189.
87. Markovitz A. A new in vivo termination function for DNA polymerase I of Escherichia coli K12. Mol Microbiol 2005; 55:1867-82; PMID:15752206; http://dx.doi.org/10.1111/j.1365-2958.2005.04513.x.
88. May EW, Craig NL. Switching from cut-and-paste to replicative Tn7 transposition. Science 1996; 272:401-4; PMID:8602527; http://dx.doi.org/10.1126/science.272.5260.401.
89. McInerney P, Johnson A, Katz F, O'Donnell M. Characterization of a triple DNA polymerase replisome. Mol Cell 2007; 27:527-38; PMID:17707226; http://dx.doi.org/10.1016/j.molcel.2007.06.019.
90. McInerney P, O'Donnell M. Functional uncoupling of twin polymerases: mechanism of polymerase dissociation from a lagging-strand block. J Biol Chem 2004; 279:21543-51; PMID:15014081; http://dx.doi.org/10.1074/jbc.M401649200.
91. McKenzie GJ, Craig NL. Fast, easy and efficient: site-specific insertion of transgenes into enterobacterial chromosomes using Tn7 without need for selection of the insertion event. BMC Microbiol 2006; 6:39; PMID:16646962; http://dx.doi.org/10.1186/1471-2180-6-39.
92. McKown RL, Orle KA, Chen T, Craig NL. Sequence requirements of Escherichia coli attTn7, a specific site of transposon Tn7 insertion. J Bacteriol 1988; 170:352-8; PMID:2826397.
93. McKown RL, Waddell CS, Arciszewska LK, Craig NL. Identification of a transposon Tn7-dependent DNA-binding activity that recognizes the ends of Tn7. Proc Natl Acad Sci USA 1987; 84:7807-11; PMID:2825163; http://dx.doi.org/10.1073/pnas.84.22.7807.
94. Mirkin EV, Mirkin SM. Replication fork stalling at natural impediments. Microbiol Mol Biol Rev 2007; 71:13-35; PMID:17347517; http://dx.doi.org/10.1128/MMBR.00030-06.
95. Mitra R, Fain-Thornton J, Craig NL. piggyBac can bypass DNA synthesis during cut and paste transposition. EMBO J 2008; 27:1097-109; PMID:18354502; http://dx.doi.org/10.1038/emboj.2008.41.
96. Mitra R, McKenzie GJ, Yi L, et al. Characterization of the TnsD-attTn7 complex that promotes site-specific insertion of Tn7. Mob DNA 2010; 1:18; PMID:20653944; http://dx.doi.org/10.1186/1759-8753-1-18.
97. Mizuuchi K. Transpositional recombination: mechanistic insights from studies of mu and other elements. Annu Rev Biochem 1992; 61:1011-51; PMID:1323232; http://dx.doi.org/10.1146/annurev.bi.61.070192.005051.
98. Mizuuchi M, Baker TA, Mizuuchi K. Assembly of phage Mu transpososomes: cooperative transitions assisted by protein and DNA scaffolds. Cell 1995; 83:375-85; PMID:8521467; http://dx.doi.org/10.1016/0092-8674(95)90115-9.
99. Namgoong SY, Harshey RM. The same two monomers within a MuA tetramer provide the DDE domains for the strand cleavage and strand transfer steps of transposition. EMBO J 1998; 17:3775-85; PMID:9649447; http://dx.doi.org/10.1093/emboj/17.13.3775.
100. Nicolas E, Lambin M, Hallet B. Target immunity of the Tn3-family transposon Tn4430 requires specific interactions between the transposase and the terminal inverted repeats of the transposon. J Bacteriol 2010; 192:4233-8; PMID:20562304; http://dx.doi.org/10.1128/JB.00477-10.
101. Nisa-Martínez R, Jimenez-Zurdo JI, Martinez-Abarca F, et al. Dispersion of the RmInt1 group II intron in the Sinorhizobium meliloti genome upon acquisition by conjugative transfer. Nucleic Acids Res 2007; 35:214-22; PMID:17158161; http://dx.doi.org/10.1093/nar/gkl1072.
102. North SH, Nakai H. Host factors that promote transpososome disassembly and the PriA-PriC pathway for restart primosome assembly. Mol Microbiol 2005; 56:1601-16; PMID:15916609; http://dx.doi.org/10.1111/j.1365-2958.2005.04639.x.
103. O'Donnell M, Jeruzalmi D, Kuriyan J. Clamp loader structure predicts the architecture of DNA polymerase III holoenzyme and RFC. Curr Biol 2001; 11:R935-46; PMID:11719243; http://dx.doi.org/10.1016/S0960-9822(01)00559-0.
104. Orle KA, Craig NL. Identification of transposition proteins encoded by the bacterial transposon Tn7. Gene 1991; 104:125-31; PMID:1655576; http://dx.doi.org/10.1016/0378-1119(91)90478-T.
105. Parks AR, Li Z, Shi Q, Owens RM, et al. Transposition into replicating DNA occurs through interaction with the processivity factor. Cell 2009; 138:685-95; PMID:19703395; http://dx.doi.org/10.1016/j.cell.2009.06.011.
106. Parks AR, Peters JE. Tn7 elements: engendering diversity from chromosomes to episomes. Plasmid 2009; 61:1-14; PMID:18951916; http://dx.doi.org/10.1016/j.plasmid.2008.09.008.
107. Parks AR, Peters JE. Transposon Tn7 is widespread in diverse bacteria and forms genomic islands. J Bacteriol 2007; 189:2170-3; PMID:17194796; http://dx.doi.org/10.1128/JB.01536-06.

108. Peters JE, Craig NL. Tn7 recognizes transposition target structures associated with DNA replication using the DNA-binding protein TnsE. Genes Dev 2001; 15:737-47; PMID:11274058; http://dx.doi.org/10.1101/gad.870201.

109. Peters JE, Craig NL. Tn7 transposes proximal to DNA double-strand breaks and into regions where chromosomal DNA replication terminates. Mol Cell 2000; 6:573-82; PMID:11030337; http://dx.doi.org/10.1016/S1097-2765(00)00056-3.

110. Peters JE, Craig NL. Tn7: smarter than we thought. Nat Rev Mol Cell Biol 2001; 2:806-14; PMID:11715047; http://dx.doi.org/10.1038/35099006.

111. Rådström P, Skold O, Swedberg G, et al. Transposon Tn5090 of plasmid R751, which carries an integron, is related to Tn7, Mu, and the retroelements. J Bacteriol 1994; 176:3257-68; PMID:8195081.

112. Ramírez MS, Bello H, Gonzalez Rocha G, et al. Tn7:In2-8 dispersion in multidrug resistant isolates of Acinetobacter baumannii from Chile. Rev Argent Microbiol 2010; 42:138-40; PMID:20589338.

113. Ramírez MS, Quiroga C, Centron D. Novel rearrangement of a class 2 integron in two non-epidemiologically related isolates of Acinetobacter baumannii. Antimicrob Agents Chemother 2005; 49:5179-81; PMID:16304199; http://dx.doi.org/10.1128/AAC.49.12.5179-5181.2005.

114. Ramírez MS, Stietz MS, Vilacoba E, et al. Increasing frequency of class 1 and 2 integrons in multidrug-resistant clones of Acinetobacter baumannii reveals the need for continuous molecular surveillance. Int J Antimicrob Agents 2011; 37:175-7; PMID:21177078; http://dx.doi.org/10.1016/j.ijantimicag.2010.10.006.

115. Ramírez MS, Vargas LJ, Cagnoni V, et al. Class 2 integron with a novel cassette array in a Burkholderia cenocepacia isolate. Antimicrob Agents Chemother 2005; 49:4418-20; PMID:16189138; http://dx.doi.org/10.1128/AAC.49.10.4418-4420.2005.

116. Rao JE, Craig NL. Selective recognition of pyrimidine motif triplexes by a protein encoded by the bacterial transposon Tn7. J Mol Biol 2001; 307:1161-70; PMID:11292332; http://dx.doi.org/10.1006/jmbi.2001.4553.

117. Rao JE, Miller PS, Craig NL. Recognition of triple-helical DNA structures by transposon Tn7. Proc Natl Acad Sci USA 2000; 97:3936-41; PMID:10737770; http://dx.doi.org/10.1073/pnas.080061497.

118. Reimmann C, Moore R, Little S, et al. Genetic structure, function and regulation of the transposable element IS21. Mol Gen Genet 1989; 215:416-24; PMID:2540414; http://dx.doi.org/10.1007/BF00427038.

119. Reyes O, Beyou A, Mignotte-Vieux C, Richaud F. Mini-Mu transduction: cis-inhibition of the insertion of Mud transposons. Plasmid 1987; 18:183-92; PMID:2832860; http://dx.doi.org/10.1016/0147-619X(87)90061-8.

120. Rice P, Mizuuchi K. Structure of the bacteriophage Mu transposase core: a common structural motif for DNA transposition and retroviral integration. Cell 1995; 82:209-20; PMID:7628012; http://dx.doi.org/10.1016/0092-8674(95)90308-9.

121. Rice PA, Baker TA. Comparative architecture of transposase and integrase complexes. Nat Struct Biol 2001; 8:302-7; PMID:11276247; http://dx.doi.org/10.1038/86166.

122. Richardson JM, Colloms SD, Finnegan DJ, Walkinshaw MD. Molecular architecture of the Mos1 paired-end complex: the structural basis of DNA transposition in a eukaryote. Cell 2009; 138:1096-108; PMID:19766564; http://dx.doi.org/10.1016/j.cell.2009.07.012.

123. Robinson MK, Bennett PM, Richmond MH. Inhibition of TnA translocation by TnA. J Bacteriol 1977; 129:407-14; PMID:318647.

124. Ronning DR, Li Y, Perez ZN, et al. The carboxy-terminal portion of TnsC activates the Tn7 transposase through a specific interaction with TnsA. EMBO J 2004; 23:2972-81; PMID:15257292; http://dx.doi.org/10.1038/sj.emboj.7600311.

125. Rose A. TnAbaR1: a novel Tn7-related transposon in Acinetobacter baumannii that contributes to the accumulation and dissemination of large repertoires of resistance genes. Bioscience Horizons 2010; 3:40-8; http://dx.doi.org/10.1093/biohorizons/hzq006.

126. Sakai J, Chalmers RM, Kleckner N. Identification and characterization of a pre-cleavage synaptic complex that is an early intermediate in Tn10 transposition. EMBO J 1995; 14:4374-83; PMID:7556079.

127. Sakai J, Kleckner N. The Tn10 synaptic complex can capture a target DNA only after transposon excision. Cell 1997; 89:205-14; PMID:9108476; http://dx.doi.org/10.1016/S0092-8674(00)80200-7.

128. Sarnovsky RJ, May EW, Craig NL. The Tn7 transposase is a heteromeric complex in which DNA breakage and joining activities are distributed between different gene products. EMBO J 1996; 15:6348-61; PMID:8947057.

129. Savilahti H, Mizuuchi K. Mu transpositional recombination: donor DNA cleavage and strand transfer in trans by the Mu transposase. Cell 1996; 85:271-80; PMID:8612279; http://dx.doi.org/10.1016/S0092-8674(00)81103-4.

130. Schumacher S, Clubb RT, Cai M, et al. Solution structure of the Mu end DNA-binding ibeta subdomain of phage Mu transposase: modular DNA recognition by two tethered domains. EMBO J 1997; 16:7532-41; PMID:9405381; http://dx.doi.org/10.1093/emboj/16.24.7532.

131. Schweidenback CT, Baker TA. Dissecting the roles of MuB in Mu transposition: ATP regulation of DNA binding is not essential for target delivery. Proc Natl Acad Sci USA 2008; 105:12101-7; PMID:18719126; http://dx.doi.org/10.1073/pnas.0805868105.

132. Seringhaus M, Kumar A, Hartigan J, et al. Genomic analysis of insertion behavior and target specificity of mini-Tn7 and Tn3 transposons in Saccharomyces cerevisiae. Nucleic Acids Res 2006; 34:e57; PMID:16648358; http://dx.doi.org/10.1093/nar/gkl184.

133. Sharpe PL, Craig NL. Host proteins can stimulate Tn7 transposition: a novel role for the ribosomal protein L29 and the acyl carrier protein. EMBO J 1998; 17:5822-31; PMID:9755182; http://dx.doi.org/10.1093/emboj/17.19.5822.

134. Sherratt DJ. Bacterial chromosome dynamics. Science 2003; 301:780-5; PMID:12907786; http://dx.doi.org/10.1126/science.1084780.

135. Shi Q, Huguet-Tapia JC, Peters JE. Tn917 targets the region where DNA replication terminates in Bacillus subtilis, highlighting a difference in chromosome processing in the firmicutes. J Bacteriol 2009; 191:7623-7; PMID:19820088; http://dx.doi.org/10.1128/JB.01023-09.

136. Shi Q, Parks AR, Potter BD, et al. DNA damage differentially activates regional chromosomal loci for Tn7 transposition in Escherichia coli. Genetics 2008; 179:1237-50; PMID:18562643; http://dx.doi.org/10.1534/genetics.108.088161.

137. Skelding Z, Queen-Baker J, Craig NL. Alternative interactions between the Tn7 transposase and the Tn7 target DNA binding protein regulate target immunity and transposition. EMBO J 2003; 22:5904-17; PMID:14592987; http://dx.doi.org/10.1093/emboj/cdg551.

138. Skelding Z, Sarnovsky R, Craig NL. Formation of a nucleoprotein complex containing Tn7 and its target DNA regulates transposition initiation. EMBO J 2002; 21:3494-504; PMID:12093750; http://dx.doi.org/10.1093/emboj/cdf347.

139. Stellwagen AE, Craig NL. Analysis of gain-of-function mutants of an ATP-dependent regulator of Tn7 transposition. J Mol Biol 2001; 305:633-42; PMID:11152618; http://dx.doi.org/10.1006/jmbi.2000.4317.

140. Stellwagen AE, Craig NL. Avoiding self: two Tn7-encoded proteins mediate target immunity in Tn7 transposition. EMBO J 1997; 16:6823-34; PMID:9362496; http://dx.doi.org/10.1093/emboj/16.22.6823.

141. Stellwagen AE, Craig NL. Gain-of-function mutations in TnsC, an ATP-dependent transposition protein that activates the bacterial transposon Tn7. Genetics 1997; 145:573-85; PMID:9055068.

142. Stellwagen AE, Craig NL. Mobile DNA elements: controlling transposition with ATP-dependent molecular switches. Trends Biochem Sci 1998; 23:486-90; PMID:9868372; http://dx.doi.org/10.1016/S0968-0004(98)01325-5.

143. Stukenberg PT, Turner J, O'Donnell M. An explanation for lagging strand replication: polymerase hopping among DNA sliding clamps. Cell 1994; 78:877-87; PMID:8087854; http://dx.doi.org/10.1016/S0092-8674(94)90662-9.

144. Surette MG, Buch SJ, Chaconas G. Transpososomes: stable protein-DNA complexes involved in the in vitro transposition of bacteriophage Mu DNA. Cell 1987; 49:253-62; PMID:3032448; http://dx.doi.org/10.1016/0092-8674(87)90566-6.

145. Suski C, Marians KJ. Resolution of converging replication forks by RecQ and topoisomerase III. Mol Cell 2008; 30:779-89; PMID:18570879; http://dx.doi.org/10.1016/j.molcel.2008.04.020.

146. Swingle B, Markel E, Costantino N, Bubunenko MG, Cartinhour S, Court DL. Oligonucleotide recombination in Gram-negative bacteria. Mol Microbiol 2010; 75:138-48; PMID:19943907; http://dx.doi.org/10.1111/j.1365-2958.2009.06976.x.

147. Swingle B, O'Carroll M, Haniford D, Derbyshire KM. The effect of host-encoded nucleoid proteins on transposition: H-NS influences targeting of both IS903 and Tn10. Mol Microbiol 2004; 52:1055-67; PMID:15130124; http://dx.doi.org/10.1111/j.1365-2958.2004.04051.x.

148. Tang M, Cecconi C, Bustamante C, Rio DC. Analysis of P element transposase protein-DNA interactions during the early stages of transposition. J Biol Chem 2007; 282:29002-12; PMID:17644523; http://dx.doi.org/10.1074/jbc.M704106200.

149. Tang Y, Cotterill S, Lichtenstein CP. Genetic analysis of the terminal 8-bp inverted repeats of transposon Tn7. Gene 1995; 162:41-6; PMID:7557414; http://dx.doi.org/10.1016/0378-1119(95)92859-6.

150. van Luenen HG, Colloms SD, Plasterk RH. The mechanism of transposition of Tc3 in C. elegans. Cell 1994; 79:293-301; PMID:7954797; http://dx.doi.org/10.1016/0092-8674(94)90198-8.

151. Waddell CS, Craig NL. Tn7 transposition: recognition of the attTn7 target sequence. Proc Natl Acad Sci USA 1989; 86:3958-62; PMID:2542960; http://dx.doi.org/10.1073/pnas.86.11.3958.

152. Waddell CS, Craig NL. Tn7 transposition: two transposition pathways directed by five Tn7-encoded genes. Genes Dev 1988; 2:137-49; PMID:2834269; http://dx.doi.org/10.1101/gad.2.2.137.
153. Wilkins BM, Lanka E. DNA processing and replication during plasmid transfer between Gram-negative bacteria. In: Clewell D, ed. Bacterial Conjugation. New York: Plenum Press, 1993:105-36
154. Wolkow CA, DeBoy RT, Craig NL. Conjugating plasmids are preferred targets for Tn7. Genes Dev 1996; 10:2145-57; PMID:8804309; http://dx.doi.org/10.1101/gad.10.17.2145.
155. Yamauchi M, Baker TA. An ATP-ADP switch in MuB controls progression of the Mu transposition pathway. EMBO J 1998; 17:5509-18; PMID:9736628; http://dx.doi.org/10.1093/emboj/17.18.5509.
156. Yanagihara K, Mizuuchi K. Progressive structural transitions within Mu transpositional complexes. Mol Cell 2003; 11:215-24; PMID:12535534; http://dx.doi.org/10.1016/S1097-2765(02)00796-7.
157. Yuzhakov A, Turner J, O'Donnell M. Replisome assembly reveals the basis for asymmetric function in leading and lagging strand replication. Cell 1996; 86:877-86; PMID:8808623; http://dx.doi.org/10.1016/S0092-8674(00)80163-4.
158. Zhou L, Mitra R, Atkinson PW, et al. Transposition of hAT elements links transposable elements and V(D)J recombination. Nature 2004; 432:995-1001; PMID:15616554; http://dx.doi.org/10.1038/nature03157.

CHAPTER 2

Mercury Resistance Transposons

Sofia Mindlin and Mayya Petrova*

Abstract

Mercury resistance transposons belong to diverse groups of mobile elements that differ in structural and functional traits. They are broadly distributed among many bacterial species residing in diverse environments such as ancient permafrost, mercury ore, pristine or contaminated soil and water and in clinics. The Tn21 subgroup transposons play an active part in the dissemination of antibiotic resistance determinants between medically important bacteria; the other transposons are chiefly distributed in natural environments. This review focuses on a set of transposons detected in both Gram-positive and Gram-negative bacteria, the molecular structure and properties of its members and the role of recombination in their diversity and evolution.

Introduction

Studies on *mer* (mercury resistance) transposons have mainly been conducted in the context of the medical and veterinary consequences of antibiotic-resistance because many mobile elements that carry antibiotic resistance determinants also contain *mer* operons.[1,2]

Mercury resistance transposons were first isolated in the mid-1970s from clinical sources (transposons Tn501, Tn21 and Tn2603). Studies on the genetic structure and transposition properties of these and related mobile elements constituted the first phase of *mer* transposon investigations.[3-6] As a result, it became evident that transposon Tn21 and many of its closest relatives carry within them an additional genetic element, an integron that encodes a site-specific integration system responsible for acquisition of antibiotic resistance cassette genes.[2,7,8] These studies highlighted the role of *mer* transposons in the distribution of antibiotic resistance determinants to clinics throughout the world and, for this reason, many laboratories were involved in their isolation and characterization. This constituted the second phase of *mer* transposon investigations, during which its molecular structure, functioning and evolution were determined. During this phase, studies were conducted on the distribution of the *mer* transposon in various environments, including contaminated soil and water, on transposons isolated from environmental and clinical sources and on the horizontal transfer of transposons.[1,9-11] At present, more than 50 functional transposons and variants that confer resistance to mercury have been isolated and completely or partially sequenced (Table 1).

Structure and Diversity of mer Operons

The best-studied mechanism of mercury resistance is the reduction of toxic Hg^{2+} to relatively nontoxic metallic mercury by the cytoplasmic flavoenzyme, mercuric reductase.[15-17] Bacterial genes determining mercury resistance are organized into well-characterized *mer* operons. The *mer* operons of Gram-negative bacteria contain a metalloregulator gene (*merR*) and three structural genes: genes that encode a transport system that delivers the toxic mercuric ions into the cells (*merTP*) and a gene that encodes the intracellular enzyme, mercuric reductase (*merA*), which

*Institute of Molecular Genetics, Russian Academy of Sciences, Moscow, Russia.
Corresponding Author: Sofia Mindlin—Email: mindlin@img.ras.ru

Bacterial Integrative Mobile Genetic Elements, edited by Adam P. Roberts and Peter Mullany.
©2013 Landes Bioscience.

Table 1. List of mercury resistance transposons

Name	Family or Subgroup	Source	Mer Genes	Presence/Absence of Integron	Size (kb)	AC or (reference)
Tn21	Tn21	NR1 from *Shigella flexneri*	RTPCADEurf2	In2	19.7	AP000342
Tn501	Tn21	pVS1 from *P. aeruginosa*	RTPADEurf2	-	8.2	Z00027
Tn502	Tn5053	pVS6 from *P. aeruginosa* RM1	RTPCADEurf2	-	9.6	EU306743
Tn512	Tn5053	*P. aeruginosa* AW54a	RTPFADE	-	8.4	EU306744
Tn1696	Tn21	R1033 from *P. aeruginosa*	RTPCADEurf2	In4	16.3	U12338
Tn1831	Tn21	Inc.P1 plasmid R702	RTPCADEurf2	In	16.9	(12)
Tn2011	Tn21	Rms213 from *E.coli*	RTPCADEurf2	In	~19	(13)
Tn2410	Tn21	R1767 from *S. typhimurium*	RTPCADEurf2	In2-related	18.5	(4)
Tn2411	Tn21	R1767 from *S. typhimurium*	RTPCADEurf2	In2-related	18.0	(4)
Tn2424	Tn21	NR79 from *Pseudomonas* sp	RTPCADEurf2	In21	25.4	AF047479 (P)
Tn2425	Tn21	Inc.FII plasmid pMH2	RTPCADEurf2	In	21.4	(12)
Tn2426	Tn21	*Shigella sonnei*	RTPCADEurf2	In		M86913
Tn2603	Tn21	RGN238 from *E. coli*	RTPCADEurf2	In8	19.6	AJ009819 (P)
Tn2608	Tn21	RGN823 from *K. pneumoniae*	RTPCADEurf2	In	13.5	(14)
Tn2613	Tn21	pCS229 from *P. mirabilis*	RTPCADEurf2	-	7.2	(14)
Tn3926	Tn21	*Yersinia enterocolitica* YE138A14	RTPCADEurf2	-	8.0	X78059, X14236 (P)
Tn5036	Tn21	*Enterobacter cloaceae* TC256	RTPCADEurf2Δ	-	8.0	Y09025
Tn5037	Tn21	*Thiobacillus ferrooxidans*	RTPA	-	6.6	AJ251743
Tn5041	Tn3	*Pseudomonas* sp KHP41	RTPCAorfYDEurf2	-	14.9	X98999
Tn5042	Tn5042	*P. fluorescens* ED94–62a	RTPCAB	-	7.0	AJ563380
Tn5044	Tn3	*Xanthomonas* sp TAP44–3	RTPCAsigYDurf858	-	10.8	Y17691
Tn5046	Tn3	*Pseudomonas* sp LS46–6	RTPCADorYurf1,2	-	10.1	Y18360
Tn5050	Tn21	*Pseudomonas* sp LS45–3	RTPCADEurf2	-	Nd	Y17719 (P)
Tn5051	Tn21	*Pseudomonas putida* HU1–6	RTPFADEurf2	-	Nd	Y17897 (P)

continued on next page

Table 1. Continued

Name	Family or Subgroup	Source	Mer Genes	Presence/Absence of Integron	Size (kb)	AC or (reference)
Tn5053	Tn5053	Xanthomonas campestris W17	RTPFADEurf2Δ	-	8.4	L40585
Tn5053v2	Tn5053	pMER05	RTPFADEurf2Δ	-	Nd	L20694, Z23093 (P)
Tn5056	Tn5053	Klebsiella sp LS13–39	R1TPAGB1D1+R2B2D2Eurf2	-	13.6	AJ302769–77 (P)
Tn5057	Tn5053	E.coli CH210	RTPFADEurf2Δ	-	Nd	AJ302763–68 (P)
Tn5058	Tn5053	Pseudomonas putida ED23–33	R1TPAGB1D1+R2B2D2Eurf2	-	12.4	Y17897
Tn5059	Tn21	Enterobacter sp	RTPCADEurf2	-	8.0	Y09026, Y10102, Y10103 (P)
Tn5060	Tn21	Pseudomonas sp A19–1	RTPCADEurf2	-	8.7	AJ551280
Tn5070	Tn3	Shewanella putrefaciens BW13	RTPA	-	6.7	Y17830
Tn5073	Tn21	Klebsiella pneumoniae M426	RTPCADEurf2	-	Nd	AF461013 (P)
Tn5074	Tn5053	Proteus morganii M567	RTPFADEurf2Δ	-	Nd	AF461012 (P)
Tn5075	Tn21	E. coli M634	RTPCADEurf2	-	11.3	AF457211
Tn5083	Tn21	Bacillus megateriumMK 64–1	Nd	-	Nd	Y18009 (P)
Tn5084	Tn21	pKLH6 from B.cereus RC607	B3R1E-likeTPAR2B2B1	-	~11.8	AF138877, Y17741, Y17748–49, AJ277277, AB036431 (P)
Tn5085	Tn21	pKLH3 from Exiguobacterium sp TC38–2b	B3R1E-likeTPAR2B2B1	-	11.8	X99457, Y08064, Y17750-Y17752 (P)
Tn5086	Tn21	enteric plasmids	RTPCADEurf2	In22	15.3	X58425 (P)
TnMER11	Tn21	chromosome of Bacillus megaterium MB1	B3R1E-likeTPAR2B2B1	-	14.5	AB022308, AB027306, AB022308 (P)
Tn5718	Tn5053	pSB102	RTPGCCABDEurf2	-	10.4	AJ304453
Tn50580	Tn5053	pTP6	R1TPAGB1+ R2B2D2E	-	12.1	AM048832
Tn6009	Tn916	Gram+ and Gram-isolates	RorIsTAborf1Δ	-	~24	E239355 (P)
TnAO22	Tn21	Achromobacter sp AO22	RTPA	-	8.2	EU696790

The vast majority of transposons presented in the table are functional transposons which have been partially or completely sequenced. Nd—not determined; (P)—partially sequenced.

converts mercuric ions into less toxic metallic mercury (for review see refs. 17–20). Thus, the minimum set of essential mercury resistance genes is *merRTPA*; other genes (*merC, merF, merD, merE, orf2, orfY*) occur in various combinations in the *mer* operons of Gram-negative bacteria and are not essential accessory genes.[16,20] The simplest of all *mer* operons are those of the transposons, Tn*5037*[21] and Tn*5070*[11]; the most complex is the *mer* operon of Tn*5718* (Table 1).[22] It should be also noted that in addition to *mer* operon-mediated inorganic mercury resistance, some *mer* operons confer organomercurial resistance. The latter operons contain the *merB* gene, which encodes organomercurial lyase,[19,20] and in some cases, the *merG* gene (Table 1), which is possibly involved in utilizing organomercurials.[23] The *mer* operons in the functional transposons of Gram-positive bacteria differ significantly from those of Gram-negative bacteria. The commonly occurring *mer* modules of *Bacillus* spp strains (Tn*5085* in Table 1) contain two *mer* operons that contain three *merB* genes that encode different organomercurial lyase enzymes[24,25] but do not contain any genes that encode the mercury transport system. In contrast, *Staphylococcus* spp *mer* operons (Tn*6009* in Table 1) contain only the *merT* transport gene and one *merB* gene.[26] Both of these *mer* operons contain the minimum set of known mercury resistance genes, including *merR* and *merA* (Table 1).

Mercury Resistance Transposons of Gram-Negative Bacteria

According to Kleckner,[27] transposons can be divided into two classes: class I transposons or composite transposons, which are flanked by two copies of an IS element; and the more homologous class II transposons, which are flanked by short inverted repeats.

All known mercury resistance transposons of Gram-negative bacteria contain two modules: (i) a mercury resistance module, and (ii) a transposition module. Some transposons incorporate additional genes or mobile elements, such as ISs and integrons. The genes of the mercury resistance modules (*mer* operons) are closely related, whereas the transposition modules belong to different families of mobile elements.

The majority of mercury resistance transposons are class II transposons. They are classed into several groups according to transposition module organization, and the main groups are the Tn*3* family and its subgroups.

Tn*21* Subgroup of Tn*3* Family Transposons

Transposons belonging to this subgroup are widespread among medically important bacteria and environmental bacteria.[11,28-30] The best-studied transposons of this subgroup are two archetypal transposons: Tn*21*, which was isolated from plasmid NR1 (R100) in a clinical strain of *Shigella flexneri* in Japan,[3] and Tn*501*, which was isolated from plasmid pVS1 in a *Pseudomonas aeruginosa* isolate in Australia (Fig. 1A).[5,31,32]

Both transposable elements are characterized by flanking inverted repeats (IRs) 38 bp in length and have two genes that are involved in transposition, *tnpA* and *tnpR*, which encode transposase TnpA and resolvase TnpR, respectively, and are transcribed in the same direction.[3,5,33] The model of the molecular mechanism of transposition[33-35] involves two stages (Fig. 1B). The first stage is the creation of a cointegrate structure of two replicons joined by two copies of the transposon. Formation of the cointegrate structure requires the *tnpA* gene product, TnpA, which interacts with the highly specific, repeated outer extremities of the transposon. The second phase is resolution of the cointegrate structure by the resolvase, TnpR, which interacts with the internal resolution site (*res* site), resulting in a copy of the transposon at its original site in the donor replicon and a new copy in the target replicon[1,3] (Fig. 1B). Thus, the transposition of Tn*21* and Tn*501* is a replicative process that results in the creation of a new transposon copy.[1,2] Transposition into the recipient molecule results in duplication of five base pairs of the target sequence (Fig. 1C).

In the Tn*21* and Tn*501* transposons, the transposase-IR interactions are interchangeable, i.e., TnpA of Tn*21* can act on the Tn*501* IRs.[36] Similarly, a tnpR mutant of Tn*21* was complemented by Tn*501*.[37] It is worth also noting that transposases of Tn*21* subgroup transposons contain a functionally important DD(35)E motif, as do many other transposases.[9,38-40]

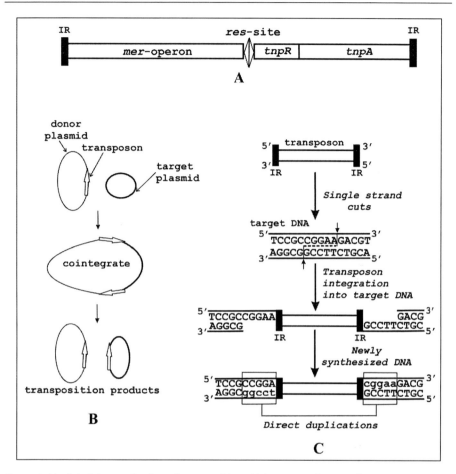

Figure 1. Model of the mechanism of transposition. A) Structure of Tn*21*-subgroup transposon. B) Replicative transposition. C) Duplication of a target DNA. See text for details.

Simple Transposons Closely Related to Tn21 and Tn501

The distribution of Tn*501*- and Tn*21*-related transposons in environmental and medically important bacterial strains has been investigated in many studies, which have shown them to be common.[14,28,29,42]

In addition, a survey of modest-sized collections of Gram-negative mercury resistant strains discovered novel functional mercury resistance transposons, which were nearly identical to the archetype transposons Tn*21* and Tn*501*, and transposons that were more distantly related to them. Their putative structures were confirmed by partial sequencing of various transposon regions (Fig. 2).

Most of the novel transposons belong to the Tn*21* branch and were isolated from environmental bacterial strains and include Tn*5036*, Tn*5059* and Tn*5060*[11,30,43] and from clinical strains including Tn*1696*, Tn*2603*, Tn*2613*, Tn*3926* and Tn*5075*.[14,42,44] Transposons characterized by mosaic structures have been identified in both groups.[11,30]

Two novel transposons of the Tn*501* branch have been analyzed in detail. Tn*5037* from *Thiobacillus ferrooxidans* G66 possesses properties of Tn*501* and Tn*21*: like Tn*501*, its *mer* operon does not contain the *merC* gene, but its *tnp* module structure is more closely related to that of Tn*21* than that of Tn*501*.[21]

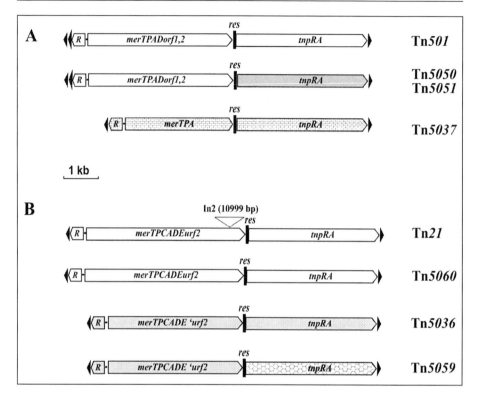

Figure 2. Genetic organization of Tn*21* subgroup transposons The location and polarity of genes and *orf*s are shown with arrows; filled arrowheads indicate the internal Tn*21*-like and terminal inverted repeats; *R, T, P, C, A* and *D* are the corresponding *mer* genes; 1 and 2 indicate *orf1(merE)* and *orf2*; *res* is the *res* site; *tnpR, tnpA*, transposition genes; different shadings denote sequences of different types; In*2* is the integron (see text for details). A) Tn*501*-branch. B) Tn*21*-branch.

A more separated Tn*5051* detected in a *P. putida* strain contained a *mer* operon that was virtually identical to that of Tn*501*; however, its *tnp* genes differed from those of Tn*501* by 30%. It was concluded that Tn*5051* is a recombinant transposon[11] (see below).

Complex Mercury Resistance Tn*21* Subgroup Transposons and Their Evolution

It has long been known that besides the simple transposons, the Tn*21* subgroup includes a set of complex transposable elements that carry different combinations of antibiotic resistance determinants[1,4,6,14,45] (Table 2). This is in general agreement with the concept that complex mercury resistance transposons evolved by insertion and deletion of DNA sequences, particularly via integrons in an ancestral mercury resistance transposon designed Tn*X*[46] (also called Tn*21Δ*).[2]

Tanaka et al.[52] first suggested that transposon Tn*2613* may be the ancestral *mer* transposon of the complex Tn*21* subgroup transposons because it consists solely of the *mer* operon, transposition module (*tnpA, tnpR, res*) and inverted repeats. However, Tn*2613* differs from Tn*21* in terms of its restriction map, suggesting that it was not the immediate predecessor of Tn*21*. No other appropriate candidates for Tn*X* were detected among the *mer* transposons studied.[30,33,42]

Subsequently, two novel functional transposons closely related to Tn*21* were identified: Tn*5075*, isolated from a pre-antibiotic era clinical strain, *Escherichia coli* M634,[44] and Tn*5060*, isolated from a permafrost bacterial strain, *Pseudomonas* sp A19–1.[43]

Table 2. *Diversity of integrons found in mercury resistance Tn21 subgroup transposons*[a]

Transposon	Cassette-Associated Resistance Genes	Resistance Phenotype[b]	Refs.
Tn*21*	*aadA1*	Sm, Sp	2
Tn*1696*	*aacC1-aadA2-cmlA1*	Gm, Sm, Sp, Cm	46
Tn*2410*	*bla(OXA-2)*	Ap	4
Tn*2424*	*aacC1-catB2-aadA1*	Gm,Cm, Sm, Sp	12
Tn*2426*	*aadB-aadA1-aacA-cat*	Gm, Sm, Sp, Ami, Cm	45
Tn*2603*	*bla(OXA)-aadA1*	Ap, Sm, Sp	14
Tn*4000*	*aadB-aadA1*	Gm, Sm, Sp	47
Tn*5086*	*drfA7*	Tp	48
Unnamed	*bla(imp13)-aacA4*	Ap, Km	49
Unnamed	*bla(vim1)-aacA4-aadA2-cmlA*	Ap, Km, Sm, Sp, Cm	50
Tn*1696*-like	*aacA4-aacC1-aadA2-cmlA1*	Km, Gm, Sm, Sp, Cm	51

[a] selected data; [b] Ami-amikacin; Ap-ampicillin; Cm- chloramphenicol; Gm-gentamycin; Km-kanamycin; Sm-streptomycin; Sp-spectinomycin; Tp-trimethoprim.

Structural analysis indicated that Tn*5060* is closely related to Tn*X*, but differs from it in terms of several substitutions.[43]

Between 1980 and 2000, it was frequently claimed that Tn*21* is ancestral to all transposons belonging to the Tn*21* subgroup and that In*2*, the integron found in Tn*21*, is ancestral to all other known integrons (see for example refs. 1 and 53). However, these views were disproved in the early 2000s when the molecular structure of Tn*1696* was revealed. It was established that Tn*1696* arose by the independent insertion of another integron (In*4*) into another backbone transposon (Tn*5036*-like), rather than via Tn*21*.[46] Subsequently, it was shown that Tn*21* is simply one of many cases in which an integron was acquired by a mercury resistance transposon. There are at least six other examples of complex transposons that originated independently via integron insertion into various simple mercury resistance transposons (Table 3). In most cases, the integron was inserted into or close to the *res*-sites of a simple backbone mercury resistance transposon,[57] which is in accordance with the known insertion specificity of integrons.[46]

Table 3. *List of independently originated complex mercury resistance transposons*

Complex Transposon	Backbone Transposon	Class I Integron	Integron Location	Refs.
Tn*21*	Tn*5060*-like	In*2*	> 350 bp from *res*-region	2
Tn*1696*	Tn*5036*-like	In*4*	between *resII* and *resI*-sites	46
Unnamed	Tn*5036*-like	In-t8	between *resII* and *resI*-sites, 9 bp from the point of insertion in Tn*1696*	54
Tn*6005*	Tn*5036*-like	In(Tn*6006*)	between *resII* and *resI*-sites, 6 bp from the point of insertion in Tn*1696*	55
Unnamed	Tn*5051*-like	unnamed	~200 bp from *res*-region	49
Unnamed	Tn*5051*-like	In*70.2*	*resI*-site	50
Unnamed	a hybrid of Tn*21* and Tn*5036*-like	In*34*-like	between *resII* and *resI*-sites, 16 bp from the point of insertion in Tn*1696*	56

The apparent prevalence of mercury resistance transposons among complex transposons isolated from medically important bacteria (Table 3) may be explained by the initial distribution of backbone mercury resistance transposons in clinics because of the use of mercury-containing compounds for treatment and disinfection.[58] After the introduction of antibiotics, these transposons likely served as bases for the formation of complex integron-containing multidrug resistance transposons. On the other hand, none of the many mercury resistance transposons recovered from permafrost samples contain integrons.[59]

Mercury Resistance Transposons Belonging to Tn*3* Family

The Tn5041 Subgroup

Transposon Tn*5041* was identified in a chromosome of *Pseudomonas* sp strain KHP41 isolated from a mercury mine in Central Asia[60] and has been sequenced completely.[10]

Tn*5041* is a Tn*3* family transposon that falls into a distinct subgroup on the "outskirts" of the Tn*3* family (Fig. 3A).[40] Tn*5041* is 14.9 kb in length and contains a typical Tn*3* family *tnpA* gene. It is bound by 47 imperfect IRs; the 38 outer base pairs of these IRs show similarity to the IRs of transposons of the Tn*3* family.[10] Transposition of Tn*5041* occurred via cointegrate formation, suggesting its replicative mechanism. However, neither of the putative resolution proteins encoded by Tn*5041* resolved the cointegrates formed during transposition.[61]

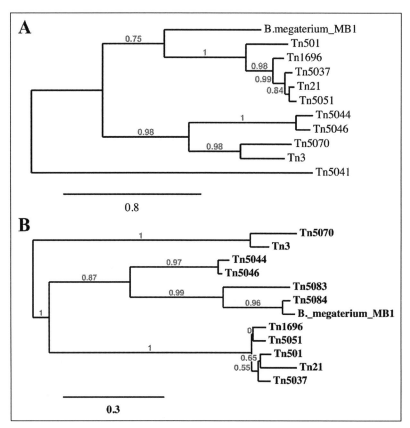

Figure 3. Neighbor–joining distance dendrogram of the TnpA (A) and TnpR (B) amino acids sequences of Tn*3* family transposons constructed by on line software (http://www.phylogeny. fr.[41]). Acs, see in Table 1.

The central part of Tn*5041* is occupied by a *mer* operon with 70% similarity to the classical *mer* operons of Tn*21* and Tn*501*. Besides the *mer* operon and the *tnpA* gene, Tn*5041* contains a 4-kb region that accommodates several apparently defective genes and mobile elements (Fig. 4A).

Twenty-four closely related variants of Tn*5041* have been characterized.[62] These variants were detected in diverse *Pseudomonas* spp strains isolated from environmental sources. Fourteen distinct types of this transposon differed from each other by 1–7 single-event DNA polymorphisms (point mutations, deletions, mosaic regions and insertions at the left arm of the transposons).

It is noteworthy that Tn*5041* can transpose only in its host strain, KHP41, and not in *P. aeruginosa* PAO-R or *Escherichia coli* K12. The host dependence of Tn*5041* transposition may be determined by certain host factors, as occurs in the closely related toluene degradation transposon, Tn*4651*.[63] Host dependence limits the distribution of Tn*5041*-like transposons among bacteria other than *Pseudomonas* spp. Tn*5041* subgroup transposons have not been detected in environmental strains of Enterobacteriaceae, in the genera *Acinetobacter* and *Xanthomonas*[10,30,61] or in medically important bacteria.[2,16]

Tn*5044* Coding for Temperature-Sensitive Mercury Resistance

Tn*5044* (10.8 kb) was discovered in *Xanthomonas campestris* TAP44-3 from the Kamchatka peninsula.[64] In addition to the standard set of mer*RTPCAD* genes, the *mer* operon of Tn*5044* contains a gene named *sigY* that encodes the RNA polymerase sigma factor-like protein (Fig. 4B). In Tn*5044*, mercury resistance is expressed at low (30°C) temperatures but not at elevated temperatures (37–41.5°C). Sequence analysis of the transposase (*tnpA*) genes places Tn*5044* and its close relatives in the Tn*3* subgroup of the Tn*3* family. However, the orientation of their resolvase and transposase genes is unusual for the Tn*3* family: *tnpR* is proximal to the end of the transposon, whereas the divergently transcribed *tnpA* is oriented inwardly (Fig. 4B). The region between the *tnpA* and *tnpR* genes is unusually large and contains two short conserved open reading frames. Tn*5044* also differs from the Tn*3* subgroup transposons in that it has unusually long-terminal inverted repeats (214 bp); the 46 outermost of these are identical to the IRs of the Tn*3* family transposons.[64]

It noteworthy that Tn*5044* was found in two other environmental strains of *X. campestris* and in one strain of *P. aeruginosa* from the same area of the Kamchatka peninsula.[64] However, no additional Tn*5044* variants containing the *sigY* gene in the *mer* operon have been detected among the 2000 mercury-resistant strains from other geographical regions.[65]

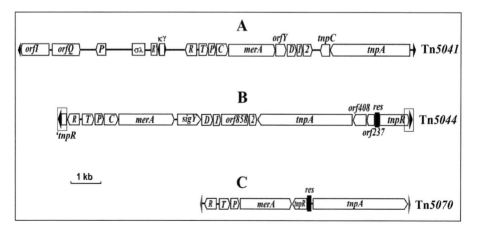

Figure 4. Genetic organization of Tn*3* subgroup transposons Tn*5041* (A), Tn*5044*(B), Tn*5070* (C). orfP, σλ, R', κγ, mobile elements or its relics within Tn*5041*. The long-terminal IRs of Tn*5044* are boxed; *orf1(1)*, *orf858*, *orf408*, *orf237* indicate genes with unknown functions. Other designations are as in Figure 2.

The Endemic Transposon, Tn5070

This transposon was found only once in an environmental bacterial strain, *Shewanella putrefaciens* BW13, which was isolated from a brook in Bethesda (USA).[11] Tn5070 6,7 kb is a functional transposon bounded by 38-bp inverted repeats, which is typical for Tn3-family transposons. The right arm of this transposon contained *tnpA* and *tnpR* genes transcribed in opposite directions and a *res* site located between these genes (Fig. 4C). The *mer* operon of Tn5070, located in its left arm, contains the minimum set of essential genes, i.e., only the *mer R, T, P* and *A* genes, and *merR* is transcribed in the same direction as the structural genes, which is typical for the *mer* operons of Gram-positive bacteria.[20,66] Attempts to detect this transposon in other samples collected from the original locality and other localities have failed. It can be thus concluded that, unlike many other mercury resistance transposons, Tn5070 is an endemic transposon or occurs in very few localities.[65]

Tn5053 Family Transposons

Similar to the mercury transposons of the Tn3 family, all members of the Tn5053 family contain the mercury resistance and the transposition modules. The mercury resistance module of archetypal transposon Tn5053 carries a *mer* operon, *merRTPFADE*, which is closely related to the classical *mer* operons, but its transposition module differs in organization and properties from that of Tn3 family transposons. The transposition module of Tn5053 and related transposons contain four genes involved in transposition (*tniABQR*) and the entire transposon is bounded by 25-bp terminal inverted repeats. Transposition of Tn5053 family transposons occurs via cointegrate formation mediated by TniA, TniB and TniQ, followed by site-specific cointegrate resolution.[9] This is catalyzed by the TniR resolvase at the *res* region, which is located upstream of the *tniR* gene (Fig. 5A). A unique feature of Tn5053 and related transposons is their transpositional behavior. These transposons exhibit a striking insertional preference for the *res* regions of natural plasmids and transposons and have been designated *res* site hunters.[57] It was also revealed that a *res* region serves as a target for Tn5053 insertions only in the presence of a functional cognate resolvase gene.[57,66] One of the most efficient site-specific recombination systems outside the Tn5053 transposon is the *res*-ParA multimer resolution system of plasmid RP1.[57,66] Tn5053 insertion events occur in clusters inside the target *res* regions and, on occasion, can occur in the vicinity of *res*. Typically, the Tn5053 related transposons demonstrate strict orientation specificity, with IRi closest to the target resolvase gene.[57,67]

It was recently shown that the Tn5053-related transposon, Tn502, can transpose by a par-independent process,[67] which confirms the previous observations of this group.[68] The alternative transposition pathway of Tn502 is a two-step replicative process leading in most cases to the formation of cointegrates that can be resolved only in the presence of the par locus via TniR- or RecA-mediated events.[67] The authors proposed that the exceptional behavior of Tn502 is due to the involvement in transposition of the *tniM* gene because this gene is present in Tn502 but not in the other transposons of the Tn5053 family.[67,69]

Transposons Tn5056, Tn5058 and Tn5718 belong to a distinct branch of the Tn5053 family and differ from Tn5053 with respect to size and structure. Tn5058 was isolated from permafrost *P. putida* strain ED23–33 and has been completely sequenced (12,372 bp).[59] The right arm contains the same set of transposition genes as that identified in Tn5053 (Fig. 5A). The left arm of Tn5058 carries two distinct *mer* operons: (i) a left broad-spectrum *mer* operon, *mer1*, in which the *merG* and *merB* genes are responsible for organomercurial resistance, and (ii) a right shortened *mer2* operon lacking the functional mercury reductase gene (Fig. 5A). Tn5058 proved to be very closely related to Tn5056 and its variants found in the European part of Russia.[11] The only substantial differences between these transposons were an extra 1.2 segment found to be inserted into the Tn5056 *mer2* locus, and the 343-bp divergent segment detected in the *merB2* gene of Tn5058[11,59] (Fig. 5A).

Tn5058 also showed a close but more complex relationship to Tn5718, which was found in a plasmid, pSB102, isolated from the rhizosphere of alfalfa.[22] Compared with Tn5058, the proximal and distal regions of the Tn5718 transposition module exhibited only a few substitutions, and its central region differed by 5–10% substitution (Fig. 5A).

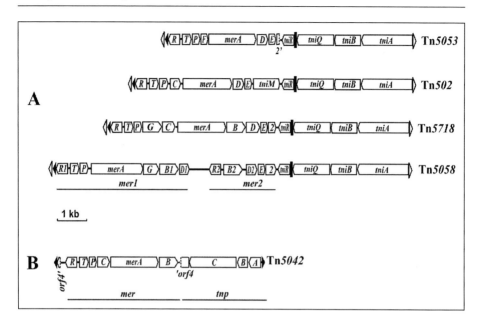

Figure 5. Genetic organization of Tn*5053* family (A) and Tn*5042* (B) transposons. *G* and *B* are the *merG* and *merB* genes controlling resistance to organomercurials; *tnpA, tnpB, tnpC* are the transposition genes of Tn*5042*; *orf4′* and *′orf4*, fragments of *orf4* frame with unknown function within transposition module of Tn*5042*. Other designations are as in Figure 2 (see text for details).

The *tni* gene diversity of transposons belonging to the Tn*5053* family is presented in Table 4.

It should be noted that Tn*5053*-related transposons are widespread in the environment[11,57,70,71] and rarely occur in clinical settings. The few exceptions are transposon Tn*5074* from the Murray collection of pre-antibiotic era enterobacteria[44] and transposons Tn*502* and Tn*512*, which were isolated from clinical *P. aeruginosa* strains.[67,68] Note that the Tn*5053* family transposons are closely related to transposons of the Tn*402* family in that they have similar transposition module structures and transpositional behavior. However, Tn*402* family transposons contain an integron recombination system instead of a mercury resistance module and their relatives are involved in dissemination of antibiotic resistance genes in clinics.[7,39,66]

Table 4. Similarity of Tn5053 tni genes with these of the other Tn5053 family transposons

| Transposon | Source | Nucleotide Identity, % | | | | |
		TniR, 614 bp	tniQ, 1220 bp	tniB, 909 bp	tniA, 1680 bp	ACs
Tn*5053* v12	*Pseudomonas* sp., pCT14	99.8	100	99.8	99.9	DQ126685
Tn*502*	*Pseudomonas aeruginosa*, RM1	85,7	78.9	88.1	85.8	EU306743
Tn*512*	*Pseudomonas* sp, AW54a	99.8	78.4	88.0	85.8	EU306744
Tn*5058*	*Pseudomonas* sp ED23–33	87.5	78.0	88.0	85.8	Y17897
Tn*50580*	Uncultured bacterium, pTP6	89.2	78.1	88.1	85.8	AM048832
Tn*5718*	pSB102	88.4	78.3	89.3	86.7	AJ304453

Tn*5042*-Like Transposons

Transposon Tn*5042*, which was found in the ancient Siberian permafrost strain, *P. fluorescens* ED94–62a,[59] belongs to a discrete type of mercury resistance transposons. Tn*5042*-like transposons are widespread among environmental bacteria. All transposons of this type are very closely related (less than 1% nucleotide substitutions). They were found with 15% frequency in collections of Gram-negative mercury-resistant bacterial strains from New Zealand, European Russia and Siberian permafrost.[59] A variant of Tn*5042* (99.3% identity) was recently discovered in plasmid pQBR103 of *P. fluorescens* SBW25 growing on sugar beet in the UK.[72]

The transposition module of Tn*5042* is about 2.7 kb long, contains three full-length open reading frames of 390, 345 and 1506 bp (Fig. 5B) and is most similar to the IS*66* family elements.[73] Tn*5042* is flanked by 20-bp imperfect inverted repeats, which showed up to 75% identity to the terminal repeats of other elements of the IS*66* family. Like the other IS*66*-related elements, Tn*5042* produced an 8-base pair duplication in the target site upon transposition.[59] There is a *mer* operon about 4.0 kb long in the left arm of Tn*5042*. It contains a standard set of *mer* genes (*merRTPCA*), followed by the *merB* gene responsible for broad-spectrum (organomercurial) resistance (Fig. 5B). The *merRTPCA* genes are most closely related to the *mer* genes of transposon Tn*5041*[10] (80–83% identical nucleotides), whereas the Tn*5042 merB* gene is most closely related to the *merB2* gene of the Tn*5058 mer* operon.[59]

It should be noted that the homology between Tn*5042* and Tn*5041* extended upstream of the *mer* operon. Interestingly enough, a similar situation has been observed with Tn*5046*,[11] which, being most similar in the *mer* operon to Tn*5041*, had more extensive similarity to Tn*5041* upstream of the *mer* operon than Tn*5042* (491- and 88-bp regions, respectively). These data are indicative of the involvement of a Tn*5041* ancestor in the origin of both Tn*5042* and Tn*5046*.

Mercury Resistance Transposons of Gram-Positive Bacteria

The first clear indication of the involvement of class II transposons in the dissemination of mercury resistance determinants in Gram-positive bacteria was obtained only in the late 1990s.[24,74,75] Soon after mercury resistance transposons Tn*5084*, Tn*5085* and Tn*MER11*—which conferred organomercurial resistance—were revealed in several *Bacillus* strains and in a strain of *Exiguobacterium* sp (TC38-2b) (Fig. 6A). Tn*5084* was localized on a plasmid of *B. cereus* RC607 isolated from Boston Harbor, USA, and on a chromosome of *B. cereus* VKM684 (ATCC10702) from the VKM collection.[25,74] Tn*5085*, which is closely related to Tn*5084*, was found on a plasmid in *Exiguobacterium* strain TC38-2b isolated from the Carpathians.[25,74] Tn*MER11*, which is closely related to Tn*5085*, was revealed in the chromosome of *B. megaterium* MB1 isolated from mercury-polluted sediments in Minamata Bay, Japan.[75]

The transposition module of all three aforementioned transposon types belongs to the Tn*21* subgroup of transposons.[1] It contains a transposase gene, *tnpA*, a resolvase gene, *tnpR*, and a *res* site upstream of *tnpR*.[75] A neighbor-joining distance dendrogram showing the position of Gram-positive bacteria transposons among other mercury resistance transposons of the Tn*3* family is presented in Figure 3.

It should be mentioned that another type of *mer* operon that confers a narrow spectrum of mercury resistance has been discovered in environmental *Bacillus* strains. However, the involvement of transposons in their distribution is still obscure.[76]

Recently, a new conjugative transposon, Tn*6009*, the first transposon to have a direct linkage between *mer* genes and the *tet*(M) gene, was described in bacteria from clinical sources.[26] Tn*6009* is a complex transposon containing a Tn*916* element directly linked to a *mer* operon with the same genetic organization as the staphylococcal *mer* operon from plasmids pI258 and pMS97.[77] In Tn*6009*, the *mer* operon is linked to the complete transposon, Tn*916*, by a unique 37-bp region (Fig. 6B). Tn*6009* variants were identified in 66 isolates from two Gram-positive and three Gram-negative bacterial genera from Nigeria and Portugal. The authors hypothesized that the carriage of Tn*6009* may explain why the Gram-positive *mer* genes were previously identified in Gram-negative mercury resistant isolates from Portuguese subjects.[78]

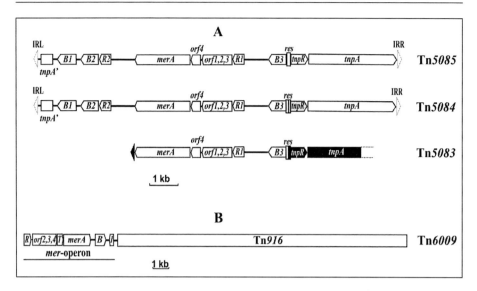

Figure 6. The structure of the transposons detected in Gram-positive bacteria. A) Tn*3* family transposons. *B1, B2, B3* are the corresponding *mer* genes encoding different organomercurial lyase enzymes. B) Conjugative transposon Tn*6009*. See text for details.

Recombinant Transposons and Their Formation Mechanisms

Close examination of the molecular organization of mercury resistance transposons indicates that it is safe to assume that many are recombinants originating from interactions between related and unrelated mobile elements. It has been established that Tn*501* has a recombinant type of structure. Presumably, it arose by insertion of a Tn*21*-like element into a related transposon followed by deletion of the Tn*21*-like transposition module[36] (Fig. 7A). A transpositional mechanism in which recombinant transposons are generated was also proposed for Tn*5053* and its relatives,[59-79] Tn*5041*[10] and other transposons.[44] Tn*5041* is thought to have acquired the *mer* operon from a Tn*21*-related transposon and transposition genes from a precursor of a related toluene degradative transposon, Tn*4651*.[10]

It has been also demonstrated that some of the Tn*21* subgroup transposons have exchanged transposition modules by site specific recombination at the *res* site. The recombinant *res* sites were found in several Tn*21* subgroup transposons,[11,21,30] in particular in Tn*5036*, Tn*5059*, Tn*5051* and Tn*5037* (Fig. 7B). Furthermore, the structures of two recombinant transposons found in Gram-positive *Bacillus* strains revealed the involvement of site specific recombination in their generation (see Fig. 6A). One of these, Tn*5084*, was derived from Tn*5085* by replacing the resolvase gene and the *res* with corresponding fragments from an unknown transposon. The other, Tn*5083*, likely resulted from site specific recombination at *res* sites between Tn*5085* and some other transposon.[25]

The formation of mercury-resistance recombinant transposons via a site-specific recombination event was also revealed by other research groups.[51,56] Thus, it can be concluded that exchange of transposition modules by recombination at the *res* site is a common event in the evolution of mercury resistance transposons.

The parental forms of several recombinant transposons have been identified. One of these, Tn*5046*, is a chimera and consists of Tn*5044* and Tn*5041* regions (Fig. 7C). It has been speculated that Tn*5046* arose by homologous recombination.[11] Homologous recombination has also been proposed as a mechanism for the evolution of other transposons.[51,56,65]

Thus, at least three discrete recombinational processes operate in the creation of mercury resistance transposon diversity: transpositional recombination, a site specific resolvase–*res* system

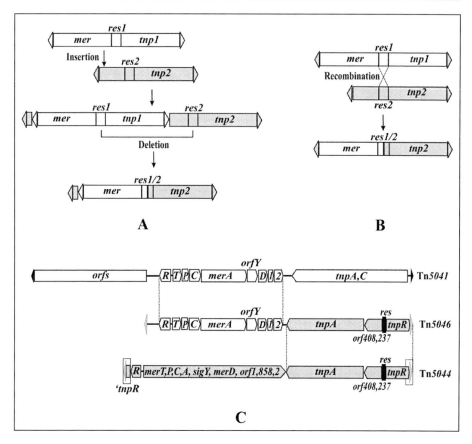

Figure 7. Recombinational processes generating the formation of recombinant transposons (scheme). A) Transpositional recombination; B) Site specific recombination at *res* sites; C) Homologous recombination. Transposons are depicted out of scale. See comments in the text.

and homologous recombination. It is evident that two or all three of these processes were involved in generating some recombinant transposons.[36,51,62]

The high frequency of recombinant transposons suggests that two transposons are often combined in one bacterial cell. This is supported by the presence of two transposons (two *mer* operons, or a transposon and a *mer* operon) in one strain or even in one plasmid.[11,16,80] For instance, the *E. coli* CH-210 strain harbors two transposons, Tn*5057* and Tn*5059*, which belong to different families and are located in the chromosome and in a plasmid, respectively.[11,30] Many strains are known to possess two plasmids with different *mer* operons or transposons.[16]

Distribution of *mer* Transposons and Their Horizontal Transfer

Horizontal gene transfer, along with mutations and genome rearrangements, is considered a major evolutionary mechanism providing for rapid and, in some cases, qualitative changes in hereditary properties of microorganisms. Epidemics, generation of new pathogenic strains by nonpathogenic bacteria, development and rapid spreading of resistance to new drugs and many other phenomena can be explained in terms of horizontal gene transfer (for review see refs. 81, 82).

The mercury resistance determinants, particularly transposons conferring resistance to mercury, can be regarded as a suitable test system for studying horizontal gene transfer in environmental and clinically important bacteria populations.

Table 5. Distribution of Tn5053 variants in natural bacterial populations

Tn5053 Variant	Host Bacterium	Source	Refs.
Tn5053 (prototype)	*Xanthomonas* sp W17	Kyrgyzia, mine	70
Tn5053vl	*Pseudomonas* sp TC24–2	Carpathians, mine	11
Tn5053v2	*P. fluorescens* (pMER05)	UK, river Mersey	57, 71
Tn5053v3	*Pseudomonas* sp HU2–9	USA, New York, river Hudson	11
Tn5053v4	*P. putida*	USA, Bethesda, a stream	11
Tn5053v5	LS47–3, unidentified	Russia, Saratov, river Volga	11
Tn5053v6	*Pseudomonas* sp. LS56–2	Russia, Sergiev Posad, water	11
Tn5053v7	*Enterobacter* sp LS50–2	Russia, Kashin, water	11
Tn5053v10	*Alcaligenes* sp (pMER330)	UK, river Mersey	57, 71
Tn5053v11	*P. fluorescens* (pMER327)	UK, river Mersey	57, 71
Tn5053v12	*Pseudomonas* sp CT14	an industrial wastewater bioreactor	84

Several authors have shown that transposons related to the archetype transposons of clinical bacteria (i.e., Tn501 and Tn21) are common in environmental strains.[20,28,29] Evidence for the distribution of identical or nearly identical elements among bacteria of different taxonomic groups was obtained in studies of the comparative molecular structure of transposons isolated from mercury resistant strains of soil and aquatic bacteria.[11,25,30] Two nearly identical transposons of the Tn21 subgroup, Tn5036 and Tn5036vl, were isolated from *Enterobacter cloacae* (Carpathians) and *Aeromonas* sp (New York) strains, respectively.[11] The transposons proved to be closely related to Tn3926 from a clinical isolate.[42] Transposons almost identical to Tn501 were found in two environmental *Pseudomonas* strains from different regions of the United States.[11] Transposons closely related to Tn5051 (Tn501 branch) initially found in a *P. putida* strain isolated from a water sample in New York were once again discovered in other genera of bacteria isolated from clinical sources[49,50] and in *Achromobacter* sp AO22 from a lead-contaminated industrial site in Australia.[83]

Several cases of horizontal transfer among Gram-negative bacteria were also observed for transposons of other subgroups and families. The global distribution of Tn5053 variants among various bacteria is interesting (Table 5). Several variants were partly sequenced and proved to differ from classical Tn5053 by less than 0.5%, suggesting relatively recent horizontal transfer of their precursor.[11,71,78] Several transposons more distantly related to Tn5053 were also detected in different geographical regions.[11] In addition, the almost identical transposons Tn5058 and Tn50580 belonging to the Tn5053 family were found in plasmids of permafrost *Pseudomonas* strain ED23-33 and present day bacteria from contaminated sediment slurry of the Nura River (Kazakhstan), respectively.[59,85]

Transposon Tn5042[59] and the some variants of Tn5041[62] were shown to be widely distributed in environmental bacterial populations of ancient and modern bacterial strains. In the early 2000s, horizontal transfer of mercury resistance transposons was detected in Gram-positive bacteria. Indeed, TnMER11 of *B. megaterium* MB1 (Japan) differed from Tn5085 found in *Exiguobacterium* sp TC38-2b (Carpathians) only by the presence of a group II intron.[24,25,74] Many other strains of *Bacillus* spp dwelling in Minamata Bay and in Boston Harbor, USA, contained transposons nearly identical to Tn5084, Tn5085 and TnMER11. It was concluded that these three types of transposons conferring resistance to organomercuriates may contribute to the worldwide distribution and horizontal dissemination of *mer* operons among *Bacillus* strains in natural environments.[86]

Conclusion

At least three distinct types of transposons are involved in a global dissemination of mercury determinants among Gram-negative and Gram-positive bacteria: (i) the Tn3 family transposons composed of different subgroups with a major Tn21 subgroup, (ii) the Tn5053 family transposons composed of several branches, and (iii) the Tn5042-like transposons, which have so far been little investigated. The Tn21 subgroup and the Tn5053 family transposons can be regarded as pandemic transposons globally disseminated between different bacterial species. The diversity in their structure persistently increases, especially by extensive recombination between various elements, both between the transposition genes and between the transposition and the *mer* genes. In contrast, Tn5041 subgroup transposable elements are found only in pseudomonads[62] and Tn5044-like transposons and Tn5070 can be regarded as endemic.[65]

Many laboratories have called attention to studies on the mechanisms of acquisition of antibiotic resistance genes by *mer* transposons and their evolution. It is commonly supposed that mercury resistance transposons formed long before the beginning of the industrial era and became widely distributed in environmental bacterial populations and were subsequently transferred into clinical bacterial strains, where they have assumed the functions of the predecessors of present-day complex transposable elements encoding multiple antibiotic resistance. The mechanisms of generation of complex transposons by insertion of integrons containing antibiotic resistance cassettes have also been thoroughly investigated.[2,46,49,59]

On analyzing the results of *mer* transposon studies, one can see that a considerable body of work on mercury resistance transposons has been performed and that the major problems have been resolved. On the other hand, several novel transposons have recently been identified. Among them are novel types of complex transposons arising from interaction between mobile elements residing in natural environments and in the clinical setting, which are potentially able to disseminate among medically important bacteria.[26,51,55,76]

Because of this, it is felt that future studies should chiefly be devoted to description of novel complex transposable elements in bacterial chromosomes and plasmids and to elucidation of the mechanisms responsible for their formation. Possibly, most attention should be devoted to investigating Gram-positive bacteria transposons, especially complex elements, which are less well understood. Regarding noncomplex transposons, future studies should also be devoted to their applied use. There is a great opportunity to use mercury resistance transposons with organomercurial resistance for industrial bioprocesses. In particular, Gram-positive *Bacillus* strains could be exploited for applications such as the detection of mercury in contaminated soils and mercury bioremediation of contaminated sites.[87,88]

Acknowledgments

This work was partly supported by the Russian Foundation for Basic Research, grants 08-04-00263-a and 11-04-01217-a.

References

1. Grinsted J, de la Cruz F, Schmitt R. The Tn21 subgroup of bacterial transposable elements. Plasmid 1990; 24:163-89. PMID:1963947 doi:10.1016/0147-619X(90)90001-S
2. Liebert CA, Hall RM, Summers AO. Transposon Tn21, flagship of the floating genome. Microbiol Mol Biol Rev 1999; 63:507-22. PMID:10477306
3. de la Cruz F, Grinsted J. Genetic and molecular characterization of Tn21, a multiple resistance transposon from R100.1. J Bacteriol 1982; 151:222-8. PMID:6282806
4. Kratz J, Schmidt F, Wiedemann B. Characterization of Tn2411 and Tn2410, two transposons derived from R-plasmid R1767 and related to Tn2603 and Tn21. J Bacteriol 1983; 155:1333-42. PMID:6309748
5. Brown NL. The structure and expression of the mercury-resistance transposon Tn501. Folia Biol (Praha) 1984; 30:7-17. PMID:6327411
6. Lett MC, Bennett PM, Vidon DJ. Characterization of Tn3926, a new mercury-resistance transposon from Yersinia enterocolitica. Gene 1985; 40:79-91. PMID:3005130 doi:10.1016/0378-1119(85)90026-5

7. Stokes HW, Hall RM. A novel family of potentially mobile DNA elements encoding site-specific gene-integration functions: integrons. Mol Microbiol 1989; 3:1669-83. PMID:2560119 doi:10.1111/j.1365-2958.1989.tb00153.x

8. Martinez E, de la Cruz F. Genetic elements involved in Tn21 site-specific integration, a novel mechanism for the dissemination of antibiotic resistance genes. EMBO J 1990; 9:1275-81. PMID:2157593

9. Kholodii GY, Mindlin SZ, Bass IA, et al. Four genes, two ends, and a res region are involved in transposition of Tn5053: a paradigm for a novel family of transposons carrying either a mer operon or an integron. Mol Microbiol 1995; 17:1189-200. PMID:8594337 doi:10.1111/j.1365-2958.1995.mmi_17061189.x

10. Kholodii GY, Yurieva OV, Gorlenko ZhM, et al. Tn5041: a chimeric mercury resistance transposon closely related to the toluene degradative transposon Tn4651. Microbiology 1997; 143:2549-56. PMID:9274008 doi:10.1099/00221287-143-8-2549

11. Mindlin S, Kholodii G, Gorlenko Z, et al. Mercury resistance transposons of Gram-negative environmental bacteria and their classification. Res Microbiol 2001; 152:811-22. PMID:11763242 doi:10.1016/S0923-2508(01)01265-7

12. Meyer JF, Nies BA, Kratz J, et al. Evolution of Tn21-related transposons: isolation of Tn2425, which harbours IS161. J Gen Microbiol 1985; 131:1123-30. PMID:2991421

13. Nakazawa H, Mitsuhashi S. Tn2011, a new transposon encoding oxacillin-hydrolyzing beta-lactamase. Antimicrob Agents Chemother 1983; 23:407-12. PMID:6303211

14. Tanaka M, Yamamoto T, Sawai T. Fine structure of transposition genes on Tn2603 and complementation of its tnpA and tnpR mutations by related transposons. Mol Gen Genet 1983; 191:442-50. PMID:6314094 doi:10.1007/BF00425761

15. Silver S, Misra TK, Laddaga RA. DNA sequence analysis of bacterial toxic heavy metal resistances. Biol Trace Elem Res 1989; 21:145-63. PMID:2484581 doi:10.1007/BF02917247

16. Liebert CA, Wireman J, Smith T, et al. Phylogeny of mercury resistance (mer) operons of Gram-negative bacteria isolated from the fecal flora of primates. Appl Environ Microbiol 1997; 63:1066-76. PMID:9055422

17. Barkay T, Miller SM, Summers AO. Bacterial mercury resistance from atoms to ecosystems. FEMS Microbiol Rev 2003; 27:355-84. PMID:12829275 doi:10.1016/S0168-6445(03)00046-9

18. Summers AO. Organization, expression, and evolution of genes for mercury resistance. Annu Rev Microbiol 1986; 40:607-34. PMID:3535655 doi:10.1146/annurev.mi.40.100186.003135

19. Hobman J, Brown NL. Bacterial mercury-resistance genes. In: Sigel A and Sigel H. eds. Metal Ions in Biological Systems. New York: Marcel Dekker, Inc, 1997:527-568.

20. Osborn AM, Bruce KD, Strike P, et al. Distribution, diversity and evolution of the bacterial mercury resistance (mer) operon. FEMS Microbiol Rev 1997; 19:239-62. PMID:9167257 doi:10.1111/j.1574-6976.1997.tb00300.x

21. Kaliaeva ES, Kholodii GIa, Bass IA, et al. Tn5037-a Tn21-like mercury transposon, detected in Thiobacillus ferrooxidans. Genetika 2001; 37:1160-4. PMID:11642118

22. Schneiker S, Keller M, Dröge M et al. The genetic organization and evolution of the broad host range mercury resistance plasmid pSB102 isolated from a microbial population residing in the rhizosphere of alfalfa. Nucleic Acids Res 2001; 29:5169-81. PMID:11812851 doi:10.1093/nar/29.24.5169

23. Kiyono M, Pan-Hou H. The merG gene product is involved in phenylmercury in resistance Pseudomonas strain K-62. J Bacteriol 1999; 181:726-30. PMID:9922233

24. Huang CC, Narita M, Yamagata T, et al. Identification of three merB genes and characterization of a broad-spectrum mercury resistance module encoded by a class II transposon of Bacillus megaterium strain MB1. Gene 1999; 239:361-6. PMID:10548738 doi:10.1016/S0378-1119(99)00388-1

25. Bogdanova E, Minakhin L, Bass I, et al. Class II broad-spectrum mercury resistance transposons in Gram-positive bacteria from natural environments. Res Microbiol 2001; 152:503-14. PMID:11446519 doi:10.1016/S0923-2508(01)01224-4

26. Soge OO, Beck NK, White TM, et al. A novel transposon, Tn6009, composed of a Tn916 element linked with a Staphylococcus aureus mer operon. J Antimicrob Chemother 2008; 62:674-80. PMID:18583328 doi:10.1093/jac/dkn255

27. Kleckner N. Transposable elements in prokaryotes. Annu Rev Genet 1981; 15:341-404. PMID:6279020 doi:10.1146/annurev.ge.15.120181.002013

28. Dahlberg C, Hermansson M. Abundance of Tn3, Tn21, and Tn501 transposase (tnpA) sequences in bacterial community DNA from marine environments. Appl Environ Microbiol 1995; 61:3051-6. PMID:7487037

29. Pearson AJ, Bruce KD, Osborn AM, et al. Distribution of class II transposase and resolvase genes in soil bacteria and their association with mer genes. Appl Environ Microbiol 1996; 62:2961-5. PMID:8702289

30. Yurieva O, Kholodii G, Minakhin L, et al. Intercontinental spread of promiscuous mercury-resistance transposons in environmental bacteria. Mol Microbiol 1997; 24:321-9. PMID:9159519 doi:10.1046/j.1365-2958.1997.3261688.x

31. Bennett PM, Grinsted J, Choi CL, et al. Characterisation of Tn501, a transposon determining resistance to mercuric ions. Mol Gen Genet 1978; 159:101-6. PMID:416334 doi:10.1007/BF00401753

32. Misra TK, Brown NL, Fritzinger DC, et al. Mercuric ion-resistance operons of plasmid R100 and transposon Tn501: the beginning of the operon including the regulatory region and the first two structural genes. Proc Natl Acad Sci USA 1984; 81:5975-9. PMID:6091128 doi:10.1073/pnas.81.19.5975

33. Brown NL, Evans LR. Transposition in prokaryotes: transposon Tn501. Res Microbiol 1991; 142:689-700. PMID:1660177 doi:10.1016/0923-2508(91)90082-L

34. Shapiro JA. Molecular model for the transposition and replication of bacteriophage Mu and other transposable elements. Proc Natl Acad Sci USA 1979; 76:1933-7. PMID:287033 doi:10.1073/pnas.76.4.1933

35. Galas DJ, Chandler M. On the molecular mechanism of transposition. Proc Natl Acad Sci USA 1981; 78:4858-62. PMID:6272280 doi:10.1073/pnas.78.8.4858

36. Grinsted J, Brown NL. A Tn21 terminal sequence within Tn501: complementation of tnpA gene function and transposon evolution. Mol Gen Genet 1984; 197:497-502. PMID:6098802 doi:10.1007/BF00329949

37. Diver WP, Grinsted J, Fritzinger DC, et al. DNA sequences of and complementation by the tnpR genes of Tn21, Tn501 and Tn1721. Mol Gen Genet 1983; 191:189-93. PMID:6312271 doi:10.1007/BF00334812

38. Kulkosky J, Jones KS, Katz RA, et al. Residues critical for retroviral integrative recombination in a region that is highly conserved among retroviral/retrotransposon integrases and bacterial insertion sequence transposases. Mol Cell Biol 1992; 12:2331-8. PMID:1314954

39. Rådström P, Sköld O, Swedberg G, et al. Transposon Tn5090 of plasmid R751, which carries an integron, is related to Tn7, Mu, and the retroelements. J Bacteriol 1994; 176:3257-68. PMID:8195081

40. Yurieva O, Nikiforov V. Catalytic center quest: comparison of transposases belonging to the Tn3 family reveals an invariant triad of acidic amino acid residues. Biochem Mol Biol Int 1996; 38:15-20. PMID:8932514

41. Dereeper A, Guignon V, Blanc G, et al. Phylogeny.fr: robust phylogenetic analysis for the non-specialist. Nucleic Acids Res 2008; 36:W465-9. PMID18424797

42. Osbourn SE, Turner AK, Grinsted J. Nucleotide sequence within Tn3926 confirms this as a Tn21-like transposable element and provides evidence for the origin of the mer operon carried by plasmid pKLH2. Plasmid 1995; 33:65-9. PMID:7753910 doi:10.1006/plas.1995.1008

43. Kholodii G, Mindlin S, Petrova M, et al. Tn5060 from the Siberian permafrost is most closely related to the ancestor of Tn21 prior to integron acquisition. FEMS Microbiol Lett 2003; 226:251-5. PMID:14553919 doi:10.1016/S0378-1097(03)00559-7

44. Essa AM, Julian DJ, Kidd SP, et al. Mercury resistance determinants related to Tn21, Tn1696, and 5053 in enterobacteria from the preantibiotic era. Antimicrob Agents Chemother 2003; 47:1115-9. PMID:12604550 doi:10.1128/AAC.47.3.1115-1119.2003

45. Zühlsdorf MT, Wiedemann B. Functional and physiological characterization of the Tn21 cassette for resistance genes in Tn2426. J Gen Microbiol 1993; 139:995-1002. PMID:8393071

46. Partridge SR, Brown HJ, Stokes HW, et al. Transposons Tn1696 and Tn21 and their integrons In4 and In2 have independent origins. Antimicrob Agents Chemother 2001; 45:1263-70. PMID:11257044 doi:10.1128/AAC.45.4.1263-1270.2001

47. Schmidt FR, Nucken EJ, Henschke RB. Nucleotide sequence analysis of 2"- aminoglycoside nucleotidyl-transferase ANT(2") from Tn4000: its relationship with AAD(3𝔘') and impact on Tn21 evolution. Mol Microbiol 1988; 2:709-17. PMID:2850441 doi:10.1111/j.1365-2958.1988.tb00081.x

48. Sundström L, Swedberg G, Skold O. Characterization of transposon Tn5086, carrying the site-specifically inserted gene dhfrVII mediating trimethoprim resistance. J Bacteriol 1993; 175:1796-805. PMID:8383666

49. Toleman MA, Biedenbach D, Bennett D, et al. Genetic characterization of a novel metallo- beta-lactamase gene, blaIMP-13, harboured by a novel Tn5051-type transposon disseminating carbapenemase genes in Europe: report from the SENTRY worldwide antimicrobial surveillance programme. J Antimicrob Chemother 2003; 52:583-90. PMID:12951335 doi:10.1093/jac/dkg410

50. Riccio ML, Pallecchi L, Docquier JD, et al. Clonal relatedness and conserved integron structures in epidemiologically unrelated Pseudomonas aeruginosa strains producing the VIM-1 metallo-{beta}-lactamase from different Italian hospitals. Antimicrob Agents Chemother 2005; 49:104-10. PMID:15616282 doi:10.1128/AAC.49.1.104-110.2005

51. Novais A, Baquero F, Machado E, et al. International spread and persistence of TEM-24 is caused by the confluence of highly penetrating enterobacteriaceae clones and an IncA/C2 plasmid containing

Tn1696:Tn1 and IS5075-Tn21. Antimicrob Agents Chemother 2010; 54:825-34. PMID:19995930 doi:10.1128/AAC.00959-09

52. Tanaka M, Yamamoto T, Sawai T. Evolution of complex resistance transposons from an ancestral mercury transposon. J Bacteriol 1983; 153:1432-8. PMID:6298184

53. Bissonnette L, Roy PH. Characterization of In0 of Pseudomonas aeruginosa plasmid pVS1, an ancestor of integrons of multiresistance plasmids and transposons of gram-negative bacteria. J Bacteriol 1992; 174:1248-57. PMID:1310501

54. Villa L, Carattoli A. Integrons and transposons on the Salmonella enterica serovar typhimurium virulence plasmid. Antimicrob Agents Chemother 2005; 49:1194-7. PMID:15728925 doi:10.1128/AAC.49.3.1194-1197.2005

55. Labbate M, Chowdhury PR, Stokes HW. A class 1 integron present in a human commensal has a hybrid transposition module compared to Tn402: evidence of interaction with mobile DNA from natural environments. J Bacteriol 2008; 190:5318-27. PMID:18502858 doi:10.1128/JB.00199-08

56. Márquez C, Labbate M, Raymondo C, et al. Urinary tract infections in a South American population: dynamic spread of class 1 integrons and multidrug resistance by homologous and site-specific recombination. J Clin Microbiol 2008; 46:3417-25. PMID:18753343 doi:10.1128/JCM.00835-08

57. Minakhina S, Kholodii G, Mindlin S, et al. Tn5053 family transposons are res site hunters sensing plasmidal res sites occupied by cognate resolvases. Mol Microbiol 1999; 33:1059-68. PMID:10476039 doi:10.1046/j.1365-2958.1999.01548.x

58. Porter FD, Silver S, Ong C, et al. Selection for mercurial resistance in hospital settings. Antimicrob Agents Chemother 1982; 22:852-8. PMID:6758691

59. Mindlin S, Minakhin L, Petrova M, et al. Present-day mercury resistance transposons are common in bacteria preserved in permafrost grounds since the Upper Pleistocene. Res Microbiol 2005; 156:994-1004. PMID:16084067 doi:10.1016/j.resmic.2005.05.011

60. Khesin RB, Karasyova EV. Mercury-resistant plasmids in bacteria from a mercury and antimony deposit area. Mol Gen Genet 1984; 197:280-5. PMID:6394954 doi:10.1007/BF00330974

61. Kholodii GIa, Mindlin SZ, Gorlenko ZhM, et al. Molecular genetic analysis of the Tn5041 transposition system. [Russian.]. Genetika 2000; 36:459-69. PMID:10822806

62. Kholodii G, Gorlenko Zh, Mindlin S, et al. Tn5041-like transposons: molecular diversity, evolutionary relationships and distribution of distinct variants in environmental bacteria. Microbiology 2002; 148:3569-82. PMID:12427948

63. Tsuda M, Iino T. Genetic analysis of a transposon carrying toluene degrading genes on a TOL plasmid pWW0. Mol Gen Genet 1987; 210:270-6. PMID:2830457 doi:10.1007/BF00325693

64. Kholodii G, Yurieva O, Mindlin S, et al. Tn5044, a novel Tn3 family transposon coding for temperature-sensitive mercury resistance. Res Microbiol 2000; 151:291-302. PMID:10875286 doi:10.1016/S0923-2508(00)00149-2

65. Gorlenko ZhM, Kaliaeva S, Bass IA, et al. Distribution of transposons Tn5044 and Tn5070 with unusual mer operons in environmental bacterial populations. Genetika 2004; 40:1717-21. PMID:15648157

66. Kamali-Moghaddam M. Sundström L. Transposon targeting determined by resolvase. FEMS Microbiol Lett 2000; 186:55-9. PMID:10779712 doi:10.1111/j.1574-6968.2000.tb09081.x

67. Petrovski S, Stanisich VA. Tn502 and Tn512 are res site hunters that provide evidence of resolvase-independent transposition to random sites. J Bacteriol 2010; 192:1865-74. PMID:20118251 doi:10.1128/JB.01322-09

68. Stanisich VA, Arwas R, Bennett PM, et al. Characterization of Pseudomonas mercury-resistance transposon Tn502, which has a preferred insertion site in RP1. J Gen Microbiol 1989; 135:2909-15. PMID:2559144

69. Petrovski S, Blackmore DW, Jackson KL, et al. Mercury(II)-resistance transposons Tn502 and Tn512, from Pseudomonas clinical strains, are structurally different members of the Tn5053 family. Plasmid 2011; 65:58-64. PMID:20800080 doi:10.1016/j.plasmid.2010.08.003

70. Kholodii GY, Yurieva OV, Lomovskaya OL, et al. Tn5053, a mercury resistance transposon with integron's ends. J Mol Biol 1993; 230:1103-7. PMID:8387603 doi:10.1006/jmbi.1993.1228

71. Hobman J, Kholodii G, Nikiforov V, et al. The sequence of the mer operon of pMER327/419 and transposon ends of pMER327/419, 330 and 05. Gene 1994; 146:73-8. PMID:8063107 doi:10.1016/0378-1119(94)90835-4

72. Tett A, Spiers AJ, Crossman LC, et al. Sequence-based analysis of pQBR103; a representative of a unique, transfer-proficient mega plasmid resident in the microbial community of sugar beet. ISME J 2007; 1:331-40. PMID:18043644

73. Mahillon J, Chandler M. Insertion sequences. Microbiol Mol Biol Rev 1998; 62:725-74. PMID:9729608

74. Bogdanova ES, Bass IA, Minakhin LS, et al. Horizontal spread of mer operons among Gram-positive bacteria in natural environments. Microbiology 1998; 144:609-20. PMID:9534232 doi:10.1099/00221287-144-3-609

75. Huang CC, Narita M, Yamagata T, et al. Structure analysis of a class II transposon encoding the mercury resistance of the Gram-positive bacterium Bacillus megaterium MB1, a strain isolated from minamata bay, Japan. Gene 1999; 234:361-9. PMID:10395910 doi:10.1016/S0378-1119(99)00184-5

76. Narita M, Chiba K, Nishizawa H, et al. Diversity of mercury resistance determinants among Bacillus strains isolated from sediment of Minamata Bay. FEMS Microbiol Lett 2003; 223:73-82. PMID:12799003 doi:10.1016/S0378-1097(03)00325-2

77. Laddaga RA, Chu L, Misra TK, et al. Nucleotide sequence and expression of the mercurial-resistance operon from Staphylococcus aureus plasmid pI258. Proc Natl Acad Sci USA 1987; 84:5106-10. PMID:3037534 doi:10.1073/pnas.84.15.5106

78. Ojo KK, Tung D, Luis H, et al. Gram-positive merA gene in Gram-negative oral and urine bacteria. FEMS Microbiol Lett 2004; 238:411-6. PMID:15358427

79. Mindlin SZ, Bass IA, Bogdanova ES, et al. Horizontal transfer of mercury resistance genes in environmental bacterial populations. Mol Biol 2002; 36:160-70 doi:10.1023/A:1015353402657.

80. Reniero D, Mozzon E, Galli E, et al. Two aberrant mercury resistance transposons in the Pseudomonas stutzeri plasmid pPB. Gene 1998; 208:37-42. PMID:9479042 doi:10.1016/S0378-1119(97)00641-0

81. Ochman H, Lawrence JG, Groisman EA. Lateral gene transfer and the nature of bacterial innovation. Nature 2000; 405:299-304. PMID:10830951 doi:10.1038/35012500

82. Koonin EV, Makarova KS, Aravind L. Horizontal gene transfer in prokaryotes: quantification and classification. Annu Rev Microbiol 2001; 55:709-42. PMID:11544372 doi:10.1146/annurev.micro.55.1.709

83. Ng SP, Davis B, Palombo EA, et al. A Tn5051-like mer-containing transposon identified in a heavy metal tolerant strain Achromobacter sp. AO22. BMC Res Notes 2009; 2:38. PMID:19284535 doi:10.1186/1756-0500-2-38

84. Bramucci M, Chen M, Nagarajan V. Genetic organization of a plasmid from an industrial wastewater Bioreactor. Appl Microbiol Biotechnol 2006; 71:67-74. PMID:16244860 doi:10.1007/s00253-005-0119-2

85. Smalla K, Haines AS, Jones K, et al. Increased abundance of IncP-1beta plasmids and mercury resistance genes in mercury-polluted river sediments: first discovery of IncP-1beta plasmids with a complex mer transposon as the sole accessory element. Appl Environ Microbiol 2006; 72:7253-9. PMID:16980416 doi:10.1128/AEM.00922-06

86. Narita M, Matsui K, Huang CC, et al. Dissemination of TnMER11-like mercury resistance transposons among Bacillus isolated from worldwide environmental samples. FEMS Microbiol Ecol 2004; 48:47-55. PMID:19712430 doi:10.1016/j.femsec.2003.12.011

87. Narita M, Yamagata T, Ishii H, et al. Simultaneous detection and removal of organomercurial compounds by using the genetic expression system of an organomercury lyase from the transposon TnMER11. Appl Microbiol Biotechnol 2002; 59:86-90. PMID:12073137 doi:10.1007/s00253-002-0946-3

88. Bontidean I, Mortari A, Leth S, et al. Biosensors for detection of mercury in contaminated soils. Environ Pollut 2004; 131:255-62. PMID:15234092 doi:10.1016/j.envpol.2004.02.019

CHAPTER 3

Integrons:
Antibiotic Resistance Evolution and Beyond

Piklu Roy Chowdhury, H.W. Stokes and Maurizio Labbate*

Abstract

Integrons include a site-specific recombination system that can capture gene cassettes. Gene cassettes are the smallest known mobilizable units of DNA and normally only comprise a single gene and a recombination site essential for the site-specific recombination event to occur. Although the site specific recombination reaction is catalyzed by an integrase protein that is a member of the tyrosine family of site specific recombinases, the biology and biochemistry of the system is unusual in a number of respects. Most notably, gene cassettes comprise a family of elements that are highly diverse both with respect to the genes and the recombination sites within them. Integrons first came to prominence as a consequence of their infiltrating pathogenic Gram negative bacteria. In this context integrons commonly possess multiple cassettes with the associated genes conferring, collectively, resistance to a wide range of clinically important antibiotics. As a consequence they are one of the single biggest contributors to the evolution of antibiotic resistant bacteria. Integrons however are ancient structures that are widely distributed among the Proteobacteria. In the broader context, cassette associated genes are remarkably diverse and highly novel. This mobile cassette "metagenome" includes a pool of novel gene cassettes potentially available to whole microbial communities. While the function of most cassette genes in this pool remains unknown, it is clear that they are rich source of innovation and novelty.

Introduction

Prokaryotes are different to eukaryotes in a number of ways. One of the most notable of these is the ability of the former to exchange DNA between individuals to an extent that is greatly in excess of that seen in the latter. Consequently, it is not possible to understand prokaryotic evolution without an understanding of Lateral Gene (or Genetic) Transfer (LGT).[1] LGT is a complex process that is dependent on mechanisms that move DNA between cells and mechanisms that provide for incorporation of DNA into the host genome so as to allow propagation with cell division. Incorporation of DNA can be by autonomous replication, as in the case of many plasmids, or by some specific integrating mechanism. Many of these integrating mechanisms are the subject of dedicated chapters in this book.

LGT is a process that shapes genomes in both the short and long-term. In the long-term, up to 30% of a genome can be acquired by LGT[2] with the movement of DNA occurring across disparate phylogenetic groups. Indeed, the promiscuous spread of DNA by LGT is important in understanding the microbial speciation and has even brought in to question the notion of creating a universal species concept.[3-5] Under conditions of strong selection, LGT can also have profound impacts in the very short-term to the point where bacteria can be seen to evolve in "real time" from a human perspective. The best example of this is the evolution of multi drug resistant bacteria

*The i3 Institute, University of Technology, Sydney. Australia.
Corresponding Author: Maurizio Labbate—Email: maurizio.labbate@uts.edu.au

Bacterial Integrative Mobile Genetic Elements, edited by Adam P. Roberts and Peter Mullany.
©2013 Landes Bioscience.

which is now seen as one of the great problems and challenges of the 21st Century. As is evident from many of the chapters in this book, what drew our attention to many integrating elements is their association with resistance genes and the havoc that has resulted.

The integron was identified as a discrete genetic element just over 20 years ago.[6] Its defining feature is a site-specific recombination system that captures discrete DNA units known as gene cassettes and, as such, contributes to LGT by providing a specific system that contributes to gene incorporation. The integron/gene cassette system contributes to LGT in both the short and long-term and this chapter will examine how it does this. In the short-term, the integron gene cassette system is a major contributor to the spread of antibiotic resistance genes in Gram-negative bacteria and it was in this context that it was first identified. The system however is an ancient one and has been a feature of genomes in many Proteobacteria for hundreds of millions of years. In this broader evolutionary context it has been influencing bacterial adaptation in ways that we are yet to fully understand.

Integrons: Emergence of a New Discrete Genetic Element

The discovery of microbes that produce antibacterial agents by Sir Alexander Fleming in 1928, followed by the commercial production of what was later shown to be penicillin in the 1940s, marked the onset of a golden era in the history of medical science - 'The antibiotic era'. However, resistance to penicillin emerged within a couple of years after it come into widespread use.[7] This led to the search for new antibiotics, a consequence of which was the evolution of a wide range of resistance mechanisms in pathogenic bacteria.[8] Over time this has culminated in the major global problem of multiple drug resistance in pathogenic bacteria. The rate at which the various resistance mechanisms evolved intrigued the scientific community and guided research to the identification of the newly discovered phenomena of LGT and its components, the mobile genetic elements including transposons and plasmids.

The initial indication of the presence of drug resistant genes at similar genetic locations on plasmids and transposons was perceived in the early 1980s, when combination of classical and advanced genetic techniques like heteroduplex analysis and restriction mapping were applied to compare closely related drug resistant mobile elements.[9-12] Quickly following on from these studies, application of DNA sequencing to resistance analysis from the mid-1980 onwards, revealed the presence of identical DNA sequences adjacent to these antibiotic resistant genes suggesting insertional hotspots of these elements.[13-16] Detailed characterization of the common regions adjacent to the resistance loci led to the identification of a novel site specific gene recombination system, similar to the phage λ system. This came to be known as the "integron."[6]

The original definition referred to a specific integron although different examples possessed diverse antibiotic resistance gene cassettes. Over time it became clear that integrons comprised a diverse family of elements and we now know that integrons are ancient genetic entities[17-19] that predate the antibiotic era and are equally prevalent in non-clinical environmental microorganisms,[20-23] where they are rarely linked to antibiotic resistant genes.[24] Thus, discovered in context of antibiotic resistance, integrons play a diverse role in the evolution of bacteria.

The Integron Site-Specific Recombination System

Functionally, integrons possess a site-specific recombination systems capable of capturing, accumulating and expressing tandem genes present in arrays.[6] The captured units are called the 'gene cassettes'[25,26] that are inserted at a specific recombination site associated with the core integron (Fig. 1A), *attI*. Based on the encoded IntI protein sequence, integrons have been characterized into different classes, of which classes 1, 2 and 3 are most clinically important since they are primarily associated with antibiotic resistance genes.[27,28] Clinically, class 1 integrons are most prevalent, detected in 22% to 59% of Gram-negative clinical isolate surveys.[29] Recently they have also started appearing in Gram-positive bacteria.[15,30,31]

Structurally, clinically-derived class 1 integrons are most frequently physically linked to the transposition genes and inverted repeats of Tn*402* transposons although, in most cases, part/s of the transposition module are deleted. The original integron definition as applied to class 1 included both the components of the site-specific recombination system and the other features between the Tn*402* inverted repeats thereby also making a class 1 integron a defective Tn*402* transposon.[6,32] The characteristic Tn*402*-like inverted repeats (IRi and IRt) therefore demarcate the boundaries of clinical class 1 integrons within which are two conserved segments (CS), the 5'-CS and the 3'-CS.[6,33] A variable region (VR), made of arrays of gene cassettes, is present between the two conserved regions (Fig. 1B). For the purposes of this review we define a clinical class 1 integron as one that is associated with the remnants of a Tn*402* transposition module and a 3'-CS. The 5'-CS of clinical class 1 integrons contains the core site-specific recombination unit made up of the *intI1* integrase gene, product of which catalyzes site specific insertion and excision of genes, when present as gene cassettes.[25] Such insertion events mostly target a recombination site called the *attI1*[34] also present within the 5'-CS. Class 1 integrons also possess a promoter (P_c) that express cassette associated genes[35,36] and is an integral part of the core integron. The 3'-CS

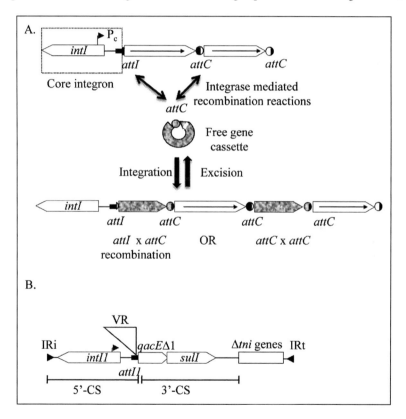

Figure 1. A) The integron/gene cassette site-specific recombination system. The integron platform consists of the *intI* gene, *attI* site and P_c promoter. The gene cassette usually consists of a single promoterless ORF and a recombination site called *attC*. In its free form, the gene cassette is circularized and is recombined into the integron by an integrase-mediated reaction between *attI* and *attC* (or *attC* and *attC* in lower frequencies). B) The genetic structure of a typical clinically derived class 1 integron. The 5'-CS contains the core integron consisting of IRi, *intI1*, *attI* and P_c. The variable region (VR) is where gene cassettes are inserted. The 3'-CS minimally consists of *qacEΔ* and *sulI* however, sometimes *orf5* and *orf6* make up part of it. Remnants of the Tn*402* transposition module (*tni*) and IRt can sometimes be found downstream of the 3'-CS.

of clinical class 1 integrons minimally consist of a truncated quaternary ammonium compound resistant gene (*qacEΔ*) and a sulphonamide resistant gene (*sul1*) while two other open reading frames, *orf5* and *orf6*, are sometimes present.[6,37] Remnants of the Tn*402* transposition module (transposition (*tni*) genes and IRt) are present beyond the 3'-CS in some clinical class 1 integrons. The antibiotic era has seen the widespread infiltration of the clinical type of class 1 integrons into diverse general environments, ranging from polluted to pristine sites and most commonly those enriched by the use of antibiotics.[38,39] Class 1 integrons are also found in environmental samples and are characterized[20,38] by specific structural differences to the clinical class 1 integrons. The most definitive ones are absence of any physical link to the defective transposition module of Tn*402* transposons and their, primarily, chromosomal location.

A recent classification based on the association of integron classes with mobile elements, like transposons and plasmids, or chromosomal locations has been used to categorise integrons into two broad groups - the mobile integrons and the chromosomal integrons. The mobile integrons broadly include the class 1s, notwithstanding that they can be occasionally found in a chromosome, as well as the class 2 and class 3 integrons.[27,40] They are primarily associated with antibiotic resistant genes, and have also been referred to as 'mobile resistance integrons' (MRI),[28] again notwithstanding that resistance cassettes can be sometimes found in chromosomal integrons. Chromosomal integrons (CI), which in fact comprise the largest number of distinct classes, apart from their chromosomal location, also are typically associated with genes of unknown functions.[28,41]

The Mobilisation of Integrons and the Spread of Antibiotic Resistance

The most frequently isolated class 1 integrons from clinical isolates conventionally retain remnants of the Tn*402* transposition module and the characteristic inverted repeats (Fig. 1B) as noted above. Thus, although defective, they are capable of transposition when the necessary essential functions are supplied in trans.[32] Transposition of this family of transposon however shows a high level of target site selection, especially toward resolution (*res*) sites that are found in many other transposons and conjugative plasmids. Consequently, Tn*402* and its close relatives belonging to the broad Tn*5053* family are referred to as '*res* site hunters'.[42,43] Such associations provide mobility to the integrons and have resulted in the rapid dissemination of resistance genes among Gram-negative bacteria. As a result, there has been widespread infiltration of the clinical-type of class 1 integrons into diverse general environments[44-48] and this is most evident among mercury resistant transposons.

Recently, four examples of class 1 integrons have been reported with functional transposition modules, different to that seen in Tn*402* type.[49] All such isolates have been recovered from clinical specimens, namely *Enterobacter cloaceae*,[50] *Pseudomonas putida*[51,52] and *Pseudomonas aeruginosa*[53] and indicate the presence of other non-Tn*402* type '*res*' hunter transposons are possibly also prevalent within clinical samples. These findings suggest that mobilization of a chromosomal class 1 integron may have occurred on multiple independent occasions.[20,38] The latter event is considered an important step in the dissemination of class 1 integrons[38] in the antibiotic era.

The IntI Subfamily of Recombinases

Integron integrases belong to the tyrosine recombinase family of site-specific recombinases, related to phage-λ integrase.[54] The family is characterized by the presence of a tyrosine residue approximately within 20 amino acids from the carboxy-terminus of the integrase protein. It is also characterized by having three short conserved regions of amino acids named patch I, II and III. The distinguishing feature of the IntI subfamily of tyrosine recombinase is the presence of 18 or 19 amino acid residues between patch II and III, which are absent in other tyrosine recombinases.[21,54] Based on the amino acid sequence homology of the IntI protein the pair wise homology between proteins derived from integrons of different classes range between 41 and 57%, including integrons isolated from environmental sources.[21,26] This is also generally true when the three major classes of mobilized integrons are compared. The class 1 IntI protein (IntI1) shares 57% amino acid identity with IntI3[27] and IntI2 has integrons 43% identity with IntI1.

Gene Cassette Structure

Gene cassettes are the smallest mobilisable units of DNA consisting of, most frequently, a promoter-less open reading frame and an associated recombination site called the 59-base element (59-be)[13] or *attC*.[55] Gene cassettes comprise the units of insertion into integrons. Given that the genes within them lack their own promoter such genes are generally dependent on a promoter, designated P_c and located within the core integron, for expression. Gene cassettes can exist in two forms, as independent free covalently closed circles within bacterial cells and in a linear integrated form when integrated into the integron. Gene cassettes are not replicons and so cannot replicate in the independent circular form. Rather, they need to be inserted in an integron for replication to occur, which in turn is dependent on the integron being located in a replicon – normally either a chromosome or a plasmid.

The most common recombination integration reaction is one between *attI* and *attC* which sees a circular cassette inserted immediately adjacent to P_c. The IntI protein can discriminate between the two *attC* strands such that the insertion sees the associated cassette gene oriented so that expression from P_c is possible (Fig. 1A).[35,36] The term 59-be is the original term for this recombination site since the first few identified gene cassettes at the time were believed to have consisted of 59-bases.[13] However, subsequently it was clear that the site could range from 55bp (*oxa118* and *aadA4* cassettes) to 141bp (*imp15* cassette) in length.[28] In 1997, Hansson et al., proposed the term *attC* to make it consistent with the other similar recombination systems[56] and this term has become accepted in the contemporary literature to describe this site.[55]

Streamlined assemblage of gene cassettes at *attI* sites of integrons results in the formation of 'cassette arrays'. Arrays can consist of up to eight or even more genes in mobile resistance integrons[28] while over 200 cassettes[57] have been reported in chromosomal integrons. The repertoire of antibiotic resistant gene cassettes is increasing over time. In a 2009 compilation, there were 132 gene cassettes conferring resistance to known antibiotics and 62 additional *gcu* cassettes (or gene cassettes of unknown functions) associated with mobile integrons.[28] A similar compilation in 2007 of the cassette pool of chromosomal integrons among Vibrionaceae identified 1677 gene cassettes, 65% of which were novel with no previous database entry.[18]

The Recombination Sites and Reaction—*attI* and *attC*

The recombination sites recognized by IntI proteins possess some structural similarities to the Int family more broadly. In particular both *attI* and *attC* sites have an identifiable 'simple site' made of a pair of inverted repeats forming the enzyme-binding sites, separated by a 6–8bp spacer region.[58] The integron gene cassette system, in contrast, consist of two different recombination sites, *attI* and *attC* and only a single strand exchange occurs in the course of integron mediated recombination reactions.[59]

The *attI1* site consists of a 'simple site' and two additional integrase binding sites known as the DR1 and the DR2, located at the 5′ end of the *attI* region are specifically present in case of class 1 integrons.[34,60,61] The simple site comprises about 20 bases and includes a pair of inverted repeat sequences each of 7bp. The inverted repeats are designated L (left) domain and R (right) domain. The recombination cross over point is located between the G and TT in a conserved 5′-GTT-3′ triplet within the R integrase-binding domain.[25]

The cassette associated *attC* sites are characterized by the presence of two simple sites, one at each of its ends and each containing a pair of conserved regions that comprise IntI binding domains made up of 7 or 8 base pairs. Individually, each simple site is an imperfect inverted repeat. In addition, each *attC* site is also an imperfect inverted repeat across its whole length (Fig. 2). In total there four IntI binding domains, two in each simple site, and are referred to as IL, 2L (left end) and 2R, 1R (right end)[62] (Fig. 2). The recombination crossover point is located at the 1R region in between the G and TT of the conserved 5′-GTT-3′ triplet. The presence of two simple sites and other palindromic sequences enables *attC* regions to display a cruciform secondary structure when present as single DNA, and which is essential for interactions of the integron-integrase[63,64] during the recombination reaction.

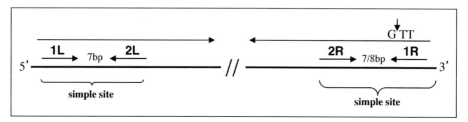

Figure 2. *attC* sequences are imperfect inverted repeats (arrows demonstrate regions of symmetry) with conserved regions of approximately 20-bp at each end. These conserved regions include a simple site (1L/2L and 1R/2R) that facilitate IntI binding and are themselves imperfect inverted repeats with a 6–8 bp spacer in between. The core-site (1R) and inverse core site (1L) are the regions of most conservation between *attC* sites and are inversely related to one another. The core site contains the highly conserved GTT with the recombination crossover point occurring between G and TT. The region between the two simples sites (marked by //) is of variable length and sequence. The diagram represents the top strand of DNA. The associated cassette gene would be to the left of the figure.

The preferred recombination site for insertion of gene cassettes into an integron backbone is the integrase-associated *attI* sites even where previously inserted cassettes present possible target sites in the form of their *attC* sites. Other rare insertion events are possible. Examples of such secondary sites insertions[65] are exemplified by presence of an *aadB* in plasmid backbones, *dfrA14* in the *strA* gene and *gcu30* cassette in *intI1* gene.[28] Notwithstanding these, the commonly encountered recombination reactions include: *attI* x *attC*, *attC* x *attC* (Fig. 1A) and, less likely, *attI* x *attI*. The first two are most frequently encountered in the case of independent cassette insertions and deletions, while the third mostly leads to co-integrate formation.[66]

In the case of an *attC-attC* recombination, a Holliday junction is established between two appropriately folded single stranded *attC*s by an integrase complex.[41] Resolution of the synapse leads to the excision of a covalently closed single stranded intermediate from the bottom strand, while the top strand remains unchanged. Upon replication one of the molecules loses a cassette while the other remains unchanged. Recombination reaction between *attI* and *attC* sites is however different as the *attI* sites are recognized by integrase while in classical double-stranded forms, in contrast to the *attC* recognition, thus, productive strand exchange occurs successfully only by the replication of the first strand exchange reaction.[63,67] The cassette insertion reaction also affects one of the daughter DNA molecules, while the other remains unchanged. Thus, one of the most significant features of an integrase mediated recombination reaction which makes it different to other similar reactions, is the ability of the integrase to recognize single stranded DNA molecules folded in specific cruciform structures that are facilitated by characteristic inverted repeats present at the primary and secondary recombination sites.[68,69]

Chromosomal Integrons

Chromosomal integrons are commonly found in the general environment with substantial diversity in integrase sequence and number or associated cassettes (Table 1). In organisms found in the broader environment beyond the clinical/commensal setting, integrons are often found in chromosomal locations and rarely carry known antibiotic resistance gene cassettes. Given this, integrons are regarded as having a more general role in evolution than simply carrying and expressing antibiotic resistance genes. Chromosomal integrons have mainly been characterized in members of the γ-Proteobacteria and β-Proteobacteria, although chromosomal intergrons have also been found among the δ-Proteobacteria, Planctomycetes and Spirochaetales. Phylogenetic analyses of integrases identified three main integron clades: the soil/freshwater Proteobacteria group, the inverted integrase group and the marine γ-Proteobacteria group with class 1 and 3 part of the soil/freshwater Proteobacteria group and class 2 part of the marine γ-Proteobacteria group.[18,41] As their

Table 1. Selected list of bacterial species containing chromosomal integrons

Bacterial Group	Bacterial Strain	Number of Gene Cassettes
γ-Proteobacteria		
Vibrionaceae	*Vibrio cholerae* N16961	179
	Vibrio vulnificus CMCP6	217
	Vibrio rotiferianus DAT722	116
	Vibrio parahemolyticus RIMD2210633	69
	Vibrio alginolyticus 12G1	51
	Vibrio fischeri ES114	38
Pseudoalteromonas	*Pseudoalteromonas tunicata* D2	7
	Pseudoalteromonas haloplanktis TAC125	5
Xanthomonadaceae	*Xanthomonas campestris* pv *campestris* ATCC 33913	22
	Xanthomonas campestris pv *vesicatoria*	3
Pseudomonadaceae	*Pseudomonas stutzeri* Q	> 7
	Pseudomonas alcaligenes	32
Shewanella	*Shewanella* sp MR-7	2
	Shewanella denitrificans OS217	0
Alteromonadaceae	*Saccharophagus degradans* 2–40	73
β-Proteobacteria		
	Nitrosomonas europaea ATCC19718	0
	Nitrosomonas eutropha C71	3
	Azoarcus sp EbN1	3
	Rubrivivax gelatinosus PM1	2
δ-Proteobacteria		
	Geobacter metallireducens GS-15	3
Planctomycetes	*Rhodopirellula baltica* SH1	0
Spirochaetales	*Treponema denticola* ATCC35405	45

name implies, the inverted integrase group have their associated *intI* gene in the inverse orientation (i.e., the *intI* reads toward the *attI*) compared with the two other integron clades (Fig. 1A). These phylogenetic studies clearly show an environmental source for the clinically important integrons.

There are characteristics that differentiate chromosomal integrons from integrons found from the class 1 integrons common in bacteria from clinical environments. In particular, they are more likely to be 'fixed' and consequently co-evolve with their host for a substantial amount of time. This can be seen with integrons from *Vibrio* spp and *Xanthomonas* spp whose integrases form a group together.[18,41] Even so, phylogenetic studies suggest that chromosomal integrons have been subject to lateral transfer at some stage in their evolutionary history.[18,70] Other characteristics of chromosomal integrons are that they can harbour large cassette arrays with some species containing over 150 cassettes (Table 1). Furthermore, they tend to harbour cassettes with higher *attC* sequence similarity when such sites are compared within a Genus or species. This observation has been made in the chromosomal arrays of multiple bacteria including *V. cholerae*, *V. fischeri*, *V. metschnikovii*, *Pseudomonas alcaligenes*, *Pseudomonas stutzeri*, *X. campestris* and *Treponema denticola*.[19,70-73] In *V. cholerae* the *attC* sites are highly conserved, differing by less than 10% nucleotide identity.

Adaptive Role of Integrons and Cassettes

Recent studies have provided new insight into the biology of integrons. It has been known for some time that insertion, excision and rearrangement of cassettes are dependent on the integrase but the conditions that induce or influence integrase transcription have only recently been explored in detail. The bacterial SOS response is a global response to DNA damage.[74] However, it is emerging as a general stress response that enhances bacterial adaptation through increased mutation and lateral gene transfer.[75] Recently, the SOS response was reported to regulate the integron-integrase in *V. cholerae* 4.5-fold and in the class 1 integron in *E. coli* by 37-fold and that this subsequently increased the level of gene cassette excision by 141-fold and 340-fold respectively.[76] The key regulatory element in the integrase promoter for SOS regulation, a LexA binding site, is conserved across the vast majority of integrase promoters indicating similar regulation in most integrons.[77] From this data it is clear that the integron is an element that assists the cell in adapting to stress. In environments that the microorganism is maladapted, the integron is likely to provide innovation through the acquisition of new gene cassettes from surrounding bacterial communities or through shuffling of non-transcribed gene cassettes distant from P_c.[76]

Guerin and colleagues[76] have indicated that P_c is the main promoter for cassette transcription and that cassettes distant from P_c are "banked" until they are shuffled to a position close to P_c. However, a recent study has demonstrated that the majority of gene cassettes in a 116-cassette vibrio array were transcribed in normal and stress growth conditions. Further analysis identified numerous promoters throughout the array. These promoters were diverse in sequence and responded to growth and stress conditions differently.[78] Thus, cassette transcription is not only dependent on P_c although if P_c is a strong promoter, movement of cassettes in front of P_c would ensure adequate transcription under conditions of strong selection and the upregulation of IntI and its impact on increasing shuffling into *attI* implies that P_c has a key role in expressing cassette genes whose expression may otherwise be problematical. The presence of additional promoters provides an extra level of innovation. Apart from allowing most of the cassettes to contribute to bacterial adaptation, they provide gene cassettes access to diverse regulatory options. Repeated rearrangement of cassettes containing genes and/or promoters within large cassette arrays has the capacity to assemble operon-like groups of co-expressed and complimentary cassettes within an array. A study using a synthetic integron has demonstrated the ability for the integron to do exactly this. Artificial gene cassettes containing the genes for tryptophan biosynthesis were delivered into a synthetic integron platform with multiple rounds of recombination and selection creating numerous arrangements of varying fitness and tryptophan production capacities.[79]

A defining feature of the SOS response is that it is activated by the presence of long-living ssDNA in the cell, a signal for DNA damage. As a result, laterally acquired ssDNA was hypothesized to induce SOS and was proven to be the case for ssDNA acquired by conjugation.[80] Further to this, conjugation was shown to induce the integron-integrase and increase cassette recombination.[80] While yet to be demonstrated, SOS is also likely to be induced by transformation and transduction. Therefore, the process of LGT itself enhances integron activity maximizing the potential for the continued acquisition of new gene cassettes and genetic diversity in bacterial populations.

Gene Cassette Diversity—Metagenomic Studies

The relative conservation of *attC* sites has permitted amplification of gene cassettes from environmental metagenomic DNA.[81] In earlier studies with small scale sequencing of amplified gene cassettes, substantial genetic diversity and novelty were found.[24,81,82] Even within small physical distances, gene cassette populations differed markedly.[82] Recent large-scale sequencing studies have reiterated this finding and provided more insight into gene cassettes in ecosystems.[23,83,84] These studies combined show that approximately 80% of gene cassettes from metagenomic samples have ORFs with no known homolog or homologs of unknown function.[18,24] Nevertheless, a theme emerging from these studies is that environments harbour a gene cassette ecotype.

The idea of a gene cassette ecotype is supported by studies looking at the functional distribution of cassette-encoded proteins from different environmental sites and from the functional identification of cassette proteins that would appear to benefit microbes in that environment. The functional distribution of cassette sequences from multiple sites in Halifax (Canada) was substantially different from that found in vibrio genomes.[83] Even within the Halifax environment, geographically separated sites subject to raw sewage were more similar than other more pristine sites.[83] The functional profile of cassettes amplified from an industrially contaminated site in Sydney (Canada) was distinct from Halifax and vibrio genomes.[83] Functional analysis of cassette-encoded proteins from polluted sites appear to be involved in the degradation of toxic compounds and pollutants consistent with the environment they are found.[22,83-85] In coral mucous communities a significant proportion of cassette-encoded proteins were implicated in biochemical processes involved in antibiotic resistance, consistent with coral microbiome ecology.[86] Finally, amplified cassettes from hydrothermal vent mussels contained ORFs consistent with the symbiont-animal host metabolic processes.[23] Overall, these data are suggestive of gene cassette ecotypes due to different selection regimes acting on these gene pools.

Gene Cassette Diversity—Functional Gene Content Studies

The large number of cassette-encoded proteins with no identifiable function has made it difficult to understand what the adaptive roles of these mobile units are in bacterial evolution. Most of what is known about cassette encoded gene function is based on bioinformatics with few studies confirming their role experimentally. There is a danger of predicting a biological role for cassette-encoded proteins because they are often general in activity. For example, putative acetyltransferase or methylases are common but determining the primary substrate they are modifying is crucial to understanding its biological role. Nevertheless, bioinformatic analyses of cassette-encoded proteins have suggested diverse functions.

From polluted sites, cassette-encoded proteins have been suggested to be involved in protection or degradation of pollutants including organochlorine, oil, dioxins, polychlorine biphenyls and phenylacetic acid[22,84] (Table 2). In one polluted environment, LysR-like transcriptional regulators were common indicating that cassette-encoded proteins may influence gene expression.[84] In another environment, cassette-encoded proteins putatively involved in the recovery of environmental nutrients including transport binding proteins and isochorismatases, involved in iron scavenging were identified.[83] In a coral mucous environment, a high proportion of cassettes predicted to be involved in antibiotic resistance were detected in *Vibrio* species.[86]

Of those cassette protein functions confirmed experimentally most are from *Vibrio* species however, diverse functions are observed including polysaccharide biosynthesis, lipases, dNTP pyrophosphohydrolases, a sulfate-binding protein, cold-shock protein and a drug-binding protein putatively involved in transcriptional regulation (Table 2). Some cassette proteins have also been implicated in virulence. In *V. cholerae* these include a heat-stable toxin gene (*sto*), a mannose-fucose resistance hemagglutinin (*mrhA*) and a chloramphenicol resistance gene (*catB9*) (Table 2). In *V. vulnificus, V. alginolyticus* and *V. parahemolyticus* is a cassette encoding the PAS factor, a small protein preferentially induced during human infection by *V. vulnificus* CMCP98K which mediates secretion of periplasmic proteins (Table 2).

A common gene cassette is that which contain toxin-antitoxin (TA) modules. TA systems are a diverse family of selfish genetic elements that prevent growth of their host if not maintained.[87] TA gene cassettes are common in large gene cassette arrays but absent in small arrays and have been demonstrated to stabilize cassette arrays by prevent large multi-cassette deletions.[88]

Table 2. Examples of experimentally confirmed functional ORFs in gene cassettes

Source of Cassette	Function	Determination of Function	Ref.
Vibrio cholerae	Sulfate-binding protein	Complementation of *E. coli* mutation	19
Vibrio cholerae OP4G	Transcriptional regulation	Crystal structure determination and drug binding assay	105
Vibrio cholerae N16961	Chloramphenicol resistance	Active when expressed in *E. coli* on medium containing chloramphenicol	106
Vibrio cholerae GP156	Heat stable enterotoxin	Active in suckling mouse assay when expressed in *E. coli*	107
Vibrio cholerae	mannose-fucose resistance hemagglutinin	Mutagenesis and testing in infant mouse model	108,109
Vibrio marinus	Psychrophilic lipase	Active when expressed in *E. coli* at 10°C	19
Vibrio vulnificus CMCP6	Cold shock	Complementation of cold shock phenotype in *E. coli*	110
Vibrio vulnificus CMCP98K	Secretion	Expression in *E. coli* mediates secretion of periplasmic proteins	111-113
Vibrio vulnificus 1003	Capsular polysaccharide biosynthesis	Transposon mutagenesis	114
Vibrio rotiferianus DAT722	dNTP pyrophosphohydrolase (iMazG)	Crystal structure determination. Expressed in *E. coli* and enzyme activity measured	115
Soil metagenomic DNA	Potential transport protein	Crystal structure determination	116
Soil metagenomic DNA	ATPase activity	Expressed in *E. coli* and enzyme activity measured	117
Soil metagenomic DNA	Methyltransferase activity	Expressed in *E. coli* and enzyme activity measured	117

Use of Chromosomal Integron-Cassette Arrays in Phylogenetics

Genome studies have shown mobile DNA to often be the defining feature that separates closely related bacterial strains, particularly within pathogenic species.[89,90] Epidemiological studies rely on simple methodologies that have the capacity to differentiate closely related strains so that the introduction, movement and evolution of pathogens can be monitored. Numerous methodologies have been developed to address this such as restriction nuclease digestion of chromosomal DNA and subsequent pulsed field gel electrophoresis (PFGE), ribotyping and PCR-based techniques targeting repetitive regions in genomes (for e.g., representative molecular methods in Foley et al.[91] for Gram-negative food pathogens).

The integron cassette array is an excellent target for strain typing. Through various studies and genome sequencing projects, it is now known as one of the most variable parts of the genome[57,72,86,92,93] and due to the presence of conserved *attC* sites, can be easily targeted by PCR. Given this, the integron cassette array can provide fine resolution in the differentiation of closely related strains using inexpensive and easily accessible technology. *attC*-PCR is one such methodology that targets the integron cassette array for strain typing.

To assist in understanding the benefits of this method, a general introduction of the evolution of *V. cholerae* is given, however we direct the reader to a recent review[94] for more comprehensive information. While more than 200 serotypes of *V. cholerae* exist, only O1 and O139 are responsible for epidemic and pandemic cholera.[95] The O1 serogroup can be further subdivided into Classical and El Tor biotypes. There have been seven pandemics of cholera with the seventh and current pandemic beginning in 1961 and driven by the El Tor biotype. The classical biotype is believed to be responsible for the previous six pandemics.[95] The O139 serogroup emerged in 1992 causing epidemics in India and Bangladesh but not replacing O1 El Tor.[96] More recently, variants of O1 El Tor have emerged and replaced prototype seventh-pandemic O1 El Tor strains called "'altered" and "hybrid" which cannot be easily typed as classical or El Tor by standard phenotypic assays.[97-99]

attC—Primer Based PCR

attC-PCR relies on primers that target the conserved *attC* sites (Fig. 3) and amplifying the variably sized gene cassettes by PCR. Since *attC* site sequences are conserved within species but deviate between species, primers are designed to be species specific. The first study to develop this method separated amplified gene cassettes on an agarose gel producing a DNA size fragment fingerprint of 5–8 DNA fragments per strain.[100] This methodology met the guidelines set by the European Study Group on Epidemiological markers of the European Society for Clinical Microbiology and Infectious Diseases on reproducibility, typeability and stability. It was used to type pathogenic and environmental strains in South America and was comparable to REP-PCR (a method that uses PCR to target repetitive sequences across the genome) and PFGE. Recent studies using this method have corroborated these findings by showing it to provide finer resolution than ERIC-PCR (a similar method to REP-PCR)[101] and in separating closely related *V. cholerae* strains.[102]

The resolution of *attC*-PCR was significantly enhanced using fluorescently labeled *attC* primers and using a DNA sequencer with high-resolution denaturing gel electrophoresis for separation and detection of the amplified PCR products.[103] As a result, approximately 80 DNA fragments were detected improving resolution more than 10-fold. This methodology was proven to have higher resolution than *recA* and multi-locus sequence phylogeny.[103] It had the capacity to separate the O1 Classical, O1 El Tor and O139 groups and further resolution within these groups which is not achievable using standard methods. Furthermore, this enhanced *attC*-PCR method was used to identify a large 40 gene cassette deletion present in a subset of the seventh pandemic strains O1 El Tor and in all O139 strains, mapping the emergence of O139 from a subset of O1 El Tor.

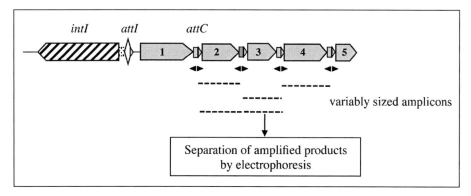

Figure 3. *attC*-PCR relies on the use of primers that target the conserved *attC* sites and amplifying the variably size gene cassettes by PCR. The resulting amplicons are sized by gel electrophoresis producing a strain-specific fingerprint. Primers are shown as black triangles and some potential amplicons as dashed lines. While not shown in the figure the protocol will also potentially amplify excised circular cassettes. Such an outcome however does not change the fingerprint pattern for a specific strain.

The emergence of O139 in 1992 and the recent emergence of "altered" and "hybrid" *V. cholerae* O1 strains as distinct from the prototype seventh pandemic O1 El Tor strains continues to emphasize the rapidity by which these strains evolve.[104] Genome sequencing has shown the integron cassette array in these new strains as a major region of variation[93,104] confirming the suitability of targeting this area of the genome for typing studies. Further studies would be well suited to use *attC*-PCR for mapping the epidemiology of *V. cholerae*. Providing the *attC* sequences are conserved, *attC*-PCR can also be applied in other bacterial species.

Conclusion

The integron gene/cassette system is clearly a useful tool for those bacteria that have it for evolution and adaptation. Information obtained about the system to date clearly shows that it is important in ways that particularly impact on human health. This includes rapid evolution to combat antibiotic usage and a more general role in niche adaptation that probably includes the ability to colonise a host and the implications for pathogenicity that flow from that. In 2007 a number of outstanding questions in relation to this system were posed by Boucher and colleagues.[18] Many of these, along with others, were recently stated or restated by Cambray and colleagues[77] reinforcing that much is still to be learned about the integron/gene cassette system. Some of these questions are fundamental. The most prominent of these is that of cassette creation. That is, how do *attC* sites and gene combine to become mobilized by this system? Despite speculation since the system was first described, the answer is still not known. Yet this is fundamental, since it impacts not just on the existing mobile pool but of the potential for static genes to be mobilized. Indeed, movement between the mobile and static gene pools could be in both directions. Thus, there may be an evolutionary advantage in fixing mobile genes in a cell line where such genes become indispensable to cell networks. The presence of chromosomal genes long ago acquired by LGT may be so derived. The argument could also be extended to bacterial operons given that the organisational structure of an operon is essentially the same as a multi cassette integron array as recently demonstrated by Bikard et al.[79] Namely, it represents the co-expression of functionally unrelated genes in a single transcriptional unit. Thus, mobile arrays could facilitate the evolution of multi protein biochemical pathways in way that does not adversely impact on a cell. Selection then may lead to fixation of such pathways by loss of *attC* sites if they proved particularly advantageous.

References

1. Ragan MA, Beiko RG. Lateral gene transfer: open issues. Philos Trans R Soc Lond B Biol Sci 2009; 364:2241-51 doi:10.1098/rstb.2009.0031. PMID:19571244
2. Koonin EV, Makarova AS, Aravind L. Horizontal gene transfer in Prokaryotes. Quantification and classification. Annu Rev Microbiol 2001; 55:709-42 doi:10.1146/annurev.micro.55.1.709. PMID:11544372
3. Konstantinidis KT, Tiedje JM. Genomic insights that advance the species definition for prokaryotes. Proc Natl Acad Sci USA 2005; 102:2567-72 doi:10.1073/pnas.0409727102. PMID:15701695
4. Rosselló-Mora R. Opinion: The species problem, can we achieve a universal concept? Syst Appl Microbiol 2003; 26:323-6 doi:10.1078/072320203322497347. PMID:14529174
5. Gogarten JP, Doolittle WF, Lawrence JG. Prokaryotic evolution in light of gene transfer. Mol Biol Evol 2002; 19:2226-38. PMID:12446813
6. Stokes HW, Hall RM. A novel family of potentially mobile DNA elements encoding site-specific gene-integration functions: integrons. Mol Microbiol 1989; 3:1669-83 doi:10.1111/j.1365-2958.1989.tb00153.x. PMID:2560119
7. Alanis AJ. Resistance to antibiotics: are we in the post-antibiotic era? Arch Med Res 2005; 36:697-705 doi:10.1016/j.arcmed.2005.06.009. PMID:16216651
8. Davies J, Davies D. Origins and evolution of antibiotic resistance. Microbiol Mol Biol Rev 2010; 74:417-33 doi:10.1128/MMBR.00016-10. PMID:20805405
9. Ward JM, Grinsted J. Physical and genetic analysis of the Inc-W group plasmids R388, Sa, and R7K. Plasmid 1982; 7:239-50 doi:10.1016/0147-619X(82)90005-1. PMID:6285397
10. Schmidt F, Klopfer-Kaul I. Evolutionary relationship between Tn21-like elements and pBP201, a plasmid from Klebsiella pneumoniae mediating resistance to gentamicin and eight other drugs. Mol Gen Genet 1984; 197:109-19 doi:10.1007/BF00327930. PMID:6096667

11. Tanaka M, Yamamoto T, Sawai T. Evolution of complex resistance transposons from an ancestral mercury transposon. J Bacteriol 1983; 153:1432-8. PMID:6298184

12. Meyer JF, Nies BA, Wiedemann B. Amikacin resistance mediated by multiresistance transposon Tn2424. J Bacteriol 1983; 155:755-60. PMID:6307980

13. Cameron FH, Groot Obbink DJ, Ackerman VP, et al. Nucleotide sequence of the AAD(2") aminoglycoside adenylyltransferase determinant aadB. Evolutionary relationship of this region with those surrounding aadA in R538-1 and dhfrII in R388. Nucleic Acids Res 1986; 14:8625-35 doi:10.1093/nar/14.21.8625. PMID:3024112

14. Hall RM, Vockler C. The region of the IncN plasmid R46 coding for resistance to beta-lactam antibiotics, streptomycin/spectinomycin and sulphonamides is closely related to antibiotic resistance segments found in IncW plasmids and in Tn21-like transposons. Nucleic Acids Res 1987; 15:7491-501 doi:10.1093/nar/15.18.7491. PMID:2821509

15. Shi L, Zheng M, Xiao Z, et al. Unnoticed spread of class 1 integrons in Gram-positive clinical strains isolated in Guangzhou, China. Microbiol Immunol 2006; 50:463-7. PMID:16785718

16. Sundström L, Radstrom P, Swedberg G, et al. Site-specific recombination promotes linkage between trimethoprim- and sulfonamide resistance genes. Sequence characterization of dhfrV and sulI and a recombination active locus of Tn21. Mol Gen Genet 1988; 213:191-201 doi:10.1007/BF00339581. PMID:3054482

17. Nemergut DR, Robeson MS, Kysela RF, et al. Insights and inferences about integron evolution from genomic data. BMC Genomics 2008; 9:261 doi:10.1186/1471-2164-9-261. PMID:18513439

18. Boucher Y, Labbate M, Koenig JE, et al. Integrons: mobilizable platforms that promote genetic diversity in bacteria. Trends Microbiol 2007; 15:301-9 doi:10.1016/j.tim.2007.05.004. PMID:17566739

19. Rowe-Magnus DA, Guerout A-M, Ploncard P, et al. The evolutionary history of chromosomal super-integrons provides an ancestry for multiresistant integrons. Proc Natl Acad Sci USA 2001; 98:652-7 doi:10.1073/pnas.98.2.652. PMID:11209061

20. Stokes HW, Nesbo CL, Holley M, et al. Class 1 integrons potentially predating the association with Tn402-like transposition genes are present in a sediment microbial community. J Bacteriol 2006; 188:5722-30 doi:10.1128/JB.01950-05. PMID:16885440

21. Nield BS, Holmes AJ, Gillings MR, et al. Recovery of new integron classes from environmental DNA. FEMS Microbiol Lett 2001; 195:59-65 doi:10.1111/j.1574-6968.2001.tb10498.x. PMID:11166996

22. Elsaied H, Stokes HW, Kitamura K, et al. Marine integrons containing novel integrase genes, attachment sites, attI, and associated gene cassettes from polluted sediments in Suez and Tokyo Bays. ISME J 2011; 5:1162-67 DOI:doi:10.1038/ismej.2010.208. PMID:21248857

23. Elsaied H, Stokes HW, Nakamura T, et al. Novel and diverse integron integrase genes and integron-like gene cassettes are prevalent in deep-sea hydrothermal vents. Environ Microbiol 2007; 9:2298-312 doi:10.1111/j.1462-2920.2007.01344.x. PMID:17686026

24. Holmes AJ, Gillings MR, Nield BS, et al. The gene cassette metagenome is a basic resource for bacterial genome evolution. Environ Microbiol 2003; 5:383-94 doi:10.1046/j.1462-2920.2003.00429.x. PMID:12713464

25. Hall RM, Brookes DE, Stokes HW. Site-specific insertion of genes into integrons: role of the 59-base element and determination of the recombination cross-over point. Mol Microbiol 1991; 5:1941-59 doi:10.1111/j.1365-2958.1991.tb00817.x. PMID:1662753

26. Hall RM, Collis CM, Kim MJ, et al. Mobile gene cassettes and integrons in evolution. Ann N Y Acad Sci 1999; 870:68-80 doi:10.1111/j.1749-6632.1999.tb08866.x. PMID:10415474

27. Recchia GD, Hall RM. Gene cassettes: a new class of mobile element. Microbiology 1995; 141:3015-27 doi:10.1099/13500872-141-12-3015. PMID:8574395

28. Partridge SR, Tsafnat G, Coiera E, et al. Gene cassettes and cassette arrays in mobile resistance integrons. FEMS Microbiol Rev 2009; 33:757-84 doi:10.1111/j.1574-6976.2009.00175.x. PMID:19416365

29. Labbate M, Case RJ, Stokes HW. The integron/gene cassette system: an active player in bacterial adaptation. Methods Mol Biol 2009; 532:103-25 doi:10.1007/978-1-60327-853-9_6. PMID:19271181

30. Nandi S, Maurer JJ, Hofacre C, et al. Gram-positive bacteria are a major reservoir of Class 1 antibiotic resistance integrons in poultry litter. Proc Natl Acad Sci USA 2004; 101:7118-22 doi:10.1073/pnas.0306466101. PMID:15107498

31. Nesvera J, Hochmannova J, Patek M. An integron of class 1 is present on the plasmid pCG4 from Gram-positive bacterium Corynebacterium glutamicum. FEMS Microbiol Lett 1998; 169:391-5 doi:10.1111/j.1574-6968.1998.tb13345.x. PMID:9868786

32. Brown HJ, Stokes HW, Hall RM. The integrons In0, In2, and In5 are defective transposon derivatives. J Bacteriol 1996; 178:4429-37. PMID:8755869

33. Hall RM, Collis CM. Mobile gene cassettes and integrons: capture and spread of genes by site-specific recombination. Mol Microbiol 1995; 15:593-600 doi:10.1111/j.1365-2958.1995.tb02368.x. PMID:7783631

34. Partridge SR, Recchia GD, Scaramuzzi C, et al. Definition of the attI1 site of class 1 integrons. Microbiology 2000; 146:2855-64. PMID:11065364
35. Collis CM, Hall RM. Expression of antibiotic resistance genes in the integrated cassettes of integrons. Antimicrob Agents Chemother 1995; 39:155-62. PMID:7695299
36. Lévesque C, Brassard S, Lapointe J, et al. Diversity and relative strength of tandem promoters for the antibiotic-resistance genes of several integrons. Gene 1994; 142:49-54 doi:10.1016/0378-1119(94)90353-0. PMID:8181756
37. Hall RM, Brown HJ, Brookes DE, et al. Integrons found in different locations have identical 5' ends but variable 3' ends. J Bacteriol 1994; 176:6286-94. PMID:7929000
38. Gillings M, Boucher Y, Labbate M, et al. The evolution of class 1 integrons and the rise of antibiotic resistance. J Bacteriol 2008; 190:5095-100 doi:10.1128/JB.00152-08. PMID:18487337
39. Gillings MR, Holley MP, Stokes HW. Evidence for dynamic exchange of qac gene cassettes between class 1 integrons and other integrons in freshwater biofilms. FEMS Microbiol Lett 2009; 296:282-8 doi:10.1111/j.1574-6968.2009.01646.x. PMID:19459951
40. Arakawa Y, Murakami M, Suzuki K, et al. A novel integron-like element carrying the metallo-beta-lactamase gene blaIMP. Antimicrob Agents Chemother 1995; 39:1612-5. PMID:7492116
41. Mazel D. Integrons: agents of bacterial evolution. Nat Rev Microbiol 2006; 4:608-20 doi:10.1038/nrmicro1462. PMID:16845431
42. Minakhina S, Kholodii G, Mindlin S, et al. Tn5053 family transposons are res site hunters sensing plasmidal res sites occupied by cognate resolvases. Mol Microbiol 1999; 33:1059-68 doi:10.1046/j.1365-2958.1999.01548.x. PMID:10476039
43. Kholodii GY, Mindlin SZ, Bass IA, et al. Four genes, two ends, and a res region are involved in transposition of Tn5053: a paradigm for a novel family of transposons carrying either a mer operon or an integron. Mol Microbiol 1995; 17:1189-200 doi:10.1111/j.1365-2958.1995.mmi_17061189.x. PMID:8594337
44. Chen S, Zhao S, White DG, et al. Characterization of multiple-antimicrobial-resistant Salmonella serovars isolated from retail meats. Appl Environ Microbiol 2004; 70:1-7 doi:10.1128/AEM.70.1.1-7.2004. PMID:14711619
45. Literák I, Vanko R, Dolejska M, et al. Antibiotic resistant Escherichia coli and Salmonella in Russian rooks (Corvus frugilegus) wintering in the Czech Republic. Lett Appl Microbiol 2007; 45:616-21 doi:10.1111/j.1472-765X.2007.02236.x. PMID:17916127
46. Schlüter A, Szczepanowski R, Puhler A, et al. Genomics of IncP-1 antibiotic resistance plasmids isolated from wastewater treatment plants provides evidence for a widely accessible drug resistance gene pool. FEMS Microbiol Rev 2007; 31:449-77 doi:10.1111/j.1574-6976.2007.00074.x. PMID:17553065
47. Van TT, Moutafis G, Istivan T, et al. Detection of Salmonella spp. in retail raw food samples from Vietnam and characterization of their antibiotic resistance. Appl Environ Microbiol 2007; 73:6885-90 doi:10.1128/AEM.00972-07. PMID:17766455
48. Gillings MR, Labbate M, Sajjad A, et al. Mobilization of a Tn402-like class 1 integron with a novel cassette array via flanking miniature inverted-repeat transposable element-like structures. Appl Environ Microbiol 2009; 75:6002-4 doi:10.1128/AEM.01033-09. PMID:19648375
49. Rådström P, Skold O, Swedberg G, et al. Transposon Tn5090 of plasmid R751, which carries an integron, is related to Tn7, Mu, and the retroelements. J Bacteriol 1994; 176:3257-68. PMID:8195081
50. Labbate M, Chowdhury PR, Stokes HW. A class 1 integron present in a human commensal has a hybrid transposition module compared to Tn402: evidence of interaction with mobile DNA from natural environments. J Bacteriol 2008; 190:5318-27 doi:10.1128/JB.00199-08. PMID:18502858
51. Juan C, Zamorano L, Mena A, et al. Metallo-beta-lactamase-producing Pseudomonas putida as a reservoir of multidrug resistance elements that can be transferred to successful Pseudomonas aeruginosa clones. J Antimicrob Chemother 2010; 65:474-8 doi:10.1093/jac/dkp491. PMID:20071364
52. Marchiaro P, Viale AM, Ballerini V, et al. First report of a Tn402-like class 1 integron carrying blaVIM-2 in Pseudomonas putida from Argentina. J Infect Dev Ctries 2010; 4:412-6. PMID:20601796
53. Lagatolla C, Edalucci E, Dolzani L, et al. Molecular evolution of metallo-beta-lactamase-producing Pseudomonas aeruginosa in a nosocomial setting of high-level endemicity. J Clin Microbiol 2006; 44:2348-53 doi:10.1128/JCM.00258-06. PMID:16825348
54. Nunes-Düby SE, Kwon HJ, Tirumalai RS, et al. Similarities and differences among 105 members of the Int family of site-specific recombinases. Nucleic Acids Res 1998; 26:391-406 doi:10.1093/nar/26.2.391. PMID:9421491
55. Rowe-Magnus DA, Guerout AM, Mazel D. Super-integrons. Res Microbiol 1999; 150:641-51 doi:10.1016/S0923-2508(99)00127-8. PMID:10673003
56. Hansson K, Skold O, Sundstrom L. Non-palindromic attL sites of integrons are capable of site-specific recombination with one another and with secondary targets. Mol Microbiol 1997; 26:441-53 doi:10.1046/j.1365-2958.1997.5401964.x. PMID:9402016

57. Chen CY, Wu KM, Chang YC, et al. Comparative genome analysis of Vibrio vulnificus, a marine pathogen. Genome Res 2003; 13:2577-87 doi:10.1101/gr.1295503. PMID:14656965
58. Grindley ND, Whiteson KL, Rice PA. Mechanisms of site-specific recombination. Annu Rev Biochem 2006; 75:567-605 doi:10.1146/annurev.biochem.73.011303.073908. PMID:16756503
59. Collis CM, Grammaticopoulos G, Briton J, et al. Site-specific insertion of gene cassettes into integrons. Mol Microbiol 1993; 9:41-52 doi:10.1111/j.1365-2958.1993.tb01667.x. PMID:8412670
60. Collis CM, Kim MJ, Stokes HW, et al. Binding of the purified integron DNA integrase IntI1 to integron- and cassette-associated recombination sites. Mol Microbiol 1998; 29:477-90 doi:10.1046/j.1365-2958.1998.00936.x. PMID:9720866
61. Gravel A, Messier N, Roy PH. Point mutations in the integron integrase IntI1 that affect recombination and/or substrate recognition. J Bacteriol 1998; 180:5437-42. PMID:9765577
62. Stokes HW, O'Gorman DB, Recchia GD, et al. Structure and function of 59-base element recombination sites associated with mobile gene cassettes. Mol Microbiol 1997; 26:731-45 doi:10.1046/j.1365-2958.1997.6091980.x. PMID:9427403
63. Francia MV, Zabala JC, de la Cruz F, et al. The IntI1 integron integrase preferentially binds single-stranded DNA of the attC site. J Bacteriol 1999; 181:6844-9. PMID:10542191
64. Johansson C, Kamali-Moghaddam M, Sundstrom L. Integron integrase binds to bulged hairpin DNA. Nucleic Acids Res 2004; 32:4033-43 doi:10.1093/nar/gkh730. PMID:15289577
65. Recchia GD, Stokes HW, Hall RM. Characterisation of specific and secondary recombination sites recognised by the integron DNA integrase. Nucleic Acids Res 1994; 22:2071-8 doi:10.1093/nar/22.11.2071. PMID:8029014
66. Shearer JE, Summers AO. Intracellular steady-state concentration of integron recombination products varies with integrase level and growth phase. J Mol Biol 2009; 386:316-31 doi:10.1016/j.jmb.2008.12.041. PMID:19135452
67. Bouvier M, Demarre G, Mazel D. Integron cassette insertion: a recombination process involving a folded single strand substrate. EMBO J 2005; 24:4356-67 doi:10.1038/sj.emboj.7600898. PMID:16341091
68. Bouvier M, Ducos-Galand M, Loot C, et al. Structural features of single-stranded integron cassette attC sites and their role in strand selection. PLoS Genet 2009; 5:e1000632 doi:10.1371/journal.pgen.1000632. PMID:19730680
69. Loot C, Bikard D, Rachlin A, et al. Cellular pathways controlling integron cassette site folding. EMBO J 2010; 29:2623-34 doi:10.1038/emboj.2010.151. PMID:20628355
70. Rowe-Magnus DA, Guerout A-M, Biskri L, et al. Comparative analysis of superintegrons: Engineering extensive genetic diversity in the Vibrionaceae. Genome Res 2003; 13:428-42 doi:10.1101/gr.617103. PMID:12618374
71. Vaisvila R, Morgan RD, Posfai J, et al. Discovery and distribution of super-integrons among Pseudomonads. Mol Microbiol 2001; 42:587-601 doi:10.1046/j.1365-2958.2001.02604.x. PMID:11722728
72. Gillings MR, Holley MP, Stokes HW, et al. Integrons in Xanthomonas: A source of species genome diversity. Proc Natl Acad Sci USA 2005; 102:4419-24 doi:10.1073/pnas.0406620102. PMID:15755815
73. Coleman N, Tetu S, Wilson N, et al. An unusual integron in Treponema denticola. Microbiology 2004; 150:3524-6 doi:10.1099/mic.0.27569-0. PMID:15528643
74. Sutton MD, Smith BT, Godoy VG, et al. The SOS response: recent insights into umuDC-dependent mutagenesis and DNA damage tolerance. Annu Rev Genet 2000; 34:479-97 doi:10.1146/annurev.genet.34.1.479. PMID:11092836
75. Aertsen A, Michiels CW. Upstream of the SOS response: figure out the trigger. Trends Microbiol 2006; 14:421-3 doi:10.1016/j.tim.2006.08.006. PMID:16934473
76. Guerin E, Cambray G, Sanchez-Alberola N, et al. The SOS response controls integron recombination. Science 2009; 324:1034 doi:10.1126/science.1172914. PMID:19460999
77. Cambray G, Guerout A, Mazel D. Integrons. Annu Rev Genet 2010; 44:141-66 doi:10.1146/annurev-genet-102209-163504. PMID:20707672
78. Michael CA, Labbate M. Gene cassette transcription in a large integron-associated array. BMC Genet 2010; 11:82 doi:10.1186/1471-2156-11-82. PMID:20843359
79. Bikard D, Julié-Galau S, Cambray G, et al. The synthetic integron: an in vivo genetic shuffling device. Nucleic Acids Res 2010; 38:e153 doi:10.1093/nar/gkq511. PMID:20534632
80. Baharoglu Z, Bikard D, Mazel D. Conjugative DNA transfer induces the bacterial SOS response and promotes antibiotic resistance development through integron activation. PLoS Genet 2010; 6:e1001165 doi:10.1371/journal.pgen.1001165. PMID:20975940
81. Stokes HW, Holmes AJ, Nield BS, et al. Gene cassette PCR: sequence-independent recovery of entire genes from environmental DNA. Appl Environ Microbiol 2001; 67:5240-6 doi:10.1128/AEM.67.11.5240-5246.2001. PMID:11679351

82. Michael CA, Gillings MR, Holmes AJ, et al. Mobile gene cassettes: a fundamental resource for bacterial evolution. Am Nat 2004; 164:1-12 doi:10.1086/421733. PMID:15266366

83. Koenig JE, Boucher Y, Charlebois RL, et al. Integron-associated gene cassettes in Halifax Harbour: assessment of a mobile gene pool in marine sediments. Environ Microbiol 2008; 10:1024-38 doi:10.1111/j.1462-2920.2007.01524.x. PMID:18190517

84. Koenig JE, Sharp C, Dlutek M, et al. Integron gene cassettes and degradation of compounds associated with industrial waste: The case of the Sydney Tar ponds. PLoS ONE 2009; 4:e5276 doi:10.1371/journal.pone.0005276. PMID:19390587

85. Nemergut DR, Martin AP, Schmidt SK. Integron diversity in heavy-metal-contaminated mine tailings and inferences about integron evolution. Appl Environ Microbiol 2004; 70:1160-8 doi:10.1128/AEM.70.2.1160-1168.2004. PMID:14766601

86. Koenig JE, Bourne DG, Curtis B, et al. Coral-mucus-associated Vibrio integrons in the Great Barrier Reef: genomic hotspots for environment adaptation. ISME J 2011; 5: 962-72 DOI:doi:10.1038/ismej.2010.193. PMID:21270840

87. Engelberg-Kulka H, Glaser G. Addiction modules and programmed cell death and antideath in bacteria cultures. Annu Rev Microbiol 1999; 53:43-70 doi:10.1146/annurev.micro.53.1.43. PMID:10547685

88. Szekeres S, Dauti M, Wilde C, et al. Chromosomal toxin-antitoxin loci can diminish large-scale genome reductions in the absence of selection. Mol Microbiol 2007; 63:1588-605 doi:10.1111/j.1365-2958.2007.05613.x. PMID:17367382

89. Wren BW. Microbial genome analysis: insights into virulence, host adaptation and evolution. Nat Rev Genet 2000; 1:30-9 doi:10.1038/35049551. PMID:11262871

90. Ochman H, Lawrence JG, Groisman EA. Lateral gene transfer and the nature of bacterial innovation. Nature 2000; 405:299-304 doi:10.1038/35012500. PMID:10830951

91. Foley SL, Lynne AM, Nayak R. Molecular typing methodologies for microbial source tracking and epidemiological investigations of Gram-negative bacterial food pathogens. Infect Genet Evol 2009; 9:430-40 doi:10.1016/j.meegid.2009.03.004. PMID:19460308

92. Feng L, Reeves PR, Lan R, et al. A recalibrated molecular clock and independent origins for the cholera pandemic clones. PLoS ONE 2008; 3:e4053 doi:10.1371/journal.pone.0004053. PMID:19115014

93. Chin CS, Sorenson J, Harris JB, et al. The origin of the Haitian cholera outbreak strain. N Engl J Med 2011; 364:33-42 doi:10.1056/NEJMoa1012928. PMID:21142692

94. Cho YJ, Yi H, Lee JH, et al. Genomic evolution of Vibrio cholerae. Curr Opin Microbiol 2010; 13:646-51 doi:10.1016/j.mib.2010.08.007. PMID:20851041

95. Kaper JB, Morris J Jr., Levine M. Cholera. Clin Microbiol Rev 1995; 8:48-86. PMID:7704895

96. Shimada T, Balakrish Nair G, Deb BC, et al. Outbreak of Vibrio cholerae non-01 in India and Bangladesh. Lancet 1993; 341(8856):1346-7 doi:10.1016/0140-6736(93)90855-B PMID: 8098474.

97. Ansaruzzaman M, Bhuiyan NA, Nair GB, et al. Cholera in Mozambique, variant of Vibrio cholerae. Emerg Infect Dis 2004; 2:2057-9. PMID:16010751

98. Nair GB, Faruque SM, Bhuiyan NA, et al. New variants of Vibrio cholerae O1 biotype El Tor with attributes of the classical biotype from hospitalized patients with acute diarrhoea in Bangladesh. J Clin Microbiol 2002; 40:3296-9 doi:10.1128/JCM.40.9.3296-3299.2002. PMID:12202569

99. Nair GB, Qadri F, Holmgren J, et al. Cholera due to altered El Tor strains of Vibrio cholerae O1 in Bangladesh. J Clin Microbiol 2006; 44:4211-3 doi:10.1128/JCM.01304-06. PMID:16957040

100. Castañeda NC, Pichel M, Orman B, et al. Genetic characterization of Vibrio cholerae isolates from Argentina by V. cholerae repeated sequences-polymerase chain reaction. Diagn Microbiol Infect Dis 2005; 53:175-83 doi:10.1016/j.diagmicrobio.2005.05.008. PMID:16249063

101. Tokunaga A, Yamaguchi H, Morita M, et al. Novel PCR-based genotyping method, using genomic variability between repetitive sequences of toxigenic Vibrio cholerae O1 El Tor and O139. Mol Cell Probes 2010; 24:99-103 doi:10.1016/j.mcp.2009.11.002. PMID:19900536

102. Chowdhury N, Asakura M, Neogi SB, et al. Development of simple and rapid PCR-fingerprinting methods for Vibrio cholerae on the basis of genetic diversity of the superintegron. J Appl Microbiol 2010; 109:304-12. PMID:20070445

103. Labbate M, Boucher Y, Joss MJ, et al. Use of chromosomal integron arrays as a phylogenetic typing system for Vibrio cholerae pandemic strains. Microbiology 2007; 153:1488-98 doi:10.1099/mic.0.2006/001065-0. PMID:17464063

104. Grim CJ, Hasan NA, Taviani E, et al. Genome sequence of hybrid Vibrio cholerae O1 MJ-1236, B-33, and CIRS101 and comparative genomics with V. cholerae. J Bacteriol 2010; 192:3524-33 doi:10.1128/JB.00040-10. PMID:20348258

105. Deshpande CN, Harrop SJ, Boucher Y, et al. Crystal structure of an integron gene cassette-associaed protein from Vibrio cholerae identified a cationic drug-binding module. PLoS ONE 2011; 6:e16934 doi:10.1371/journal.pone.0016934. PMID:21390267

106. Rowe-Magnus DA, Guerout A-M, Mazel D. Bacterial resistance evolution by recruitment of super-integron gene cassettes. Mol Microbiol 2002; 43:1657-69 doi:10.1046/j.1365-2958.2002.02861.x. PMID:11952913
107. Ogawa A, Takeda T. The gene encoding the heat-stable enterotoxin of Vibrio cholerae is flanked by 123-bp direct repeats. Microbiol Immunol 1993; 37:607-16. PMID:8246823
108. Barker A, Clark CA. Identification of VCR, a repeated sequence associated with a locus encoding a hemagglutinin in Vibrio cholerae O1. J Bacteriol 1994; 176:5450-8. PMID:8071223
109. Franzon VL, Barker A, Manning P. Nucleotide sequence encoding the mannose-fucose-resistant hemagglutinin of Vibrio cholerae O1 and construction of a mutant. Infect Immun 1993; 61:3032-7. PMID:8514410
110. Rowe-Magnus DA. Integrase-directed recovery of functional genes from genomic libraries. Nucleic Acids Res 2009; 37:e118 doi:10.1093/nar/gkp561. PMID:19596808
111. Lee JH, Yang S-T, Rho S-H, et al. Crystal structure and functional studies reveal that PAS factor from Vibrio vulnificus is a novel member of the Saposin-fold family. J Mol Biol 2006; 355:491-500 doi:10.1016/j.jmb.2005.10.074. PMID:16318855
112. Kim YR, Lee SE, Kim CM, et al. Characterization and pathogenic significance of Vibrio vulnificus antigens preferentially expressed in septicemic patients. Infect Immun 2003; 71:5461-71 doi:10.1128/IAI.71.10.5461-5471.2003. PMID:14500463
113. Tokugawa K, Kakitani M, Ishii T, et al. A novel protein secretion factor from a VIbrio species which operates in Escherichia coli. J Biotechnol 1994; 35:69-76 doi:10.1016/0168-1656(94)90190-2. PMID:7765027
114. Smith AB, Siebeling RJ. Identification of genetic loci required for capsular expression in Vibrio vulnificus. Infect Immun 2003; 71:1091-7 doi:10.1128/IAI.71.3.1091-1097.2003. PMID:12595419
115. Robinson A, Guilfoyle AP, Harrop SJ. A putative house-cleaning enzyme encoded within an integron array: 1.8 Å crystal structure defines a new MazG subtype. Mol Microbiol 2007; 66:610-21 doi:10.1111/j.1365-2958.2007.05932.x. PMID:17892463
116. Robinson A, Wu PS, Harrop SJ, et al. Integron-associated mobile gene cassettes code for folded proteins: the structure of Bal32a, a new member of the adaptable alpha+beta barrel family. J Mol Biol 2005; 346:1229-41 doi:10.1016/j.jmb.2004.12.035. PMID:15713477
117. Nield BS, Willows RD, Torda AE, et al. New enzymes from environmental cassette arrays: Functional attributes of a phosphotransferase and an RNA-methyltransferase. Protein Sci 2004; 13:1651-9 doi:10.1110/ps.04638704. PMID:15152095

CHAPTER 4

Inteins and Introns

Kristen S. Swithers and J. Peter Gogarten*

Abstract

Introns and inteins compose a group of mobile genetic elements that are ubiquitous across the tree of life. This chapter describes these parasitic genetic elements, specifically group I introns, group II introns, and inteins. These genetic elements are spliced out of either RNA or proteins after transcription or translation. Some of these molecular parasites encode homing endonucleases that help in the invasion of intein/intron free alleles. In addition, free-standing homing endonucleases are considered parasites in their own right. The molecular parasites that use a homing endonuclease have a life cycle of their own called the homing cycle and have played a role in gene and genome evolution.

Introduction

The discovery of the intron is one of the most fundamental discoveries of the twentieth century. The concept of what is now termed the intron was first introduced in 1977 by several groups. A large mRNA early transcript in the eukaryotic nucleus was observed in adenovirus 2 and in HeLa cells that appeared to be post transcriptional modified to produce a significantly smaller late mRNA transcript.[1-4] The intragenic culprit of this peculiar phenomena was later termed to be the intron.[5] These spliceosomal introns are processed out of the mRNA during maturation by the spliceosome, which can be composed of up to 200 unique proteins.[6] Spliceosomal introns are present in varying numbers in all eukaryotic lineages, and they are completely absent from prokaryotic lineages. Some supposedly deep branching lineages, such as the diplomonad *Giardia*, contain many fewer spliceosomal introns, and fewer recognized spliceosomal proteins. However, the reduced complexity of the spliceosome and the paucity of introns does not necessarily reflect the ancestral state but in part may reflect the outcome of genome streamlining.[7] The complete loss of the spliceosome might be counterselected, because it is involved in transsplicing of essential proteins.[8]

Soon after the discovery of introns, theories for the purpose of these intragenic elements were proposed; the most noteworthy being the genes in pieces idea.[5,9] This idea proposes that introns are present to accelerate protein evolution by allowing for shuffling of protein domains to generate new protein variants. The abundance of introns in eukaryotes implies that eukaryotes may have evolved more complex functions more rapidly than prokaryotes.[9] This theory was subsequently developed into the introns early theory, which postulates that introns were present in the protein coding genes of the ancestor of the eukaryotes and prokaryotes, and that spliceosomal introns were retained in the eukaryotes and lost in the prokaryotes through genome streamlining.[10-13] The opposing theory is the introns late theory. After the discovery of group II introns (discussed further in the next section), the intron late hypothesis was formulated which proposed that introns are actually selfish genetic elements that spread into the nuclear genome from the alphaproteobacterial

*Department of Molecular and Cell Biology, University of Connecticut, Storrs, Connecticut, USA.
Corresponding Author: J. Peter Gogarten—Email: jpgogarten@gmail.com

Bacterial Integrative Mobile Genetic Elements, edited by Adam P. Roberts and Peter Mullany.
©2013 Landes Bioscience.

ancestor of the mitochondria to the eukaryotic genome.[14-17] More recent evidence has provided a compelling argument for the introns late hypothesis over the intron early hypothesis,[18] although it appears that the most recent ancestor of all eukaryotes already possessed spliceosomes and spliceosomal introns.

Group II Introns

Group II introns are self-splicing mobile genetic elements. They are thought to be the predecessors of both the Eukaryotic spliceosomal intron and retrotransposons, and have had a profound role in shaping evolution across the tree of life.[19-21] Group II introns are found in all domains of life; Bacteria, Archaea and in the mitochondria and chloroplasts of fungi, plants, protists and annelid worms.[22] The RNA component of group II introns tends to be about 500 nt long and has six conserved secondary structural domains (DI-DVI). Group II introns are divided into three classes of RNA secondary structures (IIA, IIB and IIC) and different subclasses of structural variation, IIA1, IIA3 IIB1 and IIB2. Open reading frames (ORFs) are often found within the DIV domain encoding between one and four intron encoded proteins (IEPs); these include proteins with domains for reverse transcriptase (RT), maturase/splicing (X), DNA-binding (D), and endonuclease (En) activity.

Group II introns present in prokaryotes are considered to be self-splicing and mobile. Group II introns present in chloroplasts and mitochondria tend to lack open reading frames; if an ORF is present, it tends to be pseudogene-like, with premature stop codons. In addition, the RNA domains tend to be degenerate, which suggests that they are immobile and have lost their self-splicing ability. For example, of 20 group II introns present in the organelles of plants none are known to self-splice.[23,24] However, to maintain functional host genes, the introns must be spliced out. These degenerate self-splicing elements now rely on host encoded splicing factors.[25-27]

Mobility Mechanisms for Group II Introns

Similar to spliceosomal introns, group II introns splice through two transesterification reactions that lead to exon ligation and intron excision (Fig. 1 A). However, for group II introns the reactions are catalyzed by the RNA, which folds into conserved secondary and tertiary structures, creating an active site for Mg^{2+} ions to bind to and catalyze the reactions.[28-31] This ribozyme activity results in the intron forming a lariat loop that is free to move to a new intron less target site. Group II introns are mobile by at least two different self-splicing mechanisms, retrohoming and retrotransposition.[32] In these mechanisms the intron and intron encoded proteins come together and function as a ribonucleoprotein (RNP) to integrate into an intron less site. Retrohoming occurs for group II introns that have the En domain among their IEPs.[33-35] Retrohoming is a site-specific process that is 100 times more efficient than retrotransposition.[35] In the site specific copy and paste mechanism of retrohoming the RNP recognizes specific nucleotide bases or DNA structural features, and the intron reverse splices into the insertion site as a linear intron between the two DNA exons. Then the En domain cleaves the opposite DNA strand a short distance downstream and uses the 3′ OH end of the cleaved DNA as a primer to initiate reverse transcription and invoke the DNA repair and/or recombination systems of the host organism. The group II introns that lack an En domain or have a mutated En domain are still mobile but are less successful than those that do. These elements use a retrotransposition mechanism where they reverse splice into the single stranded DNA that is exposed during replication and transcription. They preferentially reverse splice to the lagging stranding where they can use the Okazaki fragments as primers to reverse transcribe into the host genome.

Evolution of Group II Introns

Group II introns may have evolved from a retroelement. This hypothesis states that the ancestor of extant group II introns was a bacterial ribozyme containing a RT domain. Phylogenetic trees based on the RNA component and ORFs of introns suggest a coevolution of both elements.[36-38] A more recent study revealed incongruence between the phylogeny of some RT proteins and

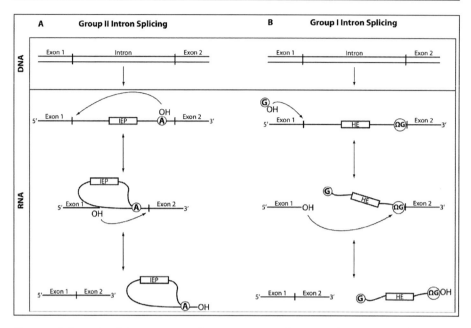

Figure 1. Schematic of the group II and group I intron splicing mechanisms. Panel A depicts the general group II intron splicing mechanism and the two transesterification reactions that occur (adapted from Toro et al.).[32] The intron encoded protein (IEP) aids in mobility and/or splicing. The final products are the ligated exons and a lariat loop RNA intermediate, which can move to an intronless allele either by retrohoming or retrotransposition. Panel B shows the general group I intron splicing mechanism and the two transesterification reactions that occur. G-OH indicates the guanosine (or GMP, GDP or GTP) bound to the catalytic site of the intron. ΩG indicates the conserved terminal guanine or adenine[63] residue at the 3' end of the intron. The open reading frame encoding the homing endonuclease (HE), which aids in group I intron mobility, is depicted in the middle of the intron. Similar to the group II intron, one of the final products are the ligated exons and in contrast to the group II intron the other product is the linear RNA intermediate.

the group II introns they reside in, possibly suggesting these ORFs move between introns.[39] One possible mechanism for this is the occurrence of twintrons.[40] These are found in Archaea and Cyanobacteria, and are created when one group II intron inserts into another group II intron.[41] The inner intron must be spliced out before the outer intron. One can imagine a scenario where the inner intron looses splicing activity and decays, leaving one of its ORFs behind resulting in an intron with an ORF from a different group II intron class.[39]

Group II introns are thought to be the ancestors of spliceosomal introns and non-LTR-retrotransposons.[19-21] Their evolutionary relationship with the spliceosomal intron is supported by the fact that group II introns and spliceosomal introns have a very similar splicing mechanism, comparable boundary sequences, and structural similarities.[42-44] It is hypothesized that group II introns evolved in bacteria and were transferred to the eukaryotic genome from the alphaproteobacterial endosymbiont ancestor of the mitochondria and the cyanobacterial endosymbiont ancestor of the plastid, then proliferated in the eukaryotic genome and evolved into the spliceosomal intron. Group II introns have played a role in shaping both eukaryote and prokaryote evolution. A recent study of group II intron proliferation in *Wolbachia*, an extant alphaproteobacteria, whose lifestyle is similar to the endosymbiont ancestor of the mitochondria, showed that mobile genetic elements are associated with genomic rearrangements and gene conversion events.[45]

Applications of Group II Introns

Many gene disruption studies using derivatives of group II introns, targetrons, have been deployed in organisms such as *Lactococcus lactis, E. coli, Staphylococcus aureus, Francilla, Clostrida*.[46-49] A targetron is a site directed mutagenesis system that uses the group II intron from the *ltrB* gene of *Lactococcus lactis* adapted to function in various hosts. Availability of this system has greatly facilitated gene inactivation studies in the genus *Clostridium*. This targetron system may also facilitate gene function studies in other organisms for which a genetic system has not yet been developed.

Group I Introns

Group I introns are a self-splicing introns that splice out after transcription via an autocatalytic two-step transesterification reaction. The first group I intron was identified in *Tetrahymena thermophila* and it was one of the first rybozymes discovered in the early 1980s.[50] They are small RNAs that range in size from 250 to 500 nucleotides and they are composed of a catalytic RNA domain and often contain a homing endonuclease (HE) encoding ORF.[51-53] The HE recognizes a large (14–40 bp) region in an intronless allele and catalyzes a double strand cut. See the section on homing endonucleases below for discussion of target site selection and the HE's role in the life cycle of the self splicing elements. Group I introns are present in the nuclear and organellar genomes of Eukaryotes, in some Eukaryotic viruses, Bacteria, and bacteriophage; to date they have not been identified in Archaea.[54] Group I introns are found within genes essential to the cell, specifically in 16S and 23S rRNA, tRNAs, ribonucleotide reductase, NADH dehydrogenase, recombinase A, DNA polymerase, thymidylate synthase, and genes involved in photosynthesis.[53,55]

Interestingly these elements are found more frequently in structural RNAs in bacteria. This could be due the fact that transcription and translation are coupled in bacteria, which might not allow the newly synthesized mRNA sufficient time before translation to form the tertiary structure necessary for splicing.[56] Group I introns tend to be more abundant in the mitochondria and chloroplasts of eukaryotes than in bacteria. The traditional explanation for this was the lack of sexual reproduction among prokaryotes. Sexual reproduction in eukaryotes may bring homologous intron-less and intron-containing alleles together more frequently, providing more opportunities for rapid intron invasion in a population. However, in light of horizontal gene transfer (HGT), which now appears to be rampant in prokaryotes,[57] HGT should provide ample opportunities for the introduction of intron containing alleles, making the traditional explanation for the disparity in group I intron abundances unsatisfactory.[57,58] The co-existence of multiple genomes in the same cell that is found after phage and virus infections of prokaryotes and in mitochondria was also proposed as a factor favoring intron movement.[58] Another possible explanation for the inequality of intron frequency in eukaryotes and bacteria is that some bacterial group I introns may inhibit growth. For example, when the group I introns of *Tetrahymena* and *Coxiella* are expressed in *E. coli*, the introns were shown to associate with ribosomes and inhibit translation, which ultimately slowed growth.[59-61]

Although group I introns are not well conserved at the sequence level they all share a conserved secondary structure, and use guanosine as a cofactor in splicing.[62,63] Based on the conserved structure, group I introns are classified into 14 subgroups. The secondary structure is labeled by paired elements P1-P10. Ikawa et al.[64] provides a detailed description of these elements and of their three dimensional arrangement. P3-P7-P9 and P4-P5-P6 form helices that make up the core of the secondary structure. Helix P3-P7-P9 forms the guanosine binding site, which on binding activates the self-splicing reaction.[64] In contrast to group II introns, the excised intron is generally linear (Fig. 1B); however, it has been observed to circularize in some instances.[65]

Inteins

In the late 1980s, the first intein was discovered in the yeast vacuolar ATPases.[66,67] The first hint at something unusual resulted from comparisons of vacuolar ATPase catalytic subunits from *Neurospora crassa*[68] and carrot[69] with a gene in yeast, initially thought to encode a calcium ion transporter.[70] Hirata et al.[67] showed that an additional sequence in the yeast homolog was retained in the processed mRNA, but absent from the functioning host protein. Inteins are present in all domains of life; as of April 2, 2011 the intein database (InBase) lists 523 inteins residing in 65 distinct insertion sites within 36 different proteins.[71] Inteins are found within proteins that are essential to cell function. These elements sit within protein coding sequences and are translated with the genes they reside in. Through an autocatalytic mechanism the intervening sequence (intein) removes itself from the protein and the flanking protein sequences (exteins) are spliced together resulting in a functional host protein.[72-75] Four conserved motifs are recognized in the general intein structure, called either Blocks A, B, F and G or N1, N3, C2 and C1.[71,76] Recently inteins have been divided into three classes based on different splicing mechanisms.[77,78] Most inteins are members of class 1, and are considered to follow the standard intein splicing mechanism. Here we will discuss only class 1 inteins (Fig. 2); for more information on class 2 and 3 inteins see the article by Tori and Perler.[78] Inteins are also divided into two groups based on their size: large and mini inteins, which are differentiated by the presence or absence of a homing endonuclease domain.[74,79]

All inteins share a low degree of sequence similarity in their splicing domain, which is formed by the amino and C-terminal parts of the intein sequence. This similarity is sufficient to identify inteins in PSI-blast searches, and suggests a common origin of the inteins splicing domain and an evolutionary relationship with the autocatalytic domain of hedgehog proteins. The latter domain cleaves the hedgehog protein and adds cholesterol to the other domain of the parent proteins.[75,80,81] Inteins found in the same positions within a protein are more similar and in phylogenetic analyses group together with strong support values; however, too little phylogenetic information is retained in the sequences to reconstruct the relationship between inteins located in different insertion sites, and between inteins and proteins which contain homologous domains.[75,82,83] The general mechanism for intein splicing is a rapid succession of four nucleophilic attacks (Fig. 2). (1) An N-S acetyl rearrangement of the peptide bond on the N-terminus of the intein. (2) A transesterification between the nucleophile on the C-extein and the thioester on the N-terminus of the intein. (3) An Asn-cyclization and peptide bond cleavage, and (4) an N-O acyl shift rearrangement of the ester bond, releasing the intein from the extein and ligating the flanking exteins together to generate a functional protein.[78]

Usually an intein is encoded as a contiguous piece of DNA that is flanked by the DNA encoding the host protein. The amino and carboxy-terminal exteins are transcribed and translated together with the intein sequences. However, in an illustration of a selectively neutral pathway toward higher complexity, the *dnaE* gene in *Synechocystis* sp PCC6803 encoding DNA polymerase III catalytic subunit, the primary polymerase in DNA replication, is broken into two separate genes, encoded in different parts of the genome.[84] The N-terminal extein, together with the N-terminal part of the intein, is transcribed independently from the C-terminal part of the intein, which is transcribed and translated together with the C-terminal extein. The two parts of the intein, produced from separate genes, associate after translation and then catalyze the splicing reaction that joins the two exteins into the functional DNA polymerase subunit. At no point along the evolutionary history, from the first invasion of the host protein by the intein to the gene breaking into two separate units, does it appear that the events were associated with a significant selective advantage or disadvantage for the cyanobacterium. A functional DNA polymerase is produced at all steps along the way; however, initially the intein was superfluous, it could be considered a molecular parasite, and the organism presumably would have survived well without the intein being present;

Figure 2, viewed on following page. Schematic of the class I intein splicing mechanism adapted from Elleuche and Poggeler.[79] The four step reaction starts with an N-S acetyl rearrangement of the peptide bond on the N-terminus (1), proceeds to a transesterification between the nucleophile on the C-extein and the thioester on the N-terminus of the intein (2), then to an Asn-cyclization and peptide bond cleavage (3), and finally an N-O acyl shift rearrangement of the ester bond, releasing the intein from the extein and ligating the flanking exteins together to generate the functional host protein (4). The excised intein can move to an inteinless allele and cleave it with the help of a homing endonuclease domain.

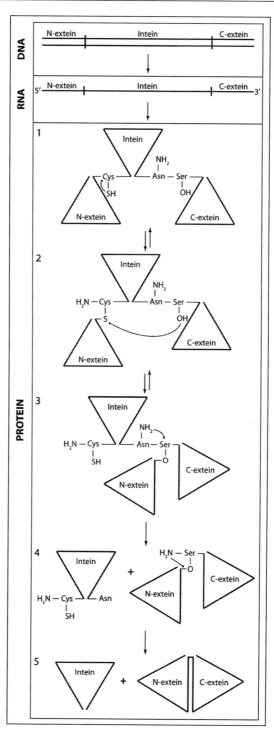

Figure 2. Please see the figure legend on the previous page.

however, in the end the intein has become an indispensible part of the machinery that synthesizes the functioning DNA polymerase. A simple deletion of the intein is no longer possible. While the intein likely has never provided a selective advantage to the host organism, it has become an essential part of a complex machinery that is under strong purifying selection, because without its splicing activity, a functioning DNA polymerase could no longer be produced. It is possible that future mechanisms might evolve that use the complex machinery to better regulate the synthesis of the DNA polymerase; however, the complex system initially evolved without providing an adaptive advantage. The split intein in DnaE has a wide distribution in cyanobacteria,[85] suggesting that this rather complex gene structure is an evolutionarily stable arrangement. Arlin Stoltzfus describes other examples for neutral pathways toward complexity, including spliceosomal introns, which might have evolved along a similar trajectory.[86] A recent genome survey found several split inteins separated only by free standing homing endonuclease gene.[87] This arrangement might be an intermediate on the path to split inteins encoded in different parts of the genome, and also might be a precursor for large inteins that integrate homing endonuclease and splicing domains into a single transcript.[87]

Applications of Inteins

A wide range of applications in biotechnology were developed based on inteins. These include kits for intein mediated protein purification that are commercially available and use different mechanisms to activate the splicing reaction leading to the release of the pure protein (see ref. 79 for a recent review). Split inteins have been used to synthesize toxic proteins from two non-toxic precursors, to synthesize cyclic proteins, and to introduce labels or synthetic peptides into proteins.[79,88]

Insertion Sites of Introns and Inteins

Understanding where in a protein inteins, group I introns, group II introns and spliceosomal introns tend to accumulate sheds light on how these parasitic genetic elements persist over long periods of time. Comparative studies have surveyed the plethora of genomic data available to attempt to answer this. Studies on intein and group I intron insertions sites have shown they are found in conserved sites, while studies on group II and spliceosomal introns did not reveal a significant conserved site preference.[55,75,89-93] This disparity between inteins, group I introns and group II introns can be explained by differences in mobility between the elements. Inteins and group I introns are generally mobile by homing endonucleases, these recognize large sequences that occur infrequently in the genome. Targeting conserved sites in conserved proteins guarantees that excision happens and is exact. Mutations in the insertion site that would allow a host protein to become resistant against the homing endonuclease, would likely result in loss of function of the host protein, and inaccurate splicing would result in small insertions or deletions in the most conserved region of the host protein. Most conserved proteins provide a vital function to the cell, so exact excision is required to maintain a functional protein, thus strong purifying selection acts on the intein to maintain accurate and efficient splicing activity. Targeting a conserved motif that occurs in different protein families might also aid the elements move to a new target site. Another reason for targeting conserved proteins is that protein will be present in a more distantly related organism, which may allow transfer and persistence in a new divergent population. The more organisms a specific group I introns is found in, the more conserved the insertion site is.[55] Although most group II introns are mobile via an endonuclease, it has a much smaller recognition site, which occurs more frequently in a genome. It appears that group II introns evolved a different strategy that relies on frequent propagation to out pace loss.[55] An exemplary comparison of target site conversation of inteins, and group I and group II introns is depicted in Figure 3. Both inteins and group I introns are found inserted in the most conserved region of the ribonucleotide reductase, whereas the group II introns do not show such a preference. See reference 55 for further examples and a statistical evaluation of target site conservation.

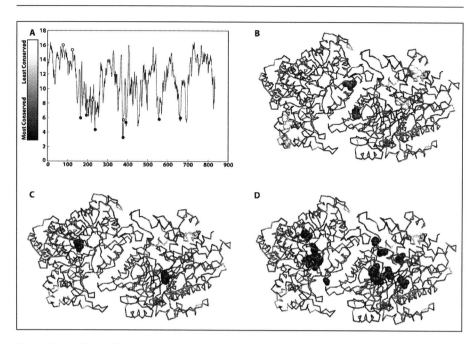

Figure 3. Positions of inteins, group I introns and group II introns along the protein sequence and in the structure of ribonucleotide reductases. Panel A shows the conservation profile of class I and class II ribonucleotide reductases. The x-axis in the amino acid positions along a multiple sequence alignment of ribonucleotide reductases and the y-axis is the conservation level in a five amino acid window centered around the indicated position. The lower the conservation score the more conserved a position is. The profile was calculated as described in[55] except the window size is five. The positions of inteins are shown as dark gray dots (or red in the online color image), the significance level for conserved site preference of inteins in the ribonucleotide reducatase is p < 0.0001; group I introns are shown as medium gray dots (or orange in the online color image), the significance level for conserved site preference of group I introns is p = 0.0256, and group II introns are shown in light gray (or blue in the online color image), with a significance level for conserved site preference of p = 0.7407. I.e., the null hypothesis of insertion independent of site conservation is rejected for inteins and group I introns, but not for group II introns. The ribonucleotide reductase hosts seven inteins, one group I intron and three group II introns. Panels B-D show the crystal structure of a dimer of the human ribonucleotide reductase R1 subunit (RNR) (PDB ID: 2WGH).[113] Each is colored according to sequence conservation. Panel B shows the crystal structure with the three group II introns insertion sites mapped to it as spheres. Panel C shows the RNR structure with the one group I intron insertion site mapped on to it as a sphere. Panel D shows the RNR crystal structure with seven intein insertion sites mapped to it as spheres. The group I intron and intein insertion sites map to conserved sites while the Group II intron do not. A color version of this image is available online at http://www.landesbioscience.com/curie.

Homing Endonucleases

Homing endonucleases (HE) are a class of site-specific endonucleases that have a large recognition site (14–40 bp) and provide mobility to group I introns and inteins.[94-96] They provide mobility to these genetic elements by targeting and cleaving intronless or inteinless alleles. The double stranded cleavage invokes the hosts DNA repair mechanisms, which uses the intron or intein containing allele as a template and copies the HE containing element into the newly cleaved target site.

How the homing endonucleases became associated with the splicing elements is an intriguing question. Among phages, freestanding HEs are common along with HE-less group I introns. The free standing HEs are thought to provide a competitive advantage for the phage when it comes in contact with competing phage, as the HE will cleave the foreign phage genome.[97] The recent sequencing of a large number of phages provides a snap shot of how HEs could have possibly associated with introns and inteins. In cyanophages a free standing HE, F-CphI, is encoded downstream of the *psbA* gene and within the *psbA* gene there is a HE free group I intron.[98] In related phages it was shown that the free standing HE can cleave the intronless intron insertion site. In T7 and T3 phages free standing HEs are able to cleave a gene encoding a DNA polymerase. Among T-phages there are some that have a free standing HE and the same HE sitting within a group I intron. Both of these elements were shown to cleave the intron insertion site.[99] Another study of split inteins and HEs in the global ocean survey (GOS) metagenomic data showed a possible progression of an HE becoming associated with inteins.[87] This progression from free standing HE to a nested HE provides a picture of how a HE could have become associated with its autocatalytic splicing element and suggests phage competition as a likely driver for the HE associating with its splicing element.

While homing endonucleases have large recognition sites, they do not require a 100% match to catalyze a double strand cut. HEs tolerate different nucleotides especially in positions that allow for synonymous substitutions of the host protein, i.e., mutations in a protein coding gene that do not change the encoded protein because of the redundancy of the genetic code.[100-102] This low precision of the HEs makes it more difficult for the host to evolve immunity to the HE through changes in the target site sequence. However, the low precision of the HEs also means that double strand cuts may be created in positions that do not correspond to the target site. This activity may result in genome rearrangements,[103] which is the process that has become the main function of the mating type switching HO endonuclease in yeast. This endonuclease evolved from a large intein whose descendents also include the intein in the yeast vacuolar ATPase catalytic subunit.[75,104]

The Homing Cycle

HEs are frequently found in inteins and group I introns, and provide a means for mobility of these parasitic genetic elements. The genetic elements using HEs go through cycles of invasion, decay and loss (compare Fig. 4). This cycle, termed the homing cycle, was first proposed by Goddard and Burt to explain the cyclic nature of group I intron gain and loss. This later was expanded to include inteins.[75,96,105] The activity of the HE and its maintenance through purifying selection depend on the availability of unoccupied target sites.

The general model starts with the invasion of a population by an HE containing mobile element that is transferred from another gene, species, population or subpopulation. When an allele with HE and an allele that does not contain an intron/intein in the target site are present in the same cell as a consequence of sex or gene transfer, the HE has the opportunity to cleave the empty target site and trigger the conversion of the empty target site into an allele with inserted HE. The resulting super Mendelian inheritance leads to an increase in frequency of the HE containing allele throughout the invaded population. However, once all target sites are occupied the parasitic genetic element has no where else to go within the local population, and two possible outcomes may occur: (1) It acquires a beneficial function for the host and remains under purifying selection or (2) The homing endonuclease begins to lose function through random mutations in the absence of purifying selection, then precise loss of the intron or intein occurs. If the latter occurs, the cycle of invasion by an allele with a functioning HE can begin again.

The relationships between the alleles with empty target site, alleles that harbor a functioning HE, and alleles with a nonfunctioning HE form a non-transitive competition network (Fig. 4). For each of the three alleles, the alleles outcompetes one of the other alleles, but loses out against the third allele. The progression of an individual target site is indicated by the central arrow. If each of the pairwise interactions leads to fixation of the winning allele, the traditional homing cycle results. However, in large populations the three types of alleles may coexist, and different subpopulations may be dominated by different alleles that propagate in the larger population similar

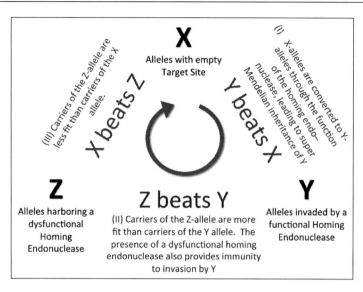

Figure 4. A molecular rock-paper-scissors game: The diagram depicts the pairwise relationships between carriers of the X allele, which does not contain a self-splicing element (intein or group I intron) nor a homing endonuclease (HE); the Y allele, containing a self-splicing element encoding a functioning HE; and the Z allele, which encodes a defective HE. The relative pairwise fitness differences are depicted by three inequalities: (I) Y beats X because the HE activity during sex or following gene transfer converts homing endonuclease free alleles into those that contain the molecular parasite; (II) Z beats Y because the defective HE produces less double strand breaks; the presence of the molecular parasite in the HE target site also provides immunity to the HE; (III) X beats Z, because its carriers have to expand less energy on replicating the molecular parasite and on synthesizing the defective HE. The three alleles also depict the main serial succession (central circular arrow) of the state of a particular homing endonuclease target site: The empty target site (X) is converted to Y through invasion by the homing endonuclease, the homing endonuclease in the Y alleles decays through mutation into the Z allele with dysfunctional homing endonuclease, and finally a precise deletion of the intein/intron restores the empty target site, i.e., Y converts to X. If each allele is successively fixed in the population, the central arrow also describes the homing cycle;[75,105] however, the different alleles may also coexist in a population for long periods of time.[96,108]

to waves in an fluid medium, without reaching fixation in the population.[96] In a large, well mixed populations the frequencies of the three alleles may follow a trajectory where higher frequency waves for each of the alleles follow one another, oscillating in state space around an equilibrium, similar to a Lotka[106]-Volterra[107] predator-prey model.[108] With the amount of metagenomic data that is being accumulated it seems possible that a quantitative and realistic model for the relations between the different alleles can be developed. One outcome of such a model would be a better understanding of gene flow within and between microbial populations.

Acquisition of New Functions

One way a HE and mobile genetic elements can avoid elimination and escape the homing cycle is to develop a function advantageous for the host cell. The catalysis of genome rearrangements has already been discussed above. Another example is the bacterial intein-like (BIL) domains, which are thought to be remnants of the HINT domain of inteins. They are found in adhesin proteins and are thought to create protein diversity.[109,110] Another example for an escape from the homing cycle is a *Naegleria* group I intron whose homing endonuclease has acquired a maturase function.[111]

Conclusion/Outlook

Many questions on the evolution and function of molecular parasites remain unanswered. For inteins, the halophillic archaea have been proposed as a model system to further understand intein propagation and the effects inteins may have on their host.[112] Utilizing the vast amount of environmental metagenomic data has provided novel insights on how homing endonucleases might have become associated with the self splicing elements.[87] Metagenomic data, deep sequencing projects, and better experimental systems promise a quantitative understanding of the host parasite interactions that govern the life cycle of HEs. Because gene transfer is a crucial step in this life cycle, these studies will also provide information on gene flow within and between populations.

Acknowledgment

This work was supported through the NASA Exobiology (NNX08AQ10G) and the National Science Foundation Assembling the Tree of Life program (DEB 0830024).

References

1. Mozer TJ, Thompson RB, Berget SM, Warner HR. Isolation and characterization of a bacteriophage T5 mutant deficient in deoxynucleoside 5'-monophosphatase activity. J Virol 1977; 24:642-50. PMID:335083
2. Chow LT, Gelinas RE, Broker TR, Roberts RJ. An amazing sequence arrangement at the 5' ends of adenovirus 2 messenger RNA. Cell 1977; 12:1-8. PMID:902310 doi:10.1016/0092-8674(77)90180-5
3. Evans RM, Fraser N, Ziff E, et al. The initiation sites for RNA transcription in Ad2 DNA. Cell 1977; 12:733-9. PMID:922890 doi:10.1016/0092-8674(77)90273-2
4. Goldberg S, Schwartz H, Darnell JE Jr. Evidence from UV transcription mapping in HeLa cells that heterogeneous nuclear RNA is the messenger RNA precursor. Proc Natl Acad Sci USA 1977; 74:4520-3. PMID:270700 doi:10.1073/pnas.74.10.4520
5. Gilbert W. Why genes in pieces? Nature 1978; 271:501. PMID:622185 doi:10.1038/271501a0
6. Zhou Z, Licklider LJ, Gygi SP, Reed R. Comprehensive proteomic analysis of the human spliceosome. Nature 2002; 419:182-5. PMID:12226669 doi:10.1038/nature01031
7. Chen XS, White WT, Collins LJ, Penny D. Computational identification of four spliceosomal snRNAs from the deep-branching eukaryote Giardia intestinalis. PLoS ONE 2008; 3:e3106. PMID:18769729 doi:10.1371/journal.pone.0003106
8. Kamikawa R, Inagaki Y, Tokoro M, et al. Split introns in the genome of Giardia intestinalis are excised by spliceosome-mediated trans-splicing. Curr Biol 2011; 21:311-5. PMID:21315596 doi:10.1016/j.cub.2011.01.025
9. Doolittle WF. Genes in pieces: were they ever together? Nature 1978; 272:581-2 doi:10.1038/272581a0.
10. Roger AJ, Doolittle WF. Molecular evolution. Why introns-in-pieces? Nature 1993; 364:289-90. PMID:8332184 doi:10.1038/364289a0
11. Doolittle WF, Stoltzfus A. Molecular evolution. Genes-in-pieces revisited. Nature 1993; 361:403. PMID:8429878 doi:10.1038/361403a0
12. Gilbert W, Glynias M. On the ancient nature of introns. Gene 1993; 135:137-44. PMID:8276250 doi:10.1016/0378-1119(93)90058-B
13. Jeffares DC, Mourier T, Penny D. The biology of intron gain and loss. Trends Genet 2006; 22:16-22. PMID:16290250 doi:10.1016/j.tig.2005.10.006
14. Cavalier-Smith T. Intron phylogeny: a new hypothesis. Trends Genet 1991; 7:145-8. PMID:2068786 doi:10.1016/0168-9525(91)90102-V
15. Stoltzfus A, Spencer DF, Zuker M, et al. Testing the exon theory of genes: the evidence from protein structure. Science 1994; 265:202-7. PMID:8023140 doi:10.1126/science.8023140
16. Logsdon JM Jr. The recent origins of spliceosomal introns revisited. Curr Opin Genet Dev 1998; 8:637-48. PMID:9914210 doi:10.1016/S0959-437X(98)80031-2
17. Logsdon JM Jr., Tyshenko MG, Dixon C, et al. Seven newly discovered intron positions in the triose-phosphate isomerase gene: evidence for the introns-late theory. Proc Natl Acad Sci USA 1995; 92:8507-11. PMID:7667320 doi:10.1073/pnas.92.18.8507
18. Koonin EV. The origin of introns and their role in eukaryogenesis: a compromise solution to the introns-early versus introns-late debate? Biol Direct 2006; 1:22. PMID:16907971 doi:10.1186/1745-6150-1-22
19. Cech TR. The generality of self-splicing RNA: relationship to nuclear mRNA splicing. Cell 1986; 44:207-10. PMID:2417724 doi:10.1016/0092-8674(86)90751-8

20. Sharp PA. On the origin of RNA splicing and introns. Cell 1985; 42:397-400. PMID:2411416 doi:10.1016/0092-8674(85)90092-3
21. Zimmerly S, Guo H, Perlman PS, Lambowitz AM. Group II intron mobility occurs by target DNA-primed reverse transcription. Cell 1995; 82:545-54. PMID:7664334 doi:10.1016/0092-8674(95)90027-6
22. Lambowitz AM, Zimmerly S. Mobile group II introns. Annu Rev Genet 2004; 38:1-35. PMID:15568970 doi:10.1146/annurev.genet.38.072902.091600
23. Watkins KP, Rojas M, Friso G, et al. APO1 Promotes the Splicing of Chloroplast Group II Introns and Harbors a Plant-Specific Zinc-Dependent RNA Binding Domain. Plant Cell 2011; 23:1082-92 PMID:21421812
24. Kroeger TS, Watkins KP, Friso G, et al. A plant-specific RNA-binding domain revealed through analysis of chloroplast group II intron splicing. Proc Natl Acad Sci USA 2009; 106:4537-42. PMID:19251672 doi:10.1073/pnas.0812503106
25. Hausner G, Olson R, Simon D, et al. Origin and evolution of the chloroplast trnK (matK) intron: a model for evolution of group II intron RNA structures. Mol Biol Evol 2006; 23:380-91. PMID:16267141 doi:10.1093/molbev/msj047
26. Turmel M, Otis C, Lemieux C. The chloroplast genome sequence of Chara vulgaris sheds new light into the closest green algal relatives of land plants. Mol Biol Evol 2006; 23:1324-38. PMID:16611644 doi:10.1093/molbev/msk018
27. Stern DB, Goldschmidt-Clermont M, Hanson MR. Chloroplast RNA metabolism. Annu Rev Plant Biol 2010; 61:125-55. PMID:20192740 doi:10.1146/annurev-arplant-042809-112242
28. Chanfreau G, Jacquier A. Catalytic site components common to both splicing steps of a group II intron. Science 1994; 266:1383-7. PMID:7973729 doi:10.1126/science.7973729
29. Sigel RK, Vaidya A, Pyle AM. Metal ion binding sites in a group II intron core. Nat Struct Biol 2000; 7:1111-6. PMID:11101891 doi:10.1038/81958
30. Gordon PM, Piccirilli JA. Metal ion coordination by the AGC triad in domain 5 contributes to group II intron catalysis. Nat Struct Biol 2001; 8:893-8. PMID:11573097 doi:10.1038/nsb1001-893
31. Gordon PM, Fong R, Piccirilli JA. A second divalent metal ion in the group II intron reaction center. Chem Biol 2007; 14:607-12. PMID:17584608 doi:10.1016/j.chembiol.2007.05.008
32. Toro N, Jimenez-Zurdo JI, Garcia-Rodriguez FM. Bacterial group II introns: not just splicing. FEMS Microbiol Rev 2007; 31:342-58. PMID:17374133 doi:10.1111/j.1574-6976.2007.00068.x
33. Huang T, Shaikh TR, Gupta K, et al. The group II intron ribonucleoprotein precursor is a large, loosely packed structure. Nucleic Acids Res 2011; 39: 2845-54. PMID:21131279
34. Cousineau B, Smith D, Lawrence-Cavanagh S, et al. Retrohoming of a bacterial group II intron: mobility via complete reverse splicing, independent of homologous DNA recombination. Cell 1998; 94:451-62. PMID:9727488 doi:10.1016/S0092-8674(00)81586-X
35. Coros CJ, Piazza CL, Chalamcharla VR, et al. Global regulators orchestrate group II intron retromobility. Mol Cell 2009; 34:250-6. PMID:19394301 doi:10.1016/j.molcel.2009.03.014
36. Fontaine JM, Goux D, Kloareg B, Loiseaux-de Goer S. The reverse-transcriptase-like proteins encoded by group II introns in the mitochondrial genome of the brown alga Pylaiella littoralis belong to two different lineages which apparently coevolved with the group II ribosyme lineages. J Mol Evol 1997; 44:33-42. PMID:9010134 doi:10.1007/PL00006119
37. Toor N, Hausner G, Zimmerly S. Coevolution of group II intron RNA structures with their intron-encoded reverse transcriptases. RNA 2001; 7:1142-52. PMID:11497432 doi:10.1017/S1355838201010251
38. Simon DM, Clarke NA, McNeil BA, et al. Group II introns in eubacteria and archaea: ORF-less introns and new varieties. RNA 2008; 14:1704-13. PMID:18676618 doi:10.1261/rna.1056108
39. Simon DM, Kelchner SA, Zimmerly S. A broadscale phylogenetic analysis of group II intron RNAs and intron-encoded reverse transcriptases. Mol Biol Evol 2009; 26:2795-808. PMID:19713327 doi:10.1093/molbev/msp193
40. Dai L, Zimmerly S. ORF-less and reverse-transcriptase-encoding group II introns in archaebacteria, with a pattern of homing into related group II intron ORFs. RNA 2003; 9:14-9. PMID:12554871 doi:10.1261/rna.2126203
41. Drager RG, Hallick RB. A complex twintron is excised as four individual introns. Nucleic Acids Res 1993; 21:2389-94. PMID:7685079 doi:10.1093/nar/21.10.2389
42. Keating KS, Toor N, Perlman PS, Pyle AM. A structural analysis of the group II intron active site and implications for the spliceosome. RNA 2010; 16:1-9. PMID:19948765 doi:10.1261/rna.1791310
43. Madhani HD, Guthrie C. A novel base-pairing interaction between U2 and U6 snRNAs suggests a mechanism for the catalytic activation of the spliceosome. Cell 1992; 71:803-17. PMID:1423631 doi:10.1016/0092-8674(92)90556-R

44. Shukla GC, Padgett RA. A catalytically active group II intron domain 5 can function in the U12-dependent spliceosome. Mol Cell 2002; 9:1145-50. PMID:12049749 doi:10.1016/S1097-2765(02)00505-1

45. Leclercq S, Giraud I, Cordaux R. Remarkable abundance and evolution of mobile group II introns in Wolbachia bacterial endosymbionts. Mol Biol Evol 2011; 28:685-97. PMID:20819906 doi:10.1093/molbev/msq238

46. Chen Y, McClane BA, Fisher DJ, et al. Construction of an alpha toxin gene knockout mutant of Clostridium perfringens type A by use of a mobile group II intron. Appl Environ Microbiol 2005; 71:7542-7. PMID:16269799 doi:10.1128/AEM.71.11.7542-7547.2005

47. Rodriguez SA, Yu JJ, Davis G, et al. Targeted inactivation of Francisella tularensis genes by group II introns. Appl Environ Microbiol 2008; 74:2619-26. PMID:18310413 doi:10.1128/AEM.02905-07

48. Yao J, Zhong J, Fang Y, et al. Use of targetrons to disrupt essential and nonessential genes in Staphylococcus aureus reveals temperature sensitivity of Ll.LtrB group II intron splicing. RNA 2006; 12:1271-81. PMID:16741231 doi:10.1261/rna.68706

49. Yao J, Zhong J, Lambowitz AM. Gene targeting using randomly inserted group II introns (targetrons) recovered from an Escherichia coli gene disruption library. Nucleic Acids Res 2005; 33:3351-62. PMID:15947133 doi:10.1093/nar/gki649

50. Kruger K, Grabowski PJ, Zaug AJ, et al. Self-splicing RNA: autoexcision and autocyclization of the ribosomal RNA intervening sequence of Tetrahymena. Cell 1982; 31:147-57. PMID:6297745 doi:10.1016/0092-8674(82)90414-7

51. Michel F, Westhof E. Modelling of the three-dimensional architecture of group I catalytic introns based on comparative sequence analysis. J Mol Biol 1990; 216:585-610. PMID:2258934 doi:10.1016/0022-2836(90)90386-Z

52. Suh SO, Jones KG, Blackwell M. A Group I intron in the nuclear small subunit rRNA gene of Cryptendoxyla hypophloia, an ascomycetous fungus: evidence for a new major class of Group I introns. J Mol Evol 1999; 48:493-500. PMID:10198116 doi:10.1007/PL00006493

53. Nielsen H, Johansen SD. Group I introns: Moving in new directions. RNA Biol 2009; 6:375-83. PMID:19667762 doi:10.4161/rna.6.4.9334

54. Haugen P, Simon DM, Bhattacharya D. The natural history of group I introns. Trends Genet 2005; 21:111-9. PMID:15661357 doi:10.1016/j.tig.2004.12.007

55. Swithers KS, Senejani AG, Fournier GP, Gogarten JP. Conservation of intron and intein insertion sites: implications for life histories of parasitic genetic elements. BMC Evol Biol 2009; 9:303. PMID:20043855 doi:10.1186/1471-2148-9-303

56. Ohman-Hedén M, Ahgren-Stalhandske A, Hahne S, Sjoberg BM. Translation across the 5′-splice site interferes with autocatalytic splicing. Mol Microbiol 1993; 7:975-82. PMID:8483423 doi:10.1111/j.1365-2958.1993.tb01189.x

57. Swithers KS, Gogarten JP, Fournier GP. Trees in the web of life. J Biol 2009; 8:54. PMID:19664165 doi:10.1186/jbiol160

58. Edgell DR, Belfort M, Shub DA. Barriers to intron promiscuity in bacteria. J Bacteriol 2000; 182:5281-9. PMID:10986228 doi:10.1128/JB.182.19.5281-5289.2000

59. Nikolcheva T, Woodson SA. Association of a group I intron with its splice junction in 50S ribosomes: implications for intron toxicity. RNA 1997; 3:1016-27. PMID:9292500

60. Raghavan R, Hicks LD, Minnick MF. Toxic introns and parasitic intein in Coxiella burnetii: legacies of a promiscuous past. J Bacteriol 2008; 190:5934-43. PMID:18606739 doi:10.1128/JB.00602-08

61. Raghavan R, Minnick MF. Group I introns and inteins: disparate origins but convergent parasitic strategies. J Bacteriol 2009; 191:6193-202. PMID:19666710 doi:10.1128/JB.00675-09

62. Vicens Q, Cech TR. Atomic level architecture of group I introns revealed. Trends Biochem Sci 2006; 31:41-51. PMID:16356725 doi:10.1016/j.tibs.2005.11.008

63. Raghavan R, Hicks LD, Minnick MF. A unique group I intron in Coxiella burnetii is a natural splice mutant. J Bacteriol 2009; 191:4044-6. PMID:19376857 doi:10.1128/JB.00359-09

64. Ikawa Y, Shiraishi H, Inoue T. Minimal catalytic domain of a group I self-splicing intron RNA. Nat Struct Biol 2000; 7:1032-5. PMID:11062558 doi:10.1038/80947

65. Nielsen H, Fiskaa T, Birgisdottir AB, et al. The ability to form full-length intron RNA circles is a general property of nuclear group I introns. RNA 2003; 9:1464-75. PMID:14624003 doi:10.1261/rna.5290903

66. Kane PM, Yamashiro CT, Wolczyk DF, et al. Protein splicing converts the yeast TFP1 gene product to the 69-kD subunit of the vacuolar H(+)-adenosine triphosphatase. Science 1990; 250:651-7. PMID:2146742 doi:10.1126/science.2146742

67. Hirata R, Ohsumi Y, Nakano A, et al. Molecular structure of a gene, VMA1, encoding the catalytic subunit of H(+)-translocating adenosine triphosphatase from vacuolar membranes of Saccharomyces cerevisiae. J Biol Chem 1990; 265:6726-33. PMID:2139027

68. Bowman EJ, Tenney K, Bowman BJ. Isolation of genes encoding the Neurospora vacuolar ATPase. Analysis of vma-1 encoding the 67-kDa subunit reveals homology to other ATPases. J Biol Chem 1988; 263:13994-4001. PMID:2971651

69. Zimniak L, Dittrich P, Gogarten JP, et al. The cDNA sequence of the 69-kDa subunit of the carrot vacuolar H+- ATPase. Homology to the beta-chain of F0F1-ATPases. J Biol Chem 1988; 263:9102-12. PMID:2897965

70. Shih CK, Wagner R, Feinstein S, et al. A dominant trifluoperazine resistance gene from Saccharomyces cerevisiae has homology with F0F1 ATP synthase and confers calcium-sensitive growth. Mol Cell Biol 1988; 8:3094-103. PMID:2905423

71. Perler FB. InBase: the Intein Database. Nucleic Acids Res 2002; 30:383-4. PMID:11752343 doi:10.1093/nar/30.1.383

72. Pietrokovski S. Intein spread and extinction in evolution. Trends Genet 2001; 17:465-72. PMID:11485819 doi:10.1016/S0168-9525(01)02365-4

73. Perler FB, Davis EO, Dean GE, et al. Protein splicing elements: inteins and exteins—a definition of terms and recommended nomenclature. Nucleic Acids Res 1994; 22:1125-7. PMID:8165123 doi:10.1093/nar/22.7.1125

74. Liu XQ. Protein-splicing intein: Genetic mobility, origin, and evolution. Annu Rev Genet 2000; 34:61-76. PMID:11092822 doi:10.1146/annurev.genet.34.1.61

75. Gogarten JP, Senejani AG, Zhaxybayeva O, et al. Inteins: structure, function, and evolution. Annu Rev Microbiol 2002; 56:263-87. PMID:12142479 doi:10.1146/annurev.micro.56.012302.160741

76. Pietrokovski S. Modular organization of inteins and C-terminal autocatalytic domains. Protein Sci 1998; 7:64-71. PMID:9514260 doi:10.1002/pro.5560070106

77. Tori K, Dassa B, Johnson MA, et al. Splicing of the mycobacteriophage Bethlehem DnaB intein: identification of a new mechanistic class of inteins that contain an obligate block F nucleophile. J Biol Chem 2010; 285:2515-26. PMID:19940146 doi:10.1074/jbc.M109.069567

78. Tori K, Perler FB. Expanding the definition of class 3 inteins and their proposed phage origin. J Bacteriol 2011; 193:2035-41. PMID:21317331

79. Elleuche S, Poggeler S. Inteins, valuable genetic elements in molecular biology and biotechnology. Appl Microbiol Biotechnol 2010; 87:479-89. PMID:20449740 doi:10.1007/s00253-010-2628-x

80. Perler FB, Olsen GJ, Adam E. Compilation and analysis of intein sequences. Nucleic Acids Res 1997; 25:1087-93. PMID:9092614 doi:10.1093/nar/25.6.1087

81. Burglin TR. The Hedgehog protein family. Genome Biol 2008; 9:241. PMID:19040769 doi:10.1186/gb-2008-9-11-241

82. Koufopanou V, Goddard MR, Burt A. Adaptation for horizontal transfer in a homing endonuclease. Mol Biol Evol 2002; 19:239-46. PMID:11861883

83. Senejani AG, Hilario E, Gogarten JP. The intein of the Thermoplasma A-ATPase A subunit: structure, evolution and expression in E. coli. BMC Biochem 2001; 2:13. PMID:11722801 doi:10.1186/1471-2091-2-13

84. Wu H, Hu Z, Liu XQ. Protein trans-splicing by a split intein encoded in a split DnaE gene of Synechocystis sp. PCC6803. Proc Natl Acad Sci USA 1998; 95:9226-31. PMID:9689062 doi:10.1073/pnas.95.16.9226

85. Caspi J, Amitai G, Belenkiy O, Pietrokovski S. Distribution of split DnaE inteins in cyanobacteria. Mol Microbiol 2003; 50:1569-77. PMID:14651639 doi:10.1046/j.1365-2958.2003.03825.x

86. Stoltzfus A. On the possibility of constructive neutral evolution. J Mol Evol 1999; 49:169-81. PMID:10441669 doi:10.1007/PL00006540

87. Dassa B, London N, Stoddard BL, et al. Fractured genes: a novel genomic arrangement involving new split inteins and a new homing endonuclease family. Nucleic Acids Res 2009; 37:2560-73. PMID:19264795 doi:10.1093/nar/gkp095

88. Aranko AS, Zuger S, Buchinger E, Iwai H. In vivo and in vitro protein ligation by naturally occurring and engineered split DnaE inteins. PLoS ONE 2009; 4:e5185. PMID:19365564 doi:10.1371/journal.pone.0005185

89. Goodwin TJ, Butler MI, Poulter RT. Multiple, non-allelic, intein-coding sequences in eukaryotic RNA polymerase genes. BMC Biol 2006; 4:38. PMID:17069655 doi:10.1186/1741-7007-4-38

90. Lazarevic V. Ribonucleotide reductase genes of Bacillus prophages: a refuge to introns and intein coding sequences. Nucleic Acids Res 2001; 29:3212-8. PMID:11470879 doi:10.1093/nar/29.15.3212

91. Dai L, Zimmerly S. Compilation and analysis of group II intron insertions in bacterial genomes: evidence for retroelement behavior. Nucleic Acids Res 2002; 30:1091-102. PMID:11861899 doi:10.1093/nar/30.5.1091

92. Toro N. Bacteria and Archaea Group II introns: additional mobile genetic elements in the environment. Environ Microbiol 2003; 5:143-51. PMID:12588294 doi:10.1046/j.1462-2920.2003.00398.x

93. Sandegren L, Sjoberg BM. Distribution, sequence homology, and homing of group I introns among T-even-like bacteriophages: evidence for recent transfer of old introns. J Biol Chem 2004; 279:22218-27. PMID:15026408 doi:10.1074/jbc.m400929200

94. Chevalier BS, Stoddard BL. Homing endonucleases: structural and functional insight into the catalysts of intron/intein mobility. Nucleic Acids Res 2001; 29:3757-74. PMID:11557808 doi:10.1093/nar/29.18.3757

95. Dujon B. Group I introns as mobile genetic elements: facts and mechanistic speculations–a review. Gene 1989; 82:91-114. PMID:2555264 doi:10.1016/0378-1119(89)90034-6

96. Gogarten JP, Hilario E. Inteins, introns, and homing endonucleases: recent revelations about the life cycle of parasitic genetic elements. BMC Evol Biol 2006; 6:94. PMID:17101053 doi:10.1186/1471-2148-6-94

97. Goodrich-Blair H, Shub DA. Beyond homing: competition between intron endonucleases confers a selective advantage on flanking genetic markers. Cell 1996; 84:211-21. PMID:8565067 doi:10.1016/S0092-8674(00)80976-9

98. Zeng Q, Bonocora RP, Shub DA. A free-standing homing endonuclease targets an intron insertion site in the psbA gene of cyanophages. Curr Biol 2009; 19:218-22. PMID:19200728 doi:10.1016/j.cub.2008.11.069

99. Bonocora RP, Shub DA. A likely pathway for formation of mobile group I introns. Curr Biol 2009; 19:223-8. PMID:19200727 doi:10.1016/j.cub.2009.01.033

100. Gimble FS. Degeneration of a homing endonuclease and its target sequence in a wild yeast strain. Nucleic Acids Res 2001; 29:4215-23. PMID:11600710 doi:10.1093/nar/29.20.4215

101. Scalley-Kim M, McConnell-Smith A, Stoddard BL. Coevolution of a homing endonuclease and its host target sequence. J Mol Biol 2007; 372:1305-19. PMID:17720189 doi:10.1016/j.jmb.2007.07.052

102. Kurokawa S, Bessho Y, Higashijima K, et al. Adaptation of intronic homing endonuclease for successful horizontal transmission. FEBS J 2005; 272:2487-96. PMID:15885098 doi:10.1111/j.1742-4658.2005.04669.x

103. Nishioka M, Fujiwara S, Takagi M, Imanaka T. Characterization of two intein homing endonucleases encoded in the DNA polymerase gene of Pyrococcus kodakaraensis strain KOD1. Nucleic Acids Res 1998; 26:4409-12. PMID:9742242 doi:10.1093/nar/26.19.4409

104. Bakhrat A, Baranes K, Krichevsky O, et al. Nuclear import of ho endonuclease utilizes two nuclear localization signals and four importins of the ribosomal import system. J Biol Chem 2006; 281:12218-26. PMID:16507575 doi:10.1074/jbc.M600238200

105. Goddard MR, Burt A. Recurrent invasion and extinction of a selfish gene. Proc Natl Acad Sci USA 1999; 96:13880-5. PMID:10570167 doi:10.1073/pnas.96.24.13880

106. Lotka AJ. Analytical Note on Certain Rhythmic Relations in Organic Systems. Proc Natl Acad Sci USA 1920; 6:410-5. PMID:16567509 doi:10.1073/pnas.6.7.410

107. Volterra V. Variations and fluctuations of the number of individuals in animal species living together in Animal Ecology. In: Chapman RN, ed. Animal Ecology. New York: McGraw–Hill; 1931.

108. Yahara K, Fukuyo M, Sasaki A, Kobayashi I. Evolutionary maintenance of selfish homing endonuclease genes in the absence of horizontal transfer. Proc Natl Acad Sci USA 2009; 106:18861-6. PMID:19837694 doi:10.1073/pnas.0908404106

109. Dori-Bachash M, Dassa B, Peleg O, et al. Bacterial intein-like domains of predatory bacteria: a new domain type characterized in Bdellovibrio bacteriovorus. Funct Integr Genomics 2009; 9:153-66. PMID:19153786 doi:10.1007/s10142-008-0106-7

110. Amitai G, Belenkiy O, Dassa B, et al. Distribution and function of new bacterial intein-like protein domains. Mol Microbiol 2003; 47:61-73. PMID:12492854 doi:10.1046/j.1365-2958.2003.03283.x

111. Wikmark OG, Einvik C, De Jonckheere JF, Johansen SD. Short-term sequence evolution and vertical inheritance of the Naegleria twin-ribozyme group I intron. BMC Evol Biol 2006; 6:39. PMID:16670006 doi:10.1186/1471-2148-6-39

112. Barzel A, Naor A, Privman E, et al. Homing endonucleases residing within inteins: evolutionary puzzles awaiting genetic solutions. Biochem Soc Trans 2011; 39:169-73. PMID:21265767 doi:10.1042/BST0390169

113. Fairman JW, Wijerathna SR, Ahmad MF, et al. Structural basis for allosteric regulation of human ribonucleotide reductase by nucleotide-induced oligomerization. Nat Struct Mol Biol 2011; 18:316-22. PMID:21336276 doi:10.1038/nsmb.2007

CHAPTER 5

Restriction-Modification Systems as Mobile Epigenetic Elements

Yoshikazu Furuta[1,2] and Ichizo Kobayashi*[1-3]

Abstract

Transfer of mobile genetic elements between prokaryotes is limited by restriction-modification systems. Restriction-modification systems consist of a modification enzyme that epigenetically methylates a specific DNA sequence, and a restriction endonuclease (restriction enzyme) that cuts DNA lacking this epigenetic mark. These elements were discovered because they attack mobile genetic elements. However, recent studies have revealed that they are themselves mobile. In some cases, the mobility of restriction-modification systems is through symbiosis with other forms of mobile elements. In other cases, movement is unlinked to other mobile elements. The systems may insert into the genome with long and variable target duplication, or into the intergenic region of an operon. Insertion of restriction-modification systems induces other genome rearrangements such as amplification and inversion. Even a domain within a protein can be the unit of mobility: some restriction-modification system subunits that recognize a target DNA sequence contain mobile amino acid sequences that can apparently move between different domains of a protein through recombination of DNA sequences encoding them. This mobility extends the biological significance of restriction-modification systems beyond defense: the systems define, and sometimes even force, epigenetic order on a genome. The multilevel conflicts involving these mobile epigenetic elements may drive prokaryotic evolution.

Introduction: Restriction-Modification Systems in Epigenetic Conflicts

DNA methyltransferases methylate specific bases in recognition sequences and generate three types of modified base: 5-methylcytosine (m5C), N4-methylcytosine (m4C), and N6-methyladenine (m6A). This is a type of epigenetic modification because it is passed on through maintenance methylation after DNA replication.[1] Epigenetic DNA methylation can be involved in gene regulation, and variation in DNA methyltransferases can potentially provide diversity in the gene expression status of the prokaryotic cell.

Many prokaryotic DNA methyltransferases are paired with restriction enzymes, which were first discovered through their ability to restrict phage infection.[2] Restriction enzymes are DNA endonucleases that recognize specific DNA sequences and introduce a break (Fig. 1A). This activity restricts the establishment of invading DNAs that lack proper DNA methylation, such as bacteriophage DNA genomes, plasmids, and DNA fragments delivered through natural

[1]Department of Medical Genome Sciences, Graduate School of Frontier Sciences, University of Tokyo, Tokyo, Japan; [2]Institute of Medical Science, University of Tokyo, Minato-ku, Tokyo, Japan; [3]Department of Biophysics and Biochemistry, Graduate School of Science, University of Tokyo, Tokyo, Japan.
*Corresponding Author: Ichizo Kobayashi—Email: ikobaya@ims.u-tokyo.ac.jp

Bacterial Integrative Mobile Genetic Elements, edited by Adam P. Roberts and Peter Mullany.
©2013 Landes Bioscience.

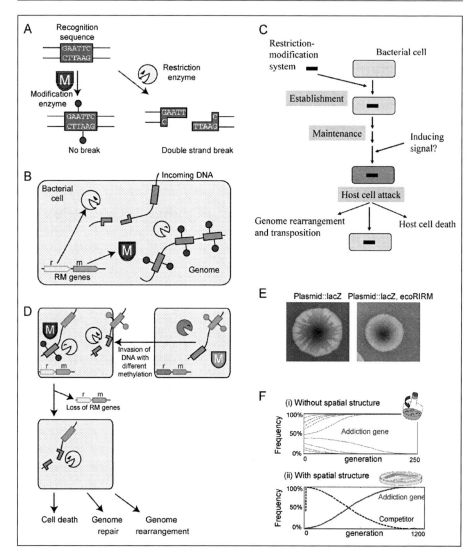

Figure 1. Type II restriction–modification system. A) A modification enzyme and a restriction endonuclease of an RM system. B) Attack on incoming DNA. C) A hypothetical life cycle. D) Epigenetic conflicts and postsegregational killing involving RM systems. DNA methylated at the red site but unmethylated at the green site cannot invade a cell with a green-yellow RM system because of restriction. Loss of the green-yellow RM system leads genome DNA breakage because the modification enzyme cannot methylate all recognition sites after genome replication and protein dilution by cell division. The breakage leads to cell death, genome repair, or genome rearrangement. E) A bacterial colony showing postsegregational killing. Left, an unstable plasmid with the *lacZ* gene; right, the plasmid with insertion of the EcoRI RM system. Unpublished results by Noriko Takahashi and Ichizo Kobayashi. F) Spread of a postsegregational killing gene (addiction gene) in the presence of spatial structure (theoretical work). Modified from Mochizuki et al. Genetics 2006; 172:1309-23;[82] with permission of the Genetics Society of America. A color version of this figure is available online at http://www.landesbioscience.com/curie.

transformation machinery (Fig. 1B). The potentially lethal cleavage of cellular DNA in cells that harbor a restriction enzyme is prevented by epigenetic DNA methylation by the cognate DNA methyltransferase (Fig. 1A,B). Genes encoding the restriction enzyme and the methyltransferase are often located next to each other and form a unit called a restriction–modification (RM) system. RM systems are classified into four types, Type I, II, III, and IV, based on their genetic and biochemical characteristics.

The primary biological significance of RM systems is often assumed to be their activity as a defense system for host cells against invading DNA such as bacteriophages. However, an RM system will not defend a bacterial cell from invasion of DNA from a bacterial cell carrying the same RM system. An RM system will attack invading DNA only when it lacks its epigenetic mark. The essence of the RM phenomena is conflict involving an epigenetic system rather than defense against invading DNA. In response to violation of epigenetic status by invasion of foreign DNA or loss of an RM system, RM systems induce DNA breakage and, consequently, reactions such as DNA damage repair, cell death and genome rearrangements (Fig. 1C,D).

RM systems are themselves mobile and can therefore be designated epigenetic mobile elements. This review introduces the mobility of RM systems first and then the conflicts between RM systems and their host cells and between RM systems themselves. These conflicts, which can involve host cell death, affect the conservation or change of the host epigenetic status.

Abundance of RM Systems

RM gene homologs that have been identified in completely sequenced bacterial genomes are in the REBASE database.[3] Some genomes—for example, those from *Hemophilus influenzae*, *Methanococcus jannaschii*, *Helicobacter pylori*, *Neisseria meningitidis*, *Neisseria gonorrhoeae*, and *Xylella fastidiosa*—have large numbers of RM gene homologs, although many RM gene homologs are specific to a single strain within a given species. No RM system is native to eukaryotes, although a family of viruses that use the unicellular eukaryotic alga Chlorella as a host produce RM systems.[4]

Types of Restriction Systems

RM systems are currently classified into four types (I-IV), each with a unique mechanism for target recognition (Fig. 2).[2] Type I systems consist of R, M and specificity (S) subunit genes. Formation of a multisubunit structure is necessary for modification (SM) and restriction (SMR) activities.[5] Sequence recognition is determined by the target recognition domains (TRDs) in the S subunit. Each of the two TRDs, TRD1 and TRD2, recognizes one half of a bipartite target sequence.[6] DNA methylation takes place within the recognition sequence, whereas the cleavage site is at a variable distance from the recognition sequence. After binding to an unmodified recognition sequence, restriction enzyme complexes are thought to translocate DNA toward themselves from both directions, in a reaction coupled to ATP hydrolysis. DNA is cleaved where two restriction enzyme complexes meet.[7]

Type II systems consist of separate R and M enzymes, which independently recognize a target sequence and catalyze reactions.[2,8] M proteins have amino acid sequence motifs that are common to DNA methyltransferases and are well conserved, and their target recognition domain can be easily identified.[9] R proteins share much less similarity.[10]

RM systems of the Type IIG subgroup are defined by the presence of a single polypeptide that apparently results from a fusion of R and M proteins. This Type IIG subgroup can be further classified into two subgroups by sequence similarity to either Type II or Type I systems.[11] The class similar to Type II (Fig. 2C) has the same TRD structure as typical Type II M enzymes within an RM polypeptide.[2,11-13] The other class similar to Type I, such as AloI,[14] has a fused S subunit in addition to the RM fusion polypeptide (Fig. 2D). As in Type I systems, the S subunit counterpart has two TRDs. Although R and M are always fused in Type IIG systems, the S subunit is sometimes separated from the RM fused gene (Fig. 2E).[15]

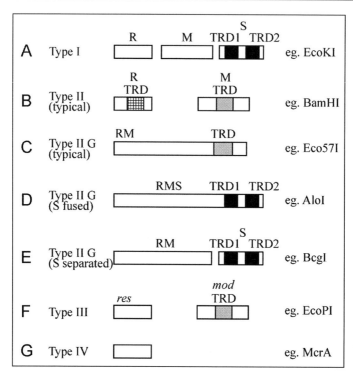

Figure 2. Gene organization of various types of restriction–modification systems. See text for explanation. TRD: target recognition domain.

Type III systems consist of *res* and *mod* genes (Fig. 2F). The *mod* gene product has modification activity by itself, while the complex of the two gene products has restriction enzyme activity.[16] The Mod subunit is responsible for target recognition and its TRD can be easily identified, similar to M enzymes of Type II systems.[17-19]

Type IV systems contain a class of enzymes that cleave DNA only when the recognition site is methylated (Fig. 2F).[2] McrA, McrBC, and Mrr are prototype *Escherichia coli* enzymes in this class that show different restriction spectra.[1,20-23]

Gene Organization

There are many variations to the simplified schemes above.[3] A Type II system may have two different modification enzymes for the top and bottom strands of the recognition sequence. It may carry two different nucleases for the different strands. Some RM systems have a gene for a regulatory protein (C protein, Fig. 3A, also see below).

The relative direction of the constituent genes can vary. In the simplest case of a Type II system with one R gene and one M gene, the two genes may lie in the same orientation, or in opposite orientations.

In most cases, genes in an RM system are clustered into a single locus, but in some cases they are unlinked or in a more complex context.[24] In a Type I system of *Staphylococcus aureus*, one pair of M and S genes lie on one genomic island, another M and S pair are on another genomic island, and an R gene is outside of both islands (Fig. 3B). The hypothesis is that the R gene product can associate with either MS pair to form a two-faced restriction enzyme that recognizes two different sequences.[25] *Lactococcus lactis* has plasmids that carry an S gene but not an R or an M gene (Fig. 3C). The plasmid-encoded S gene products form active Type I RM systems with R and M products

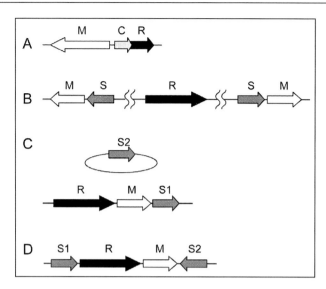

Figure 3. Gene order in restriction–modification systems. A) PvuII RM system, which has a control (C) gene in addition to R and M genes.[44] B) Type I RM system in *Staphylococcus aureus*.[25] A single R gene may interact with two sets of M and S genes. C) Type I RM system in *Lactococcus lactis*.[26] A second copy of S gene is on a plasmid. D) Type I RM system in *Mycoplasma pulmonis*.[27] R and M genes are flanked by two S genes in the opposite orientation.

from genes on the chromosome.[26] In *Mycoplasma pulmonis*, two S genes in opposing orientation flank an R gene and an M gene in the inverted orientation (Fig. 3D). Recombination between the two S genes is discussed below.[27]

Mobility of RM Systems Revealed from Molecular Evolutionary Analyses

Evolutionary analyses suggest that RM genes have undergone extensive horizontal transfer between different groups of microorganisms.[28,29] Early studies found that close homologs occur in distantly related organisms such as Eubacteria and Archaea (archaebacteria).[30] Extensive sequence alignment and phylogenetic tree construction now provide strong support for this point.[31-34] Additional evidence for extensive horizontal transfer of RM genes comes from incongruencies between methyltransferase phylogenetic trees and rRNA gene trees from the same species.[31] Moreover, the GC content and/or codon usage of RM genes often differs from the majority of other genes in the genome.[31,35-37] This indicates that some RM genes may have joined the genome relatively recently through horizontal transfer from distantly related bacteria.

RM Systems on Mobile Genetic Elements

Genome comparisons revealed that the RM systems are mobile and involved in genome rearrangements. In some cases, mobility is acquired by carriage on other mobile elements (Table 1) such as plasmids,[38-45] phages,[45-52] conjugative elements/genomic islands,[25,53-57] transposons,[58-60] and integrons.[61,62] Indeed, some RM systems are located immediately adjacent to mobility-related genes such as those for transposases, integrases, or resolvases.[63-66] RM systems are also found that form a composite transposon by insertion of insertion sequence (IS) to both sides.[67]

The mobile elements themselves can be lost from the genome. Carrying a Type II RM system may allow their stable maintenance through postsegregational killing, as demonstrated for plasmids.[68,69] This may represent the symbiosis of two genetic elements that provides mutual benefits of maintenance and mobility.

Table 1. RM systems on mobile genetic elements

(i) Plasmid

ecoRI	*Escherichia coli*	pMB1[42]
paeR7I	*Pseudomonas aeruginosa*	pMG7[41]
ecoRII	*Escherichia coli*	N3[43]
ecoRV	*Escherichia coli*	pLG74[40]
ssoII	*Shigella sonnei*	P4[38]
bsp6I	*Bacillus sp* strain RFL6	pXH13[39]
ecoP15I	*Escherichia coli*	p15B[45]
pvuII	*Proteus vulgaris*	pPvu1[44]

See REBASE[3] for other examples.

(ii) Bacteriophage

ecoP1	*Escherichia coli*	P1[45]
ecoT38I	*Escherichia coli*	P2[51]
ecoO109I	*Escherichia coli*	P4[52]
bsuM	*Bacillus subtilis*	Prophage 3[49]
hindIII	*Hemophilus influenzae*	phi-flu[50]
sau42I	*Staphylococcus aureus*	phi-42[48]
dam	*Escherichia coli*	T2[47]
PHG11b_1 (putative)	*Flavobacterium*	phi-11b[46]

(iii) Integrative Conjugative Element/Genomic Island

sth368I	*Streptococcus thermophilus*	ICESt1[56]
llaI	*Lactococcus lactis*	pTR2030[55]
ORF29	*Providencia rettgeri*	R391[57]
Type II M	*Streptococcus pneumoniae*	Tn5252[54]
Type I MS	*Staphylococcus aureus*	Genomic islands[25]
Type II RM	*Streptococcus pyogenes*	Chimeric element phi-10394.4[53]

(iv) Transposon

Type I RMS	*Sinorhizobium meliloti*	Flanked by ISRm1 and ISRm11[59]
R	*Acidithiobacillus caldus*	Flanked by ISAtc1-like[58]
rle39BI	*Rhizobium leguminosarum*	Flanked by ISRle39[60]

See other examples in reference 74.

(v) Integron

xbaI	*Xanthomonas campestris*	Reference 62
hphI	*Vibrio metschnikovii*	Reference 62
M.Vch01I	*Vibrio cholerae*	Reference 61

(vi) Linked to Mobility-Related Genes

bglII	*Bacillus globigii*	Transposon resolvase homolog[66]
accI	*Weeksella zoohelcum*	Transposase and integrase[64]
yenI	*Yersinia enterocolitica*	IS1222[63]
sptAI	*Salmonella enterica*	Integrase[65]

RM Systems as Mobile Genetic Elements

In other cases, RM systems are found inserted in the genome, but not linked to a mobile element (Table 2). They sometimes insert into an operon-like gene cluster (Fig. 4A (i))[70-72] by a recombination process that could be related to the DNA cleavage activity of the restriction

Table 2. Genome rearrangements associated with RM systems

(i) Simple Insertion into an Operon-Like Gene Cluster

Hemophilus aegyptius	*haeII* insertion between *mucE* and *mucF*[72]
Neisseria meningitidis	*nmeBI* insertion between *pheS* and *pheT*[71]
Streptococcus suis	*ssuDAT1I* insertion between *purH* and *purD*[70]

(ii) Substitution

Helicobacter pylori	Simple substitution of Type I, II, and III RM genes[73,100]
Pyrococcus species	Simple substitution of Type II *pabI* RM system[137]

See other examples in reference 74.

(iii) Allelic Diversity

Streptococcus suis	Two isoschizomers of R genes in *ssuDat1I* system[70]
Campylobacter jejuni	Diversity in S genes and neighbor genes, which are between R and M genes, at *hsdIAB* system[76]
Escherichia coli	Variation in ICR (Immigration control center, Type I RM system)[75]

See other examples in reference 74.

(iv) Insertion with Long Target Duplication

Helicobacter pylori	Insertion of Type II RM systems with duplication of 36- to 369-bp region in target genome[73,74]
Hemophilus influenzae	Insertion of Type I RM systems with duplication of 39-bp region in target genome[74]
Campylobacter jejuni	Insertion of Type III RM systems with duplication of 107- to 208-bp region in target genome[74]
Burkholderia species	Insertion of Type IV R with duplication of 263- to 270-bp region in target genome[74]

See other examples in reference 74.

(v) Insertion with Neighboring Inversion

Helicobacter pylori	10-kb inversion next to Type II RM system[100]
Pyrococcus species	108-kb inversion next to Type I RM system and 5-kb inversion next to predicted RM system[137]

(vi) Transposition

Helicobacter pylori	Type II RM system at different loci[100]
Pyrococcus species	Predicted M gene at different loci[137]
Xanthomonas oryzae	Transposon-like structure without transposase[74]
Neisseria gonorrhoeae	Transposon-like structure without transposase[74]

enzyme.[73] Insertion might give a competitive advantage to the operon as to the mobile elements. Some RM systems are substituted by another RM system at the same locus.[70,74-76]

In *Helicobacter pylori*, insertion of several RM systems has occurred with the duplication of a long target sequence (Fig. 4A (iii)).[73] The generality of this mode was confirmed by systematic comparison analysis of RM systems on the completely sequenced genomes.[74] This survey also led to the discovery of RM systems with a transposon-like structure where RM genes replaced the transposase gene (Fig. 4B).[74]

Attack on the Host Bacterial Genome: Type II RM Systems

Works have demonstrated that RM systems also act as a watchdog maintaining epigenetic order in cells (Fig. 5). Alteration in the epigenetic status might lead to double-strand breakage of the self

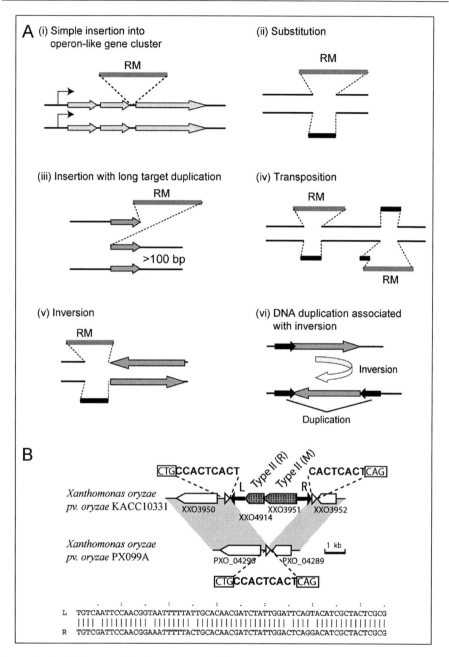

Figure 4. Restriction-modification systems and genome rearrangements. A) Various patterns of genome rearrangements linked to an RM system (or a DNA methyltransferase gene (vi)). B) A transposon-like RM system. A Type II RM gene pair flanked by long (65 bp), imperfect inverted repeats (L and R) has inserted into a genome with a short (8 bp) target duplication. The 5′ CTGCCAG (boxed) apparently split by this target sequence is the predicted recognition sequence of the RM system. Modified from Furuta et al. Nucleic Acids Res 2010; 38:2428-43;[74] Oxford University Press.

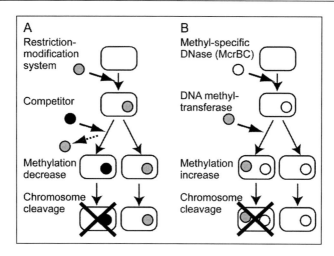

Figure 5. Host attack by restriction systems in conflict with epigenetic systems. A) Postsegregational killing by Type II systems. When a resident RM gene complex is replaced by a competitor genetic element, a decrease in the modification enzyme level exposes newly replicated chromosomal restriction sites to lethal cleavage by the remaining restriction enzyme molecules (Fig. 1D). Intact genome copies survive in uninfected and unaltered neighboring clonal cells. B) A methyl-specific Type IV restriction enzyme (McrBC). When a DNA methylation system enters a cell and begins methylating chromosomal recognition sites, McrBC triggers cell death by chromosomal cleavage. Intact genome copies survive in uninfected and unaltered neighboring clonal cells. Modified from Fukuda et al. Genome Biol 2008; 9:R163;[34] Biomed Central.

genome by the restriction enzyme,[34,68,77,78] which can result in cell death or genome rearrangements. This may eliminate unstable cells and maintain the epigenome status.

Some Type II RM systems cause chromosomal cleavage of their host cells when their genes are eliminated, for example, by a competitor genetic element (Figs. 1D and 5A).[68] When an RM system is stably maintained in a cell, the restriction enzyme does not cleave the genomic DNA because of protection through epigenetic methylation by the cognate methyltransferases. However, when the RM system is lost, the concentration of the restriction and modification enzymes is decreased through cell division,[79] resulting in undermethylated sites on newly replicated chromosomes.[80] The remaining restriction enzyme molecules cleave the unmethylated recognition sequence and cause cell death. The net result is survival of cells that were not invaded by the competitor. This process is called "postsegregational killing" or "genetic addiction."[24] The capability of an RM system in forcing maintenance on its host can become stronger by a mutation in its methyltransferase.[81]

Figure 1E visualizes the effect of postsegregational killing during the formation of bacterial colony. An unstable plasmid in the bacterial cell is lost during colony formation and leads to formation of papillae (Fig. 1E, left). However, when the EcoRI RM system is present on the plasmid, no papillae are formed because the plasmid-free cells are killed (Fig. 1E, right).

Postsegregational killing occurs because of a conflict between the RM system (or the plasmid) and the host bacteria and is an example of intragenomic conflicts. A theoretical work demonstrated that starting from very few copies, a postsegregational killing gene can increase in a population in the presence of a spatial structure (Fig. 1F (ii)). In the absence of the spatial structure, the gene is quickly lost unless it is abundant at the beginning (Fig. 1F (i)).[82]

Post-segregational cell killing by one RM system is inhibited by the presence of another RM system recognizing the same DNA sequence, because the M protein of the latter system protects the genome from cleavage by the R protein of the former system.[83] This indicates the presence of competition for recognition sequences between RM systems. Thus, a recognition sequence of RM systems defines an incompatibility group. This competition explains the individual specificity and collective diversity in RM systems' sequence recognition. The competition may be one-sided when the recognition sequence of one RM is included in the recognition sequence of the other RM.[84] Another incompatibility relationship between RM systems is found in a regulatory protein that delays expression of the R protein upon entry of an RM system into a new host bacterial cell.[85]

Recent studies revealed a common pathway of stress-induced cell death in bacteria.[86,87] Transcriptome analysis during postsegregational killing by a Type II RM system revealed its similarity to killing by several bacteriocidal antibiotics.[78] Thus, RM systems switch on the death pathway intrinsic to the host bacterial cells. Gene products that program bacterial cell death, such as the restriction enzymes discussed here, are likely to work upstream of the common cell death pathway.

Attack on the Host Bacterial Genome: Type IV Restriction Systems

Several studies demonstrated that phages or plasmids carrying a DNA methyltransferase gene cannot be propagated in an strain of *E. coli* which harbors *mcrBC*, a methylated DNA specific restriction enzyme.[88] Whether the block to propagation is caused by repeated methylation and subsequent cleavage of the introduced DNA,[88] or to host genome methylation and its cleavage was not known. Fukuda et al. demonstrated that McrBC inhibits establishment of the gene for the DNA methyltransferase PvuII (M.PvuII, 5′CAGm4CTG) in *E. coli*,[34] even when the gene is on a plasmid lacking its recognition sequence. This result suggests that the transferred DNA does not need to have methylated sites for McrBC-dependent inhibition,[34] suggesting that host genome cleavage accompanied by cell death inhibits the establishment of the methyltransferase gene (Fig. 5B). The underlying mechanism of cell death was revealed by observing *E. coli* chromosomal DNA infected with lambda phage carrying the M.PvuII gene.[34] Accumulation of large linear DNAs corresponding to broken chromosomes, and smaller DNAs of variable size was observed, which likely reflected chromosomal degradation. The *mcrBC*-dependence strongly suggests that M.PvuII-mediated chromosomal methylation triggers chromosomal cleavage by McrBC, followed by chromosomal degradation. This, in turn, indicates that inhibition of phage multiplication (restriction) is caused by host death.[34] This type of conflict between DNA methyltransferase genes carried by bacteriophages and methyl-specific restriction enzymes is biologically relevant because DNA methyltransferase genes are often found in bacteriophage genomes.[89-92]

In addition to M.PvuII, the M.SinI (GGW^m5CC) and M.MspI (^m5CCGG) cause McrBC-dependent cell death, whereas M.SsoII (C^m5CNGG) does not. These results are consistent with the R^mC sequence specificity of McrBC observed in vitro.[93] McrBC has the potential to act as a defense system against many DNA methyltransferases with an appropriate specificity. Such conflicts between McrBC and invading epigenetic DNA methylation systems may have driven diversification of sequence recognition by the methyltransferases and by the McrBC family.[94]

Type II systems cause cell death when a particular mode of epigenetic DNA methylation decreases, while this Type IV system causes cell death when an epigenetic DNA methylation mode increases. Nonetheless, the result of both systems is the maintenance of an epigenetic order defined by DNA methylation (Fig. 5).

The DNA replication fork may be the site of action of McrBC, which can cleave a model DNA replication fork in vitro.[77] Cleavage of a fork requires methylation on both arms and results in removal of one or both arms. Most cleavage events remove the methylated sites from the fork. This suggests that acquisition of even rare modification patterns will be recognized and rejected efficiently by modification-dependent restriction systems that recognize two sites.

Attack on the Host Bacterial Genome: Type I RM Systems

Restriction alleviation is a phenotypic decrease in restriction activity against invading DNA that can be induced by DNA-damaging agents; this also occurs constitutively in some bacterial mutants. The underlying mechanisms vary among the restriction system types.[95-98] Restriction alleviation is proposed to be a mechanism for protecting chromosomes from restriction at newly generated replication forks that produce unmethylated restriction sites.[5]

A recent work demonstrated that a Type I restriction enzyme cleaves model replication forks at their branch point in vitro.[77] Cleavage was dependent on a recognition sequence on one of the arms and was inhibited when the site was hemi-methylated. The results suggested that the enzyme binds to DNA at the recognition sequence and tracks along the DNA, cleaving when it encounters a branch point.

Fork cleavage may take place on chromosomal DNA under conditions of extra replication initiation. From an unmethylated recognition sequence, the restriction enzyme tracks on the DNA. If the fork is moving forward during replication, DNA cleavage might not occur. However, when the enzyme meets an arrested replication fork, one arm is cleaved, possibly leading to cell death or a round of repair through recombination and replication. This mechanism might lead to elimination of cells with an unstable genome and to maintenance of an intact genome.[26] This hypothesized function is similar to that of programmed cell death in multicellular organisms.

Anti-Restriction Systems

The anti-restriction modification systems evolved by bacteriophages represent examples of co-evolution with the host bacterium.[16] Host bacteria cells have evolved anti-restriction features, such as solitary methyltransferases that protect restriction sites from lethal attack by an RM system.[81,99] Another sign of adaptation, restriction avoidance, is discussed below.

RM Systems and Genome Rearrangements

The attack of RM systems on the host genome induces repair by recombination and replication and may induce genome rearrangements (Fig. 4A). Systematic genome comparison shows the involvement of RM systems in genome rearrangements (Table 2). RM systems can be inserted into the genomes with other mobile elements (see above) or inserted by themselves with long target duplication (Fig. 4A (iii)).[73,74] A large genome inversion event is seen in the neighborhood of RM insertion in *Helicobacter pylori* (Fig. 4A (v)).[100] Activity of an RM system likely induced unequal homologous recombination at IS3 sequences in *E. coli*, causing genome-wide rearrangements.[80] A methyltransferase gene in *Helicobacter pylori* is linked to an event of DNA duplication associated with inversion (Fig. 4A (vi)), a mechanism formally similar to replicative inversion in several DNA transposons.[101] A Type II restriction-modification system accelerates genome and phenotype changes in bacterial experimental evolution.[102]

We found that the *bamHI* gene complex, flanked by long direct repeats, amplifies in the *Bacillus subtilis* chromosome in its restriction activity dependent manner.[103] These results led us to propose that RM gene complexes increase in frequency in the cell population in a life cycle similar to that of a DNA virus (Fig. 1C). Insertion with long target duplication (Fig. 4A (iii), discussed above) results in formation of direct repeats flanking an RM system, giving it the potential for amplification.

Impact on Genome Evolution

The presence of an RM gene complex[104] and the action of a restriction enzyme[105] induce the SOS response, as does the action of a methylated DNA-specific endonuclease.[106] Global mutagenesis generates heterogeneity in the cell population, similar to other stressful conditions,[107] and may help the survival of the genome. These mechanisms could contribute to the evolution of restriction avoidance (discussed below) and to the inactivation of a RM gene complex. Genomes show signs of strong selection against restriction sites. Systematic avoidance of potential restriction sites (palindromes) in bacterial genomes is called restriction avoidance and has been characterized by informatics.[108-110] This is likely the result of selection after host attack by RM systems. (However, evaluating the contribution, if any, of selection by restriction attacks on groups of transferred genomic DNAs is difficult.) In many genomes, restriction avoidance is more pronounced in the genome proper than in prophages.[109,110] This is probably because the genome has experienced long-term selection by RM systems, while the prophages, as newcomers, have not yet undergone selection.[109,110] This restriction avoidance would lead to a decrease in the virulence of the particular

RM system against the genome, which is almost always beneficial for the host and can be beneficial to the RM system itself under some conditions.

DNA methylation may locally increase mutation. 5-methylcytosine shows a higher rate of deamination than cytosine. Its deamination at the C/G pair in duplex DNA results in the formation of thymine and hence the generation of a T/G mispair.[111,112] Very short patch mismatch correction by a *vsr* gene linked to a solitary cytosine methyltransferase *dcm* in *E. coli*, can repair the mispair generated by methylation by *dcm* product and restore the mC/G pair.[112] A homolog of the *vsr* gene is linked to several RM genes that produce 5-methylcytosine,[3] some of which are active. Local mutagenesis induced by the action of the RM systems leads to loss of the restriction sites in the genome, which can be beneficial to the host. The anti-mutagenesis activity of Vsr homologs linked to RM systems will prevent this loss and may be of immediate benefit to the RM system.

Domain Sequence Movement in the Specificity Subunit

The above results revealed that RM systems have many features as mobile genetic elements. Recent work demonstrates that parts of RM system genes are variable and even mobile.

DNA sequence recognition of the Type I RM system specificity subunit is mediated by TRD1 and TRD2. These TRDs show diversity among strains, consistent with the diversification of recognition sequences.[113-117] In some cases, diversification occurred by a specific genome rearrangement mechanism. In *Mycoplasma pulmonis*, two S genes flank the R and M genes, which are in an inverted orientation and prone to recombine with each other, resulting in TRD shuffling.[27] In *Lactococcus lactis*, two copies of the S genes are on different plasmids, interacting through homologous recombination to create two chimeric S genes for one RM system with shuffled recognition sequences.[118] S genes tolerates exchange of the sequences between TRD1 and TRD2 by circular permutation.[119] Weak sequence similarity (36% identity in amino acid sequence) was detected between TRD1 and TRD2 of from different species, that is TRD1 of StySKI from *Salmonella enterica* and TRD2 of EcoR124I from *Escherichia coli*.[120]

Surprisingly, several sequences are shared by TRD1 and TRD2 genes at the same locus in several bacteria: these domain sequences appear to have moved between two positions within a single protein (Fig. 6).[121] The gene/protein organization can be represented as x-(TRD1)-y-x-(TRD2)-y, where x and y are repeated. Movement probably occurs by recombination at these flanking DNA repeats. Lateral domain movements within a protein, which we have designated DOMO (*do*main *mo*vement), represent novel routes for the diversification of proteins.

Gene Regulation

RM systems possess mechanisms that tightly regulate their gene expression to suppress the potential of a lethal attack on the host bacteria. When RM systems enter a new host bacterial cell with a genome that lacks proper methylation, they avoid killing the cell by expressing the modification enzyme first (Fig. 1C).[122] The restriction endonuclease and modification enzyme activities must be carefully regulated not only during RM system establishment in a new host, but also during maintenance. When RM genes are lost from a cell or the epigenetic status is disturbed, the restriction enzyme will attack the chromosome through postsegregational killing, leading to cell death or genome repair by recombination and replication. DNA fragments encoding RM systems may be released into the environment after cell death, invading other cells and establishing in the genome of a new host. This life cycle is similar to other mobile elements such as lysogenic phages and DNA transposons.

The regulatory machinery of mobile RM systems is expected to be host-independent, to bypass differences in the host factors affecting their establishment, maintenance and host attack. Thus far, regulatory mechanisms have been studied mainly at the transcriptional level,

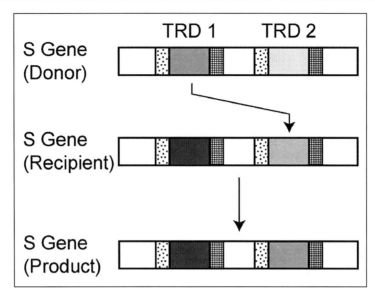

Figure 6. Domain movement. Movement of an amino-acid sequence from one protein domain to another domain of the same protein. The underlying mechanism is movement of the corresponding DNA sequence through recombination at sequences flanking the domain sequences. Modified from Furuta et al. PLoS ONE 2011; 6:e18819;[34] Public Library of Science.

as three main modes of regulation. One employs C-proteins,[123] which specifically bind a DNA operator sequence through a helix-turn-helix (HTH) motif to temporally control expression of the restriction enzyme, the modification enzyme or both.[124,125] This tight, finely tuned regulation operates via transcriptional feedback circuits.[126] Moreover, C proteins can efficiently delay expression of the restriction enzyme during establishment in a new host cell.[85,122] In the second type of regulation, the modification enzyme represses transcription of its own gene and occasionally stimulates restriction gene expression by DNA binding via its HTH domain.[127-129] In the third mode, the coordinated expression of R-M systems depends on the methylation status of a cognate recognition site(s) in their promoter region.[130,131]

Recent analyses revealed regulation by intragenic reverse promoters from which antisense RNAs are transcribed.[132-135]

Conclusion and Perspective

RM systems, originally found to be a barrier to gene mobility, turn out to be mobile elements themselves. They are mobile epigenetic elements because they define, and sometimes even force an epigenetic status on a genome. Multilevel conflicts involving these epigenetic systems may drive prokaryotic evolution.

This model is based on laboratory experiments and genome comparisons and needs to be examined through additional experimental and theoretical studies.[136] The expanding accumulation of bacterial genome sequences, especially within a species, may allow more detailed analysis. The concept of conflicts between epigenetic systems may provide information for understanding eukaryotic evolution and the origin of life.

Acknowledgments

We thank Noriko Takahashi for providing Figure 1E.

References

1. Ishikawa K, Fukuda E, Kobayashi I. Conflicts targeting epigenetic systems and their resolution by cell death: novel concepts for methyl-specific and other restriction systems. DNA Res 2010; 17:325-42. PMID:21059708 doi:10.1093/dnares/dsq027

2. Roberts RJ, Belfort M, Bestor T, et al. A nomenclature for restriction enzymes, DNA methyltransferases, homing endonucleases and their genes. Nucleic Acids Res 2003; 31:1805-12. PMID:12654995 doi:10.1093/nar/gkg274

3. Roberts RJ, Vincze T, Posfai J, et al. REBASE–a database for DNA restriction and modification: enzymes, genes and genomes. Nucleic Acids Res 2010; 38(Database issue):D234-6. PMID:19846593 doi:10.1093/nar/gkp874

4. Van Etten JL, Meints RH. Giant viruses infecting algae. Annu Rev Microbiol 1999; 53:447-94. PMID:10547698 doi:10.1146/annurev.micro.53.1.447

5. Murray NE. Type I restriction systems: sophisticated molecular machines (a legacy of Bertani and Weigle). Microbiol Mol Biol Rev 2000; 64:412-34. PMID:10839821 doi:10.1128/MMBR.64.2.412-434.2000

6. Gough JA, Murray NE. Sequence diversity among related genes for recognition of specific targets in DNA molecules. J Mol Biol 1983; 166:1-19. PMID:6304321 doi:10.1016/S0022-2836(83)80047-3

7. Studier FW, Bandyopadhyay PK. Model for how type I restriction enzymes select cleavage sites in DNA. Proc Natl Acad Sci USA 1988; 85:4677-81. PMID:2838843 doi:10.1073/pnas.85.13.4677

8. Pingoud A, Fuxreiter M, Pingoud V, et al. Type II restriction endonucleases: structure and mechanism. Cell Mol Life Sci 2005; 62:685-707. PMID:15770420 doi:10.1007/s00018-004-4513-1

9. Malone T, Blumenthal RM, Cheng X. Structure-guided analysis reveals nine sequence motifs conserved among DNA amino-methyltransferases, and suggests a catalytic mechanism for these enzymes. J Mol Biol 1995; 253:618-32. PMID:7473738 doi:10.1006/jmbi.1995.0577

10. Orlowski J, Bujnicki JM. Structural and evolutionary classification of Type II restriction enzymes based on theoretical and experimental analyses. Nucleic Acids Res 2008; 36:3552-69. PMID:18456708 doi:10.1093/nar/gkn175

11. Morgan RD, Dwinell EA, Bhatia TK, et al. The MmeI family: type II restriction-modification enzymes that employ single-strand modification for host protection. Nucleic Acids Res 2009; 37:5208-21. PMID:19578066 doi:10.1093/nar/gkp534

12. Janulaitis A, Petrusyte M, Maneliene Z, et al. Purification and properties of the Eco57I restriction endonuclease and methylase–prototypes of a new class (type IV). Nucleic Acids Res 1992; 20:6043-9. PMID:1334260 doi:10.1093/nar/20.22.6043

13. Nakonieczna J, Kaczorowski T, Obarska-Kosinska A, et al. Functional analysis of MmeI from methanol utilizer Methylophilus methylotrophus, a subtype IIC restriction-modification enzyme related to type I enzymes. Appl Environ Microbiol 2009; 75:212-23. PMID:18997032 doi:10.1128/AEM.01322-08

14. Cesnaviciene E, Petrusyte M, Kazlauskiene R, et al. Characterization of AloI, a restriction-modification system of a new type. J Mol Biol 2001; 314:205-16. PMID:11718555 doi:10.1006/jmbi.2001.5049

15. Kong H. Analyzing the functional organization of a novel restriction modification system, the BcgI system. J Mol Biol 1998; 279:823-32. PMID:9642063 doi:10.1006/jmbi.1998.1821

16. Dryden DT, Murray NE, Rao DN. Nucleoside triphosphate-dependent restriction enzymes. Nucleic Acids Res 2001; 29:3728-41. PMID:11557806 doi:10.1093/nar/29.18.3728

17. Fox KL, Dowideit SJ, Erwin AL, et al. Haemophilus influenzae phasevarions have evolved from type III DNA restriction systems into epigenetic regulators of gene expression. Nucleic Acids Res 2007; 35:5242-52. PMID:17675301 doi:10.1093/nar/gkm571

18. Srikhanta YN, Dowideit SJ, Edwards JL, et al. Phasevarions mediate random switching of gene expression in pathogenic Neisseria. PLoS Pathog 2009; 5:e1000400. PMID:19390608 doi:10.1371/journal.ppat.1000400

19. Srikhanta YN, Fox KL, Jennings MP. The phasevarion: phase variation of type III DNA methyltransferases controls coordinated switching in multiple genes. Nat Rev Microbiol 2010; 8:196-206. PMID:20140025 doi:10.1038/nrmicro2283

20. Kelleher JE, Raleigh EA. A novel activity in Escherichia coli K-12 that directs restriction of DNA modified at CG dinucleotides. J Bacteriol 1991; 173:5220-3. PMID:1830580

21. Waite-Rees PA, Keating CJ, Moran LS, et al. Characterization and expression of the Escherichia coli Mrr restriction system. J Bacteriol 1991; 173:5207-19. PMID:1650347

22. Mulligan EA, Dunn JJ. Cloning, purification and initial characterization of E. coli McrA, a putative 5-methylcytosine-specific nuclease. Protein Expr Purif 2008; 62:98-103. PMID:18662788 doi:10.1016/j.pep.2008.06.016

23. Mulligan EA, Hatchwell E, McCorkle SR, et al. Differential binding of Escherichia coli McrA protein to DNA sequences that contain the dinucleotide m5CpG. Nucleic Acids Res 2010; 38:1997-2005. PMID:20015968 doi:10.1093/nar/gkp1120

24. Kobayashi I. Restriction-modification systems as minimal forms of life. Restriction endonucleases 2004:19-62.
25. Kuroda M, Ohta T, Uchiyama I, et al. Whole genome sequencing of meticillin-resistant Staphylococcus aureus. Lancet 2001; 357:1225-40. PMID:11418146 doi:10.1016/S0140-6736(00)04403-2
26. Schouler C, Gautier M, Ehrlich SD, et al. Combinational variation of restriction modification specificities in Lactococcus lactis. Mol Microbiol 1998; 28:169-78. PMID:9593305 doi:10.1046/j.1365-2958.1998.00787.x
27. Dybvig K, Sitaraman R, French CT. A family of phase-variable restriction enzymes with differing specificities generated by high-frequency gene rearrangements. Proc Natl Acad Sci USA 1998; 95:13923-8. PMID:9811902 doi:10.1073/pnas.95.23.13923
28. Kobayashi I, Nobusato A, Kobayashi-Takahashi N, et al. Shaping the genome–restriction-modification systems as mobile genetic elements. Curr Opin Genet Dev 1999; 9:649-56. PMID:10607611 doi:10.1016/S0959-437X(99)00026-X
29. Kobayashi I. Behavior of restriction-modification systems as selfish mobile elements and their impact on genome evolution. Nucleic Acids Res 2001; 29:3742-56. PMID:11557807 doi:10.1093/nar/29.18.3742
30. Nölling J, de Vos WM. Characterization of the archaeal, plasmid-encoded type II restriction-modification system MthTI from Methanobacterium thermoformicicum THF: homology to the bacterial NgoPII system from Neisseria gonorrhoeae. J Bacteriol 1992; 174:5719-26. PMID:1512204
31. Nobusato A, Uchiyama I, Kobayashi I. Diversity of restriction-modification gene homologues in Helico-bacter pylori. Gene 2000; 259:89-98. PMID:11163966 doi:10.1016/S0378-1119(00)00455-8
32. Jeltsch A, Kroger M, Pingoud A. Evidence for an evolutionary relationship among type-II restriction endonucleases. Gene 1995; 160:7-16. PMID:7628720 doi:10.1016/0378-1119(95)00181-5
33. Bujnicki JM, Radlinska M. Molecular phylogenetics of DNA 5mC-methyltransferases. Acta Microbiol Pol 1999; 48:19-30. PMID:10467693
34. Fukuda E, Kaminska KH, Bujnicki JM, et al. Cell death upon epigenetic genome methylation: a novel function of methyl-specific deoxyribonucleases. Genome Biol 2008; 9:R163. PMID:19025584 doi:10.1186/gb-2008-9-11-r163
35. Jeltsch A, Pingoud A. Horizontal gene transfer contributes to the wide distribution and evolution of type II restriction-modification systems. J Mol Evol 1996; 42:91-6. PMID:8919860 doi:10.1007/BF02198833
36. Mrázek J, Karlin S. Detecting alien genes in bacterial genomes. Ann N Y Acad Sci 1999; 870:314-29. PMID:10415493 doi:10.1111/j.1749-6632.1999.tb08893.x
37. Parkhill J, Achtman M, James KD, et al. Complete DNA sequence of a serogroup A strain of Neisseria meningitidis Z2491. Nature 2000; 404:502-6. PMID:10761919 doi:10.1038/35006655
38. Karyagina AS, Lunin VG, Degtyarenko KN, et al. Analysis of the nucleotide and derived amino acid sequences of the SsoII restriction endonuclease and methyltransferase. Gene 1993; 124:13-9. PMID:7916706 doi:10.1016/0378-1119(93)90756-S
39. Lubys A, Janulaitis A. Cloning and analysis of the plasmid-borne genes encoding the Bsp6I restriction and modification enzymes. Gene 1995; 157:25-9. PMID:7607501 doi:10.1016/0378-1119(94)00795-T
40. Glatman LI, Moroz AF, Yablokova MB, et al. A novel plasmid-mediated DNA restriction-modification system in E. coli. Plasmid 1980; 4:350-1. PMID:6261280 doi:10.1016/0147-619X(80)90072-4
41. Hinkle NF, Miller RV. pMG7-mediated restriction of Pseudomonas aeruginosa phage DNAs is determined by a class II restriction endonuclease. Plasmid 1979; 2:387-93. PMID:113797 doi:10.1016/0147-619X(79)90022-2
42. Greene PJ, Gupta M, Boyer HW, et al. Sequence analysis of the DNA encoding the Eco RI endonuclease and methylase. J Biol Chem 1981; 256:2143-53. PMID:6257703
43. Kosykh VG, Buryanov YI, Bayev AA. Molecular cloning of EcoRII endonuclease and methylase genes. Mol Gen Genet 1980; 178:717-8. PMID:6248737 doi:10.1007/BF00337884
44. Calvin Koons MD, Blumenthal RM. Characterization of pPvu1, the autonomous plasmid from Proteus vulgaris that carries the genes of the PvuII restriction-modification system. Gene 1995; 157:73-9. PMID:7607530 doi:10.1016/0378-1119(94)00618-3
45. Hümbelin M, Suri B, Rao DN, et al. Type III DNA restriction and modification systems EcoP1 and EcoP15. Nucleotide sequence of the EcoP1 operon, the EcoP15 mod gene and some EcoP1 mod mutants. J Mol Biol 1988; 200:23-9. PMID:2837577 doi:10.1016/0022-2836(88)90330-0
46. Borriss M, Lombardot T, Glockner FO, et al. Genome and proteome characterization of the psychrophilic Flavobacterium bacteriophage 11b. Extremophiles 2007; 11:95-104. PMID:16932843 doi:10.1007/s00792-006-0014-5
47. Miner Z, Hattman S. Molecular cloning, sequencing, and mapping of the bacteriophage T2 dam gene. J Bacteriol 1988; 170:5177-84. PMID:3053648
48. Dempsey RM, Carroll D, Kong H, et al. Sau42I, a BcgI-like restriction-modification system encoded by the Staphylococcus aureus quadruple-converting phage Phi42. Microbiology 2005; 151:1301-11. PMID:15817797 doi:10.1099/mic.0.27646-0

49. Ohshima H, Matsuoka S, Asai K, et al. Molecular organization of intrinsic restriction and modification genes BsuM of Bacillus subtilis Marburg. J Bacteriol 2002; 184:381-9. PMID:11751814 doi:10.1128/JB.184.2.381-389.2002
50. Hendrix RW, Smith MC, Burns RN, et al. Evolutionary relationships among diverse bacteriophages and prophages: all the world's a phage. Proc Natl Acad Sci USA 1999; 96:2192-7. PMID:10051617 doi:10.1073/pnas.96.5.2192
51. Kita K, Kawakami H, Tanaka H. Evidence for horizontal transfer of the EcoT38I restriction-modification gene to chromosomal DNA by the P2 phage and diversity of defective P2 prophages in Escherichia coli TH38 strains. J Bacteriol 2003; 185:2296-305. PMID:12644501 doi:10.1128/JB.185.7.2296-2305.2003
52. Kita K, Tsuda J, Kato T, et al. Evidence of horizontal transfer of the EcoO109I restriction-modification gene to Escherichia coli chromosomal DNA. J Bacteriol 1999; 181:6822-7. PMID:10542186
53. Euler CW, Ryan PA, Martin JM, et al. M.SpyI, a DNA methyltransferase encoded on a mefA chimeric element, modifies the genome of Streptococcus pyogenes. J Bacteriol 2007; 189:1044-54. PMID:17085578 doi:10.1128/JB.01411-06
54. Sampath J, Vijayakumar MN. Identification of a DNA cytosine methyltransferase gene in conjugative transposon Tn5252. Plasmid 1998; 39:63-76. PMID:9473447 doi:10.1006/plas.1997.1316
55. O'Sullivan DJ, Zagula K, Klaenhammer TR. In vivo restriction by LlaI is encoded by three genes, arranged in an operon with llaIM, on the conjugative Lactococcus plasmid pTR2030. J Bacteriol 1995; 177:134-43. PMID:7528201
56. Burrus V, Bontemps C, Decaris B, et al. Characterization of a novel type II restriction-modification system, Sth368I, encoded by the integrative element ICESt1 of Streptococcus thermophilus CNRZ368. Appl Environ Microbiol 2001; 67:1522-8. PMID:11282600 doi:10.1128/AEM.67.4.1522-1528.2001
57. Böltner D, MacMahon C, Pembroke JT, et al. R391: a conjugative integrating mosaic comprised of phage, plasmid, and transposon elements. J Bacteriol 2002; 184:5158-69. PMID:12193633 doi:10.1128/JB.184.18.5158-5169.2002
58. van Zyl LJ, Deane SM, Louw LA, et al. Presence of a family of plasmids (29 to 65 kilobases) with a 26-kilobase common region in different strains of the sulfur-oxidizing bacterium Acidithiobacillus caldus. Appl Environ Microbiol 2008; 74:4300-8. PMID:18515486 doi:10.1128/AEM.00864-08
59. Stiens M, Becker A, Bekel T, et al. Comparative genomic hybridisation and ultrafast pyrosequencing revealed remarkable differences between the Sinorhizobium meliloti genomes of the model strain Rm1021 and the field isolate SM11. J Biotechnol 2008; 136:31-7. PMID:18562031 doi:10.1016/j.jbiotec.2008.04.014
60. Rochepeau P, Selinger LB, Hynes MF. Transposon-like structure of a new plasmid-encoded restriction-modification system in Rhizobium leguminosarum VF39SM. Mol Gen Genet 1997; 256:387-96. PMID:9393436 doi:10.1007/s004380050582
61. Heidelberg JF, Eisen JA, Nelson WC, et al. DNA sequence of both chromosomes of the cholera pathogen Vibrio cholerae. Nature 2000; 406:477-83. PMID:10952301 doi:10.1038/35020000
62. Rowe-Magnus DA, Guerout AM, Ploncard P, et al. The evolutionary history of chromosomal super-integrons provides an ancestry for multiresistant integrons. Proc Natl Acad Sci USA 2001; 98:652-7. PMID:11209061 doi:10.1073/pnas.98.2.652
63. Antonenko V, Pawlow V, Heesemann J, et al. Characterization of a novel unique restriction-modification system from Yersinia enterocolitica O:8 1B. FEMS Microbiol Lett 2003; 219:249-52. PMID:12620628 doi:10.1016/S0378-1097(03)00047-8
64. Brassard S, Paquet H, Roy PH. A transposon-like sequence adjacent to the AccI restriction-modification operon. Gene 1995; 157:69-72. PMID:7607529 doi:10.1016/0378-1119(94)00734-A
65. Naderer M, Brust JR, Knowle D, et al. Mobility of a restriction-modification system revealed by its genetic contexts in three hosts. J Bacteriol 2002; 184:2411-9. PMID:11948154 doi:10.1128/JB.184.9.2411-2419.2002
66. Anton BP, Heiter DF, Benner JS, et al. Cloning and characterization of the Bg/II restriction-modification system reveals a possible evolutionary footprint. Gene 1997; 187:19-27. PMID:9073062 doi:10.1016/S0378-1119(96)00638-5
67. Takahashi N, Ohashi S, Sadykov MR, et al. IS-Linked Movement of a Restriction-Modification System. PLoS ONE 2011; 6:e16554. PMID:21305031 doi:10.1371/journal.pone.0016554
68. Naito T, Kusano K, Kobayashi I. Selfish behavior of restriction-modification systems. Science 1995; 267:897-9. PMID:7846533 doi:10.1126/science.7846533
69. Naito Y, Naito T, Kobayashi I. Selfish restriction modification genes: resistance of a resident R/M plasmid to displacement by an incompatible plasmid mediated by host killing. Biol Chem 1998; 379:429-36. PMID:9628334 doi:10.1515/bchm.1998.379.4-5.429
70. Sekizaki T, Otani Y, Osaki M, et al. Evidence for horizontal transfer of SsuDAT1I restriction-modification genes to the Streptococcus suis genome. J Bacteriol 2001; 183:500-11. PMID:11133943 doi:10.1128/JB.183.2.500-511.2001

71. Claus H, Friedrich A, Frosch M, et al. Differential distribution of novel restriction-modification systems in clonal lineages of Neisseria meningitidis. J Bacteriol 2000; 182:1296-303. PMID:10671450 doi:10.1128/JB.182.5.1296-1303.2000
72. Stein DC, Gunn JS, Piekarowicz A. Sequence similarities between the genes encoding the S.NgoI and HaeII restriction/modification systems. Biol Chem 1998; 379:575-8. PMID:9628358
73. Nobusato A, Uchiyama I, Ohashi S, et al. Insertion with long target duplication: a mechanism for gene mobility suggested from comparison of two related bacterial genomes. Gene 2000; 259:99-108. PMID:11163967 doi:10.1016/S0378-1119(00)00456-X
74. Furuta Y, Abe K, Kobayashi I. Genome comparison and context analysis reveals putative mobile forms of restriction-modification systems and related rearrangements. Nucleic Acids Res 2010; 38:2428-43. PMID:20071371 doi:10.1093/nar/gkp1226
75. Sibley MH, Raleigh EA. Cassette-like variation of restriction enzyme genes in Escherichia coli C and relatives. Nucleic Acids Res 2004; 32:522-34. PMID:14744977 doi:10.1093/nar/gkh194
76. Miller WG, Pearson BM, Wells JM, et al. Diversity within the Campylobacter jejuni type I restriction-modification loci. Microbiology 2005; 151:337-51. PMID:15699185 doi:10.1099/mic.0.27327-0
77. Ishikawa K, Handa N, Kobayashi I. Cleavage of a model DNA replication fork by a Type I restriction endonuclease. Nucleic Acids Res 2009; 37:3531-44. PMID:19357093 doi:10.1093/nar/gkp214
78. Asakura Y, Kobayashi I. From damaged genome to cell surface: transcriptome changes during bacterial cell death triggered by loss of a restriction-modification gene complex. Nucleic Acids Res 2009; 37:3021-31. PMID:19304752 doi:10.1093/nar/gkp148
79. Ichige A, Kobayashi I. Stability of EcoRI restriction-modification enzymes in vivo differentiates the EcoRI restriction-modification system from other postsegregational cell killing systems. J Bacteriol 2005; 187:6612-21. PMID:16166522 doi:10.1128/JB.187.19.6612-6621.2005
80. Handa N, Nakayama Y, Sadykov M, et al. Experimental genome evolution: large-scale genome rearrangements associated with resistance to replacement of a chromosomal restriction-modification gene complex. Mol Microbiol 2001; 40:932-40. PMID:11401700 doi:10.1046/j.1365-2958.2001.02436.x
81. Ohno S, Handa N, Watanabe-Matsui M, et al. Maintenance forced by a restriction-modification system can be modulated by a region in its modification enzyme not essential for methyltransferase activity. J Bacteriol 2008; 190:2039-49. PMID:18192396 doi:10.1128/JB.01319-07
82. Mochizuki A, Yahara K, Kobayashi I, et al. Genetic addiction: selfish gene's strategy for symbiosis in the genome. Genetics 2006; 172:1309-23. PMID:16299387 doi:10.1534/genetics.105.042895
83. Kusano K, Naito T, Handa N, et al. Restriction-modification systems as genomic parasites in competition for specific sequences. Proc Natl Acad Sci USA 1995; 92:11095-9. PMID:7479944 doi:10.1073/pnas.92.24.11095
84. Chinen A, Naito Y, Handa N, et al. Evolution of sequence recognition by restriction-modification enzymes: selective pressure for specificity decrease. Mol Biol Evol 2000; 17:1610-9. PMID:11070049
85. Nakayama Y, Kobayashi I. Restriction-modification gene complexes as selfish gene entities: roles of a regulatory system in their establishment, maintenance, and apoptotic mutual exclusion. Proc Natl Acad Sci USA 1998; 95:6442-7. PMID:9600985 doi:10.1073/pnas.95.11.6442
86. Kohanski MA, Dwyer DJ, Hayete B, et al. A common mechanism of cellular death induced by bactericidal antibiotics. Cell 2007; 130:797-810. PMID:17803904 doi:10.1016/j.cell.2007.06.049
87. Kohanski MA, Dwyer DJ, Collins JJ. How antibiotics kill bacteria: from targets to networks. Nat Rev Microbiol 2010; 8:423-35. PMID:20440275 doi:10.1038/nrmicro2333
88. Blumenthal RM, Cotterman MM. Isolation of mutants in a DNA methyltransferase through mcrB-mediated restriction. Gene 1988; 74:271-3. PMID:2854810 doi:10.1016/0378-1119(88)90301-0
89. Iida S, Meyer J, Bachi B, et al. DNA restriction–modification genes of phage P1 and plasmid p15B. Structure and in vitro transcription. J Mol Biol 1983; 165:1-18. PMID:6302279 doi:10.1016/S0022-2836(83)80239-3
90. Noyer-Weidner M, Walter J, Terschuren PA, et al. M.phi 3TII: a new monospecific DNA (cytosine-C5) methyltransferase with pronounced amino acid sequence similarity to a family of adenine-N6-DNA-methyltransferases. Nucleic Acids Res 1994; 22:5517-23. PMID:7816649 doi:10.1093/nar/22.24.5517
91. Lange C, Noyer-Weidner M, Trautner TA, et al. M.H2I, a multispecific 5C-DNA methyltransferase encoded by Bacillus amyloliquefaciens phage H2. Gene 1991; 100:213-8. PMID:2055471 doi:10.1016/0378-1119(91)90369-M
92. Behrens B, Noyer-Weidner M, Pawlek B, et al. Organization of multispecific DNA methyltransferases encoded by temperate Bacillus subtilis phages. EMBO J 1987; 6:1137-42. PMID:3109889
93. Sutherland E, Coe L, Raleigh EA. McrBC: a multisubunit GTP-dependent restriction endonuclease. J Mol Biol 1992; 225:327-48. PMID:1317461 doi:10.1016/0022-2836(92)90925-A

94. Warren RA. Modified bases in bacteriophage DNAs. Annu Rev Microbiol 1980; 34:137-58. PMID:7002022 doi:10.1146/annurev.mi.34.100180.001033
95. Doronina VA, Murray NE. The proteolytic control of restriction activity in Escherichia coli K-12. Mol Microbiol 2001; 39:416-28. PMID:11136462 doi:10.1046/j.1365-2958.2001.02232.x
96. Makovets S, Doronina VA, Murray NE. Regulation of endonuclease activity by proteolysis prevents breakage of unmodified bacterial chromosomes by type I restriction enzymes. Proc Natl Acad Sci USA 1999; 96:9757-62. PMID:10449767 doi:10.1073/pnas.96.17.9757
97. Makovets S, Powell LM, Titheradge AJ, et al. Is modification sufficient to protect a bacterial chromosome from a resident restriction endonuclease? Mol Microbiol 2004; 51:135-47. PMID:14651617 doi:10.1046/j.1365-2958.2003.03801.x
98. Seidel R, Bloom JG, van Noort J, et al. Dynamics of initiation, termination and reinitiation of DNA translocation by the motor protein EcoR124I. EMBO J 2005; 24:4188-97. PMID:16292342 doi:10.1038/sj.emboj.7600881
99. Takahashi N, Naito Y, Handa N, et al. A DNA methyltransferase can protect the genome from postdisturbance attack by a restriction-modification gene complex. J Bacteriol 2002; 184:6100-8. PMID:12399478 doi:10.1128/JB.184.22.6100-6108.2002
100. Alm RA, Ling LS, Moir DT, et al. Genomic-sequence comparison of two unrelated isolates of the human gastric pathogen Helicobacter pylori. Nature 1999; 397:176-80. PMID:9923682 doi:10.1038/16495
101. Furuta Y, Kawai M, Yahara K, et al. Birth and death of genes linked to chromosomal inversion. Proc Natl Acad Sci USA 2011; 108:1501-6. PMID:21212362 doi:10.1073/pnas.1012579108
102. Asakura Y, Kojima H, Kobayashi I. Evolutionary genome engineering using a restriction-modification system. Nucleic Acids Res 2011; Aug 12. [Epub ahead of print] PMID:21785135
103. Sadykov M, Asami Y, Niki H, et al. Multiplication of a restriction-modification gene complex. Mol Microbiol 2003; 48:417-27. PMID:12675801 doi:10.1046/j.1365-2958.2003.03464.x
104. Handa N, Ichige A, Kusano K, et al. Cellular responses to postsegregational killing by restriction-modification genes. J Bacteriol 2000; 182:2218-29. PMID:10735865 doi:10.1128/JB.182.8.2218-2229.2000
105. Heitman J, Model P. SOS induction as an in vivo assay of enzyme-DNA interactions. Gene 1991; 103:1-9. PMID:1908806 doi:10.1016/0378-1119(91)90383-M
106. Heitman J, Model P. Site-specific methylases induce the SOS DNA repair response in Escherichia coli. J Bacteriol 1987; 169:3243-50. PMID:3036779
107. Higgins NP. Death and transfiguration among bacteria. Trends Biochem Sci 1992; 17:207-11. PMID:1323887 doi:10.1016/0968-0004(92)90376-K
108. Gelfand MS, Koonin EV. Avoidance of palindromic words in bacterial and archaeal genomes: a close connection with restriction enzymes. Nucleic Acids Res 1997; 25:2430-9. PMID:9171096 doi:10.1093/nar/25.12.2430
109. Rocha EP, Danchin A, Viari A. Evolutionary role of restriction/modification systems as revealed by comparative genome analysis. Genome Res 2001; 11:946-58. PMID:11381024 doi:10.1101/gr.GR-1531RR
110. Rocha EP, Viari A, Danchin A. Oligonucleotide bias in Bacillus subtilis: general trends and taxonomic comparisons. Nucleic Acids Res 1998; 26:2971-80. PMID:9611243 doi:10.1093/nar/26.12.2971
111. Friedberg EC, Walker GC, Siede W. DNA repair and mutagenesis: Amer Society for Microbiology; 1995.
112. Lieb M. Spontaneous mutation at a 5-methylcytosine hotspot is prevented by very short patch (VSP) mismatch repair. Genetics 1991; 128:23-7. PMID:1829427
113. Fuller-Pace FV, Bullas LR, Delius H, et al. Genetic recombination can generate altered restriction specificity. Proc Natl Acad Sci USA 1984; 81:6095-9. PMID:6091134 doi:10.1073/pnas.81.19.6095
114. Gubler M, Braguglia D, Meyer J, et al. Recombination of constant and variable modules alters DNA sequence recognition by type IC restriction-modification enzymes. EMBO J 1992; 11:233-40. PMID:1740108
115. Gann AA, Campbell AJ, Collins JF, et al. Reassortment of DNA recognition domains and the evolution of new specificities. Mol Microbiol 1987; 1:13-22. PMID:2838725 doi:10.1111/j.1365-2958.1987.tb00521.x
116. Tsuru T, Kawai M, Mizutani-Ui Y, et al. Evolution of paralogous genes: Reconstruction of genome rearrangements through comparison of multiple genomes within Staphylococcus aureus. Mol Biol Evol 2006; 23:1269-85. PMID:16601000 doi:10.1093/molbev/msk013
117. Waldron DE, Lindsay JA. SauI: a novel lineage-specific type I restriction-modification system that blocks horizontal gene transfer into Staphylococcus aureus and between S. aureus isolates of different lineages. J Bacteriol 2006; 188:5578-85. PMID:16855248 doi:10.1128/JB.00418-06
118. O'Sullivan D, Twomey DP, Coffey A, et al. Novel type I restriction specificities through domain shuffling of HsdS subunits in Lactococcus lactis. Mol Microbiol 2000; 36:866-75. PMID:10844674 doi:10.1046/j.1365-2958.2000.01901.x

119. Janscak P, Bickle TA. The DNA recognition subunit of the type IB restriction-modification enzyme EcoAI tolerates circular permutions of its polypeptide chain. J Mol Biol 1998; 284:937-48. PMID:9837717 doi:10.1006/jmbi.1998.2250

120. Thorpe PH, Ternent D, Murray NE. The specificity of sty SKI, a type I restriction enzyme, implies a structure with rotational symmetry. Nucleic Acids Res 1997; 25:1694-700. PMID:9108149 doi:10.1093/nar/25.9.1694

121. Furuta Y, Kawai M, Uchiyama I, et al. Domain movement within a gene: A novel evolutionary mechanism for protein diversification. PLoS ONE 2011; 6:e18819. PMID:21533192 doi:10.1371/journal.pone.0018819

122. Mruk I, Blumenthal RM. Real-time kinetics of restriction-modification gene expression after entry into a new host cell. Nucleic Acids Res 2008; 36:2581-93. PMID:18334533 doi:10.1093/nar/gkn097

123. Tao T, Bourne JC, Blumenthal RM. A family of regulatory genes associated with type II restriction-modification systems. J Bacteriol 1991; 173:1367-75. PMID:1995588

124. Knowle D, Lintner RE, Touma YM, et al. Nature of the promoter activated by C.PvuII, an unusual regulatory protein conserved among restriction-modification systems. J Bacteriol 2005; 187:488-97. PMID:15629920 doi:10.1128/JB.187.2.488-497.2005

125. Vijesurier RM, Carlock L, Blumenthal RM, et al. Role and mechanism of action of C. PvuII, a regulatory protein conserved among restriction-modification systems. J Bacteriol 2000; 182:477-87. PMID:10629196 doi:10.1128/JB.182.2.477-487.2000

126. Mruk I, Rajesh P, Blumenthal RM. Regulatory circuit based on autogenous activation-repression: roles of C-boxes and spacer sequences in control of the PvuII restriction-modification system. Nucleic Acids Res 2007; 35:6935-52. PMID:17933763 doi:10.1093/nar/gkm837

127. Som S, Friedman S. Regulation of EcoRII methyltransferase: effect of mutations on gene expression and in vitro binding to the promoter region. Nucleic Acids Res 1994; 22:5347-53. PMID:7816624 doi:10.1093/nar/22.24.5347

128. Karyagina A, Shilov I, Tashlitskii V, et al. Specific binding of sso II DNA methyltransferase to its promoter region provides the regulation of sso II restriction-modification gene expression. Nucleic Acids Res 1997; 25:2114-20. PMID:9153310 doi:10.1093/nar/25.11.2114

129. Fedotova EA, Protsenko AS, Zakharova MV, et al. SsoII-like DNA-methyltransferase Ecl18kI: interaction between regulatory and methylating functions. Biochemistry (Mosc) 2009; 74:85-91. PMID:19232054 doi:10.1134/S0006297909010131

130. Beletskaya IV, Zakharova MV, Shlyapnikov MG, et al. DNA methylation at the CfrBI site is involved in expression control in the CfrBI restriction-modification system. Nucleic Acids Res 2000; 28:3817-22. PMID:11000275 doi:10.1093/nar/28.19.3817

131. Christensen LL, Josephsen J. The methyltransferase from the LlaDII restriction-modification system influences the level of expression of its own gene. J Bacteriol 2004; 186:287-95. PMID:14702296 doi:10.1128/JB.186.2.287-295.2004

132. Liu Y, Ichige A, Kobayashi I. Regulation of the EcoRI restriction-modification system: Identification of ecoRIM gene promoters and their upstream negative regulators in the ecoRIR gene. Gene 2007; 400:140-9. PMID:17618069 doi:10.1016/j.gene.2007.06.006

133. Mruk I, Liu Y, Ge L, et al. Antisense RNA associated with biological regulation of a restriction-modification system. Nucleic Acids Res 2011.

134. Nagornykh M, Zakharova M, Protsenko A, et al. Regulation of gene expression in restriction-modification system Eco29kI. Nucleic Acids Res 2011.

135. Liu Y, Kobayashi I. Negative regulation of the EcoRI restriction enzyme gene is associated with intragenic reverse promoters. J Bacteriol 2007; 189:6928-35. PMID:17616602 doi:10.1128/JB.00127-07

136. Yahara K, Fukuyo M, Sasaki A, et al. Evolutionary maintenance of selfish homing endonuclease genes in the absence of horizontal transfer. Proc Natl Acad Sci USA 2009; 106:18861-6. PMID:19837694 doi:10.1073/pnas.0908404106

137. Chinen A, Uchiyama I, Kobayashi I. Comparison between Pyrococcus horikoshii and Pyrococcus abyssi genome sequences reveals linkage of restriction-modification genes with large genome polymorphisms. Gene 2000; 259:109-21. PMID:11163968 doi:10.1016/S0378-1119(00)00459-5

CHAPTER 6

Mobile Genetic Elements in the Genus Bacteroides, and Their Mechanism(s) of Dissemination

Mai Nguyen[1] and Gayatri Vedantam*,[2,3]

Abstract

*B*acteroides* spp organisms, the predominant commensal bacteria in the human gut have become increasingly resistant to many antibiotics. They are now also considered to be reservoirs of antibiotic resistance genes due to their capacity to harbor and disseminate these genes via mobile transmissible elements that occur in bewildering variety. Gene dissemination occurs within and from *Bacteroides* spp primarily by conjugation, the molecular mechanisms of which are still poorly understood in the genus, even though the need to prevent this dissemination is urgent. One current avenue of research is thus focused on interventions that use non-antibiotic methodologies to prevent conjugation-based DNA transfer.

Introduction

It has been estimated that the human microflora collectively number up to 100 trillion cells, 10-fold the number of human cells.[1,2] The majority of these normal flora reside in the gastrointestinal tract, and comprise over 500 bacterial species that exert a profound influence on human physiology.[3] Of these bacteria, 99% are anaerobes.[4] Anaerobic bacteroidetes are the predominant class, accounting about 30% of all bacteria in the human gut.[5,6]

Bacteroides spp are bile-resistant, non-spore-forming, gram-negative, rod-shaped anaerobes that are a dominant bacterial genus in the human colon and are less abundant in the intestines of other animals and in the environment.[7,8] *Bacteroides* sp are passed from mother to child during vaginal birth and thus become part of the human flora in the earliest stages of life.[9] The C+G nucleotide composition of *Bacteroides* genome is in the range of 40–48%. Its membranes contain sphingolipids, which are unusual in bacteria.

As an integral part of the normal human gut flora, *Bacteroides* spp play a number of roles, such as providing energy for the host in the form of short-chain fatty acids and sugars, recycling bile acids, and aiding in the development of the host immune system.[6] They also exhibit unique adaptations to successfully colonize the gut such as the ability to change their cell surface architecture,[10] ability to stimulate host expression of fucosylated glycoproteins as well as synthesize them,[11] and the ability to tolerate and use oxygen.[12]

However, when *Bacteroides* spp escape the gut due to surgery, trauma or disease, they can cause life-threatening infections such as peritonitis and intra-abdominal sepsis.[6,8,13] *Bacteroides*

[1]Section of Digestive Diseases and Nutrition, University of Illinois, Chicago, Illinois, USA; [2]Deptartment of Veterinary Science and Microbiology, University of Arizona, Tucson, Arizona, USA; [3]Southern Arizona VA Healthcare System, Tucson, Arizona, USA.
*Corresponding Author: Gayatri Vedantam—Email: gayatri@email.arizona.edu

Bacterial Integrative Mobile Genetic Elements, edited by Adam P. Roberts and Peter Mullany.
©2013 Landes Bioscience.

spp rarely cause endocarditis, inflammation of the inner layer of the heart, but when it does occur, it can be serious with a mortality rate of 21–43%,[14] and increase hospital stays by up to 15 d.[15] These organisms can also be associated with other infections such as those of the skin, soft tissue, joints (septic arthritis), and brain (abscesses and meningitis).[6] Among more than 20 species of the genus, *B. fragilis* is most frequently isolated from clinical specimens, followed by *B. ovatus* and *B. thetaiotaomicron*.[6,16,17] Enterotoxigenic *Bacteroides fragilis* (ETBF; a sub-group of *B. fragilis*) has also been implicated in inflammatory bowel disease (IBD)[18,19] and colon cancer.[20] Recently, a 5 year study in Japan revealed that *Bacteroides* bacteremia was the primary cause of colorectal carcinoma in a group of patients.[21] *Bacteroides spp.* are thus among the most commonly isolated anaerobic pathogens,[19] and it is now appreciated that this is likely due to the expression of diverse virulence factors including surface polysaccharide capsules, outer membrane vesicles, toxins and β-lactamases.[22] In addition, the capacity of *B. fragilis* to tolerate nanomolar concentrations of oxygen allows this species to predominate in infections of the peritoneal cavity.

Antimicrobial resistance in the Bacteroides further complicates the clinical picture. Many *Bacteroides* spp are resistant to aminoglycosides (gentamicin, kanamycin, streptomycin), tetracycline (nearly 85% of clinical isolates), β-lactam antibiotics (penicillin, ampicillin, cephalosporins, cefoxitin, cephamycins and carbapenems), metronidazole and the macrolide-lincosamide-streptogramin (MLS) group of antibiotics (erythromycin and clindamycin).[7,8] Of concern, all of these resistance traits have been found on transmissible genetic elements obviously contributing to resistance gene dissemination.[8] Over the past decades, carriage of the tetracycline resistance gene, *tetQ*, has increased from about 30% to more than 80% in clinical strains.[8] Specifically, over the past 10 y, resistance of *Bacteroides* spp has increased dramatically worldwide, especially to commonly prescribed antibiotics such as clindamycin and the cephalosporins.[15,23-26] *Bacteroides* spp resistance to fluoroquinolones has also increased from 1.5–12% during the past few years.[27,28] Metronidazole-resistant strains of *B. fragilis* have been reported in many countries, including Brazil,[29] India,[30] United States,[31] Hungary[32] and Poland,[33] and some Indian strains show very high levels of resistance (MIC 512mg/L).[34] Therefore, anaerobic infections have reemerged as a serious health threat, and *Bacteroides spp.* are now recognized as important pathogens.

Mobile Genetic Elements in the *Bacteroides*

Bacteroides spp harbor many conjugative and mobilizable elements (Table 1). Conjugative elements are autonomous and self-transferable, i.e., they encode all functions for DNA transfer, including those resulting in nucleic acid processing (DNA transfer initiation) as well as conjugation channel/portal assembly and DNA translocation (mating apparatus formation). Mobilizable elements encode only DNA processing functions (Fig. 1), and use (or 'hijack") the mating apparatus of co-resident conjugative element(s) to transfer to recipient bacteria. Interestingly, in the Bacteroides, a wide variety of mobilizable elements appear to be able to translocate through the same mating apparatus encoded by a single conjugative element. A significant proportion of Bacteroides mobile elements also harbor antibiotic resistance genes, and are thus responsible for the widespread dissemination of those genes. Both conjugative and mobilizable elements can be transposons or plasmids.

Plasmids

Plasmids are very common in *Bacteroides* spp and are found in 20% to 50% of strains.[6] Plasmids can replicate as independent elements in the host bacteria, and some can integrate into the host genome.[35] Many plasmids also have an origin of transfer (*oriT*) and a *trans*-acting mobilization gene, which allow them to be transferred by conjugation.[6]

Antibiotic resistance genes have been found on plasmids in *Bacteroides* spp. Genes whose products confer resistance to metronidazole, chloramphenicol, carbapenems, clindamycin and erythromycin have been found on mobile plasmids from *Bacteroides* spp clinical isolates worldwide. Resistance genes *nimA-nimF*, encoding metronidazole resistance, have also been identified on transferrable plasmids and recovered worldwide.[36] The *cfiA* gene, conferring resistance to carbapenems, has also been found in a plasmid in clinical isolates.[37]

Table 1. Mobile genetic elements found in Bacteroides spp

I. Conjugative Elements	II. Mobilizable Elements
• Self-transmissible	• Not self-transmissible
• Encode DNA-processing functions	• Encode DNA-processing functions
• Encode conjugation apparatus (CA) or mating channel for completely autonomous transfer	• Do not encode CA. Transfer is depenedent on CA formed by co-resident conjugative element(s)
• Almost always harbor antibiotic resistance genes	• May harbor antibiotic resistance genes
I.a. Conjugative Plasmids:	**II.a. Mobilizable Plasmids:**
Have *oriT* and *trans*-acting mobilization gene(s)	Have *oriT* and *trans*-acting mobilization gene(s)
Can replicate independently	Can replicate independently
May integrate into the recipient chromosome	May integrate into the recipient chromosome
Examples: pBF4,[125] pBI136[126]	*Examples:* pBFTM10[39]
I.b. Conjugative Transposons:	**II.b. Mobilizable Transposons:**
Located on chromosome	Located on chromosome
Do not replicate extra-chromosomally	Do not replicate extra-chromosomally
52 kb to 150 kb	~4 kb to 15 kb
Also referred to as "Tet elements" since most carry the tetracycline resistance gene *tetQ*.	
Examples: BTF-37,[54] CTnDOT,[55] CTnERL,[56] Tcr Emr DOT,[57] Tcr Emr 7853,[58] CTnBST,[59] CTnGERM1,[60] CTn86[62] and CTn9343.[62]	*Examples:* Tn*4399*,[45] Tn*5520*,[46] cLV25,[47] NBU1, NBU2,[48] Tn*4555*[49]

Conjugative and Mobilizable Plasmids

To date, two conjugative plasmids have been identified: the *B. fragilis* 41kb pBF4 plasmid and *B. ovatus* 80.6kb pBI136 plasmid. DNA sequencing of the transfer regions of these plasmids has been limited due to A-T rich tracts that confound assembly and analyses. Only one gene, *bctA*, encoding a 110kD protein that localizes to the membrane, has been identified to be required in mating process.[38] Multiple mobilizable plasmids have been identified, that can only be transferred via the mating channel formed by other co-resided conjugative elements; these include *B. fragilis* pLV22a and pBFTM10.[39]

Cryptic Plasmids

Bacteroides spp are also known for harboring small molecular weight cryptic plasmids at high frequency (50%).[40] These small plasmids are characterized in size classes: class I, 2.7 kb, class IIA, 4.2 kb, class IIB, 5.0kb, class IIC, 7.9 kb and class III, 5.6 kb.[40,41] The majority of small plasmids are found in class I, IIA and III. Class IIB and IIC are rare among both normal flora and clinical isolates.[40] These small molecular weight plasmids are called cryptic because, beside basic plasmid maintenance functions like replication, mobilization and in some cases, stability, they do not encode for any other obviously biological useful traits such as antibiotic resistance or production of virulence-associated proteins. For example, pBI143, a 2.7 kb plasmid, and pB8–51, a 4.2 kb plasmid, have been found to encode only replication and mobilization functions.[42,43,44] pBF35, a representative of the most frequent class III plasmids, and originating in Hungary, was also found to encode only replication, mobilization and stability functions.[41]

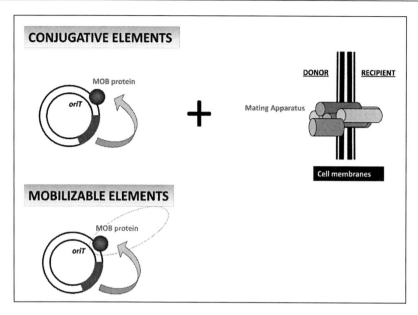

Figure 1. Schematic highlighting the differences between conjugative and mobilizable elements. In the former, both DNA processing as well as mating apparatus functions are encoded, whereas, in the latter, only DNA-processing functions are elaborated.

At first glance, it may appear that cryptic plasmids have no clinical importance because they do not carry resistance genes. However, they can have important clinical effects since they are capable of acquiring one or more antibiotic resistance genes and becoming mobile via integration with other conjugative or mobilizable transposons (or even other plasmids), resulting in further dissemination of antibiotic resistance genes. In addition, the abundance and diversity of these cryptic plasmids is cause for concern.

Transposons

Transposons, both mobilizable and conjugative, are most often located on the bacterial genome, do not replicate independently, and are copied along with the chromosomal DNA.

Mobilizable Transposons

Mobilizable transposons, like mobilizable plasmids, cannot transfer autonomously, but can be disseminated from donor bacteria via a co-resident "helper" element.[6] They use the mating apparatus of a co-resident conjugative element such as a conjugative plasmid or transposon for transfer to a recipient cell. Mobilizable transposons are invariably smaller than conjugative transposons and carry genes whose products are required for excision, DNA processing and integration of the element. However, they do not encode conjugal apparatus components and have to depend on co-resident conjugal elements like conjugative transposons. Some well-characterized mobilizable transposons in the bacteroidetes are the *B. fragilis* 9.6kb Tn*4399*,[45] *B. fragilis* 4.69kb Tn*5520*,[46] *B. fragilis* 15.3kb cLV25,[47] *B. uniformis* 10.3kb NBU1, *B. uniformis* 11.1kb NBU2[48] and *B. vulgatus* 12.5kb Tn*4555*.[49] Tn*4399* requires the gene products encoded by *mocA* (a predicted relaxase MocA), and *mocB*, for mobilization.[50] During transposition, Tn*4399* creates a 3-bp target site repeat and inserts an extra 5bp between the right inverted repeat and the target site repeat.[51] However, other known mobilizable transposons require just one gene, encoding a relaxase, for mobilization. These are Tn*5520* (BmpH[46,52]), Tn*4555* (MobA[53]), NBU1 (*MobN1*), and NBU2 (*MobN2*[48]). To date, the BmpH protein, encoded by the smallest known mobilizable transposon Tn*5520*, is the best characterized *B. fragilis* relaxase.[52]

Conjugative Transposons

Conjugative transposons (CTn's) are frequently found in *Bacteroides* spp More than 80% of *Bacteroides* strains contain at least one conjugative transposon.[8] Conjugative transposons in Bacteroides are often called "tetracycline resistance factors," and many of them can be stimulated to transfer via tetracycline exposure. They range in size from 52kb - 150kb, and include *B. fragilis* BTF-37 (37kb),[54] *B. thetaiotaomicron* CTnDOT (65kb),[55] *B. thetaiotaomicron* CTnERL (52kb),[56] *B. thetaiotaomicron* TcrEmrDOT (70kb),[57] *B. thetaiotaomicron* TcrEmr7853 (70kb),[58] *B. thetaiotaomicron* CTnBST (100kb),[59] *B. thetaiotaomicron* CTnGERM1 (75kb),[60] *B. vulgatus* CTn341 (52kb),[61] *B. fragilis* CTn86 (57kb)[62] and *B. fragilis* CTn9343 (64kb).[62] Of these, CTnDOT (65kb) from *B. thetaiotaomicron* is the best described. CTn's have also been referred to as "Tet elements" since most, but not all, carry a tetracycline resistance gene (usually *tetQ*).[35,63] Many CTns also carry the *rteABC* gene cluster, whose products are involved in the regulation of conjugal transfer.[64,65] *rteA* and *rteB* genes encode a tetracycline inducible two-component regulatory system, which controls *rteC* expression.[65] *rteC*, in turn, controls expression of genes required for excision of transmissible elements. Transcription of *tetQ*, *rteA and rteB* is constitutive but translation of these genes is elevated during exposure to tetracycline due to a translational attenuation mechanism. As a result, very low (sub-inhibitory) levels of tetracycline or its analogs can markedly elevate conjugal transfer of Tet elements and other co-resident factors by 1,000-to 10,000-fold even upon brief exposure.[64,66] It should be noted that this induction of transfer, while notable in some elements, occurs alongside the constitutive transfer events that are hallmarks of all these elements, and that have likely been occurring for millenia. Many *B. fragilis* conjugative transposons also carry erythromycin resistance genes such as *ermF* (cTnDOT),[67] *ermB* (cTnBST)[68] or *ermG* (cTnGERM1).[60]

Conjugative transposons are mainly responsible for the spread of tetracycline and erythromycin resistance in clinical isolates of *Bacteroides* spp.[8] They are not only responsible for the dissemination of the antibiotic resistance genes which they themselves carry, but also for the transfer of antibiotic resistance genes harbored by co-resident mobilizable elements, via stimulation of the excision and transfer of those mobile elements. RteA and RteB encoded within the central regulatory region of the CTnDOT/ERL family of conjugative transposons regulate the excision and mobilization of the NBU mobilizable plasmids.[69]

To date, CTnDOT (65kb) from *B. thetaiotaomicron* is the best studied conjugative transposon. Since it was first recovered from a patient with a *Bacteroides* infection,[8] CTnDOT has served as a model to study the transfer mechanism of conjugative transposons in *Bacteroides* spp. It contains an excision region, a central regulatory region and a transfer region, and its excision, integration and regulation have been extensively studied. Excision is the first step in CTnDOT transfer from the chromosome, and results in the formation of a non-replicating circular intermediate.[7] An operon containing genes required for excision (*orf2c, orf2d*, and *exc*) was identified and is regulated at the transcriptional level by the tetracycline-inducible regulatory proteins RteA, RteB and RteC.[65,70-72] Integration is the final step, resulting in the recombination of CTnDOT with the recipient bacterial genome. The integration reaction requires IntDOT, a CTnDOT-encoded protein that is a member of the tyrosine recombinase family, as well as an uncharacterized *Bacteroides* host factor.[70,73,74] Interestingly, and somewhat surprisingly, the excision proteins Orf2C, Orf2d and Exc also appear to be involved in integration.[70] However, little is known about the organization and regulation of the transfer genes responsible for the formation of the conjugation apparatus that allows the transfer of co-resident mobilizable elements. Expression of the transfer genes (*traA* to *traQ*) is activated directly by the excision proteins, and independently of RteC.[75]

Chimeric Transposons

Due to their ability to integrate into DNA, it should not be surprising to find chimeric transposons, ie. a transposon integrated into another transposon. However, to date, there are only two chimeric transposons in which a CTn inserted into another CTn, have been reported: a CTn in gram-positive bacteria, the streptococcal Tn*5253*[76]; and a CTn in gram negative bacteria, the *Bacteroides* CTn12256.[77] Although the 188kb CTn12256 was isolated from a clinical isolate in 1977, only recently has its chimeric nature been studied in more detail.[78] By cloning parts of CTn12256 into fosmids and performing a PCR walking survey, Wang, et al. found that the element was a chimera formed by the integration of a CTnDOT type element (CTnDOT2) into another CTn, CTn3Bf, that is most similar to a putative CTn found in *B. fragilis* YCH46.[78] It is interesting that although CTnDOT2 has 98% identity to CTnDOT, the transfer of the chimeric transposon CTn12256 is not dependent on tetracycline stimulation like CTnDOT and most other *Bacteroides* sp transposons. Further, the *traG* homolog on CTnBf3, but not the *traG* of CTnDOT2, is essential for transfer. However, and interestingly, neither CTn alone can transfer independently, nor can one CTn control the transfer of the other. DNA analysis has revealed that the final chimeric combination of CTnDOT2 and CTnBf3 results in a large deletion of sequence originally present in CTnDOT2, suggesting that genes present in the deleted region may have played important roles in transfer/mobilization of the CTnDOT2 element.

How are Mobile Elements Transferred within and from *Bacteroides* Spp?

The primary mechanism responsible for the dissemination of genetic elements in *Bacteroides* spp is conjugation, one of the most important mechanisms of horizontal gene transfer in prokaryotes. However, the molecular mechanism(s) of this process is poorly understood in *Bacteroides* spp

Conjugation, a subtype of the bacterial Type IV secretion system (T4SS), is defined as the uni-directional transfer of a single-stranded DNA molecule from a bacterial donor cell to a recipient cell, in a process requiring cell-to-cell contact.[79] During conjugation, one copy of the DNA strand is transferred to, and replicated in, the recipient cell. The parent DNA is retained and replicated in the donor cell. Transferred DNA molecules, which can be either plasmids or transposons, are of two types: conjugative and mobilizable. As described above, conjugative plasmids and transposons are autonomously- or self-transmissible elements, encoding all components necessary for transfer. Mobilizable plasmids and transposons are non-self-transmissible elements. Their transfers require the assistance of a co-resident conjugative transfer element. Conjugative elements tend to be large (> 30kb), while mobilizable elements are small (< 15kb).[80]

All transfer elements contain a cis-acting origin of transfer (*oriT*) sequence where transfer is initiated. *oriT*s are specific sequences, about 30–500bp in length, most often located adjacent to the transfer initiation genes known as mobilization ("Mob") genes, and forming a compact mobilization region.[81] A common feature of the *oriT* is the presence of inverted repeats juxtaposed near a DNA sequence that is nicked during the transfer process.[82] The nick (*nic*) site, a short stretch of about 10 nucleotides, is the site for recognition by the relaxase, required for the DNA transfer process.

Conjugation involves two major sets of events: Initiation (DNA processing) and conjugal apparatus formation as described below, and diagrammed in Figure 2.

Initiation (DNA Processing)

DNA processing includes binding, nicking and unwinding of the DNA, and these reactions are independent of conjugal apparatus formation.[81,83] Processing occurs via the relaxosome, a nucleoprotein complex composed of specific proteins (mobilization proteins), one of which is the relaxase.[84,85] This critical mobilization protein nicks the DNA to be transferred in a site- and strand-specific manner at the origin of transfer (*oriT*),[81,83] and then covalently associates with the 5′-end of the nicked DNA via a phosphotyrosyl linkage. The nicked DNA is further unwound from the parent molecule, and transmitted in single-stranded fashion with 5′-3′ polarity from the donor to the recipient.[81] Single-stranded copies in both the donor and the recipient are then re-circularized and restored to the double-stranded form.[81] The passage from the donor to the recipient occurs through a specialized membrane-traversing channel called the conjugal apparatus (discussed below).

Relaxase proteins, the major mobilization proteins of the relaxosomes, are usually multifunctional, and thus contain two or more protein domains. The nicking domain is always located at the N-terminus of the protein.[80] At the C-terminus, a DNA helicase, DNA primase or other domain of unknown function is almost always found.[80] Crystal structures of some relaxases have been obtained, including that of the F plasmid TraI nicking domain, with and without a bound DNA substrate,[86,87] the nicking domain of TrwC from plasmid R388 with bound DNA[88,89]

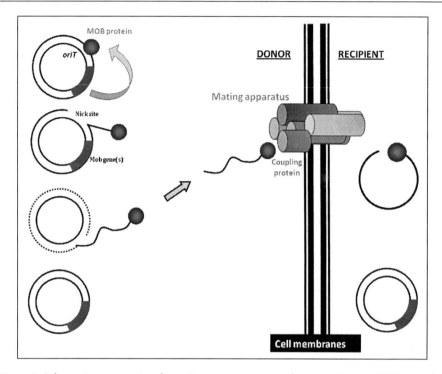

Figure 2. Schematic representing the major events occurring during conjugative DNA transfer. Cell membranes separating donor and recipient bacteria are depicted as solid black lines. The transferring element (plasmid-like in this example) is shown harboring an origin of conjugative transfer (*oriT*) as well as a mobilization (MOB) protein-encoding segment (Mob genes). As described in the text, a single-stranded DNA molecule is generated during the conjugation process, and translocated to the mating apparatus, of which one member is the coupling protein.

and that of MobA from Inc.Q plasmid R1162.[90] In many cases, the nicking domain itself contains at least three conserved protein motifs. Motif I contains the active site tyrosine, which creates a single-stranded 5' DNA nick through a trans-esterification reaction similar to that of type I topoisomerase.[83] This reaction involves the nucleophilic attack on the DNA-phosphate backbone by the hydroxyl group of the tyrosine, resulting in a reversible covalent phosphodiester bond.[81,91] Motif II is likely responsible for the recognition and noncovalent binding of the relaxase with the end of the trailing region 3' to the *nic* site. Motif III is histidine-rich and is called HUH (His-hyrophobic residue-His) or HHH (His-His-His). This motif likely facilitates the cleavage reaction (trans-esterification) by abstracting a proton from the terminal tyrosine hydroxyl, allowing the oxygen moiety to act as a nucleophile.[81] The termination of strand-transfer occurs via a second cleavage reaction, releasing a single-stranded DNA molecule in the recipient cell.[81]

Most relaxases require the activity of accessory proteins to alter the conformation of the DNA to facilitate relaxase binding. The *E. coli* plasmid RP4-encoded relaxosome requires three proteins (two cognate proteins and the host encoded integration host factor, IHF) to achieve the conformation required for the relaxase to bind and nick the DNA.[91] Similarly, relaxase activity of the TrwC relaxase of the R388 plasmid system also requires assistance from TrwA.[92] In the F plasmid system, TraY enhances the relaxase/helicase activity of TraI.[93] In *Bacteroides spp*, to date, the mobilization region of CTn341[94] and Tn5520[52] have been elaborated in detail. In CTn341 system, activity of the relaxase mobB required the accessory protein mobA.[94] However, in Tn5520, the relaxase BmpH is the first relaxase in *Bacteroides spp* reported to not require any accessory protein.[52]

Conjugation Apparatus Formation

The second major process in conjugation is the formation of the conjugal- or mating apparatus. The conjugal apparatus (CA) is a multi-protein channel that is assembled across donor and recipient cell membranes during conjugation, through which the DNA strand is transferred.[95,96] A pilus or other surface filament or proteins(s) may also be produced to facilitate adhesion and contact between two cells.[97,98]

Although the formation of the conjugal apparatus has been well studied in *Agrobacterium tumefaciens* Ti plasmids and *E.coli* F, RP4 and R388 plasmids, little is known about its structure and function in *Bacteroides* spp In *E. coli* and *A. tumefaciens*, this membrane channel is formed by 10–12 proteins (Table 2).[95,99-101] In the *E. coli* RP4 plasmid system, the mating channel is composed of 10 mating-pair gene products, a TraF pilin support protein and the coupling protein TraG.[95,102] In *E. coli* F plasmid system, the channel is composed of 11 proteins including the coupling protein TraD.[103] In *A. tumefaciens*, each of 12 proteins named VirB1 to VirB11 and VirD4 has been extensively characterized.[104,105] Recently, a cryo-electron microscopic structure of the core complex of the conjugal apparatus encoded by the *E. coli* conjugative plasmid pKM101 showed that the CA complex is 108Å wide and spanned the inner and outer membranes.[106] However, in *Bacteroides* spp the nature and function of the CA in even the best studied elements is still poorly understood. To date, the only detailed description of CA-encoding genes in *Bacteroides* spp is from a study of the *B. vulgatus* CTn341 element, where the requirement of each CA gene was assessed via the generation of gene mutants.[61]

Specificity (Coupling Protein)

One of the CA's most important components is the coupling protein (CP). CPs appear to be encoded by all conjugative plasmids and transposons, but not by mobilizable elements. The CP is unique to conjugation and is considered to be the first point of contact that the relaxosome and/or tDNA makes with the CA. The best characterized CPs are TrwB of plasmid R388 (Inc.W group), TraD of F plasmids (Inc.F group), TraG of RP4 plasmids (Inc.P group), and VirD4 of *A. tumefaciens* Ti plasmids.[107] These proteins share the following characteristics: (1) they are composed of transmembrane α-helices in their N-terminal region that mediate anchoring to the inner membrane. Indeed, they are integral inner membrane proteins.[108-110] They also typically have a cytoplasmic C-terminal domain.[95,108,111,112] Thus, their location is the link between a cytoplasmic system and the membrane complex. (2) CPs have a nucleotide binding motif, and can bind both single- and double-stranded DNA, suggesting a specific role in DNA transfer.[109] (3) A CP has a Walker box domain, and cytoplasmic domains that interact with the relaxosome.[113] The presence of Walker box motifs suggest that they use ATP hydrolysis as an energy source. It is speculated that when CP interacts with the relaxosome, the Walker-box mediates ATP hydrolysis to provide energy to "pump" the relaxosome through the CA and into the recipient cell.[114,115] With this role, CPs are considered to be "gatekeepers" of conjugation. The *A. tumefaciens* CA has two of these CP "gatekeepers," VirD4 and its required partner VirB4, both of which have nucleotide binding activity.[113,116,117] However, in *E. coli*, one CP has been described for each conjugative system. (4) CPs are often multimeric proteins. The crystal structure of the soluble cytoplasmic domain of TrwB, the CP of the Inc.W plasmid R388, shows that it is a hexameric protein, resembles a ring helicase.[118]

Most importantly, and in many transfer systems, including those elaborated by the *E. coli* F plasmids and *A. tumefaciens* Ti plasmids, CPs are highly selective for the cognate relaxosome.[119,120] Thus, a CA will only allow transfer of the plasmid that encodes it, as well as co-resident plasmids belonging to the same incompatibility group. All other plasmids and mobile elements are strictly excluded. However, and interestingly, in *B. fragilis*, putative CPs do not appear to harbor this selectivity. Conjugative transposons facilitate their own transfer, as well as that of all co-resident mobile elements, and "incompatibility" does not seem to play any role in the selection of molecules destined for dissemination. Thus, mobile elements can be transferred within, and from *B. fragilis* even to bacteria from other genera. This promiscuous transfer may be one significant driving force behind *Bacteroides* spp as reservoirs of transferable antibiotic-resistance conferring elements.

Table 2. Comparison of conjugation apparatus components of A. tumefacien, E. coli and Bacteroides spp mating systems

Mating System	Number of CA Proteins[a]	Energetic Components[b]	Core Channel Components	Pilus Components	Specific for Own Cognate Relaxosome
A. tumefaciens VirB/D system[127]	12	VirD4 (CP), VirB11, VirB4	VirB3, VirB8, VirB10, VirB6, VirB7, VirB9, VirB1	VirB2, VirB5	Yes[119]
E. coli RP4 plasmid[95]	12	TraG (CP)	TrbB, TrbC, TrbD, TrbE, TrbF, TrbG, TrbH, TrbI, TrbJ, TrbL	TraF	Yes (to closely related plasmids)[119]
E. coli F factor[103]	11	TraD (CP), TraC	TraL, TraE, TraK, TraB, TraV, TraG, TraW, TrbC,	TraA	Yes[120]
B. thetaoimicron CTnDOT[8]	17 putative gene products	Putative CP: TraG (OrfG)	N/A	N/A	No
B. fragilis BTF37[54]	N/A.	Putative CP: BctA[128]	ORF7 (TraM), ORF8 (TraN)[c]	N/A	No
B. vulgatus CTn341[61]	17 putative gene products	Putative CP: TraG (OrfG)	N/A	N/A	No

[a]Number of CA proteins includes energy-providing components, core channel components and pilus components. [b]Energetic components include coupling protein and other ATPase proteins. [c]Data from the Hecht and Vedantam laboratories.

Combating Antibiotic Resistance by Targeting Bacterial Conjugation

With the alarming rise and spread of antibiotic resistance to even new generations of antibiotics, the need to prevent the dissemination of antibiotic resistance is more urgent than ever before. Obviously, it is biologically impractical to use approaches that will eradicate all *Bacteroides spp* organisms, antibiotic-resistant or not. Since conjugation is the primary means by which *Bacteroides spp.* disseminate antibiotic resistance (and other) genes, interventions that target this process present an attractive approach to preventing unwanted horizontal DNA transfer—and only in the subset of organisms that are competent to do so. Currently, multiple groups are exploring non-antibiotic-based methodologies to prevent conjugation-based DNA transfer.[15,121-124] Different methodologies to inhibit the conjugative relaxase have been tested. Antibody libraries against the TrwC relaxase of conjugative plasmid R388 have been used to block its nicking activity within recipient cells.[122] Other studies report being able to disrupt the conjugation process by using specific inhibitors to the F plasmid relaxase.[121] Interestingly, a short palindromic repeat (CRISPR) focusing on RNA interference can also limit gene transfer—this has been tested in staphylococci by targeting relaxase genes.[123] Thus, one approach in the search for interventions to mitigate antibiotic resistance gene spread will likely focus on conjugation system-based targets.

Conclusion

Bacteroides spp., the most prominent group of bacteria in the gut, can be both beneficial symbionts, as well as dangerous opportunistic pathogens. Arguably, one of the most significant traits that characterizes the *Bacteroides* as important clinical pathogens is their capacity to harbor a plethora of transmissible genetic elements carrying many antibiotic resistance genes. Since these genes are efficiently disseminated and acquired, *Bacteroides spp.* are now considered to be reservoirs of these antibiotic resistance traits. Despite some of the advances described in this review, it should be emphasized that the majority of the mechanistic aspects of conjugative transfer—the pre-eminent process that drives the spread of antibiotic resistance in the genus—still remain to be elucidated. In-depth studies on *Bacteroides spp* conjugational components and mechanisms will thus provide insights to designing effective interventions that might focus on conjugation-system targets to prevent antibiotic resistance gene transfer.

Acknowledgments

Research in the GV laboratory is supported by grants from the US Department of Veterans Affairs and the USDA-CSREES Hatch Program.

This chapter was reproduced in part from: Nguyen M, Vedantam G. Mobile genetic elements in the genus Bacteroides, and their mechanism(s) of dissemination. Mob Genet Elements 2011; 1(3):187-196.

References

1. Ley RE, Peterson DA, Gordon JI. Ecological and evolutionary forces shaping microbial diversity in the human intestine. Cell 2006; 124:837-48; PMID:16497592; http://dx.doi.org/10.1016/j.cell.2006.02.017.
2. Xu J, Chiang HC, Bjursell MK, Gordon JI. Message from a human gut symbiont: sensitivity is a prerequisite for sharing. Trends Microbiol 2004; 12:21-8; PMID:14700548; http://dx.doi.org/10.1016/j.tim.2003.11.007.
3. Mai V, Morris JG Jr. Colonic bacterial flora: changing understandings in the molecular age. J Nutr 2004; 134:459-64; PMID:14747689.
4. Guarner F, Malagelada JR. Gut flora in health and disease. Lancet 2003; 361:512-9; PMID:12583961; http://dx.doi.org/10.1016/S0140-6736(03)12489-0.
5. Salyers AA. Bacteroides of the human lower intestinal tract. Annu Rev Microbiol 1984; 38:293-313; PMID:6388494; http://dx.doi.org/10.1146/annurev.mi.38.100184.001453.
6. Wexler HM. Bacteroides: the good, the bad, and the nitty-gritty. Clin Microbiol Rev 2007; 20:593-621; PMID:17934076; http://dx.doi.org/10.1128/CMR.00008-07.

7. Whittle G, Shoemaker NB, Salyers AA. The role of Bacteroides conjugative transposons in the dissemination of antibiotic resistance genes. Cell Mol Life Sci 2002; 59:2044-54; PMID:12568330; http://dx.doi.org/10.1007/s000180200004.

8. Shoemaker NB, Vlamakis H, Hayes K, Salyers AA. Evidence for extensive resistance gene transfer among Bacteroides spp. and among Bacteroides and other genera in the human colon. Appl Environ Microbiol 2001; 67:561-8; PMID:11157217; http://dx.doi.org/10.1128/AEM.67.2.561-568.2001.

9. Reid G. When microbe meets human. Clin Infect Dis 2004; 39:827-30; PMID:15472815; http://dx.doi.org/10.1086/423387.

10. Krinos CM, Coyne MJ, Weinacht KG, et al. Extensive surface diversity of a commensal microorganism by multiple DNA inversions. Nature 2001; 414:555-8; PMID:11734857; http://dx.doi.org/10.1038/35107092.

11. Coyne MJ, Reinap B, Lee MM, Comstock LE. Human symbionts use a host-like pathway for surface fucosylation. Science 2005; 307:1778-81; PMID:15774760; http://dx.doi.org/10.1126/science.1106469.

12. Baughn AD, Malamy MH. The strict anaerobe Bacteroides fragilis grows in and benefits from nanomolar concentrations of oxygen. Nature 2004; 427:441-4; PMID:14749831; http://dx.doi.org/10.1038/nature02285.

13. Waters VL. Conjugative transfer in the dissemination of beta-lactam and aminoglycoside resistance. Front Biosci 1999; 4:D433-56; PMID:10228095; http://dx.doi.org/10.2741/Waters.

14. Brook I. Endocarditis due to anaerobic bacteria. Cardiology 2002; 98:1-5; PMID:12373039; http://dx.doi.org/10.1159/000064684.

15. Vedantam G. Antimicrobial resistance in Bacteroides spp.: occurrence and dissemination. Future Microbiol 2009; 4:413-23; PMID:19416011; http://dx.doi.org/10.2217/fmb.09.12.

16. Shinagawa N, Osanai H, Hirata K, et al. Bacteria isolated from surgical infections and its susceptibilities to antimicrobial agents--special references to bacteria isolated between April 2009 and March 2010. Jpn J Antibiot 2011; 64:125-69; PMID:21861307.

17. Papaparaskevas J, Katsandri A, Pantazatou A, et al. Epidemiological characteristics of infections caused by Bacteroides, Prevotella and Fusobacterium species: a prospective observational study. Anaerobe 2011; 17:113-7; PMID:21664284; http://dx.doi.org/10.1016/j.anaerobe.2011.05.013.

18. Basset C, Holton J, Bazeos A, et al. Are Helicobacter species and enterotoxigenic Bacteroides fragilis involved in inflammatory bowel disease? Dig Dis Sci 2004; 49:1425-32; PMID:15481314; http://dx.doi.org/10.1023/B:DDAS.0000042241.13489.88.

19. Prindiville TP, Sheikh RA, Cohen SH, et al. Bacteroides fragilis enterotoxin gene sequences in patients with inflammatory bowel disease. Emerg Infect Dis 2000; 6:171-4; PMID:10756151; http://dx.doi.org/10.3201/eid0602.000210.

20. Toprak NU, Yagci A, Gulluoglu BM, et al. A possible role of Bacteroides fragilis enterotoxin in the aetiology of colorectal cancer. Clin Microbiol Infect 2006; 12:782-6; PMID:16842574.

21. Yoshino Y, Kitazawa T, Ikeda M, et al. Clinical features of Bacteroides bacteremia and their association with colorectal carcinoma. [In press.]. Infection 2011; PMID:21773761; http://dx.doi.org/10.1007/s15010-011-0159-8.

22. Polk BF, Kasper DL. Bacteroides fragilis subspecies in clinical isolates. Ann Intern Med 1977; 86:569-71; PMID:322563.

23. Koeth LM, Good CE, Appelbaum PC, et al. Surveillance of susceptibility patterns in 1297 European and US anaerobic and capnophilic isolates to co-amoxiclav and five other antimicrobial agents. J Antimicrob Chemother 2004; 53:1039-44; PMID:15132729; http://dx.doi.org/10.1093/jac/dkh248.

24. Snydman DR, Jacobus NV, McDermott LA, et al. National survey on the susceptibility of Bacteroides fragilis group: report and analysis of trends in the United States from 1997 to 2004. Antimicrob Agents Chemother 2007; 51:1649-55; PMID:17283189; http://dx.doi.org/10.1128/AAC.01435-06.

25. Betriu C, Culebras E, Gómez M, et al. Resistance trends of the Bacteroides fragilis group over a 10-year period, 1997 to 2006, in Madrid, Spain. Antimicrob Agents Chemother 2008; 52:2686-90; PMID:18474575; http://dx.doi.org/10.1128/AAC.00081-08.

26. Hedberg M, Nord CE; ESCMID Study Group on Antimicrobial Resistance in Anaerobic Bacteria. Antimicrobial susceptibility of Bacteroides fragilis group isolates in Europe. Clin Microbiol Infect 2003; 9:475-88; PMID:12848722; http://dx.doi.org/10.1046/j.1469-0691.2003.00674.x.

27. Stein GE, Goldstein EJ. Fluoroquinolones and anaerobes. Clin Infect Dis 2006; 42:1598-607; PMID:16652318; http://dx.doi.org/10.1086/503907.

28. Betriu C, Rodríguez-Avial I, Gómez M, et al. Changing patterns of fluoroquinolone resistance among Bacteroides fragilis group organisms over a 6-year period (1997-2002). Diagn Microbiol Infect Dis 2005; 53:221-3; PMID:16243476; http://dx.doi.org/10.1016/j.diagmicrobio.2005.06.012.

29. B D Vieira JM, Boente RF, Rodrigues Miranda K, et al. Decreased susceptibility to nitroimidazoles among Bacteroides species in Brazil. Curr Microbiol 2006; 52:27-32; PMID:16391998; http://dx.doi.org/10.1007/s00284-005-0068-0.

30. Chaudhry R, Mathur P, Dhawan B, Kumar L. Emergence of metronidazole-resistant Bacteroides fragilis, India. Emerg Infect Dis 2001; 7:485-6; PMID:11384542.

31. Schapiro JM, Gupta R, Stefansson E, Fang FC, Limaye AP. Isolation of metronidazole-resistant Bacteroides fragilis carrying the nimA nitroreductase gene from a patient in Washington State. J Clin Microbiol 2004; 42:4127-9; PMID:15364999; http://dx.doi.org/10.1128/JCM.42.9.4127-4129.2004.

32. Nagy E, Sóki J, Urban E, et al. Occurrence of metronidazole and imipenem resistance among Bacteroides fragilis group clinical isolates in Hungary. Acta Biol Hung 2001; 52:271-80; PMID:11426861; http://dx.doi.org/10.1556/ABiol.52.2001.2-3.11.

33. Wójcik-Stojek B, Bulanda M, Martirosian G, et al. In vitro antibiotic susceptibility of Bacteroides fragilis strains isolated from excised appendix of patients with phlegmonous or gangrenous appendicitis. Acta Microbiol Pol 2000; 49:171-5; PMID:11093680.

34. Dubreuil L, Odou MF. Anaerobic bacteria and antibiotics: What kind of unexpected resistance could I find in my laboratory tomorrow? Anaerobe 2010; 16:555-9; PMID:20971200; http://dx.doi.org/10.1016/j.anaerobe.2010.10.002.

35. Salyers AA, Shoemaker NB, Stevens AM, Li LY. Conjugative transposons: an unusual and diverse set of integrated gene transfer elements. Microbiol Rev 1995; 59:579-90; PMID:8531886.

36. Löfmark S, Fang H, Hedberg M, Edlund C. Inducible metronidazole resistance and nim genes in clinical Bacteroides fragilis group isolates. Antimicrob Agents Chemother 2005; 49:1253-6; PMID:15728943; http://dx.doi.org/10.1128/AAC.49.3.1253-1256.2005.

37. Nakano V, Padilla G, do Valle Marques M, Avila-Campos MJ. Plasmid-related beta-lactamase production in Bacteroides fragilis strains. Res Microbiol 2004; 155:843-6; PMID:15567279; http://dx.doi.org/10.1016/j.resmic.2004.06.011.

38. Morgan RM, Macrina FL. bctA: a novel pBF4 gene necessary for conjugal transfer in Bacteroides spp. Microbiology 1997; 143:2155-65; PMID:9245805; http://dx.doi.org/10.1099/00221287-143-7-2155.

39. Smith CJ, Tribble GD, Bayley DP. Genetic elements of Bacteroides species: a moving story. Plasmid 1998; 40:12-29; PMID:9657930; http://dx.doi.org/10.1006/plas.1998.1347.

40. Callihan DR, Young FE, Clark VL. Identification of three homology classes of small, cryptic plasmids in intestinal Bacteroides species. Plasmid 1983; 9:17-30; PMID:6300942; http://dx.doi.org/10.1016/0147-619X(83)90028-8.

41. Sóki J, Wareham DW, Rátkai C, et al. Prevalence, nucleotide sequence and expression studies of two proteins of a 5.6kb, class III, Bacteroides plasmid frequently found in clinical isolates from European countries. Plasmid 2010; 63:86-97; PMID:20026106; http://dx.doi.org/10.1016/j.plasmid.2009.12.002.

42. Smith CJ, Rollins LA, Parker AC. Nucleotide sequence determination and genetic analysis of the Bacteroides plasmid, pBI143. Plasmid 1995; 34:211-22; PMID:8825374; http://dx.doi.org/10.1006/plas.1995.0007.

43. Valentine PJ, Shoemaker NB, Salyers AA. Mobilization of Bacteroides plasmids by Bacteroides conjugal elements. J Bacteriol 1988; 170:1319-24; PMID:3343220.

44. Novicki TJ, Hecht DW. Characterization and DNA sequence of the mobilization region of pLV22a from Bacteroides fragilis. J Bacteriol 1995; 177:4466-73; PMID:7635830.

45. Hecht DW, Malamy MH. Tn4399, a conjugal mobilizing transposon of Bacteroides fragilis. J Bacteriol 1989; 171:3603-8; PMID:2544548.

46. Vedantam G, Novicki TJ, Hecht DW. Bacteroides fragilis transfer factor Tn5520: the smallest bacterial mobilizable transposon containing single integrase and mobilization genes that function in Escherichia coli. J Bacteriol 1999; 181:2564-71; PMID:10198023.

47. Bass KA, Hecht DW. Isolation and characterization of cLV25, a Bacteroides fragilis chromosomal transfer factor resembling multiple Bacteroides sp. mobilizable transposons. J Bacteriol 2002; 184:1895-904; PMID:11889096; http://dx.doi.org/10.1128/JB.184.7.1895-1904.2002.

48. Li LY, Shoemaker NB, Wang GR, et al. The mobilization regions of two integrated Bacteroides elements, NBU1 and NBU2, have only a single mobilization protein and may be on a cassette. J Bacteriol 1995; 177:3940-5; PMID:7608064.

49. Smith CJ, Parker AC. Identification of a circular intermediate in the transfer and transposition of Tn4555, a mobilizable transposon from Bacteroides spp. J Bacteriol 1993; 175:2682-91; PMID:8386723.

50. Murphy CG, Malamy MH. Characterization of a "mobilization cassette" in transposon Tn4399 from Bacteroides fragilis. J Bacteriol 1993; 175:5814-23; PMID:8397185.

51. Hecht DW, Thompson JS, Malamy MH. Characterization of the termini and transposition products of Tn4399, a conjugal mobilizing transposon of Bacteroides fragilis. Proc Natl Acad Sci U S A 1989; 86:5340-4; PMID:2546154; http://dx.doi.org/10.1073/pnas.86.14.5340.

52. Vedantam G, Knopf S, Hecht DW. Bacteroides fragilis mobilizable transposon Tn5520 requires a 71 base pair origin of transfer sequence and a single mobilization protein for relaxosome formation during conjugation. Mol Microbiol 2006; 59:288-300; PMID:16359335; http://dx.doi.org/10.1111/j.1365-2958.2005.04934.x.

53. Smith CJ, Parker AC. A gene product related to TraI is required for the mobilization of Bacteroides mobilizable transposons and plasmids. Mol Microbiol 1996; 20:741-50; PMID:8793871; http://dx.doi. org/10.1111/j.1365-2958.1996.tb02513.x.
54. Vedantam G, Hecht DW. Isolation and characterization of BTF-37: chromosomal DNA captured from Bacteroides fragilis that confers self-transferability and expresses a pilus-like structure in Bacteroides spp. and Escherichia coli. J Bacteriol 2002; 184:728-38; PMID:11790742; http://dx.doi.org/10.1128/ JB.184.3.728-738.2002.
55. Bonheyo G, Graham D, Shoemaker NB, Salyers AA. Transfer region of a bacteroides conjugative transposon, CTnDOT. Plasmid 2001; 45:41-51; PMID:11319931; http://dx.doi.org/10.1006/plas.2000.1495.
56. Bonheyo GT, Hund BD, Shoemaker NB, Salyers AA. Transfer region of a Bacteroides conjugative transposon contains regulatory as well as structural genes. Plasmid 2001; 46:202-9; PMID:11735369; http://dx.doi.org/10.1006/plas.2001.1545.
57. Li LY, Shoemaker NB, Salyers AA. Location and characteristics of the transfer region of a Bacteroides conjugative transposon and regulation of transfer genes. J Bacteriol 1995; 177:4992-9; PMID:7665476.
58. Nikolich MP, Shoemaker NB, Wang GR, Salyers AA. Characterization of a new type of Bacteroides conjugative transposon, Tcr Emr 7853. J Bacteriol 1994; 176:6606-12; PMID:7961412.
59. Gupta A, Vlamakis H, Shoemaker N, Salyers AA. A new Bacteroides conjugative transposon that carries an ermB gene. Appl Environ Microbiol 2003; 69:6455-63; PMID:14602600; http://dx.doi.org/10.1128/ AEM.69.11.6455-6463.2003.
60. Wang Y, Wang GR, Shelby A, et al. A newly discovered Bacteroides conjugative transposon, CTnGERM1, contains genes also found in gram-positive bacteria. Appl Environ Microbiol 2003; 69:4595-603; PMID:12902247; http://dx.doi.org/10.1128/AEM.69.8.4595-4603.2003.
61. Bacic M, Parker AC, Stagg J, et al. Genetic and structural analysis of the Bacteroides conjugative transposon CTn341. J Bacteriol 2005; 187:2858-69; PMID:15805532; http://dx.doi.org/10.1128/ JB.187.8.2858-2869.2005.
62. Buckwold SL, Shoemaker NB, Sears CL, Franco AA. Identification and characterization of conjugative transposons CTn86 and CTn9343 in Bacteroides fragilis strains. Appl Environ Microbiol 2007; 73:53-63; PMID:17071793; http://dx.doi.org/10.1128/AEM.01669-06.
63. Shoemaker NB, Barber RD, Salyers AA. Cloning and characterization of a Bacteroides conjugal tetracycline-erythromycin resistance element by using a shuttle cosmid vector. J Bacteriol 1989; 171:1294-302; PMID:2646276.
64. Shoemaker NB, Salyers AA. Tetracycline-dependent appearance of plasmidlike forms in Bacteroides uniformis 0061 mediated by conjugal Bacteroides tetracycline resistance elements. J Bacteriol 1988; 170:1651-7; PMID:2832373.
65. Stevens AM, Shoemaker NB, Li LY, Salyers AA. Tetracycline regulation of genes on Bacteroides conjugative transposons. J Bacteriol 1993; 175:6134-41; PMID:8407786.
66. Rashtchian A, Dubes GR, Booth SJ. Tetracycline-inducible transfer of tetracycline resistance in Bacteroides fragilis in the absence of detectable plasmid DNA. J Bacteriol 1982; 150:141-7; PMID:7061390.
67. Whittle G, Hund BD, Shoemaker NB, Salyers AA. Characterization of the 13-kilobase ermF region of the Bacteroides conjugative transposon CTnDOT. Appl Environ Microbiol 2001; 67:3488-95; PMID:11472924; http://dx.doi.org/10.1128/AEM.67.8.3488-3495.2001.
68. Wesslund NA, Wang GR, Song B, et al. Integration and excision of a newly discovered bacteroides conjugative transposon, CTnBST. J Bacteriol 2007; 189:1072-82; PMID:17122349; http://dx.doi. org/10.1128/JB.01064-06.
69. Stevens AM, Sanders JM, Shoemaker NB, Salyers AA. Genes involved in production of plasmidlike forms by a Bacteroides conjugal chromosomal element share amino acid homology with two-component regulatory systems. J Bacteriol 1992; 174:2935-42; PMID:1569023.
70. Cheng Q, Sutanto Y, Shoemaker NB, et al. Identification of genes required for excision of CTnDOT, a Bacteroides conjugative transposon. Mol Microbiol 2001; 41:625-32; PMID:11532130; http://dx.doi. org/10.1046/j.1365-2958.2001.02519.x.
71. Park J, Salyers AA. Characterization of the Bacteroides CTnDOT regulatory protein RteC. J Bacteriol 2011; 193:91-7; PMID:21037014; http://dx.doi.org/10.1128/JB.01015-10.
72. Moon K, Shoemaker NB, Gardner JF, Salyers AA. Regulation of excision genes of the Bacteroides conjugative transposon CTnDOT. J Bacteriol 2005; 187:5732-41; PMID:16077120; http://dx.doi. org/10.1128/JB.187.16.5732-5741.2005.
73. Malanowska K, Salyers AA, Gardner JF. Characterization of a conjugative transposon integrase, IntDOT. Mol Microbiol 2006; 60:1228-40; PMID:16689798; http://dx.doi.org/10.1111/ j.1365-2958.2006.05164.x.
74. Wood MM, Dichiara JM, Yoneji S, Gardner JF. CTnDOT integrase interactions with attachment site DNA and control of directionality of the recombination reaction. J Bacteriol 2010; 192:3934-43; PMID:20511494; http://dx.doi.org/10.1128/JB.00351-10.

75. Jeters RT, Wang GR, Moon K, et al. Tetracycline-associated transcriptional regulation of transfer genes of the Bacteroides conjugative transposon CTnDOT. J Bacteriol 2009; 191:6374-82; PMID:19700528; http://dx.doi.org/10.1128/JB.00739-09.

76. Ayoubi P, Kilic AO, Vijayakumar MN. Tn5253, the pneumococcal omega (cat tet) BM6001 element, is a composite structure of two conjugative transposons, Tn5251 and Tn5252. J Bacteriol 1991; 173:1617-22; PMID:1847905.

77. Bedzyk LA, Shoemaker NB, Young KE, Salyers AA. Insertion and excision of Bacteroides conjugative chromosomal elements. J Bacteriol 1992; 174:166-72; PMID:1309516.

78. Wang GR, Shoemaker NB, Jeters RT, Salyers AA. CTn12256, a chimeric Bacteroides conjugative transposon that consists of two independently active mobile elements. Plasmid 2011; 66:93-105; PMID:21777612; http://dx.doi.org/10.1016/j.plasmid.2011.06.003.

79. Willetts N, Wilkins B. Processing of plasmid DNA during bacterial conjugation. Microbiol Rev 1984; 48:24-41; PMID:6201705.

80. Garcillán-Barcia MP, Francia MV, de la Cruz F. The diversity of conjugative relaxases and its application in plasmid classification. FEMS Microbiol Rev 2009; 33:657-87; PMID:19396961; http://dx.doi.org/10.1111/j.1574-6976.2009.00168.x.

81. Lanka E, Wilkins BM. DNA processing reactions in bacterial conjugation. Annu Rev Biochem 1995; 64:141-69; PMID:7574478; http://dx.doi.org/10.1146/annurev.bi.64.070195.001041.

82. Pansegrau W, Balzer D, Kruft V, et al. In vitro assembly of relaxosomes at the transfer origin of plasmid RP4. Proc Natl Acad Sci U S A 1990; 87:6555-9; PMID:2168553; http://dx.doi.org/10.1073/pnas.87.17.6555.

83. Pansegrau W, Schröder W, Lanka E. Relaxase (TraI) of IncP alpha plasmid RP4 catalyzes a site-specific cleaving-joining reaction of single-stranded DNA. Proc Natl Acad Sci U S A 1993; 90:2925-9; PMID:8385350; http://dx.doi.org/10.1073/pnas.90.7.2925.

84. Pansegrau W, Ziegelin G, Lanka E. The origin of conjugative IncP plasmid transfer: interaction with plasmid-encoded products and the nucleotide sequence at the relaxation site. Biochim Biophys Acta 1988; 951:365-74; PMID:2850014.

85. Pansegrau W, Lanka E. Mechanisms of initiation and termination reactions in conjugative DNA processing. Independence of tight substrate binding and catalytic activity of relaxase (TraI) of IncPalpha plasmid RP4. J Biol Chem 1996; 271:13068-76; PMID:8662726; http://dx.doi.org/10.1074/jbc.271.22.13068.

86. Larkin C, Datta S, Nezami A, et al. Crystallization and preliminary X-ray characterization of the relaxase domain of F factor TraI. Acta Crystallogr D Biol Crystallogr 2003; 59:1514-6; PMID:12876370; http://dx.doi.org/10.1107/S0907444903012964.

87. Larkin C, Haft RJ, Harley MJ, et al. Roles of active site residues and the HUH motif of the F plasmid TraI relaxase. J Biol Chem 2007; 282:33707-13; PMID:17890221; http://dx.doi.org/10.1074/jbc.M703210200.

88. Boer R, Russi S, Guasch A, et al. Unveiling the molecular mechanism of a conjugative relaxase: The structure of TrwC complexed with a 27-mer DNA comprising the recognition hairpin and the cleavage site. J Mol Biol 2006; 358:857-69; PMID:16540117; http://dx.doi.org/10.1016/j.jmb.2006.02.018.

89. Guasch A, Lucas M, Moncalián G, et al. Recognition and processing of the origin of transfer DNA by conjugative relaxase TrwC. Nat Struct Biol 2003; 10:1002-10; PMID:14625590; http://dx.doi.org/10.1038/nsb1017.

90. Monzingo AF, Ozburn A, Xia S, et al. The structure of the minimal relaxase domain of MobA at 2.1 A resolution. J Mol Biol 2007; 366:165-78; PMID:17157875; http://dx.doi.org/10.1016/j.jmb.2006.11.031.

91. Byrd DR, Matson SW. Nicking by transesterification: the reaction catalysed by a relaxase. Mol Microbiol 1997; 25:1011-22; PMID:9350859; http://dx.doi.org/10.1046/j.1365-2958.1997.5241885.x.

92. Moncalián G, de la Cruz F. DNA binding properties of protein TrwA, a possible structural variant of the Arc repressor superfamily. Biochim Biophys Acta 2004; 1701:15-23; PMID:15450172.

93. Lum PL, Rodgers ME, Schildbach JF. TraY DNA recognition of its two F factor binding sites. J Mol Biol 2002; 321:563-78; PMID:12206773; http://dx.doi.org/10.1016/S0022-2836(02)00680-0.

94. Peed L, Parker AC, Smith CJ. Genetic and functional analyses of the mob operon on conjugative transposon CTn341 from Bacteroides spp. J Bacteriol 2010; 192:4643-50; PMID:20639338; http://dx.doi.org/10.1128/JB.00317-10.

95. Grahn AM, Haase J, Bamford DH, Lanka E. Components of the RP4 conjugative transfer apparatus form an envelope structure bridging inner and outer membranes of donor cells: implications for related macromolecule transport systems. J Bacteriol 2000; 182:1564-74; PMID:10692361; http://dx.doi.org/10.1128/JB.182.6.1564-1574.2000.

96. Samuels AL, Lanka E, Davies JE. Conjugative junctions in RP4-mediated mating of Escherichia coli. J Bacteriol 2000; 182:2709-15; PMID:10781537; http://dx.doi.org/10.1128/JB.182.10.2709-2715.2000.

97. Christie PJ, Atmakuri K, Krishnamoorthy V, et al. Biogenesis, architecture, and function of bacterial type IV secretion systems. Annu Rev Microbiol 2005; 59:451-85; PMID:16153176; http://dx.doi.org/10.1146/annurev.micro.58.030603.123630.

98. Collins RF, Frye SA, Balasingham S, et al. Interaction with type IV pili induces structural changes in the bacterial outer membrane secretin PilQ. J Biol Chem 2005; 280:18923-30; PMID:15753075; http://dx.doi.org/10.1074/jbc.M411603200.

99. Anthony KG, Klimke WA, Manchak J, Frost LS. Comparison of proteins involved in pilus synthesis and mating pair stabilization from the related plasmids F and R100-1: insights into the mechanism of conjugation. J Bacteriol 1999; 181:5149-59; PMID:10464182.

100. Li PL, Everhart DM, Farrand SK. Genetic and sequence analysis of the pTiC58 trb locus, encoding a mating-pair formation system related to members of the type IV secretion family. J Bacteriol 1998; 180:6164-72; PMID:9829924.

101. Li PL, Hwang I, Miyagi H, et al. Essential components of the Ti plasmid trb system, a type IV macromolecular transporter. J Bacteriol 1999; 181:5033-41; PMID:10438776.

102. Haase J, Lurz R, Grahn AM, et al. Bacterial conjugation mediated by plasmid RP4: RSF1010 mobilization, donor-specific phage propagation, and pilus production require the same Tra2 core components of a proposed DNA transport complex. J Bacteriol 1995; 177:4779-91; PMID:7642506.

103. Lawley TD, Klimke WA, Gubbins MJ, Frost LS. F factor conjugation is a true type IV secretion system. FEMS Microbiol Lett 2003; 224:1-15; PMID:12855161; http://dx.doi.org/10.1016/S0378-1097(03)00430-0.

104. Chen I, Christie PJ, Dubnau D. The ins and outs of DNA transfer in bacteria. Science 2005; 310:1456-60; PMID:16322448; http://dx.doi.org/10.1126/science.1114021.

105. Christie PJ. Type IV secretion: the Agrobacterium VirB/D4 and related conjugation systems. Biochim Biophys Acta 2004; 1694:219-34; PMID:15546668; http://dx.doi.org/10.1016/j.bbamcr.2004.02.013.

106. Fronzes R, Schäfer E, Wang L, et al. Structure of a type IV secretion system core complex. Science 2009; 323:266-8; PMID:19131631; http://dx.doi.org/10.1126/science.1166101.

107. Gomis-Rüth FX, Solà M, de la Cruz F, Coll M. Coupling factors in macromolecular type-IV secretion machineries. Curr Pharm Des 2004; 10:1551-65; PMID:15134575; http://dx.doi.org/10.2174/1381612043384817.

108. Llosa M, Bolland S, de la Cruz F. Genetic organization of the conjugal DNA processing region of the IncW plasmid R388. J Mol Biol 1994; 235:448-64; PMID:8289274; http://dx.doi.org/10.1006/jmbi.1994.1005.

109. Moncalián G, Cabezón E, Alkorta I, et al. Characterization of ATP and DNA binding activities of TrwB, the coupling protein essential in plasmid R388 conjugation. J Biol Chem 1999; 274:36117-24; PMID:10593894; http://dx.doi.org/10.1074/jbc.274.51.36117.

110. Okamoto S, Toyoda-Yamamoto A, Ito K, et al. Localization and orientation of the VirD4 protein of Agrobacterium tumefaciens on the cell membrane. Mol Gen Genet 1991; 228:24-32; PMID:1909421; http://dx.doi.org/10.1007/BF00282443.

111. Das A, Xie YH. Construction of transposon Tn3phoA: its application in defining the membrane topology of the Agrobacterium tumefaciens DNA transfer proteins. Mol Microbiol 1998; 27:405-14; PMID:9484895; http://dx.doi.org/10.1046/j.1365-2958.1998.00688.x.

112. Lee MH, Kosuk N, Bailey J, et al. Analysis of F factor TraD membrane topology by use of gene fusions and trypsin-sensitive insertions. J Bacteriol 1999; 181:6108-13; PMID:10498725.

113. Rabel C, Grahn AM, Lurz R, Lanka E. The VirB4 family of proposed traffic nucleoside triphosphatases: common motifs in plasmid RP4 TrbE are essential for conjugation and phage adsorption. J Bacteriol 2003; 185:1045-58; PMID:12533481; http://dx.doi.org/10.1128/JB.185.3.1045-1058.2003.

114. Llosa M, Gomis-Rüth FX, Coll M, de la Cruz Fd F. Bacterial conjugation: a two-step mechanism for DNA transport. Mol Microbiol 2002; 45:1-8; PMID:12100543; http://dx.doi.org/10.1046/j.1365-2958.2002.03014.x.

115. Tato I, Matilla I, Arechaga I, et al. The ATPase activity of the DNA transporter TrwB is modulated by protein TrwA: implications for a common assembly mechanism of DNA translocating motors. J Biol Chem 2007; 282:25569-76; PMID:17599913; http://dx.doi.org/10.1074/jbc.M703464200.

116. Middleton R, Sjölander K, Krishnamurthy N, et al. Predicted hexameric structure of the Agrobacterium VirB4 C terminus suggests VirB4 acts as a docking site during type IV secretion. Proc Natl Acad Sci U S A 2005; 102:1685-90; PMID:15668378; http://dx.doi.org/10.1073/pnas.0409399102.

117. Draper O, Middleton R, Doucleff M, Zambryski PC. Topology of the VirB4 C terminus in the Agrobacterium tumefaciens VirB/D4 type IV secretion system. J Biol Chem 2006; 281:37628-35; PMID:17038312; http://dx.doi.org/10.1074/jbc.M606403200.

118. Gomis-Rüth FX, Moncalián G, Pérez-Luque R, et al. The bacterial conjugation protein TrwB resembles ring helicases and F1-ATPase. Nature 2001; 409:637-41; PMID:11214325; http://dx.doi.org/10.1038/35054586.

119. Hamilton CM, Lee H, Li PL, et al. TraG from RP4 and TraG and VirD4 from Ti plasmids confer relaxosome specificity to the conjugal transfer system of pTiC58. J Bacteriol 2000; 182:1541-8; PMID:10692358; http://dx.doi.org/10.1128/JB.182.6.1541-1548.2000.

120. Sastre JI, Cabezón E, de la Cruz F. The carboxyl terminus of protein TraD adds specificity and efficiency to F-plasmid conjugative transfer. J Bacteriol 1998; 180:6039-42; PMID:9811665.

121. Lujan SA, Guogas LM, Ragonese H, et al. Disrupting antibiotic resistance propagation by inhibiting the conjugative DNA relaxase. Proc Natl Acad Sci U S A 2007; 104:12282-7; PMID:17630285; http://dx.doi.org/10.1073/pnas.0702760104.

122. Garcillán-Barcia MP, Jurado P, González-Pérez B, et al. Conjugative transfer can be inhibited by blocking relaxase activity within recipient cells with intrabodies. Mol Microbiol 2007; 63:404-16; PMID:17163977; http://dx.doi.org/10.1111/j.1365-2958.2006.05523.x.

123. Marraffini LA, Sontheimer EJ. CRISPR interference limits horizontal gene transfer in staphylococci by targeting DNA. Science 2008; 322:1843-5; PMID:19095942; http://dx.doi.org/10.1126/science.1165771.

124. Filutowicz M, Burgess R, Gamelli RL, et al. Bacterial conjugation-based antimicrobial agents. Plasmid 2008; 60:38-44; PMID:18482767; http://dx.doi.org/10.1016/j.plasmid.2008.03.004.

125. Shoemaker NB, Guthrie EP, Salyers AA, Gardner JF. Evidence that the clindamycin-erythromycin resistance gene of Bacteroides plasmid pBF4 is on a transposable element. J Bacteriol 1985; 162:626-32; PMID:2985540.

126. Smith CJ, Macrina FL. Large transmissible clindamycin resistance plasmid in Bacteroides ovatus. J Bacteriol 1984; 158:739-41; PMID:6725207.

127. Alvarez-Martinez CE, Christie PJ. Biological diversity of prokaryotic type IV secretion systems. Microbiol Mol Biol Rev 2009; 73:775-808; PMID:19946141; http://dx.doi.org/10.1128/MMBR.00023-09.

128. Hecht DW, Kos IM, Knopf SE, Vedantam G. Characterization of BctA, a mating apparatus protein required for transfer of the Bacteroides fragilis conjugal element BTF-37. Res Microbiol 2007; 158:600-7; PMID:17720457; http://dx.doi.org/10.1016/j.resmic.2007.06.004.

CHAPTER 7

Mobilisable Genetic Elements from the Clostridia

Vicki Adams, Priscilla A. Johanesen, Julian I. Rood and Dena Lyras*

Abstract

Mobilisable elements are genetic entities such as genomic islands, plasmids and transposons that are capable of intercellular movement, but are not in themselves conjugative. These elements rely on co-resident conjugative elements to provide in trans factors that facilitate their transfer to a recipient cell. A number of mobilisable transposons have been identified in the genus *Clostridium* and these fall into one of three broad categories. The first category consists of elements that encode enzymes responsible for their own transposition and mobilisation, exemplified by the Tn*4451/3* family from *Clostridium perfringens* and *Clostridium difficile*. The second category defines elements that encode enzymes for their own transposition but which do not encode any mobilisation proteins, such as tIS*Cpe8*, an IS*1595*-like element from *C. perfringens*. The third and last category consists of elements that do not contain any discernable transposase, recombinase or mobilisation genes, as found with Tn*5398* from *C. difficile*. Although each of these elements is unique, they have common features that include an origin of transfer site (*oriT*), which is an essential feature of mobilisable elements, and the presence of antibiotic resistance genes. These diverse elements are all capable of movement between bacterial cells when co-resident with conjugative elements and demonstrate that a plethora of diverse genetic elements are capable of disseminating antibiotic resistance determinants in this group of medically important pathogens. An understanding of how these elements evolve and spread is therefore important in a world where antibiotic resistance and a lack of treatment options represents an ever increasing threat to public health.

Introduction

The Disease Causing Clostridia

The genus *Clostridium* comprises a diverse collection of Gram positive, anaerobic, endospore forming bacilli, several species of which cause significant disease in both humans and animals, partly because of their ubiquitous environmental distribution.[1,2] Disease causing members are responsible for a wide range of neurotoxic, histotoxic, and enterotoxaemic diseases that result from the production of various toxins. Some species are prolific toxin producers, including *Clostridium perfringens*, different strains of which potentially produce up to 16 different toxins.[3] Another member of the genus, *Clostridium difficile*, encodes two members of the large clostridial toxin (LCT) family, Toxin A (TcdA) and Toxin B (TcdB).[4] Toxins from this group, which also include the hemorrhagic (TcsH) and lethal (TcsL) toxins of *Clostridium sordellii*

*Department of Microbiology, School of Biomedical Sciences, Monash University, Victoria, Australia.
Corresponding Author: Dena Lyras—Email: dena.lyras@monash.edu

Bacterial Integrative Mobile Genetic Elements, edited by Adam P. Roberts and Peter Mullany.
©2013 Landes Bioscience.

and α-toxin (Tcnα) from *Clostridium novyi*, are very large (in excess of 200 kDa) and have a conserved mechanism of action.[5]

The diseases caused by the clostridia depend on the species involved. Neurotoxic diseases such as botulism and tetanus are due to the release of potent neurotoxins produced by *C. botulinum* and *C. tetani*, respectively, leading to flaccid or spastic paralysis that is generally fatal without medical intervention.[6] Histotoxic diseases include gas gangrene (both spontaneous and traumatic), wound infections, post-operative and post-abortive infections. *C. sordellii* and *Clostridium septicum* are often determined to be the cause of spontaneous gas gangrene and post abortive infections, respectively, and recent work has demonstrated that α toxin from *C. septicum* is responsible for disease symptoms.[7] *C. perfringens* is the most prolific toxin producer, with a potential reservoir of toxin genes that exceeds 16 in number.[3] While not all are expressed in any one strain, there is evidence that in *C. perfringens*-mediated traumatic gas gangrene α-toxin is essential for virulence, but that it works in synergy with perfringolysin O to destroy host tissues and allow proliferation of the organism .[8]

Enterotoxaemic diseases are caused predominantly by *C. perfringens* and *C. difficile*. *C. difficile* causes a number of gastrointestinal disease syndromes that range from mild self-limiting diarrhea to more severe life-threatening pseudomembranous colitis, which can often be fatal.[9] Notably, *C. difficile* disease is associated with recent or current treatment with antibiotics, which is an important predisposing factor in disease.[9] *C. difficile* infection has become one of the leading causes of nosocomial diarrhea worldwide, with a dramatic increase in the rate and prevalence of *C. difficile* infections in many parts of the world occurring over the last decade, primarily due to the emergence of the so-called "hypervirulent" isolates belonging to the BI/NAP1/027 group.[10,11] The consequences of enteric diseases caused by *C. perfringens* are less severe than those of *C. difficile* and depend greatly on the toxins produced by particular strains.[12] Human food poisoning is very common and develops after the ingestion of large numbers of vegetative cells that have grown in the food. Sporulation in the gut leads to enterotoxin (CPE) production and subsequent diarrhea, and is generally a self-limiting disease.[12] By contrast, human necrotising enteritis occurs rarely and results from the consumption of improperly prepared food contaminated with a type C strain of *C. perfringens* that encodes β-toxin, another pore-formimg toxin, and is often fatal.[13] Animals also suffer from enteric *C. perfringens* infections. Depending on the animal, and the strain of *C. perfringens* causing the infection, many different enteritis and enetotoxaemic syndromes can occur.[14,15]

Antibiotic Resistance in the Clostridia

Like many bacterial pathogens the emergence of antibiotic resistance in clostridial species appears to be the result of antibiotic use to treat infection and/or the sub-therapeutic use of antibiotics in animal feeds.[16,17] Most of the resistance determinants that have been characterized are from *C. perfringens* and *C. difficile*, with the most prevalent phenotypes including resistance to tetracycline and erythromycin, the latter being a member of the macrolide-lincosamide-Streptogramin B or MLS group of antibiotics.[18] Resistance to other antimicrobials such as chloramphenicol, lincomycin, rifampin (or rifampicin) and fluoroquinolones has also been observed.[18-20] While resistance to antibiotics such as rifampin and fluoroquinolones has been found to be associated with chromosomal mutations in the *rpo(B)* and *gyrA* genes,[20-22] respectively, other resistance phenotypes are associated with the presence of specific determinants. More often than not these resistance determinants are associated with mobile genetic elements.

Resistance to tetracycline has been characterized in both *C. perfringens* and *C. difficile*. In *C. perfringens*, resistance can either be chromosomally encoded or associated with plasmids, all of which show similarity to the prototype 47 kb conjugative plasmid pCW3.[23] The tetracycline resistance determinant encoded by these plasmids is unique in that it consists of two overlapping genes, *tetA*(P) and *tetB*(P), which mediate resistance by different mechanisms.[24] The *tetA*(P) gene encodes a protein that acts as an active tetracycline efflux protein while *tetB*(P) encodes a protein that confers resistance by a ribosomal protection mechanism.[24] In other *C. perfringens* strains tetracycline resistance is encoded by a *tet*(M) gene.[25] This gene, which also confers resistance via a ribosomal protection mechanism, has been shown to be associated with a defective Tn*916*-like element, CW459*tet*(M), in one *C. perfringens* strain and in *C. difficile* is associated with a conjugative Tn*916*-like transposon, Tn*5397*.[26]

Erythromycin resistance in *C. perfringens* was first reported in the mid-1970s and was found to be due to the resistance determinant *erm*(B), which encodes a 23S rRNA methylase, located on a large, non-conjugative plasmid pIP402.[27] Subsequent studies showed that *erm*(B)-mediated erythromycin resistance in *C. perfringens* is relatively uncommon and that erythromycin resistance in this bacterium is most commonly associated with another class of erythromycin gene, *erm*(Q).[28] In *C. difficile* erythromycin resistance is often encoded by an *erm*(B) gene, which can be located on the mobilisable genetic element Tn*5398*,[29] however, genetic organization of these genes varies significantly.[30,31]

Lincomycin is an antibiotic that is often used for the treatment of infections caused by anaerobic bacteria[32] and resistance to this antibiotic has been reported in *C. perfringens*.[33,34] Lincomycin resistance in *C. perfringens* is common, but is usually conferred as MLS resistance by *erm*(B) or *erm*(Q) genes.[27] However, recent studies identified the *lnu*(A) and *lnu*(B) lincomycin resistance genes from *C. perfringens* chicken isolates which confer resistance by encoding a lincosamide nucleotidyltransferase that acts by directly inactivating the antibiotic.[33] Other studies on *C. perfringens* strains of animal origin detected the *lnuP* resistance gene, also encoding a lincosamide nucleotidyltransferase, which was located on a transposable genetic element, tIS*Cpe8*, that in turn was located on a conjugative plasmid, pJIR2774.[34]

Chloramphenicol resistance in *C. perfringens* and *C. difficile* is not common. In both species this phenotype is generally mediated by the *catP* gene, which encodes a chloramphenicol acetyltransferase.[35,36] The *catP* (originally called *catD* in *C. difficile*) gene in both *C. perfringens* and *C. difficile* is located on mobilisable transposons of the Tn*4451/3* family.[35,36] In *C. perfringens* chloramphenicol resistance may also be encoded by the chromosomal *catQ* gene, which does not appear to be associated with a mobile genetic element.[37]

Mobile Elements in the Clostridia

There are a number of mobile elements that have been characterized in the pathogenic clostridia. These include bacteriophage or bacteriophage-like elements, plasmids and transposons. Conjugative plasmids and integrative conjugative elements (ICEs or conjugative transposons) are elements that are capable of movement from donor to recipient cells in a process that requires cell-to-cell contact.[38,39] The bacterial conjugation machinery typically involves proteins required for the formation of the mating bridge, for DNA processing and for DNA translocation to the recipient cell. Many conjugative plasmids and transposons are also capable of mobilising co-resident plasmids and integrative mobilisable elements (IMEs or mobilisable transposons) in trans.[38,39] Typically, the role of the conjugative element in this instance is to provide the mating bridge by which the DNA is transferred. The mobilisable transposon or plasmid contains an *oriT* site, which is essential for conjugative transfer, and usually also encodes a cognate mobilisation protein (or proteins) that facilitates the mobilisation of the element to the recipient cell.[38] Mobilisable elements in the clostridia are usually transposons that carry antibiotic resistance genes.[19] The presence of these elements therefore has implications for the spread of antibiotic resistance determinants and potentially other virulence traits, such as toxins.[18]

Clostridial mobilisable transposons are diverse, involving distinct mobilisation mechanisms, some of which remain to be defined. They can be categorised into three broad categories: elements that encode enzymes responsible for their own transposition and mobilisation, exemplified by the Tn*4451/3* family found in both *C. perfringens* and *C. difficile*, elements that encode transposition enzymes, but not mobilisation proteins, such as tIS*Cpe8*, an IS*1595*-like element from *C. perfringens,* and elements that do not contain any discernable transposase/recombinase or mobilisation genes, as found in Tn*5398* from *C. difficile*. These three groups of elements will be the subject of this chapter.

The Tn*4451/3* Family of Mobilisable Transposons

*Identification of the Tn*4451/3 *Family of Elements*

Transfer of tetracycline and chloramphenicol resistance was first described in two *C. perfringens* strains, CP590 and CP600, which were found to contain plasmids conferring tetracycline and chloramphenicol resistance or clindamycin and erythromycin resistance.[40] The tetracycline and chloramphenicol resistance genes were shown to be encoded on the same conjugative plasmid, pIP401, in strain CP590. Chloramphenicol resistance was frequently lost during conjugative transfer, a phenomenon that was even more marked during transfer experiments performed in axenic mice in which transfer of chloramphenicol resistance could not be detected.[40] Strain CP600 transferred tetracycline and chloramphenicol resistance in a similar way to pIP401.[41] Analysis of the tetracycline resistant, chloramphenicol sensitive transconjugants indicated that the loss of chloramphenicol resistance coincided with the loss of approximately 6 kb of DNA from pIP401, pJIR25 (a CP600-derived tetracycline and chloramphenicol resistance plasmid) and pJIR27 (derived from another tetracycline and chloramphenicol resistant *C. perfringens* strain).[40,41] As a result of these experiments it was suggested that the 6 kb region encoding chloramphenicol resistance represented a discrete transposon.[40,41]

The chloramphenicol resistance elements from pIP401 and pJIR27 were subsequently cloned in *E. coli* and designated Tn*4451* and Tn*4452*, respectively.[42] Both elements demonstrated an unstable phenotype in *E. coli*, spontaneously excising from multicopy plasmids; in addition, very low level of transposition of these elements was also observed.[42] Heteroduplex analysis indicated that Tn*4451* and Tn*4452* were very similar, with only a 400 bp disparity being noted approximately 250 bp from one end of the 6 kb elements.[42] Further analysis of Tn*4451* showed that excision was precise in both *E. coli* and *C. perfringens* and the target sites identified from *E. coli* resembled the ends of the elements.[35] Subsequently, six similar elements were identified in *C. perfringens*[43] (Adams, Lyras and Rood, unpublished data) and five in *C. difficile*.[43,44] Interestingly, all the *C. difficile* strains shown to carry *catD* (now known as *catP*)[36] appear to carry duplicated elements that are located on restriction fragments of a similar size, suggesting that two copies of the element are located in the same chromosomal position in each of the *C. difficile* strains studied.[44] This observation suggests either that the isolates are clonal, although this seems unlikely given the geographically diverse origin of each of these strains, or that there may be a hot spot for insertion of this transposon in the *C. difficile* genome.[44]

*Molecular Characterization of the Tn*4451/3 *Family*

Tn*4451* and Tn*4453a* have been sequenced and consist of 6340 bp that encodes six genes, *tnpX*, *tnpV*, *catP*, *tnpY*, *tnpZ* and *tnpW*, with an overall nucleotide identity of 89% between the two elements[35,45] (Fig. 1). The *tnpV* and *catP* genes from each element are identical with the *tnpV* gene overlapping *tnpX* by 73 bp (Fig. 1 and Table 1).[35,45] Amino acid differences between the Tn*4451* or Tn*4453a* encoded TnpX, TnpZ and TnpY proteins are comparable, ranging between from 8% to 12%, while the TnpW gene is the most divergent, with a 24% amino acid sequence difference between the two elements (Table 1).[45]

The *tnpX* gene encodes a functional site-specific recombinase belonging to the serine recombinase family.[35,46] CatP confers chloramphenicol resistance while the TnpZ protein, discussed in greater detail below, interacts with the *oriT* or RS$_A$ site located upstream of *tnpZ* and which is conserved between Tn*4451* and Tn*4453a*.[36,47] This *oriT* site is located between the *tnpY* and *tnpZ* genes and consists of an inverted repeat that shows similarity to other Mob/Pre family systems.[47] The function of the TnpV, TnpY and TnpW proteins is unknown although deletion analysis has shown that they are not essential for transposition.[45,48] TnpV was suggested to act as an excisionase promoting excision,[35] however, *tnpV* deletion studies performed with Tn*4451* suggest this is not the case.[48] The TnpY protein contains Walker A and B boxes, which are motifs involved in ATP metabolism, and is expected to bind ATP.[35] However, deletion studies have indicated that the TnpY protein is not essential for transposition, although the presence of this protein during mobilisation changes the dynamics of the transfer process, but not the mobilisation rate.[48] Finally, the *tnpW* gene encodes a very small protein of 62 amino acids, which also has no known function.[48] Deletion of the *tnpW*

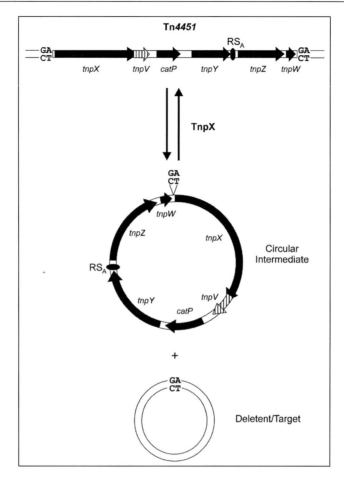

Figure 1. Genetic organization of Tn*4451* and model of excision and insertion. Tn*4451* encodes six genes, *tnpX, tnpV, catP, tnpY, tnpZ* and *tnpW*, as indicated. The *tnpV* gene overlaps the *tnpX* gene by 73 bp.[35] The transposon is flanked by directly repeated GA dinucleotides. Excision and integration is mediated by TnpX. Excision of Tn*4451* leads to the formation of a circular form of the transposon, which possesses one of the dinucleotides at the joint, and a deletion plasmid, in which one dinucleotide remains at the deletion site. Insertion involves the TnpX-mediated recombination of the circular molecule with a chromosomal target site. Reproduced from Lyras D, Rood JI. Mol Microbiol 2000; 38:588-601;[45] with permission from Society for General Microbiology.

gene resulted in a slight decrease in transposition, but this deficiency could not be complemented by the provision of a wild-type *tnpW* gene in trans. This result suggested that the decrease in transposition activity probably resulted from changes to the end of the transposon that occurred as a consequence of the deletion process, rather than loss of activity of the TnpW protein itself.[48]

Members of the Tn4451/3 Family
Tn*4451* and Tn*4452* were identified from *C. perfringens* strains CP590 and 1329, respectively.[42] Several other chloramphenicol resistant *C. perfringens* isolates were also shown by hybridization analysis to encode a *catP* gene.[43] Subsequent PCR analysis has indicated that these three strains, along with a more recently identified chloramphenicol resistant isolate also encode Tn*4451*-like elements (Adams, Lyras and Rood, unpublished). Two distinct elements from *C. difficile* strain W1

Table 1. *Comparison of amino acid sequences of proteins encoded by Tn4453a to homologous proteins identified from genome sequencing projects*

Element	Tn4453a					
	TnpX	TnpV	CatP	TnpY	TnpZ	TnpW
Tn4451#	88%	100%	100%	92%	92%	76%
C. nexile DSM1787#	88%	100%	100%	92%	92%	76%
C. difficile NAP08	93%	82%	~5 kb*	97%	96%	98%
Coprococcus sp ART55/1	93%	86%	~2 kb*	94%	94%	95%

#Tn4451 and the *C. nexile* element are identical to each other at the nucleotide sequence level. Accession numbers are as follows: Tn4451 (U15027), Tn4453 (AF226276), *C. nexile* DSM1787 (contained in NZ_ABWO00000000), *C. difficile* NAP 08 (contained in NZ_ADNX00000000) and *Coprococcus* sp ART55/1 FP929039.
*The *catP* gene was not identified in these strains, instead unrelated DNA sequences of the indicated sizes were present.

were also cloned and designated Tn4453a and Tn4453b.[36] Similar elements appear to be present in other chloramphenicol resistant *C. difficile* strains.[44]

Recent bioinformatic analysis of whole genome sequences has indicated that there are likely to be many other members of the Tn4451/3 family, especially in bacterial species found in the normal human gut (Adams and Lyras, unpublished). These species include *Clostridium nexile* and *Coprococcus* sp, which along with a *C. difficile* NAP08 strain, carry elements that show a high degree of sequence conservation to Tn4451 and Tn4453a (Table 1). Furthermore, comparative analysis suggests that two distinct element lineages exist, those more closely related to Tn4453a, including putative elements from a *C. difficile* NAP08 strain and *Coprococcus* sp ART55/1, and those more closely related to Tn4451, such as a putative *C. nexile* element which is identical to Tn4451 at the nucleotide level and putative elements from *Ruminococcus bromii* L2–63, *Phascolarctobacterium* sp YIT12067 and *Eubacterium siraeum* V10Sc8a (Adams and Lyras, unpublished). The sequence data suggests that these elements are likely to be transposition and mobilisation proficient and therefore capable of transfer between members of the human gut microbiota, which is the source of most of the genome sequences. The elements identified in the *C. difficile* NAP08 strain and the *Coprococcus* sp strain ART55/1 no longer encode chloramphenicol resistance and instead have acquired other DNA sequences, such as putative aminoglycoside resistance genes in the former strain and transcriptional regulators in the latter (Adams and Lyras, unpublished).

Mechanism of Transposition of Tn4451/3

The transposition mechanism of Tn4451, Tn4453a and, to a lesser extent, Tn4453b has been extensively studied and the transposition modules are functionally interchangeable, confirming that these elements are closely related.[36,46,45,48] This family of elements are non-replicative and excise precisely from the surrounding DNA.[35,49] They can exist in two forms, as a linear molecule, integrated into either the host chromosome or a co-resident plasmid, or as a circular, non-replicating transposition intermediate (Fig. 1). In the linear form the transposon is bound by directly repeated GA residues and is maintained and replicated together with the host molecule.[35,46,49] The second form of the transposon, the circular intermediate, results from an excision event whereby the left and right ends are abutted, with a GA dinucleotide found to be located between the two transposon ends.[35,46] The circular form has been shown to be the transposition intermediate.[45]

The linear form of the Tn4451/3 transposons is cleaved by TnpX at both the left (proximal to *tnpX*) and right (*tnpW*) ends. This process involves two base-pair staggered cuts that occur at the GA residues that flank the linear molecule.[46] Cleavage at the transposon ends leads to the formation of the circular intermediate, where the left and right ends are joined by a central GA dinucleotide, with the second GA dinucleotide remaining at the now empty target site (Fig. 1).[45,46] The circular intermediate does not replicate, but acts as a substrate either for TnpZ-mediated mobilisation to a conjugation recipient or for TnpX-mediated insertion into a replicating DNA molecule, such as the chromosome or a plasmid (Fig. 1).[45,46] If mobilisation of the circular intermediate to another bacterial cell occurs, TnpX-mediated insertion into a replicating DNA molecule is required for rescue and stable maintenance of the element in the new host.[45,48]

Regulation of the transposition process is achieved primarily through the expression of the *tnpX* gene and the presence of an upstream *tnpX* promoter. The *tnpX* gene is located at the extreme left end of the element, which does not contain an intact *tnpX*-specific transcriptional promoter, with only a -10 box being located within the transposon sequence (Fig. 1).[45] Expression of *tnpX* therefore depends on the presence of an appropriately spaced -35 box in the left-end flanking DNA sequence for the formation of a promoter that can drive expression of *tnpX* from the linear form of the transposon.[45] Alternatively, transposon insertion immediately downstream of a functional promoter also would lead to *tnpX* expression. Upon excision of the element and formation of the circular intermediate, a new *tnpX* promoter, P_{CI}, is formed by the almost perfect juxtaposition of a -35 box located at the right end of the element with the -10 box located upstream of the *tnpX* gene at the left end (Fig. 1),[45] leading to a high level of expression of *tnpX*. The formation of a strong *tnpX* promoter in the non-replicating circular intermediate is likely to have evolved as a survival mechanism since high levels of TnpX would ensure the integration of the element at a new site within a replicating molecule, particularly following mobilisation to a new host cell. Insertion would then serve to disengage the strong P_{CI} promoter resulting in a more stable, integrated form of the transposon.[45]

The TnpX protein belongs to the serine recombinase family,[46] a group of enzymes that includes resolvases encoded by Tn3-like elements.[50] This group of proteins is characterized by having a conserved catalytic domain, usually located at the N-terminus, that catalyzes DNA cleavage through a covalent linkage to the DNA substrate via the hydroxyl group of the catalytic serine residue.[51] The catalytic serine residue of TnpX is residue S15[46] and modeling of the catalytic domain suggests that the structure of TnpX in this region is similar to that of gamma delta and Tn3 resolvase proteins.[52] This similarity extends to the presence of many TnpX-encoded amino acids that have been shown to play a critical role in catalysis in the latter group of enzymes.[52]

Although TnpX is very similar to resolvase proteins of the Tn3 family in the N-terminal catalytic region, the rest of the protein is very different. It is four times larger and can catalyze DNA rearrangements that the smaller enzymes cannot promote, such as insertion of the encoding element.[53] Studies using purified protein have shown that the 707 amino acid TnpX protein is arranged into three distinct domains.[54] The first domain of approximately 105 amino acids encompasses the catalytic pocket that is responsible for DNA cleavage while the second domain contains residues involved in DNA binding.[53,54] The second domain also interacts with the third domain, which has been shown to contain a major DNA binding motif as well as a second, non-essential DNA binding motif[53,54] that is located at the C-terminus of the protein.[54] The TnpX protein is a dimer in solution and the dimerization site is located within the third domain of the protein.[52-54] In addition to dimerization, the TnpX recombination synapse must be able to facilitate an additional protein-protein interaction; specifically, the interaction required to bring the two DNA recombination sites together, to which a TnpX dimer will be bound.[52] The location and exact nature of this interaction site remains to be elucidated.

DNA binding studies have demonstrated that TnpX recognizes and binds to the left and right ends of the element (*attL* and *attR*) and to the joint of the circular intermediate (*attCI*).[53] Kinetic studies have shown that TnpX binds with greatest affinity to *attCI*, but binds very poorly to several insertion or target sites (*attT*). TnpX binds strongly to both *attL* and *attR*, although at lower efficiency than to *attCI*.[53] During excisive recombination, *attL* and *attR* are brought together in a synapse, which allows DNA strand exchange to occur. Since DNA binding to these two sites is equally strong it is postulated that separate dimers of TnpX bind to these sites and are subsequently brought together for synapse formation.[53] Conversely, TnpX has very different binding affinities for the insertion substrates *attCI* and *attT*, suggesting that the synapses formed for insertion and excision are distinct and that these reactions are not merely the reverse of one another.[52,53]

Studies using an in vivo genetic approach showed that TnpV, TnpY and TnpW are not required for transposition (ie excision followed by insertion). In vitro excision experiments also demonstrated that purified TnpX alone is capable of excising the integrated element and forming the circular intermediate molecule.[48] However, transposon insertion could not be detected in the in vitro experiments even though insertion occurs readily in vivo in the presence of just TnpX.[48,53] The in vitro excision reactions reach an apparent maximum substrate conversion of 50%[52] and the in vivo excision assays can achieve excision levels approaching 100%.[48] These data further support the notion that there are inherent differences between the requirements for excision and insertion and also suggest that while other transposon and host-encoded proteins are not essential for transposition, they are likely to play some role in the transposition process.

The Mechanism of Tn4451/3 Mobilisation

The TnpZ protein shows most similarity to relaxases belonging to the pMV158 or MOB$_V$ superfamily,[55,56] which encompasses enzymes that are responsible for DNA processing during conjugative transfer of plasmids and other mobile elements.[38] Very little biochemical or mutational analysis has been performed on relaxases from this family and knowledge surrounding the mechanism involved in mobilisation is limited to the observation that the relaxase is covalently attached to the DNA after nicking at a specific site, defined as the *oriT* or RS$_A$ site.[55]

The *tnpZ* gene is preceded by an RS$_A$ site that is also related to transfer sites associated with the MOB$_V$ family[35,56] and has been shown to be essential for mobilisation to occur.[47] In addition, it has been demonstrated that mobilisation does not occur without a functional *tnpZ* gene and that TnpZ functions both in cis and in trans.[47] As found with other MOB$_V$ family mobile elements, the transfer machineries of both Tn916 and RP4 have been demonstrated to facilitate the conjugative mobilisation of plasmids carrying an RS$_A$ site and the *tnpZ* gene in both *E. coli* and *C. perfringens*.[47] The mobilisation cassettes encoded by three members of the Tn4451/3 family, Tn4451, Tn4453a and Tn4453b, are functional, although the Tn4453a cassette appears to function less efficiently.[36] The difference in mobilisation levels is likely to be the result of differences in the TnpZ proteins since the RS$_A$ sites are conserved.[36] Finally, mobilisation of the Tn4451/3 family of elements is likely to have played an important role in the dissemination of *catP*-mediated chloramphenicol resistance. Of particular interest in this regard is the finding that the *catP* gene, together with the upstream *tnpV* gene, has been detected in clinical isolates of *Neisseria meningitidis*.[57,58]

The Lincomycin Resistance Element tIS*Cpe8*

Identification of tIS*Cpe8*

Resistance to lincomycin in *C. perfringens* and *C. difficile* is relatively common and is typically conferred by genetically related variants of the MLS resistance determinant Erm(B). Studies in Belgian chickens indicated that there was an increase in low-level lincomycin resistance in animal-associated *C. perfringens* strains.[33] In two of the isolates lincomycin resistance was found to be conferred by the *lnu*(A) or *lnu*(B) genes which encode lincosamide nucleotidyltransferases that act by directly inactivating lincomycin[33]. This was a significant finding as it was the first discovery of *C. perfringens* isolates that were resistant to lincomycin that did not carry an *erm* resistance gene.

Subsequently, analysis of several multiply antibiotic resistant *C. perfringens* isolates of animal origin identified several strains that were lincomycin resistant but were susceptible to erythromycin, suggesting the presence of non-*erm* lincomycin resistance determinants.[34] Two of these isolates, strains UAZ196 and 95–949, were able to transfer lincomycin resistance to other *C. perfringens* isolates by conjugation. Detailed genetic analysis of strain 95–949 showed that resistance was encoded by the lincomycin resistance gene *lnuP*, which encodes a lincosamide nucleotidyltransferase.[34] The *lnuP* gene was located on an element with similarities to an IS1595-family element from *Streptococcus agalactiae*, tIS*Sag10*, and was termed tIS*Cpe8*. This element was located on a conjugative R-plasmid, pJIR2774, which is similar to the conjugative plasmid pCW3, but does not confer tetracycline resistance.[34]

Phenotypic and Genetic Characterization of tISCpe8

tIS*Cpe8* is a 1,964 bp element that includes 24-bp imperfect inverted repeats at the termini, with 22 of the 24 bases conserved between the two ends. It encodes two genes, *InuP*, which encodes lincomycin resistance, and *tnp*, which encodes a putative 348 amino acid transposase (Fig. 2).[34] The genetic organization of tIS*Cpe8* is almost identical to that of tIS*Sag10* from S. *agalactiae*, a mobilisable transposon of the IS*Pna2* group in the IS*1595* family of transposon-like insertion elements.[59] The similarity between the two elements suggests that they may have originated from a common ancestor and that DNA transfer may have occurred between *C. perfringens* and *S. agalactiae*, possibly via intermediate hosts.[34] Comparative analysis of the tIS*Cpe8*-encoded transposase with other proteins shows that it has significant sequence identity to putative transposases from *Clostridium bartlettii* (99% identity; encoded by tIS*Cba1*), *Clostridium phytofermentans* (32% identity; encoded by tIS*Clph1*), S. *agalactiae* (36% identity; encoded by tIS*Sag10*), and *Campylobacter jejuni* (36% identity).[34] Note that at the 3'-end of the *lnuP* gene there is an *oriT*-like region which is similar to the *oriT* region found in the 3'-end of the *lnuC* gene from tIS*Sag10*,[34] an element that can be mobilisised by Tn*916*.

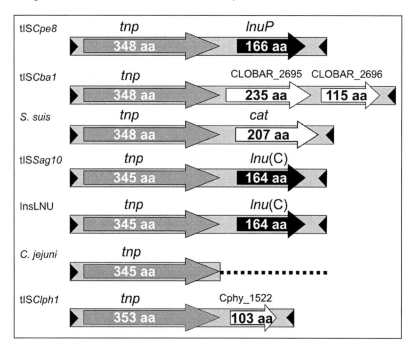

Figure 2. Genetic organization of the lincomycin resistance transposon tIS*Cpe8* and comparison to orthologous gene regions. tIS*Cpe8* encodes a putative transposase gene (*tnp*) and a linco- mycin resistance gene (*lnuP*). The direction of transcription, and the inverted repeats located at the ends of the element, are indicated by the arrows. The numbers of amino acids (aa) in the encoded proteins are indicated. The elements related to tIS*Cpe8* include tIS*Sag10* from S. *agalactiae*.[64] The remaining elements were identified by searching existing genome sequences and included tIS*Cba1* from *C. bartlettii* (accession number NZ_ABEZ02000022), an element from *S. suis* (YP_003028723.1), an element from *C. jejuni* (AY701528), InsLNU from *H. parasuis* (NC_012661.1), and tIS*Clph1* from *C. phytofermentans* (NC_010001). Related transposase genes are indicated (gray arrows), conserved inverted repeats are indicated by arrowheads, and related resistance genes are also indicated (black arrows). Open arrows indicate ORFs with unknown function and the chloramphenicol acetyltransferase gene (*cat*). Reproduced from Lyras D et al. J Bacteriol 2009; 191:6345-51;[34] with permission from American Society for Microbiology.

Transposition of tISCpe8

tIS*Cpe8* inserts into many different sites and can exist in multiple copies within the genome of the host cell, suggesting that it is capable of transposing efficiently in *C. perfringens*.[34] Evidence to support this hypothesis was obtained from experiments in which a tIS*Cpe8*–containing suicide plasmid was introduced into *C. perfringens* strain JIR325 via electroporation, resulting in the isolation of numerous independent lincomycin-resistant transformants. PCR and Southern hybridization analysis confirmed that these isolates carried tIS*Cpe8* on the chromosome independently of the suicide plasmid, suggesting that transposition of the element had occurred in the host strain. Southern hybridization analysis of tIS*Cpe8* derivatives suggested that in strains JIR325 there was a hot spot for the insertion of tIS*Cpe8*, whereas in strains CW504 and 95–949 multiple insertions were detected, implying a more random insertion pattern.[34]

Analysis of tIS*Cpe8* insertion sites showed that tIS*Cpe8* duplicates an 8-bp target sequence upon insertion, like other members of the IS*Pna2* group in the IS*1595* family.[34] The transposase encoded by tIS*Cpe8* also has several regions in common with transposases of the IS*Pna2* group, regions that have been shown to be important for protein function, including a zinc finger motif, a putative helix-turn-helix motif, and a catalytic tetrad.[34,59] Taken together these observations suggest that the tIS*Cpe8* transposition mechanism is similar to that from the IS*Pna2* group of elements. The structural and functional similarity between these transposons suggests that they have a common ancestor.[34]

Mobilisation of tISCpe8

tIS*Cpe8* does not encode any proteins with similarity to known mobilisation proteins.[34] However, at the 3'-end of the *lnuP* gene it does contain an *oriT*-like region, which is similar to that found in tIS*Sag10*. This *oriT* site is functional and allows tIS*Cpe8* to be mobilised by Tn*916* into recipient cells during conjugation.[34]

This observation that IS*Cpe8* encodes an *oriT* region within an antibiotic resistance gene is an important one and reinforces the hypothesis that conjugation plays an important role in the dissemination of genetic elements that are not obviously mobile since they do not contain conjugation or mobilisation genes. This process is likely to play a significant role in bacterial adaptation and evolution, particularly with respect to antibiotic resistance.[34]

The Erythromycin Resistance Element Tn5398

Discovery and Mobilisation of Tn5398

Early studies showed that the Erm B determinant carried by *C. difficile* strain 630, which also carries the tetracycline resistance element Tn*5397*, was transferable to *C. difficile* recipient strains by a conjugation-like mechanism.[60] This resistance determinant was subsequently shown to transfer into a *Staphylococcus aureus* recipient strain[61] as well as into *Bacillus subtilis* and back into *C. difficile* from *B. subtilis* transconjugants.[62] Transfer was not associated with plasmid mobilisation, a conclusion confirmed by hybridization studies, which showed that integration of the Erm B determinant occurred in a site-specific manner into the *C. difficile* chromosome.[62] By comparison, integration occurred non-specifically into the *B. subtilis* chromosome.[62] As a result of these observations it was concluded that the erythromycin resistance determinant is likely to reside on a conjugative transposon, which was defined as Tn*5398*.[62] However, isolation and characterization of the element showed that Tn*5398* did not encode any genes associated with characterized conjugative transposons and it was subsequently shown to be a mobilisable, non-conjugative element or IME.[30] Despite the fact that Tn*5398* does not appear to encode any mobilisation genes, it is clearly mobilisable, by an unknown element in strain 630.[30] It is likely that mobilisation involves excision of the element followed by the formation of a circular intermediate, but Tn*5398* does not appear to mediate its own excision.[30] Tn*5398* may be the derivative of a progenitor element that carried conjugation or mobilisation genes as well as a site-specific recombinase gene.[30]

Phenotypic and Genotypic Characterization of Tn5398

Tn*5398* (9630 bp) contains six complete genes (Fig. 3),[30] with two directly repeated *erm*(B) genes comprising the Erm B determinant, which is flanked by the *orf13*, *effR*, *effD* and *orf9* genes.[30] Tn*5398* also contains one incomplete gene, *orf7* (Fig. 3). None of the Tn*5398* genes encode proteins that have any homology to proteins known to be involved in mobilisation or conjugative transfer.[30] The *effD* and *effR* genes are thought to encode a protein that act as a potential efflux pump and its associated regulator, respectively. The *orf13* and *orf9* genes and the incomplete *orf7* gene encode proteins that show similarity to proteins encoded on the conjugative transposon Tn*916*, however, no function has been determined for any of these genes.[30]

Not all erythromycin resistant *C. difficile* strains carry two copies of the *erm*(B) gene. Considerable genetic variation exists within the *erm*(B)-carrying isolates, with some strains having one *erm*(B) gene and others having various derivatives of the Tn*5398*-encoded directly repeated sequence while some carry two copies of the *erm*(B) genes associated with Tn*5398*.[29,30,63] Recent bioinformatic analysis has also identified a putative mobilisable transposon in strain ATCC43255 that has similarity to Tn*5398*.[63] The Erm B determinant is not present in the ATCC43255 element, having been replaced by a gene encoding a putative secreted protein of unknown function.[63]

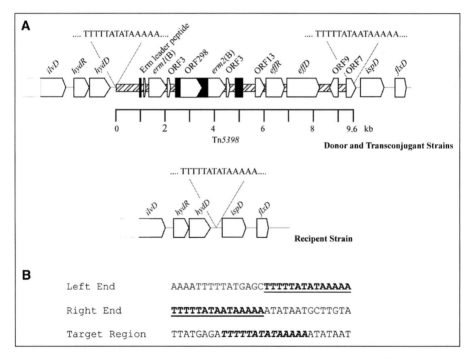

Figure 3. Genetic organization of Tn*5398*. A) A genetic map of Tn*5398* from *C. difficile* donor and transconjugant strains, and the target region in the *C. difficile* recipient strain CD37, are shown. The region encompassing the Tn*5398* element is backed by a cross-hatched box. The element is 9.6 kb in length and contains 6 complete genes and one incomplete gene, *orf7*, that are shown as open block arrows. The sequences of the ends of the transposon in the donor and transconjugant strains and the target region in the recipient strain are indicated. Regions that encompass direct repeat (DR) sequences associated with the Erm B determinant are shown by the black boxes B) Nucleotide sequences of the ends of Tn*5398* and the target region. Nucleotides representing the ends of Tn*5398* are underlined and shown in bold. Nucleotides comprising the target site in the *C. difficile* recipient strain CD37 are shown in bold italics. Modified from Farrow KA et al. Microbiology 2001; 147:2717-28;[30] with permission from the Society for General Microbiology.

Conclusion

The emergence of pathogens encoding multiple antibiotic resistance phenotypes is of significant concern and one of the main contributing factors to the emergence of these strains is the presence of resistance genes on mobile genetic elements. Mobilisable elements of the clostridia often carry antibiotic resistance determinants and are structurally and mechanistically diverse. The Tn*4451/3* family of elements encode their own mobilisation protein and cognate recognition and cleavage site, therefore requiring some compatibility between the Tn*4451/3* mobilisation complex and the mating pair formation complex encoded by the conjugative element that mediates cell-to-cell contact. By contrast, tIS*Cpe8* utilizes both the mobilisation and conjugation machinery encoded by Tn*916* to facilitate its *oriT*-dependent conjugative transfer. The mechanism of transfer utilized by Tn*5398* is less well understood but may occur using a mechanism similar to that of tIS*Cpe8*, whereby an *oriT*-like sequence, perhaps recognized by one of the conjugative elements present within strain 630, facilitates cell-to-cell movement. Clearly, the detailed analysis of these mobilisable clostridial elements has shown that their evolution and dissemination is a complex process that involves the interaction of genetic elements with very different properties and that these elements are likely to play a very significant role in clostridial adaptation and evolution.

Acknowledgments

This work was supported by Project Grants from the Australian National Health and Medical Research Council.

References

1. McClane BA. Progress in understanding clostridial toxins and their genes could reduce the damage they cause. ASM News 1995; 61:465-8.
2. Songer JG. Clostridial diseases of animals. In: Rood JI, McClane BA, Songer JG, Titball RW, eds. The clostridia: molecular biology and pathogenesis. London: Academic Press, 1997:153-82.
3. Rood JI. Virulence genes of Clostridium perfringens. Annu Rev Microbiol 1998; 52:333-60; PMID:9891801; http://dx.doi.org/10.1146/annurev.micro.52.1.333.
4. Just I, Selzer J, Wilm M, et al. Glucosylation of Rho proteins by Clostridium difficile toxin B. Nature 1995; 375:500-3; PMID:7777059; http://dx.doi.org/10.1038/375500a0.
5. Aktories K. Bacterial protein toxins that modify host regulatory GTPases. Nat Rev Microbiol 2011; 9:487-98; PMID:21677684; http://dx.doi.org/10.1038/nrmicro2592.
6. Johnson EA. Neurotoxigenic clostridia. In: Fischetti VA, Ferretti JJ, Portnoy DA, Novick RP, Rood JI, eds. Gram positive pathogens. Washington: ASM Press, 2000:540-50.
7. Kennedy CL, Krejany EO, Young LF, et al. The alpha-toxin of Clostridium septicum is essential for virulence. Mol Microbiol 2005; 57:1357-66; PMID:16102005; http://dx.doi.org/10.1111/j.1365-2958.2005.04774.x.
8. Awad MM, Ellemor DM, Boyd RL, et al. Synergistic effects of alpha-toxin and perfringolysin O in Clostridium perfringens-mediated gas gangrene. Infect Immun 2001; 69:7904-10; PMID:11705975; http://dx.doi.org/10.1128/IAI.69.12.7904-7910.2001.
9. Borriello SP. Pathogenesis of Clostridium difficile infection. J Antimicrob Chemother 1998; 41(Suppl C):13-9; PMID:9630370; http://dx.doi.org/10.1093/jac/41.suppl_3.13.
10. Warny M, Pepin J, Fang A, et al. Toxin production by an emerging strain of Clostridium difficile associated with outbreaks of severe disease in North America and Europe. Lancet 2005; 366:1079-84; PMID:16182895; http://dx.doi.org/10.1016/S0140-6736(05)67420-X.
11. McDonald LC, Killgore GE, Thompson A, et al. An epidemic, toxin gene-variant strain of Clostridium difficile. N Engl J Med 2005; 353:2433-41; PMID:16322603; http://dx.doi.org/10.1056/NEJMoa051590.
12. McClane BA. Clostridium perfringens enterotoxin and intestinal tight junctions. Trends Microbiol 2000; 8:145-6; PMID:10754565; http://dx.doi.org/10.1016/S0966-842X(00)01724-8.
13. Uzal FA, McClane BA. Recent progress in understanding the pathogenesis of Clostridium perfringens type C infections. Vet Microbiol 2011; 153:37-43; PMID:21420802; http://dx.doi.org/10.1016/j.vetmic.2011.02.048.
14. Uzal FA, Fisher DJ, Saputo J, et al. Ulcerative enterocolitis in two goats associated with enterotoxin- and beta2 toxin-positive Clostridium perfringens type D. J Vet Diagn Invest 2008; 20:668-72; PMID:18776108; http://dx.doi.org/10.1177/104063870802000526.

15. Van Immerseel F, Rood JI, Moore RJ, Titball RW. Rethinking our understanding of the pathogenesis of necrotic enteritis in chickens. Trends Microbiol 2009; 17:32-6; PMID:18977143; http://dx.doi.org/10.1016/j.tim.2008.09.005.
16. Rood JI. Antibiotic resistance determinants of Clostridium perfringens. In: Sebald M, ed. Genetics and molecular biology of anaerobic bacteria. New York: Springer-Verlag, 1993:141-55.
17. Rood JI, Maher EA, Somers EB, et al. Isolation and characterization of multiply antibiotic-resistant Clostridum perfringens strains from porcine feces. Antimicrob Agents Chemother 1978; 13:871-80; PMID:208463.
18. Lyras D, Rood JI. Transposons and antibiotic resistance determinants from Clostridium perfringens and Clostridium difficile. In: Rood J, McClane B, Songer J, Titball R, eds. The clostridia: molcular biology and pathogenesis. London: Academic Press, 1997:73-92.
19. Adams V, Lyras D, Farrow KA, Rood JI. The clostridial mobilisable transposons. Cell Mol Life Sci 2002; 59:2033-43; PMID:12568329; http://dx.doi.org/10.1007/s000180200003.
20. Drudy D, Kyne L, O'Mahony R, Fanning S. gyrA mutations in fluoroquinolone-resistant Clostridium difficile PCR-027. Emerg Infect Dis 2007; 13:504-5; PMID:17552115; http://dx.doi.org/10.3201/eid1303.060771.
21. O'Connor JR, Johnson S, Gerding DN. Clostridium difficile infection caused by the epidemic BI/NAP1/027 strain. Gastroenterology 2009; 136:1913-24; PMID:19457419; http://dx.doi.org/10.1053/j.gastro.2009.02.073.
22. Curry SR, Marsh JW, Shutt KA, et al. High frequency of rifampin resistance identified in an epidemic Clostridium difficile clone from a large teaching hospital. Clin Infect Dis 2009; 48:425-9; PMID:19140738; http://dx.doi.org/10.1086/596315.
23. Bannam TL, Teng WL, Bulach D, et al. Functional identification of conjugation and replication regions of the tetracycline resistance plasmid pCW3 from Clostridium perfringens. J Bacteriol 2006; 188:4942-51; PMID:16788202; http://dx.doi.org/10.1128/JB.00298-06.
24. Sloan J, McMurry LM, Lyras D, et al. The Clostridium perfringens TetP determinant comprises two overlapping genes: tetA(P), which mediates active tetracycline efflux, and tetB(P), which is related to the ribosomal protection family of tetracycline-resistance determinants. Mol Microbiol 1994; 11:403-15; PMID:8170402; http://dx.doi.org/10.1111/j.1365-2958.1994.tb00320.x.
25. Lyras D, Rood JI. Genetic organization and distribution of tetracycline resistance determinants in Clostridium perfringens. Antimicrob Agents Chemother 1996; 40:2500-4; PMID:8913453.
26. Roberts AP, Johanesen PA, Lyras D, et al. Comparison of Tn5397 from Clostridium difficile, Tn916 from Enterococcus faecalis and the CW459tet(M) element from Clostridium perfringens shows that they have similar conjugation regions but different insertion and excision modules. Microbiology 2001; 147:1243-51; PMID:11320127.
27. Berryman DI, Rood JI. The closely related ermB-ermAM genes from Clostridium perfringens, Enterococcus faecalis (pAM beta 1), and Streptococcus agalactiae (pIP501) are flanked by variants of a directly repeated sequence. Antimicrob Agents Chemother 1995; 39:1830-4; PMID:7486927.
28. Berryman DI, Lyristis M, Rood JI. Cloning and sequence analysis of ermQ, the predominant macrolide-lincosamide-streptogramin B resistance gene in Clostridium perfringens. Antimicrob Agents Chemother 1994; 38:1041-6; PMID:8067735.
29. Farrow KA, Lyras D, Rood JI. The macrolide-lincosamide-streptogramin B resistance determinant from Clostridium difficile 630 contains two erm(B) genes. Antimicrob Agents Chemother 2000; 44:411-3; PMID:10639372; http://dx.doi.org/10.1128/AAC.44.2.411-413.2000.
30. Farrow KA, Lyras D, Rood JI. Genomic analysis of the erythromycin resistance element Tn5398 from Clostridium difficile. Microbiology 2001; 147:2717-28; PMID:11577151.
31. Spigaglia P, Mastrantonio P. Analysis of macrolide-lincosamide-streptogramin B (MLS(B)) resistance determinant in strains of Clostridium difficile. Microb Drug Resist 2002; 8:45-53; PMID:12002649; http://dx.doi.org/10.1089/10766290252913755.
32. Spízek J, Novotná J, Rezanka T. Lincosamides: chemical structure, biosynthesis, mechanism of action, resistance, and applications. Adv Appl Microbiol 2004; 56:121-54; PMID:15566978; http://dx.doi.org/10.1016/S0065-2164(04)56004-5.
33. Martel A, Devriese LA, Cauwerts K, et al. Susceptibility of Clostridium perfringens strains from broiler chickens to antibiotics and anticoccidials. Avian Pathol 2004; 33:3-7; PMID:14681061; http://dx.doi.org/10.1080/0307945031000163291.
34. Lyras D, Adams V, Ballard SA, et al. tISCpe8, an IS1595-family lincomycin resistance element located on a conjugative plasmid in Clostridium perfringens. J Bacteriol 2009; 191:6345-51; PMID:19684139; http://dx.doi.org/10.1128/JB.00668-09.

35. Bannam TL, Crellin PK, Rood JI. Molecular genetics of the chloramphenicol-resistance transposon Tn4451 from Clostridium perfringens: the TnpX site-specific recombinase excises a circular transposon molecule. Mol Microbiol 1995; 16:535-51; PMID:7565113; http://dx.doi.org/10.1111/j.1365-2958.1995. tb02417.x.

36. Lyras D, Storie C, Huggins AS, et al. Chloramphenicol resistance in Clostridium difficile is encoded on Tn4453 transposons that are closely related to Tn4451 from Clostridium perfringens. Antimicrob Agents Chemother 1998; 42:1563-7; PMID:9660983.

37. Bannam TL, Rood JI. Relationship between the Clostridium perfringens catQ gene product and chloramphenicol acetyltransferases from other bacteria. Antimicrob Agents Chemother 1991; 35:471-6; PMID:2039197.

38. de la Cruz F, Davies J. Horizontal gene transfer and the origin of species: lessons from bacteria. Trends Microbiol 2000; 8:128-33; PMID:10707066; http://dx.doi.org/10.1016/S0966-842X(00)01703-0.

39. Wozniak RA, Waldor MK. Integrative and conjugative elements: mosaic mobile genetic elements enabling dynamic lateral gene flow. Nat Rev Microbiol 2010; 8:552-63; PMID:20601965; http://dx.doi. org/10.1038/nrmicro2382.

40. Brefort G, Magot M, Ionesco H, Sebald M. Characterization and transferability of Clostridium perfringens plasmids. Plasmid 1977; 1:52-66; PMID:220651; http://dx.doi.org/10.1016/0147-619X(77)90008-7.

41. Abraham LJ, Rood JI. Molecular analysis of transferable tetracycline resistance plasmids from Clostridium perfringens. J Bacteriol 1985; 161:636-40; PMID:2857166.

42. Abraham LJ, Rood JI. Identification of Tn4451 and Tn4452, chloramphenicol resistance transposons from Clostridium perfringens. J Bacteriol 1987; 169:1579-84; PMID:2881919.

43. Rood JI, Jefferson S, Bannam TL, et al. Hybridization analysis of three chloramphenicol resistance determinants from Clostridium perfringens and Clostridium difficile. Antimicrob Agents Chemother 1989; 33:1569-74; PMID:2554801.

44. Wren BW, Mullany P, Clayton C, Tabaqchali S. Molecular cloning and genetic analysis of a chloramphenicol acetyltransferase determinant from Clostridium difficile. Antimicrob Agents Chemother 1988; 32:1213-7; PMID:2847649.

45. Lyras D, Rood JI. Transposition of Tn4451 and Tn4453 involves a circular intermediate that forms a promoter for the large resolvase, TnpX. Mol Microbiol 2000; 38:588-601; PMID:11069682; http:// dx.doi.org/10.1046/j.1365-2958.2000.02154.x.

46. Crellin PK, Rood JI. The resolvase/invertase domain of the site-specific recombinase TnpX is functional and recognizes a target sequence that resembles the junction of the circular form of the Clostridium perfringens transposon Tn4451. J Bacteriol 1997; 179:5148-56; PMID:9260958.

47. Crellin PK, Rood JI. Tn4451 from Clostridium perfringens is a mobilizable transposon that encodes the functional Mob protein, TnpZ. Mol Microbiol 1998; 27:631-42; PMID:9489674; http://dx.doi. org/10.1046/j.1365-2958.1998.00712.x.

48. Lyras D, Adams V, Lucet I, Rood JI. The large resolvase TnpX is the only transposon-encoded protein required for transposition of the Tn4451/3 family of integrative mobilizable elements. Mol Microbiol 2004; 51:1787-800; PMID:15009902; http://dx.doi.org/10.1111/j.1365-2958.2003.03950.x.

49. Abraham LJ, Rood JI. The Clostridium perfringens chloramphenicol resistance transposon Tn4451 excises precisely in Escherichia coli. Plasmid 1988; 19:164-8; PMID:2901770; http://dx.doi.org/10.1016/0147-619X(88)90055-8.

50. Sherrat D. Tn3 and related transposable elements: site-specific recombination and transposition. In: Berg DE, Howe MM, eds. Mobile DNA. Washington: American Society for Microbiology, 1989:163-84.

51. Smith MC, Thorpe HM. Diversity in the serine recombinases. Mol Microbiol 2002; 44:299-307; PMID:11972771; http://dx.doi.org/10.1046/j.1365-2958.2002.02891.x.

52. Adams V, Lucet IS, Tynan FE, et al. Two distinct regions of the large serine recombinase TnpX are required for DNA binding and biological function. Mol Microbiol 2006; 60:591-601; PMID:16629663; http://dx.doi.org/10.1111/j.1365-2958.2006.05120.x.

53. Adams V, Lucet IS, Lyras D, Rood JI. DNA binding properties of TnpX indicate that different synapses are formed in the excision and integration of the Tn4451 family. Mol Microbiol 2004; 53:1195-207; PMID:15306021; http://dx.doi.org/10.1111/j.1365-2958.2004.04198.x.

54. Lucet IS, Tynan FE, Adams V, et al. Identification of the structural and functional domains of the large serine recombinase TnpX from Clostridium perfringens. J Biol Chem 2005; 280:2503-11; PMID:15542858; http://dx.doi.org/10.1074/jbc.M409702200.

55. Francia MV, Varsaki A, Garcillán-Barcia MP, et al. A classification scheme for mobilization regions of bacterial plasmids. FEMS Microbiol Rev 2004; 28:79-100; PMID:14975531; http://dx.doi. org/10.1016/j.femsre.2003.09.001.

56. Garcillán-Barcia MP, Francia MV, de la Cruz F. The diversity of conjugative relaxases and its application in plasmid classification. FEMS Microbiol Rev 2009; 33:657-87; PMID:19396961; http://dx.doi. org/10.1111/j.1574-6976.2009.00168.x.
57. Galimand M, Gerbaud G, Guibourdenche M, et al. High-level chloramphenicol resistance in Neisseria meningitidis. N Engl J Med 1998; 339:868-74; PMID:9744970; http://dx.doi.org/10.1056/ NEJM199809243391302.
58. Shultz TR, Tapsall JW, White PA, et al. Chloramphenicol-resistant Neisseria meningitidis containing catP isolated in Australia. J Antimicrob Chemother 2003; 52:856-9; PMID:14563894; http://dx.doi. org/10.1093/jac/dkg452.
59. Siguier P, Gagnevin L, Chandler M. The new IS1595 family, its relation to IS1 and the frontier between insertion sequences and transposons. Res Microbiol 2009; 160:232-41; PMID:19286454; http://dx.doi. org/10.1016/j.resmic.2009.02.003.
60. Wüst J, Hardegger U. Transferable resistance to clindamycin, erythromycin, and tetracycline in Clostridium difficile. Antimicrob Agents Chemother 1983; 23:784-6; PMID:6870225.
61. Hächler H, Berger-Bächi B, Kayser FH. Genetic characterization of a Clostridium difficile erythromycin-clindamycin resistance determinant that is transferable to Staphylococcus aureus. Antimicrob Agents Chemother 1987; 31:1039-45; PMID:2821888.
62. Mullany P, Wilks M, Tabaqchali S. Transfer of macrolide-lincosamide-streptogramin B (MLS) resistance in Clostridium difficile is linked to a gene homologous with toxin A and is mediated by a conjugative transposon, Tn5398. J Antimicrob Chemother 1995; 35:305-15; PMID:7759394; http://dx.doi. org/10.1093/jac/35.2.305.
63. Brouwer MS, Warburton PJ, Roberts AP, et al. Genetic organisation, mobility and predicted functions of genes on integrated, mobile genetic elements in sequenced strains of Clostridium difficile. PLoS One 2011; 6:e23014; PMID:21876735; http://dx.doi.org/10.1371/journal.pone.0023014.
64. Achard A, Villers C, Pichereau V, Leclercq R. New lnu(C) gene conferring resistance to lincomycin by nucleotidylation in Streptococcus agalactiae UCN36. Antimicrob Agents Chemother 2005; 49:2716-9; PMID:15980341; http://dx.doi.org/10.1128/AAC.49.7.2716-2719.2005.

CHAPTER 8

pSAM2, a Paradigm for a Family of Actinomycete Integrative and Conjugative Elements

Emilie Esnault, Alain Raynal and Jean-Luc Pernodet*

Abstract

PSAM2 is a self-transmissible integrative and replicative mobile genetic element isolated from *Streptomyces ambofaciens*. It belongs to the large family of Actinomycete integrative and conjugative elements (AICEs). These elements have the ability to mobilize chromosomal DNA regions and thus play a role in horizontal gene transfer in this group of bacteria. As *Streptomyces* and other related actinobacteria are important producers of secondary metabolites of medical interest, these AICEs could play a role in the flux of secondary metabolites biosynthetic or resistance genes. pSAM2 is among the best characterized AICEs, as its excision/integration, replication and transfer have been characterized. In particular, pSAM2 transfer, as the one of other *Streptomyces* mobile genetic elements, involves double-stranded DNA, contrarily to the classical conjugation of single-stranded DNA molecules. A bioinformatics analysis allowed the identification of pSAM2-like elements in several suborders of *Actinomycetales*.

Introduction

Prokaryotic genomes are highly dynamic, due to internal rearrangements and horizontal gene transfer (HGT). One of the important mechanisms for HGT is conjugation. Two kinds of mobile elements are able to conjugate, plasmids and ICEs (Integrative Conjugative Elements). The term integrating and conjugative element was used for the first time by Hagège et al.[1] to describe pSAM2, a *Streptomyces ambofaciens* conjugative element that is integrated in the chromosome and excises and replicates before conjugation. Then Burrus et al.[2] proposed to extend the term applied to pSAM2 to all mobile genetic elements that excise by site-specific recombination into a circular form, self-transfer by conjugation and integrate into the host genome, whatever the specificity and the mechanism of integration and conjugation could be.

A subgroup of ICE, termed AICE (Actinomycete Integrative and Conjugative Elements) was defined by te Poele et al.[3] They represent a special class of ICEs, not only because they are found in *Actinomycetales* but also because, unlike most other ICEs, they have the ability to replicate autonomously like plasmids. *Actinomycetales* are important producers of bioactive secondary metabolites.[4] Because AICEs are able to mobilize non-conjugative plasmids and chromosomal regions, they could play an important role in horizontal transfer. In particular they could be involved in the transfer of secondary metabolite biosynthetic or resistance genes, thus being a driving force in the evolution of secondary metabolism.[5] Like other ICEs, AICEs have a modular organization.[6] Each module includes all the genes involved in a given biological function and

*Université Paris-Sud, CNRS, UMR8621, Institut de Génétique et Microbiologie, Orsay, France.
Corresponding Author: Jean-Luc Pernodet—Email: jean-luc.pernodet@igmors.u-psud.fr

Bacterial Integrative Mobile Genetic Elements, edited by Adam P. Roberts and Peter Mullany.
©2013 Landes Bioscience.

modules are combined to form a functional backbone.[7] The backbone of AICEs is composed of four modules involved in site-specific recombination, replication, conjugation and regulation. Several AICEs have been described, but few of them have been functionally characterized.

In this chapter, we present the knowledge available on the different functions of pSAM2, discuss the distribution of pSAM2-like elements in sequenced genomes and describe genetic tools derived from pSAM2.

Overview of pSAM2 Functions

pSAM2 was first isolated as a covalently closed circular DNA molecule when the plasmid contents of five *Streptomyces ambofaciens* strains were investigated.[8] These five strains corresponded to two different geographic isolates (strains ATCC23877 and DSM40697), and to three mutant derivatives of the wild type strain ATCC23877 (strains ATCC15154, JI3212 and RP181110). A large circular plasmid, pSAM1 (80 kb), previously described[9,10] was found in ATCC23877 and all its derivative. This plasmid is absent from strain DSM40697. A second type of covalently closed circular DNA molecule, called pSAM2 (11 kb), was isolated from the strains JI3212 and RP181110, but was absent from the parental strain ATCC23877 and from the ATCC15154 derivative. However, when pSAM2 was used as a probe in hybridization experiments with total genomic DNA from all five strains, it became clear that a copy of pSAM2 was integrated in the chromosome of the strain ATCC23877 and of all its derivatives (ATCC15154, JI3212, RP181110) and that this integrated form coexisted with the free circular form in strains JI3212 and RP181110. In all four strains, pSAM2 was integrated in the same way at the same locus. No pSAM2 sequence could be detected in strain DSM40697, either free or integrated. The *S. ambofaciens* strains harboring pSAM2 were mated with *Streptomyces lividans*. Most of the *S. lividans* exconjugants contained pSAM2. The status of pSAM2 (integrated only or free and integrated) in the *S. lividans* recipient was the same as in the donor strain, demonstrating that this was determined by pSAM2 and not by the host strain. The integration of pSAM2 in *S. lividans* occurred at a specific chromosomal location. The *S. lividans* strains which had acquired pSAM2 through mating could behave as donors in subsequent mating, indicating that all the genes required for transfer were carried by pSAM2 or highly conserved in *Streptomyces*. When protoplasts of *S. lividans* or other *Streptomyces* species were transformed with the covalently closed circular form of pSAM2, the resulting transformants contained pSAM2 integrated at a specific locus and free in the cytoplasm, showing that the free form had the ability to integrate. From these observations, pSAM2 appeared as a mobile genetic element able to integrate at a specific locus in the genome of its host, to exist also in the free circular state and to transfer from its host to a new one during mating. Further studies combining functional genetics with the analysis of pSAM2 sequence led to the identification of the genes involved in the different function of pSAM2 and in the regulation of their expression. As observed for other mobile genetic elements, these genes are organized in modules (Fig. 1). From these studies, the picture that emerged is that pSAM2 has a prophage-like mode of maintenance, being integrated into the host chromosome and replicated along with it. In response to yet uncharacterized signal(s) linked to the contact of the mycelium of the donor strain, harboring pSAM2, with the mycelium of a recipient strain, devoid of pSAM2, pSAM2 is able to excise, to replicate into the donor mycelium, to transfer, replicate and spread into the recipient mycelium and finally to integrate into a specific site of the recipient chromosome. A schematic representation of these different steps, necessary for pSAM2 transfer, is presented in Figure 2. The genes and proteins involved in the various functions excision/integration, replication, transfer and regulation are described in detail below.

Site-Specific Excision/Integration

The excision/integration system of pSAM2, and more generally of AICEs, is similar to the site-specific recombination system of temperate phages like λ. Functional analysis showed that a region of about 2.5 kb was sufficient for the integration into the chromosome at a specific locus.[11-13] This region contains the genes *xis* and *int* encoding respectively an excisionase (Xis) and a site-specific recombinase of the integrase family (Int). These proteins are similar to the ones involved

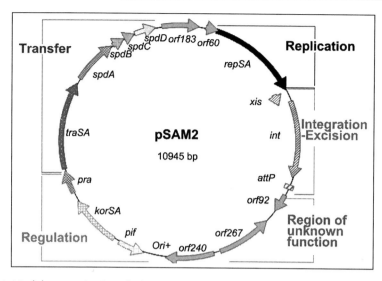

Figure 1. Modular organization of pSAM2. Genes are represented by arrows. The filing of the arrows indicates the function in which a gene is involved: dots for transfer, grid for regulation, black for replication, stripes for site-specific recombination. Genes with unknown function are represented in gray.

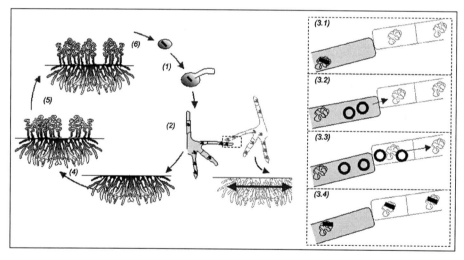

Figure 2. pSAM2: mode and kinetic of transfer in *Streptomyces*. *Streptomyces* are soil bacteria that display a complex morphological differentiation. 1) In favorable condition the spore, containing one copy of the genome germinates. 2) The mycelium grows by tip elongation into the substrate and form the vegetative mycelium which consists of septate, multinucleoid branched hyphae. 3) At this stage, physical contact between potential partners could occur randomly during natural growth. 3.1) Before mating, pSAM2 is only present in the donor strain in its integrated form. 3.2) The presence of a recipient strain, devoid of pSAM2, leads to the excision of pSAM2 in the donor cell and to the initiation of transfer. 3.3) After 9 h of mating, only pSAM2 free forms are present in the donor and in the recipient. pSAM2 spreads in the recipient mycelium. 3.4) After 48h, the transfer is finished, pSAM2 is found integrated in donor and recipient strains.[25] 4) Upon nutrient starvation, hair-like aerial hyphea emerge and differentiate into chains of unigeomic spores. 5,6) Resulting spores will all possess an integrated copy of pSAM2.

in site-specific recombination in some temperate phages. The two genes are cotranscribed and the stop codon of *xis* overlaps the start codon of *int*.[14] Int is sufficient to promote the integration in the host chromosome. Int and Xis are both required for the excision of the integrated form of pSAM2 (Fig. 3). The pSAM2 integrase belongs to the sub-family of tyrosine integrases that utilize a catalytic tyrosine to mediate strand cleavage. An efficient pSAM2 site-specific recombination system was reconstituted in *Escherichia coli* expressing Int or Xis and Int and harboring the sequences required for recombination.[15] This suggests that recombination does not require other proteins encoded by pSAM2 or the host bacteria in addition to Xis and Int.

To identify precisely the sites involved in pSAM2 integration, the sequences of these sites on the circular form of pSAM2 and on the chromosome of two different strains, *S. ambofaciens* and *S. lividans*, before and after pSAM2 integration were analyzed and compared. This led to the identification of a common region of 58 base pairs, called the identity segment, in which recombination takes place, leading to the integration/excision of pSAM2.[16] The site involved in recombination were respectively called *attP* on pSAM2 and *attB* on the chromosome; after

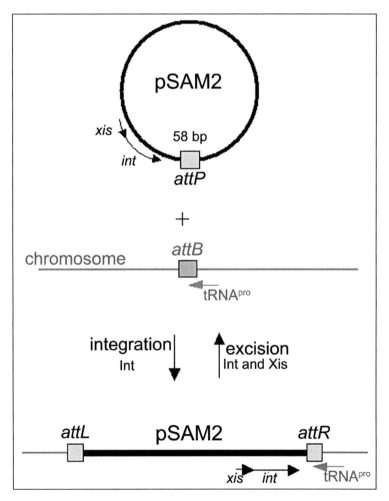

Figure 3. pSAM2 integration excision mechanism. pSAM2 integrates and excises by recombination between specific sites (*attP* and *attB* for integration, *attL* and *attR* for excision). Gray squares represent the 58-bp identity segment shared by the *att* sites.

integration pSAM2 is flanked by the *attL* and *attR* sites, as represented in Figure 3. In the chromosomal *attB* site, the 58 bp identity segment overlaps the 3' end of a tRNA[Pro] gene over 42 bp, extending from the region encoding the anticodon loop after the one encoding the 3' part of the acceptor stem (Fig. 4A).[14,17] Thus, the identity segment allows the reconstitution of a functional tRNA[Pro] gene after pSAM2 integration. The sequence downstream of the gene is modified but it should be noticed that the inverted repeats following the tRNA gene in the chromosome before integration are replaced by other inverted repeats after integration, replacing a probable terminator structure by another one. The pSAM2 site-specific recombination system reconstituted in *Escherichia coli* was used to study in detail the *att* sequences. This allowed defining the minimal *attB* site as a 26-bp sequence corresponding to the region encoding the anticodon stem loop.[15] The sequence of *attB* suggests that 9-bp imperfect inverted repeats, separated by 5 bp constitute the *attB* site, as is the case for the bacteriophage λ (Fig. 4B). The *attP* site was also characterized in detail. Functional studies defined the minimal *attP* site as a region of about 300 bp centered on the identity segment. Foot-print experiments with purified integrase identified two Int-binding sites in *attP*. These sites consist 17nt direct repeats with a single mismatch (5' GTCACGCAG(A/T) TAGACAC 3') and are located approximately 100 bp upstream and downstream of the core site where recombination takes place (Fig. 4C).[18] Two direct repeats corresponding to the first 9 nucleotides of the Int binding sequence are present in the first third of the *int* gene. Foot-print experiments showed that the integrase could bind in vitro to these sequences (Raynal, unpublished). These sequences were modified by site-directed mutagenesis without changing the Int protein sequence. The binding of Int in vitro on these mutated sequences was reduced. The integration efficiency of pSAM2 derivatives containing the mutant form of the *int* gene was lower than that of the wild-type (Raynal, unpublished). These observations might indicate that the integrase has the capacity to bind to its own gene to modulate its expression. This potential regulatory mechanism has not been described for other integrase genes.

Replication

The study of pSAM2 mode of replication showed that it replicates by a rolling-circle replication (RCR) mechanism, generating strand specific single-stranded DNA intermediates.[19] The direction of replication was determined and, although replication is only transient for pSAM2, most of the genes are transcribed in the same direction as leading strand replication, as observed for RCR plasmids.

In RCR, the replication initiator protein, or replicase, generates a site- and strand-specific nick at the leading-or double-strand origin (*ori* or *dso*).[20] Functional analysis established that two regions of pSAM2 were involved in replication.[21] One of these regions contains the gene *repSA* encoding the replicase.[1] Conserved domains involved in the recognition of the *dso* sequence and in the generation of the nick have been identified in replicases and these domains were used to classify replicases.[20,22] Tyrosine residues present in one conserved domain are involved in the nicking activity. On the basis of this classification, RepSA, which possesses two (putative) active tyrosine residues appears more similar to some phage replicases than to the ones of RCR Streptomyces plasmids.[1] In the other region of pSAM2 required for replication, a fragment of 580 bp was shown to act as a replication origin. More precisely a region of about 400 bp with no identified gene but numerous short direct repeats, and where a long hairpin secondary structure can be predicted, is proposed to contain the *dso*. In this region, a putative nick site was identified by similarity with the nick site of plasmids of the pC194 family. In RCR plasmids the *dso* is located within or in the immediate vicinity of the replicase gene. This might allow interference of the replicase-*dso* interaction with the expression of the replicase gene, which could play a role in the regulation of replication and control of the copy number. This is not the case for pSAM2 where the *dso* lies about 4 kb from the *repSA* gene. A minimal replicon containing *repSA* and *dso* was shown to replicate autonomously in *S. lividans*, confirming their involvement and role in replication.[1]

Replication is not required for the maintenance of pSAM2, which is most of the time integrated in the chromosome, but is required during transfer. Replication and transfer are linked as shown

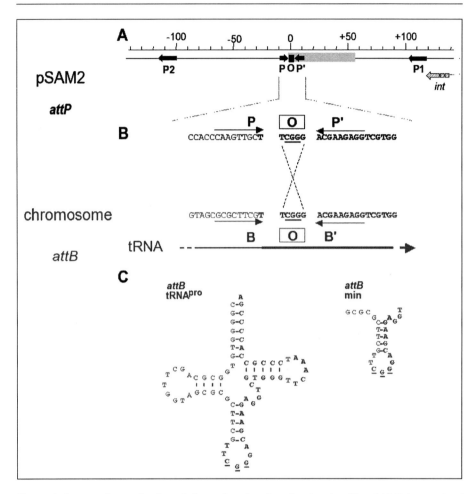

Figure 4. Structural organization of the *attP* and *attB* region involved in pSAM2 integration/excision. A) Schematic representation of the region including the *attP* site downstream the *int* gene. The central nucleotide of the potential crossover region (O) was chosen as position 0. The *int* arrow represents the 3' end of the gene. The position and the relative orientation of the inverted repeats P and P' and the arm-type repeats P1 and P2 are indicated by black arrows. The gray bar represents the 58 bp identity segment. B) DNA sequences of the POP' and BOB' regions. Thin arrows represent the 9-nt-long imperfect inverted repeats. Nucleotides belonging to the identity segment are in bold. The sequence of anticodon is underlined. C) Cloverleaf representation of the putative Pro-tRNA gene overlapping the pSAM2 *attB* site in *Streptomyces ambofaciens* and of the minimal functional *attB* site.

by the fact that no transfer is observed for pSAM2 derivatives with a mutation in the *repSA* gene or in the origin of replication.[1,21] It should be noted that the lack of replication should affect dramatically the detection of transfer as no multiplication of pSAM2 occurs in the recipient strain and thus no spreading. One possible hypothesis explaining the importance of replication during transfer could be the necessity of a high pSAM2 copy number to increase the probability of interaction between pSAM2 molecules and the conjugative apparatus. No mechanism controlling pSAM2 copy number was identified, whereas negative regulation of replication is often present in plasmids. It should also be noted that the *repSA* gene is cotranscribed with *xis* and *int*, the genes required for excision/integration.[1,23] Thus the replication is linked to transfer, the first step of which is excision and the co-transcription of the genes involved in these two functions allows the co-regulation of their expression. All these observations suggest that, as soon as the free form of pSAM2 is present, there will be a burst of replication, leading to a high pSAM2 copy number, allowing maximum transfer efficiency.

Transfer

pSAM2 transfers horizontally by conjugation. The mechanism of pSAM2 conjugation is similar to that of *Streptomyces* conjugative plasmid but very different from classical conjugation via single stranded DNA. Conjugation requires two successive events: mating pair formation, i.e., the formation of a complex were the donor and recipient cells connect physically, and then entry of the DNA into recipient cell. In classical conjugation, the first step involves specific structures like pili or filaments that facilitate cell contact.[24] A complex apparatus is then required for transferring DNA as single stranded DNA molecules generated by a relaxase. For pSAM2, AICEs and *Streptomyces* conjugative plasmids, the conjugative apparatus is simpler and double stranded DNA is transferred. The physical contact between potential partners occurs probably randomly during growth as no gene seems to be required to promote the formation of specific mating structure. A few genes are sufficient for the transfer of double stranded DNA molecules and their spreading all along the recipient mycelium.[25-28] Like the conjugation of *Streptomyces* plasmids, the conjugation of pSAM2 induces a delay in the development of the newly infected mycelium. This is visible on plate by the formation of a circular growth retardation zone called pock or lethal zygosis reaction (Fig. 5).[29]

Functional analysis of pSAM2 allowed the identification of a unique 2.8 kb region involved in transfer and pock formation. This region contains *traSA* and the four *spd* genes allowing the intermycelial and the intramycelial transfer.[19,21]

The Main Transfer Protein TraSA

pSAM2 derivatives carrying a deletion of the *traSA* gene are unable to form pocks and to transfer.[21] Although smaller, TraSA is similar to other *Streptomyces* plasmid transfer proteins and more generally to double-stranded DNA translocases.[30] These DNA translocases possess a transmembrane domain (TMD) in their N-terminal part, an FtsK/SpoIIIE domain in their central part, followed by a winged-Helix-Turn-Helix (wHTH) fold region (Fig. 5B). Sequence analysis indicates that TraSA could be associated with the membrane through a TMD positioned in its N-terminal part. An FtsK/SpoIIIE domain is also predicted in TraSA. This domain functions like a pump which actively transports double-stranded DNA. Thus, FtsK of *E. coli* segregates chromosomes into the daughter cell and SpoIIIE of *B. subtilis* pumps the DNA into the prespore.[31] The presence of this domain suggested that double-stranded DNA could be transferred during *Streptomyces* conjugation. Indeed, Possoz et al. demonstrated the intracellular transfer of unprocessed double-stranded pSAM2 molecules during conjugation, using a system based on differential restriction-modification sensitivity of single- and double-stranded DNA.[25] The conjugal transfer of double-stranded DNA seems to be shared by *Streptomyces* circular conjugative plasmid.[26-28] Recent biochemical experiments showed that TraB$_{pSVH1}$ translocase from *Streptomyces venezuelae* plasmid assembles into an hexameric ring structure with a central pore large enough to translocate double-stranded DNA, and forms pores in artificial membranes.[32]

The wHTH region from chromosomal or conjugative plasmid translocases is involved in sequence-specific DNA binding.[32] This non-covalent binding allows the directional pumping of the DNA molecule. The sequence recognized by the wHTH domain is called KOPS for FtsK and the cis-acting locus for transfer (*clt*) for conjugative plasmid and consists of 8 bp repeated sequences.[33,34] The means by which TraSA recognizes and binds to DNA have not yet been elucidated, as no wHTH domain is predicted in TraSA and as the *clt* from pSAM2 has not been identified.

Concerning TraSA expression, little is known. It has been demonstrated that *traSA* is not cotranscribed with the upstream gene *pra* encoding a regulator and that the identified regulators (Pra and KorSA) do not directly control *traSA* expression.[35]

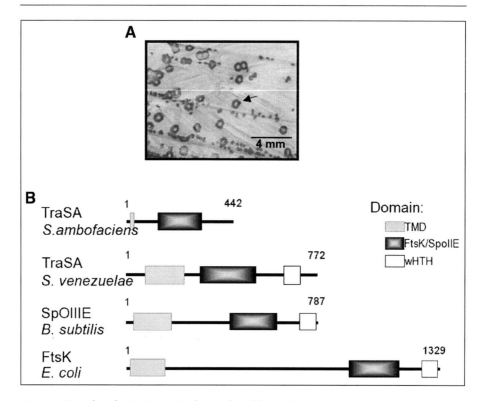

Figure 5. Transfer of pSAM2. A) Pocks produced by pSAM2-containing clones on a lawn of plasmid-free *Streptomyces lividans* mycelium. The arrow indicates a pock, a circular zone of growth inhibition. The growth inhibition is accompanied by the induction of the synthesis of the red pigmented antibiotic actinorhodine. B) Comparison of TraSA, the major transfer protein of pSAM2, with the transfer protein TraB from pSVH1 and ds DNA translocases. Schematic representation of TraSA, TraB from pSVH1, SpoIIIE and FtsK, with their predicted conserved domains (TMD: TransMembrane Domain; FtsK/SpoIIIE domain; wHTH: wigged Helix-Turn-Helix). TraSA domains were predicted using TMPred and HMMPfam.

One of these translocases, TraB from the conjugative plasmid pSG5 from *Streptomyces ghanaensis*, was expressed under the control of its native promoter as a fusion protein with eGFP. The fused protein was found to be localized at the hyphal tips of *Streptomyces lividans*. This result strongly suggests that conjugation takes place at the tips of the mating mycelium.[28]

Spread Proteins

Mutations in a region containing four genes and localized downstream *traSA* affect transfer efficiency and pock morphology. As the reduced pock size could correspond to reduced intramycelial transfer,[36] these genes were designed as spread genes and named *spdA, spdB, spdC* and *spdD*.[30] *spd* genes of pSAM2 have an organization similar to other *Steptomyces* conjugative non-integrative plasmid. These are read in the same direction, separated by only few nucleotides and since no significant hairpin structures were found downstream of them, all these genes could be cotranscribed. Moreover, except for SpdB, all these proteins are predicted to be transmembrane proteins.[30] Their precise role is not known. SpdA is homologous to SpdB2 from pSVHI, a transmembrane protein involved in DNA binding and interacting with the major transfer protein TraB.[37]

Mobilization

During conjugation, pSAM2 promotes not only its own transfer, but also the transfer of non-conjugative plasmid and chromosomal segments. The presence of pSAM2 in *S. ambofaciens* increased the frequency of chromosomal locus transfer by a factor 100 when present in one of the parents of a cross.[38] It also increased by a factor 100 to 10000 the frequency of transfer of non-conjugative plasmid cloning vectors (Esnault, unpublished). This ability to increase genetic exchange is common to others AICEs and *Streptomyces* conjugative plasmids.

Regulation

In pSAM2, a regulatory mechanism enables the transient activation of replication and site-specific integration/excision functions. A cascade of two proteins participate in this regulation: Pra activates the replication, excision and integration of pSAM2[23,35] and KorSA represses the transcription of *pra* in the absence of transfer.[39] A third protein, Pif (pSAM2 immunity factor), plays an important role and could be considered as an identity marker because it participates in the donor-recipient recognition.

Pra, a pSAM2 Specific Regulatory Protein

In the absence of transfer, the wild type pSAM2 is only integrated, but pSAM2 mutant (called pSAM2$_{B3}$ and pSAM2$_{B4}$) were found simultaneously integrated and free (five to 10 copies per genome).[8] The mutations responsible for the co-existence of the free forms of pSAM2$_{B3}$ and pSAM2$_{B4}$ with the integrated one were localized in the promoter region of the gene *pra*.[35] These two independent point mutations increase the strength of the *pra* promoter and constitutive expression of Pra provoked the appearance of the free forms of the wild type pSAM2, in the absence of conjugation. As *pra* is not directly involved in replication (absent from the minimal replicon[1]), this indicates a regulatory role for Pra. Further analyses showed that Pra was not only involved in the activation of pSAM2 replication but also activates integration and excision.[23] The three genes *repSA*, *xis* and *int* constitute an operon directly or indirectly activated by Pra. Pra deficient mutants have lost the ability to replicate, transfer efficiently and to form pocks.

The exact mechanism by which Pra regulates the expression of these genes is still unknown. A MalE-Pra fusion protein produced in *E. coli* was not able to bind any pSAM2 DNA region and the search for proteins interacting with Pra using a bacterial two hybrid system did not yield any result so far (our unpublished data). No specific domain has been revealed by sequence analysis and no Pra homologs were found in any other exhaustively studied ICEs. Therefore, Pra seems to be specific to the pSAM2 regulation system and it could be the first member of a new family of regulatory proteins. Compared with plasmid regulation, where the copy number is often limited through a negative control of the replication, Pra behaves as a positive activator that transiently activates replication when the strain is engaged in conjugation. Thus, the control of replication in pSAM2 is quite different from that found in plasmids.

KorSA, the General Transcriptional Repressor

The gene *korSA* encodes a protein belonging to the GntR family of transcriptional repressors. KorSA was shown to bind to two regions in pSAM2: the promoter of its own gene and the *pra* gene promoter.[39] Its binding site (5′ TNACTCATNTANANGAG 3′) was identified by footprint analysis. There are two binding sites upstream of *pra* and one upstream of *korSA*. This could explain the higher affinity of KorSA for the *pra* promoter. It should be noted that the mutation in the *pra* promoter of pSAM2$_{B4}$ is located in one of the KorSA binding sites. This could decrease the binding of KorSA, and, together with the increased strength of the mutant promoter, participate in the increased expression of Pra. A pSAM2 derivative in which *korSA* is partially deleted is found only in the free form. The introduction of a copy of *korSA* in trans restored the presence of the integrated form. KorSA is constantly transcribed, even during conjugation. Activation of the excision and replication of pSAM2, which are required for transfer are under the control of the positive regulator Pra, the expression of which is repressed by KorSA. Thus, a transient release of KorSA repression is necessary to initiate the events leading to pSAM2 transfer. This mechanism is not yet identified. KorSA could be considered as the main negative transcriptional regulator of pSAM2. Its role is to prevent integration, excision and replication through the control of *pra* expression.

Transcriptional repressors of the GntR family are often found in *Streptomyces* mobile genetic elements, plasmids or AICEs. In these elements, the product of the *kor* genes represses the expression of the transfer genes whose expression is deleterious (*kil* genes). This is not the case for KorSA, which does not seem to repress *traSA*. Nevertheless, the presence of a pSAM2 derivative deleted for *korSA* was found to be deleterious for the host, suggesting the existence of an attenuated Kil phenotype.[39]

Pif, a Protein Conferring Immunity

pSAM2 transfer is initiated when a mycelium harboring pSAM2 is in close contact with a mycelium devoid of pSAM2. The appearance of pSAM2 free form is an indication of transfer initiation. When both mycelia in contact harbor pSAM2, the pSAM2 free form is not observed and transfer is reduced by at least three orders of magnitude.[25] Therefore, pSAM2 carries a mechanism that prevents redundant plasmid exchange or self-transfer. This mechanism involves the gene *pif* (pSAM2 *i*mmunity *f*actor). The presence of *pif* is sufficient to confer a status of donor to the host.[40] The protein Pif contains a Nudix hydrolase motif that is required for the immunity activity. Nudix proteins catalyze the hydrolysis of nucleoside diphosphates linked to other moieties. The role usually assigned to these proteins is to sanitize the nucleotide pools because they have the ability to degrade potentially mutagenic, oxidised nucleotides thereby preventing their incorporation during replication or transcription.[41] Nudix proteins have also been shown to modify enzyme activity by removing mononucleotide moieties from cofactor.[42] The mechanism by which Pif functions remains unknown but the inactivation of the Nudix motif by changing two amino acids decreases the immunity level by a factor 100 and provokes the constant presence of the free forms of pSAM2.[40] Moreover, a *S. lividans* strain harboring a pSAM2 *pif*-minus derivative exhibits a blood-red color and poor sporulation as the mycelium in the pocks area. This suggests that there is continuously transfer in this mycelium. No DNA-binding motif or transmembrane domain, which could suggest a role of Pif as a transcriptional repressor or as a bacterial surface protein, is predicted in Pif. Possoz et al. suggested that Pif prevents the entry of pSAM2 by modifying a host component in the donor strain, thus signaling the presence of pSAM2.[25]

Nudix hydrolase proteins are not specific to pSAM2, this protein family is commonly found in AICEs, but their role remains unknown.[43]

From these studies, a general view of pSAM2 regulation emerged. pSAM2 is maintained in the integrated state due to the constitutive expression of KorSA which represses the transcription of *pra*. In response to unknown signals, linked to the contact with a mycelium devoid of pSAM2, i.e., not expressing Pif, the repression by KorSA is relieved; Pra is expressed and activates pSAM2 excision and replication in the donor mycelium. Transfer can then occur and pSAM2 can spread and integrate into the recipient mycelium. The expression of Pif in the recipient mycelium will lead to the disappearance of pSAM2 free forms and only the integrated forms will be observed. In laboratory conditions, transfer is initiated about 2 h after the initial contact between the donor and recipient and is finished after 48 h.[25] It should be noted that when the establishment of pSAM2 is prevented in the recipient strain (e.g., by restriction barrier), the signal for transfer initiation persists for a longer time, as pSAM2 free forms are detected in the donor after 72 h. The regulatory proteins allow the coordinated expression of the genes involved in the different conjugation steps and prevent their unnecessary expression to avoid redundant transfer. This is important as transfer seems to be associated with growth delay. Regulatory genes are always found on AICEs or conjugative *Streptomyces* plasmids, but they differ and no general model of regulation has been described.

Other Genes

pSAM2 sequence analysis predicts the existence of several genes to which no functions have been assigned. Two predicted genes are present between the transfer module and the replication module. The others are clustered in a region extending from *attP* to *dso*.

Genes of Unknown Function Localized between Transfer Module and Replication Module

orf183 and *orf60* are localized upstream *repSA* and are transcribed in the same direction as *repSA*. The deletion of *orf183* reduces transfer efficiency but this pSAM2 deleted derivative is still able to replicate, and to maintain in an integrated state (Raynal, unpublished data). A GGDEF domain is predicted in Orf183. This domain is known to be involved in signal transduction by catalyzing the synthesis or hydrolysis of cyclic diguanylate. Recent studies present cyclic diguanylate as an important second messenger in a signaling system that appears to be ubiquitous in bacteria.[44] Homologs of *orf183* are found in a few AICEs.[43] *orf60* is a small gene positioned between *orf183* and RepSA. It encodes a 60 amino acid protein with homologs encoded by some AICEs (see pSAM2-like elements).

A Region Containing Genes of Unknown Function

A region extending from *attP* to *dso* may contain genes. The predicted genes present in this region are *orf92*, *orf267* and *orf240*. The deduce Orf92 protein has homologs in a few uncharacterized AICEs. One homologous protein (Runsl-0366) is found for Orf267 in *Runella slithyformis* DSM 19594. For Orf240 no predicted domains and no homologs have been found in databases. Further biological analysis need to be done to confirm or invalidate the existence of these genes. The region containing these predicted genes is characterized by a GC content (58,5%) lower than the rest of the pSAM2 sequence (70%). The deletion of this region has no effect on site-specific recombination, replication and transfer, at least in laboratory conditions. Therefore, it could be subject to insertion or deletion of accessory elements of diverse origin.

Host Range

pSAM2 is able to conjugate from *S. ambofaciens* ATCC23877 to other *Streptomyces* such as *S. lividans*.[11] Experiments have also shown that pSAM2 or pSAM2 derived vectors were able to integrate site-specifically in a wide range of *Streptomyces*[12,13] but also in *Mycobacterium smegmatis*.[45] This suggests that the host range of pSAM2 could be quite broad. This is not always the case for AICEs, for instance pMEA100 and pMEA300 seem to have a narrow host range.[3] As integration into the host chromosome is required for stable maintenance of pSAM2, the specificities of the pSAM2 integration system are important for the host range. pSAM2 integrates into a tRNA gene; tRNA genes are essential and their sequences are highly conserved.[46] Therefore a functional *attB* site is likely to be present in a wide range of (actino)bacteria. In addition, the pSAM2 integrase seems to be sufficient for site-specific integration and no host factor is required.[15] These properties contribute to broaden the host range of pSAM2.

Several factors can reduce the transfer and stable establishment of pSAM2 into a new host. Among them are the general defense mechanisms developed by bacteria to withstand the constant exposure to exogenous nucleic acids. These mechanisms include restriction modification systems. As pSAM2 is transferred among *Streptomyces* as double stranded DNA, it is highly sensitive to restriction. The presence in the *Streptomyces* recipient strain of a restriction modification system, for which there are restriction sites on pSAM2, can considerably reduce the establishment of pSAM2 in the recipient.[25] Another factor reducing interspecific pSAM2 transfer could be the presence in the recipient of related AICEs. It was shown that for the AICEs SLP1, pIJ110 and pIJ408, the presence of one of them in the recipient prevented the transfer of the two others, leading to conclusion that these three AICEs belonged to the same family.[47] pSAM2 could still transfer to recipient strains containing one of these three AICEs, indicating that it belonged to another family of AICEs.[8] But we observed that the expression in the recipient strain of *pif* homologs could reduce transfer efficiency (our unpublished data). Therefore the mechanism preventing redundant transfer could also reduce the transfer of related AICEs.

pSAM2-Like Elements

Conservation and Diversity among AICEs

A few AICEs have been characterized structurally and functionally to different extents. Among them, besides pSAM2, there is pSA1.1 from *S. cyaneus*,[48] pMR2 from *Micromonospora rosaria*[49] pMEA100 from *Amycolatopsis mediterranei U32*[50] and SLP1 from *Streptomyces coelicolor (A3)2*.[51] All of them integrate into a tRNA gene by site-specific recombination. For all of them, it is possible to identify functional modules containing the genes involved in regulation, transfer, replication and site-specific integration/excision. In each of these AICEs, a region called variable is specific for each element taken individually. These regions contain several genes with no (predicted) assignment. The general organization of these regions and modules is conserved between these AICEs (Fig. 6). However, a closer examination of the individual genes present in these modules reveals the diversity of these AICEs and allows defining families among them. For instance, the genes involved in the replication of pSAM2, pMEA100 and SLP1 are quite different. pMEA100 replicates most probably by a rolling circle replication mechanism, as pSAM2, but its replicase RepAM lack significant sequence similarity with other characterized prokaryotic Rep proteins and the putative origin of replication is located in the *repAM* coding sequence.[52] The originality of RepAM led te Poele and coworkers to bring together some AICEs, identified by sequence analysis of actinomycete genomes, in a family of pMEA elements.[43,53] The mechanism of SLP1 replication is not fully characterized but is different from the one of pSAM2 or pMEA100. It requires a region containing a putative ATP-binding protein (SCO4617) and a protein weakly related to DNA primase/polymerase (SCO4618).[3,54,55] SLP1 sequence analysis shows that the region between *SCO4617* and *xis* contains several repeats and regions of dyad symmetry that may correspond to the origin of replication. Homologs of *repSA* are present in the replication module of pSA1.1 from *S. cyaneus* and pMR2 from *M. rosaria*. Moreover, in these elements the genes of the transfer module present the same organization as those from pSAM2: the *traSA* homolog is localized upstream of the spread genes and not downstream, as for pMEA100 and SLP1. In addition, the *pra* gene, which has no homolog in pMEA-elements or in SLP1 has homologs in the regulatory module of pSA1.1 and pMR2. This led us to define pSAM2-like elements as ICEs possessing homologs of *pra*, *traSA* and *repSA*, as summarized in Table 1. pSAM2-like elements were identified in sequenced genomes by Blast searches for Pra, TraSA and RepSA homologs.

Distribution and Diversity and of pSAM2-Like Elements

Forty pSAM2-like AICEs were identified in 35 genomes. All bacteria harboring pSAM2-like elements belong to the *Actinomycetales* order. The *Actinomycetales* belongs to the branch of high G+C Gram-positive bacteria. Many of them form branching mycelium and produce spores. The *Actinomycetales* are taxonomically divided in 13 suborders. pSAM2-like elements have been found in the suborders *Streptomycineae, Micromonosporineae, Pseudonocardineae, Streptosporangineae, Corynebacterineae* and *Frankineae*. Therefore pSAM2-like elements are widely distributed, contrarily to pMEA-elements, which are only found in the genus *Amycolatopsis* and the closely related genus *Saccharopolyspora*.[3] For each *Actinomycetales* suborder, at least one pSAM2-like element is presented in Figure 6c. None of them resemble exactly pSAM2. Nevertheless, in most cases, and for all those presented in Figure 6c, they are inserted into a tRNA gene and flanked by direct repeats. They all contain a low GC variable region, often containing transposase genes. The regulatory, transfer, replication and integration/excision modules are in the same order. The regulatory modules characterized by the presence of *pra* homologs also often contain genes encoding GntR and Nudix proteins. In some transfer modules, no spread genes could be identified. The region between *traSA* and *repSA* allowed variability since the nature and number of genes varies from one element to the other. Concerning the excision/integration module, if an integrase gene is always found, the *xis* gene could be absent.. The similarity between the integrases from pSAM2-like elements is weak compared with the similarity of other conserved proteins. This could be related to the fact that pSAM2-like elements are inserted into various tRNA genes.

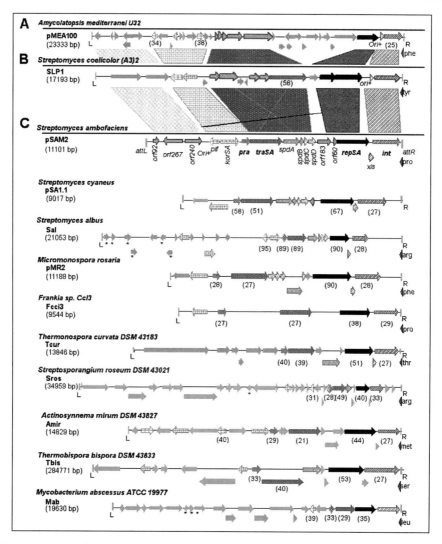

Figure 6. Structural organization of AICEs. A) pMEA100, a representative of the pMEA family, B) the SLP1 element, C) pSAM2 family: pSAM2, pSA1.1, pMR2 and examples of newly identified pSAM2-like elements:Sal (genes SalbJ 010100019238 to SalbJ 010100019358), Fcci3 (genes Francci3 4270 to Francci3 4274), Tcur (genes Tcur 4523 to Tcur 4536), Sros (genes Sros 8691 to Sros 8658), Amir (genes Amir 1277 to Amir 1294), Tbis (genes Tbis 0092 to Tbis 0078), Mab (genes MAB 2072 to MAB 2102). The position and orientation of genes are indicated by arrows. The pSAM2 gene names are indicated under the corresponding arrow. Specific representation is assigned to differentiate each functional module: girds for genes involved in regulation, dots for genes involved in transfer, black color for genes involved in replication and stripes for genes which belong to the site-specific recombination module. Genes which belong to the variable region or with no homologs in pSAM2 are represented in gray. For pSAM2-like elements, representatives found in the different suborders are shown. Genes with sequence similarity with a pSAM2 gene are represented with the same code. When indicated, the identity percent with the pSAM2 homolog is given in brackets. A star (*) indicates a transposase gene. tRNA are represented by an arrow and the corresponding amino acids are indicated. Full bar represent repeated sequences R and L surrounding the mobile element.

Table 1. AICE family specificity

| AICE Family | AICE Name | Modules | | |
		Regulation	Transfer	Replication	
pSAM2-like elements	pSAM2 pSA1.1 pMR2	pra	Major transfer protein gene upstream spread genes	*repSA*	*Ori* far from replicase gene
pMEA-elements	pMEA 100	Not determined	Major transfer protein gene downstream spread genes	*repAM*	*Ori* within replicase gene
SLP1- elements	SLP1			Primase -polymerase	*Ori* near replicase gene

Some hints concerning the mechanisms allowing the evolution of AICEs can be obtained from the organization of regions containing pSAM2 in *S. ambofaciens* ATCC23877.[14] pSAM2 is flanked by the 58-bp repeats included in *attL* and *attR*. A third 58-bp repeat (*attX*) is present in the same region and the sequence of 42 kb between *attX* and *attL*, which has been called xSAM1, could correspond to a prophage (our unpublished data). This whole region of 53 kb (*attX-xSAM1-attL-pSAM2-attR*) is absent from *S. ambofaciens* DSM40697, where only the *attB* site is present.[14] This suggests that pSAM2 is part of a larger composite element, flanked by *attX* and *attR*, which could have arisen through site-specific accretion. Similarly, it was observed in *Streptomyces ghanaensis* that a vector using the integration system of the actinophage VWB integrates into a tRNA^Arg which overlaps the *attR* site of a pSAM2 related AICE.[56] Therefore site-specific accretion of AICEs or prophages followed by deletion could be a mechanism allowing the acquisition or exchange of modules and the evolution of AICEs, as proposed for ICEs in other bacteria.[57]

Genetic Tools Derived from pSAM2

The site-specific recombination module of pSAM2 was used to build several types of genetic tools. First, integrative vectors carrying the *int* gene and the *attP* sequence were built.[13,58] As the chromosomal integration site *attB* overlaps a tRNA gene conserved in actinobacteria, and as Int alone is sufficient to promote integration, these vectors could be used in wide range of actinobacteria, belonging to the genus *Streptomyces* but also *Mycobacterium*.[13,45,58-60] These integrating vectors provided stable integration of the cloned genes without any of the deleterious effects (in particular on antibiotic production) which are sometimes linked to the introduction of plasmid cloning vectors. The usefulness of pSAM2-derived integrating vectors was illustrated by their use in industrial *Streptomyces pristinaespiralis* strains for the improvement of pristinamycin production.[60]

Second, the knowledge obtained on the sequence required for efficient recombination between the *att* sites was used to build excisable antibiotic resistance cassettes.[61] These cassettes contain an antibiotic resistance gene flanked by minimal *attL* and *attR* sites. They are used in gene replacement experiments to generate deletion mutants. These cassettes can then be precisely excised by transiently expressing the Xis and Int proteins through the introduction of a highly unstable replicating vector carrying the *xis* and *int* genes. After excision a scar of 33, 34 or 35 nucleotides (depending on the cassette used), without any stop codon in any reading frame, is left. Thus in-frame deletion mutants can be easily obtained. This allows also the construction of unmarked mutants strains and the recycling of antibiotic resistance markers.

New vectors and genetic tools are still needed to manipulate various industrially important actinobacteria. The recently identified pSAM2-like elements constitute a source of new site-specific recombination modules that could be used to build new tools with different integration sites in *Streptomyces* chromosomes. Another application for the site-specific recombination system of pSAM2, or more generally of AICEs, could be *Streptomyces* chromosome engineering, i.e., the deletion or inversion of large chromosomal regions, as was done in *E. coli* with the integrase and *att* sites of lambda.[62,63]

Conclusion

pSAM2 is one of the AICEs in which the various functions of site-specific recombination, replication and transfer are well characterized. Some elements of the complex regulation of these functions have also been elucidated. From this knowledge, it seems that most of the pSAM2 genes are not expressed when pSAM2 is integrated in the chromosome. Their expression is induced during transfer. In particular, one of the purposes of the activation by Pra is to obtain a burst of replication, in the donor strain as well as in the recipient, in order to allow efficient transfer and the invasion of the recipient mycelium by pSAM2. Due to the apical growth of the mycelium, this step of spreading is essential for pSAM2 to be present in the aerial mycelium that will differentiate into spore chains. The ability to replicate is a distinctive characteristic of AICEs. This could be related to the fact that AICEs are mostly found in mycelium-forming actinobacteria, where replication could be required for efficient transfer and spreading. Our knowledge about pSAM2 and more generally AICEs is far from complete. For instance, in pSAM2, the signal for transfer, the mechanisms by which Pif confers immunity or by which Pra activates replication and site-specific recombination are still unknown. Their study might reveal new mechanisms of regulation, as the study of *Streptomyces* conjugation revealed an original mechanism of interspecific transfer of double stranded DNA. Comparative studies of the pSAM2-like elements and other AICEs should also provide information on the biology of AICEs, their evolution and role in horizontal gene transfer.

References

1. Hagège J, Boccard F, Smokvina T, et al. Identification of a gene encoding the replication initiator protein of the Streptomyces integrating element, pSAM2. Plasmid 1994; 31:166-83; PMID:8029324; http://dx.doi.org/10.1006/plas.1994.1018.
2. Burrus V, Pavlovic G, Decaris B, Guedon G. The ICESt1 element of Streptococcus thermophilus belongs to a large family of integrative and conjugative elements that exchange modules and change their specificity of integration. Plasmid 2002; 48:77-97; PMID:12383726; http://dx.doi.org/10.1016/S0147-619X(02)00102-6.
3. te Poele EM, Bolhuis H, Dijkhuizen L. Actinomycete integrative and conjugative elements. Antonie van Leeuwenhoek 2008; 94:127-43; PMID:18523858; http://dx.doi.org/10.1007/s10482-008-9255-x.
4. Bérdy J. Bioactive microbial metabolites. J Antibiot (Tokyo) 2005; 58:1-26; PMID:15813176; http://dx.doi.org/10.1038/ja.2005.1.
5. Ginolhac A, Jarrin C, Gillet B, et al. Phylogenetic analysis of polyketide synthase I domains from soil metagenomic libraries allows selection of promising clones. Appl Environ Microbiol 2004; 70:5522-7; PMID:15345440; http://dx.doi.org/10.1128/AEM.70.9.5522-5527.2004.
6. Toussaint A, Merlin C. Mobile elements as a combination of functional modules. Plasmid 2002; 47:26-35; PMID:11798283; http://dx.doi.org/10.1006/plas.2001.1552.
7. Burrus V, Waldor MK. Shaping bacterial genomes with integrative and conjugative elements. Res Microbiol 2004; 155:376-86; PMID:15207870; http://dx.doi.org/10.1016/j.resmic.2004.01.012.
8. Pernodet JL, Simonet JM, Guerineau M. Plasmids in different strains of Streptomyces ambofaciens: free and integrated form of plasmid pSAM2. Mol Gen Genet 1984; 198:35-41; PMID:6596483; http://dx.doi.org/10.1007/BF00328697.
9. Omura S, Ikeda H, Kitao C. The detection of a plasmid in Streptomyces ambofaciens KA-1028 and its possible involvement in spiramycin production. J Antibiot (Tokyo) 1979; 32:1058-60; PMID:528367.
10. Omura S, Ikeda H, Tanaka H. Extraction and characterization of plasmids from macrolide antibiotic-producing streptomycetes. J Antibiot (Tokyo) 1981; 34:478-82; PMID:7275829.
11. Boccard F, Pernodet J, Friedmann A, Guerineau M. Site-specific integration of plasmid pSAM2 in Streptomyces lividans and S. ambofaciens. Mol Gen Genet 1988; 212:432-39; http://dx.doi.org/10.1007/BF00330847.

12. Kuhstoss S, Richardson MA, Rao RN. Site-specific integration in Streptomyces ambofaciens: localization of integration functions in S. ambofaciens plasmid pSAM2. J Bacteriol 1989; 171:16-23; PMID:2536654.

13. Smokvina T, Mazodier P, Boccard F, et al. Construction of a series of pSAM2-based integrative vectors for use in actinomycetes. Gene 1990; 94:53-9; PMID:2227452; http://dx.doi.org/10.1016/0378-1119(90)90467-6.

14. Boccard F, Smokvina T, Pernodet JL, et al. The integrated conjugative plasmid pSAM2 of Streptomyces ambofaciens is related to temperate bacteriophages. EMBO J 1989; 8:973-80; PMID:2721504.

15. Raynal A, Tuphile K, Gerbaud C, et al. Structure of the chromosomal insertion site for pSAM2: functional analysis in Escherichia coli. Mol Microbiol 1998; 28:333-42; PMID:9622358; http://dx.doi.org/10.1046/j.1365-2958.1998.00799.x.

16. Boccard F, Smokvina T, Pernodet JL, et al. Structural analysis of loci involved in pSAM2 site-specific integration in Streptomyces. Plasmid 1989; 21:59-70; PMID:2657820; http://dx.doi.org/10.1016/0147-619X(89)90087-5.

17. Mazodier P, Thompson C, Boccard F. The chromosomal integration site of the Streptomyces element pSAM2 overlaps a putative tRNA gene conserved among actinomycetes. Mol Gen Genet 1990; 222:431-4; PMID:1703270; http://dx.doi.org/10.1007/BF00633850.

18. Raynal A, Friedmann A, Tuphile K, et al. Characterization of the attP site of the integrative element pSAM2 from Streptomyces ambofaciens. Microbiology 2002; 148:61-7; PMID:11782499.

19. Hagège J, Pernodet JL, Friedmann A, Guerineau M. Mode and origin of replication of pSAM2, a conjugative integrating element of Streptomyces ambofaciens. Mol Microbiol 1993; 10:799-812; PMID:7934842; http://dx.doi.org/10.1111/j.1365-2958.1993.tb00950.x.

20. Khan SA. Rolling-circle replication of bacterial plasmids. Microbiol Mol Biol Rev 1997; 61:442-55; PMID:9409148.

21. Smokvina T, Boccard F, Pernodet JL, et al. Functional analysis of the Streptomyces ambofaciens element pSAM2. Plasmid 1991; 25:40-52; PMID:1852016; http://dx.doi.org/10.1016/0147-619X(91)90005-H.

22. Ilyina TV, Koonin EV. Conserved sequence motifs in the initiator proteins for rolling circle DNA replication encoded by diverse replicons from eubacteria, eucaryotes and archaebacteria. Nucleic Acids Res 1992; 20:3279-85; PMID:1630899; http://dx.doi.org/10.1093/nar/20.13.3279.

23. Sezonov G, Duchene AM, Friedmann A, et al. Replicase, excisionase, and integrase genes of the Streptomyces element pSAM2 constitute an operon positively regulated by the pra gene. J Bacteriol 1998; 180:3056-61; PMID:9620953.

24. Clarke M, Maddera L, Harris RL, Silverman PM. F-pili dynamics by live-cell imaging. Proc Natl Acad Sci USA 2008; 105:17978-81; PMID:19004777; http://dx.doi.org/10.1073/pnas.0806786105.

25. Possoz C, Ribard C, Gagnat J, et al. The integrative element pSAM2 from Streptomyces: kinetics and mode of conjugal transfer. Mol Microbiol 2001; 42:159-66; PMID:11679075; http://dx.doi.org/10.1046/j.1365-2958.2001.02618.x.

26. Ducote MJ. An in vivo assay for conjugation-mediated recombination yields novel results for streptomyces plasmid pIJ101. Plasmid 2006; 55:242-8; PMID:16388851; http://dx.doi.org/10.1016/j.plasmid.2005.11.002.

27. Ducote MJ, Parkash S, Pettis GS. Minimal and contributing sequence determinants of the cis-acting locus of transfer (clt) of streptomycete plasmid pIJ101 occur within an intrinsically curved plasmid region. J Bacteriol 2000; 182:6834-41; PMID:11073933; http://dx.doi.org/10.1128/JB.182.23.6834-6841.2000.

28. Reuther J, Gekeler C, Tiffert Y, et al. Unique conjugation mechanism in mycelial streptomycetes: a DNA-binding ATPase translocates unprocessed plasmid DNA at the hyphal tip. Mol Microbiol 2006; 61:436-46; PMID:16776656; http://dx.doi.org/10.1111/j.1365-2958.2006.05258.x.

29. Kieser T, Hopwood DA, Wright HM, Thompson CJ. pIJ101, a multi-copy broad host-range Streptomyces plasmid: functional analysis and development of DNA cloning vectors. Mol Gen Genet 1982; 185:223-8; PMID:6283316; http://dx.doi.org/10.1007/BF00330791.

30. Hagège J, Pernodet JL, Sezonov G, et al. Transfer functions of the conjugative integrating element pSAM2 from Streptomyces ambofaciens: characterization of a kil-kor system associated with transfer. J Bacteriol 1993; 175:5529-38; PMID:8366038.

31. Bath J, Wu LJ, Errington J, Wang JC. Role of Bacillus subtilis SpoIIIE in DNA transport across the mother cell-prespore division septum. Science 2000; 290:995-7; PMID:11062134; http://dx.doi.org/10.1126/science.290.5493.995.

32. Vogelmann J, Ammelburg M, Finger C, et al. Conjugal plasmid transfer in Streptomyces resembles bacterial chromosome segregation by FtsK/SpoIIIE. EMBO J 2011; 30:2246-54; PMID:21505418; http://dx.doi.org/10.1038/emboj.2011.121.

33. Pettis GS, Cohen SN. Transfer of the pIJ101 plasmid in Streptomyces lividans requires a cis-acting function dispensable for chromosomal gene transfer. Mol Microbiol 1994; 13:955-64; PMID:7854128; http://dx.doi.org/10.1111/j.1365-2958.1994.tb00487.x.

34. Bigot S, Saleh OA, Lesterlin C, et al. KOPS: DNA motifs that control E. coli chromosome segregation by orienting the FtsK translocase. EMBO J 2005; 24:3770-80; PMID:16211009; http://dx.doi. org/10.1038/sj.emboj.7600835.

35. Sezonov G, Hagege J, Pernodet JL, et al. Characterization of pra, a gene for replication control in pSAM2, the integrating element of Streptomyces ambofaciens. Mol Microbiol 1995; 17:533-44; PMID:8559072; http://dx.doi.org/10.1111/j.1365-2958.1995.mmi_17030533.x.

36. Kataoka M, Seki T, Yoshida T. Regulation and function of the Streptomyces plasmid pSN22 genes involved in pock formation and inviability. J Bacteriol 1991; 173:7975-81; PMID:1720772.

37. Tiffert Y, Gotz B, Reuther J, et al. Conjugative DNA transfer in Streptomyces: SpdB2 involved in the intramycelial spreading of plasmid pSVH1 is an oligomeric integral membrane protein that binds to dsDNA. Microbiology 2007; 153:2976-83; PMID:17768240; http://dx.doi.org/10.1099/ mic.0.2006/005413-0.

38. Smokvina T, Francou F, Luzzati M. Genetic analysis in Streptomyces ambofaciens. J Gen Microbiol 1988; 134:395-402; PMID:3171544.

39. Sezonov G, Possoz C, Friedmann A, et al. KorSA from the Streptomyces integrative element pSAM2 is a central transcriptional repressor: target genes and binding sites. J Bacteriol 2000; 182:1243-50; PMID:10671443; http://dx.doi.org/10.1128/JB.182.5.1243-1250.2000.

40. Possoz C, Gagnat J, Sezonov G, et al. Conjugal immunity of Streptomyces strains carrying the integrative element pSAM2 is due to the pif gene (pSAM2 immunity factor). Mol Microbiol 2003; 47:1385-93; PMID:12603742; http://dx.doi.org/10.1046/j.1365-2958.2003.03380.x.

41. McLennan AG. The Nudix hydrolase superfamily. Cell Mol Life Sci 2006; 63:123-43; PMID:16378245; http://dx.doi.org/10.1007/s00018-005-5386-7.

42. Kloosterman H, Vrijbloed JW, Dijkhuizen L. Molecular, biochemical, and functional characterization of a Nudix hydrolase protein that stimulates the activity of a nicotinoprotein alcohol dehydrogenase. J Biol Chem 2002; 277:34785-92; PMID:12089158; http://dx.doi.org/10.1074/jbc.M205617200.

43. te Poele EM, Samborskyy M, Oliynyk M, et al. Actinomycete integrative and conjugative pMEA-like elements of Amycolatopsis and Saccharopolyspora decoded. Plasmid 2008; 59:202-16; PMID:18295883; http://dx.doi.org/10.1016/j.plasmid.2008.01.003.

44. Römling U, Gomelsky M, Galperin MY. C-di-GMP: the dawning of a novel bacterial signalling system. Mol Microbiol 2005; 57:629-39; PMID:16045609; http://dx.doi.org/10.1111/j.1365-2958.2005.04697.x.

45. Martín C, Mazodier P, Mediola MV, et al. Site-specific integration of the Streptomyces plasmid pSAM2 in Mycobacterium smegmatis. Mol Microbiol 1991; 5:2499-502; PMID:1665195; http://dx.doi. org/10.1111/j.1365-2958.1991.tb02095.x.

46. Williams KP. Integration sites for genetic elements in prokaryotic tRNA and tmRNA genes: sublocation preference of integrase subfamilies. Nucleic Acids Res 2002; 30:866-75; PMID:11842097; http://dx.doi. org/10.1093/nar/30.4.866.

47. Hopwood DA, Hintermann G, Kieser T, Wright HM. Integrated DNA sequences in three streptomycetes form related autonomous plasmids after transfer to Streptomyces lividans. Plasmid 1984; 11:1-16; PMID:6369354; http://dx.doi.org/10.1016/0147-619X(84)90002-7.

48. Miyoshi YK, Ogata S, Hayashida S. Multicopy derivative of pock-forming plasmid pSA1 in Streptomyces azureus. J Bacteriol 1986; 168:452-4; PMID:3759910.

49. Hosted TJ Jr, Wang T, Horan AC. Characterization of the Micromonospora rosaria pMR2 plasmid and development of a high G+C codon optimized integrase for site-specific integration. Plasmid 2005; 54:249-58; PMID:16024079; http://dx.doi.org/10.1016/j.plasmid.2005.05.004.

50. Moretti P, Hintermann G, Hutter R. Isolation and characterization of an extrachromosomal element from Nocardia mediterranei. Plasmid 1985; 14:126-33; PMID:2999850; http://dx.doi.org/10.1016/0147-619X(85)90072-1.

51. Bibb MJ, Ward JM, Kieser T, et al. Excision of chromosomal DNA sequences from Streptomyces coelicolor forms a novel family of plasmids detectable in Streptomyces lividans. Mol Gen Genet 1981; 184:230-40; PMID:6948998.

52. te Poele EM, Kloosterman H, Hessels GI, et al. RepAM of the Amycolatopsis methanolica integrative element pMEA300 belongs to a novel class of replication initiator proteins. Microbiology 2006; 152:2943-50; PMID:17005975; http://dx.doi.org/10.1099/mic.0.28746-0.

53. te Poele EM, Habets MN, Tan GY, et al. Prevalence and distribution of nucleotide sequences typical for pMEA-like accessory genetic elements in the genus Amycolatopsis. FEMS Microbiol Ecol 2007; 61:285-94; PMID:17535299; http://dx.doi.org/10.1111/j.1574-6941.2007.00334.x.

54. Brasch MA, Pettis GS, Lee SC, Cohen SN. Localization and nucleotide sequences of genes mediating site-specific recombination of the SLP1 element in Streptomyces lividans. J Bacteriol 1993; 175:3067-74; PMID:8387993.

55. Grant SR, Lee SC, Kendall K, Cohen SN. Identification and characterization of a locus inhibiting extrachromosomal maintenance of the Streptomyces plasmid SLP1. Mol Gen Genet 1989; 217:324-31; PMID:2770697; http://dx.doi.org/10.1007/BF02464900.

56. Ostash B, Makitrinskyy R, Walker S, Fedorenko V. Identification and characterization of Streptomyces ghanaensis ATCC14672 integration sites for three actinophage-based plasmids. Plasmid 2009; 61:171-5; PMID:19167423; http://dx.doi.org/10.1016/j.plasmid.2008.12.002.

57. Pavlovic G, Burrus V, Gintz B, et al. Evolution of genomic islands by deletion and tandem accretion by site-specific recombination: ICESt1-related elements from Streptococcus thermophilus. Microbiology 2004; 150:759-74; PMID:15073287; http://dx.doi.org/10.1099/mic.0.26883-0.

58. Kuhstoss S, Richardson MA, Rao RN. Plasmid cloning vectors that integrate site-specifically in Streptomyces spp. Gene 1991; 97:143-6; PMID:1995427; http://dx.doi.org/10.1016/0378-1119(91)90022-4.

59. Eiglmeier K, Honore N, Cole ST. Towards the integration of foreign DNA into the chromosome of Mycobacterium leprae. Res Microbiol 1991; 142:617-22; PMID:1683711; http://dx.doi.org/10.1016/0923-2508(91)90074-K.

60. Sezonov G, Blanc V, Bamas-Jacques N, et al. Complete conversion of antibiotic precursor to pristinamycin IIA by overexpression of Streptomyces pristinaespiralis biosynthetic genes. Nat Biotechnol 1997; 15:349-53; PMID:9094136; http://dx.doi.org/10.1038/nbt0497-349.

61. Raynal A, Karray F, Tuphile K, et al. Excisable cassettes: new tools for functional analysis of Streptomyces genomes. Appl Environ Microbiol 2006; 72:4839-44; PMID:16820478; http://dx.doi.org/10.1128/AEM.00167-06.

62. Valens M, Penaud S, Rossignol M, et al. Macrodomain organization of the Escherichia coli chromosome. EMBO J 2004; 23:4330-41; PMID:15470498; http://dx.doi.org/10.1038/sj.emboj.7600434.

63. Esnault E, Valens M, Espeli O, Boccard F. Chromosome structuring limits genome plasticity in Escherichia coli. PLoS Genet 2007; 3:e226; PMID:18085828; http://dx.doi.org/10.1371/journal.pgen.0030226.

CHAPTER 9

The Tn*916*/Tn*1545* Family of Conjugative Transposons

Lena Ciric,[1] Azmiza Jasni,[1] Lisbeth Elvira de Vries,[2,3] Yvonne Agersø,[2] Peter Mullany[1] and Adam P. Roberts*[1]

Abstract

The conjugative transposon Tn*916* was first discovered in the late 1970s and is, together with the related conjugative transposon Tn*1545*, the paradigm of a large family of related conjugative transposons known as the Tn*916*/Tn*1545* family, which are found in an extremely diverse range of bacteria. With the huge increase in bacterial genomic sequence data available, due to the widespread use of next generation sequencing, more putative conjugative transposons belonging to the Tn*916*/Tn*1545* family are being reported. Many of these are capable of excision, integration and conjugation. Nearly all of the Tn*916*/Tn*1545*-like elements discovered to date encode tetracycline resistance however, increasingly resistance to other antimicrobials is being found. Some of the members of the Tn*916*/Tn*1545* family of elements are composite structures which contain smaller mobile genetic elements which are also capable of transposition. Tn*916*/Tn*1545*-like elements themselves are also found within larger and more complex elements. This review will give an overview of the current knowledge of the Tn*916*/Tn*1545* family of conjugative transposons highlighting recently characterized composite elements carrying additional and novel resistance genes.

Introduction

The Conjugative Transposon Tn916

The existence of chromosomally, as opposed to plasmid encoded transferable resistance was first suspected in the late 1970s. When *Enterococcus faecalis* strain DS16 was mated with the plasmid-free *E. faecalis* strain JH2–2, some transconjugants resistant to tetracycline contained the Tn*916* determinant linked to the co-resident plasmid pAD1 which had also transferred from DS16. In addition, derivatives of DS16 devoid of pAD1 were capable of transferring tetracycline resistance to recipient strains. Transconjugants (plasmid-free) from such matings could subsequently act as donors in the transfer of tetracycline resistance. Further work showed that tetracycline resistance was conferred by an integrative element which was called a conjugative transposon and was designated Tn*916*.[1] This was the first conjugative transposon found to carry an antibiotic resistance gene and it was hypothesized that it may explain the widespread presence of tetracycline resistance among streptococci at the time.[1]

[1]Department of Microbial Diseases, UCL Eastman Dental Institute, University College London, London, UK; [2]Department of Microbiology and Risk Assessment, National Food Institute, Technical University of Denmark, Lyngby, Denmark; [3]Department of Veterinary Disease Biology, University of Copenhagen, Frederiksberg, Denmark.
*Corresponding Author: Adam P. Roberts—adam.roberts@ucl.ac.uk

Bacterial Integrative Mobile Genetic Elements, edited by Adam P. Roberts and Peter Mullany.
©2013 Landes Bioscience.

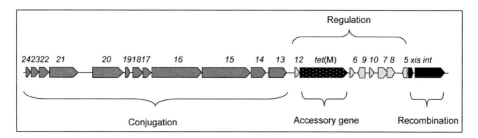

Figure 1. Schematic of Tn*916* showing the four functional modules: conjugation (dark gray/blue); regulation (light gray/green); recombination (black/red) and the accessory gene *tet*(M) (gray/black with white dots). Schematic adapted from: Roberts AP, Mullany P. Trends Microbiol 2009; 17:251-8;[4] ©2009 with permission from Elsevier.

The genetic organization of Tn*916*, and many other mobile genetic elements (MGEs), is modular.[2-4] The modules are involved in conjugation, excision and integration (recombination), regulation and accessory functions which are not involved in mobility or regulation (Fig. 1).[4] Each of these functional modules will be considered separately and in detail below.

The Tn*916* Family of Conjugative Transposons; An Ever Expanding Family of MGEs

In the past few decades, numerous mobile genetic elements with similarities to Tn*916* have been characterized (Fig. 2). While the conjugation and the regulatory genes are generally conserved the genes encoding the excisionases, integrases or recombinases and the accessory genes vary (Fig. 2). A number of elements within the family also contain insertions of smaller MGEs encoding resistance to other antimicrobials e.g., macrolides[5,6] and mercury.[7] Group II introns and IS elements have also been found in some of the Tn*916*/Tn*1545* family of MGEs (Fig. 2).[8-10] Tn*916*-like elements are found in a wide range of bacterial species belonging to at least 36 genera spanning six phyla (Fig. 3), giving the Tn*916*/Tn*1545* family an exceptionally broad host range.[4]

Here we present an overview of the current knowledge of the biological functions of Tn*916* encoded proteins and explore their genetic diversity.

The Functions of the Transposon Encoded Proteins by Module

Recombination

All of the Tn*916*/Tn*1545* family of MGEs possess a recombination module which is located at one end of the element with the direction of transcription leading out of that end of the element (Fig. 1).[11-14] In Tn*916*, this consists of two genes: encoding a tyrosine integrase, IntTn, and an excisionase, XisTn. Excision begins with the introduction of staggered endonucleolytic cuts made at each end of the element generating single-stranded, non-complementary hexanucleotides at each end of the element termed the coupling sequences.[13,15] The coupling sequences then join forming a covalent bond creating a circular intermediate molecule with a heteroduplex at the joint, while the target site from which the element has excised is also ligated. On integration into a target site, heteroduplex regions are produced on either side of the conjugative transposon which are then resolved by DNA repair or replication. One study has shown that Tn*916* is also capable of inversion in its target site. PCRs performed on DNA extracted from a broth culture of *Enterococcus faecium* DPC3675 which carries one copy of Tn*916* showed the ends of the transposon in both orientations within the target site,[16] work in our lab has shown that this may be a general property of Tn*916*/Tn*1545*-like elements (Roberts et al., unpublished).

Tn*916* and Tn*1545* both encode a tyrosine recombinase which are highly related to each other.[14,17] There is variation however between tyrosine recombinases associated with different Tn*916*-like elements e.g., Int*6000*[18] is more related to the tyrosine integrases of staphylococcal

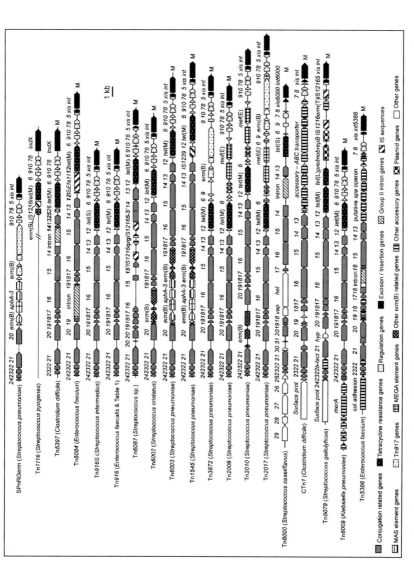

Figure 2. The structure of various members of the Tn916/Tn1545 family is shown. Functional modules are represented as shown in the key. The organisms from which the elements were isolated are shown to the left in brackets. Mobility is denoted by a capital M on the right. Schematic adapted from: Roberts AP, Mullany P. Trends Microbiol 2009; 17:251-8;[4] ©2009 with permission from Elsevier.

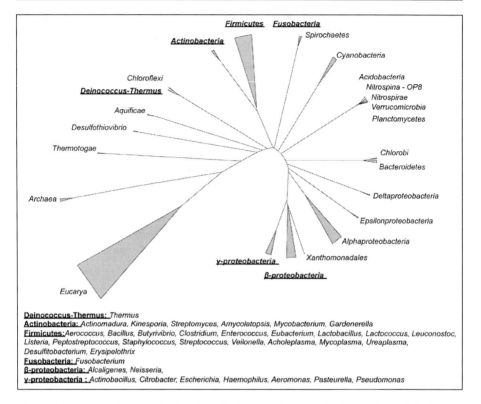

Figure 3. Taxonomic tree of life showing all of the phyla from which members of the Tn*916*/ Tn*1545* family have been isolated. The genera within each phylum are shown below. The tree was produced by aligning all Bacterial Domain sequences from the Ribosomal Database Project using the ARB package[59] and the January 2004 database.

pathogenicity islands than it is to the integrases of Tn*916* and Tn*1545*, probably reflecting different recombination events between different MGEs which have led to the formation of the different elements (Fig. 4). In some members of the Tn*916*/Tn*1545* family integration and excision is mediated by large serine recombinase (these proteins do not require an excisonase in the recombination reactions) these include Tn*5397* and Tn*1116*.[8,19]

Target Sites of Tn916

The Tn*916* IntTn protein can use multiple target sites and these have been shown to be A:T-rich. A recent study has characterized 123 insertion sites in the genome of *Butyrivibrio proteoclasticus* strain B316[20] and has shown that the consensus sequence TTTTT *TATATA* AAAAA is used (the hexanucleotide in italics is variable and forms the coupling sequences). In addition we have recently performed a similar study in *Clostridium difficile* and have shown a nearly identical consensus sequence based on almost 200 insertion sites in two different strains (Mullany et al., unpublished). Interestingly however Tn*916* has a preferred insertion site in *C. difficile* strain CD37. Here the target insertion site also consists of an A:T-rich region but is preferentially used in this strain.[21]

Conjugation

Knowledge of the specific mechanism of conjugation among the Tn*916* family members is somewhat limited, however early Tn*5* mutagenesis indicated that ORFs *24* to *13* are involved in this process (Fig. 1).[22] However none of the Tn*5* insertions were complemented so polar effects

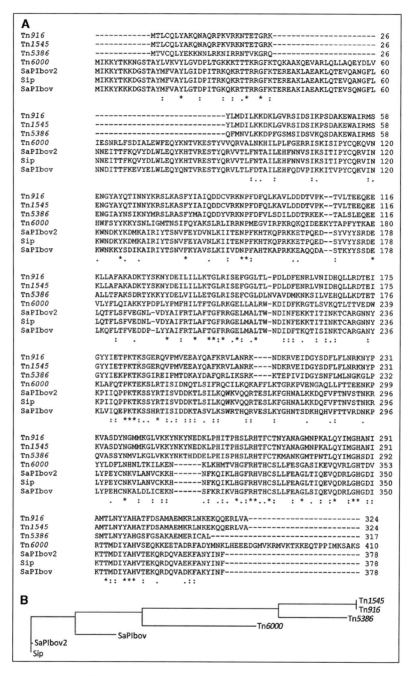

Figure 4. A) Alignment of the integrase genes of Tn916 (U09422), Tn1545 (X61025), Tn5386 (DQ321786), Tn6000 (FN555436), SaPIbov (AAG29618), SaPIbov2 (AAP55251) and Sip (AAP51267). Sequence accession numbers are shown in brackets above. "*"—identical amino acids; ":"—conserved amino acid substitutions; "."—semiconserved amino acid substitutions. B) A phylogenetic tree of the amino acid sequences is shown. Figures generated using ClustalW2 (http://www.ebi.ac.uk/Tools/msa/clustalw2/).

cannot be ruled out. The specific functions of some of these proteins, or homologs of these proteins, have been experimentally proven and are shown in Table 1. Orf20 is a relaxase nicking at the origin of transfer (*ori*T), constituting the first step of the conjugation process.[23] Tn*916* IntTn is a specificity factor for this reaction and is responsible for both the strand and sequence selection of Orf20. *Ori*T itself spans a 466 bp region containing a number of inverted repeats and is positioned between *orf20* and *orf21* of the transposon.[24] *Orf18* encodes a putative ArdA homologue which is responsible for the transposon's immunity to DNA restriction modification following conjugation by mimicking the DNA substrate for restriction enzymes. This likely contributes to the broad host range of Tn*916*.[25] The putative product of *orf14* shows some similarity to the NPL/p60 family of proteins which are associated with virulence in *Listeria monocytogenes*.[26,27]

Regulation

Our knowledge of the regulation of Tn*916*/Tn*1545*-like elements is almost completely limited to Tn*916*. Su et al. (28), proposed that the regulatory system of Tn*916* comprises of *orf12*, *orf9*, *orf7* and *orf8* (Fig. 1). This region is conserved in nearly all Tn*916*-like elements which suggest that it is extremely important for the function and / or maintenance of the elements.

The presence of several inverted repeat sequences within *orf12* are key to the proposed regulatory mechanism. It has been proposed that regulation of Tn*916* involves transcriptional attenuation and is regulated by tetracycline.[28,29] The proposed regulatory region consists of stem-loop structures, 5S:6S and 7:8 followed by a series of uracil residues in the RNA, predicted to be the transcriptional terminators ("T" on Fig. 5). In the presence of tetracycline most ribosomes are inactivated by the reversible binding of tetracycline, resulting in a build-up of charged t-RNA molecules due to a lower rate of protein synthesis. At this stage, a few ribosomes are thought to be protected by the low and basal level of Tet(M). Accumulation of charged t-RNA enables the more rapid translation of *orf12* by the protected ribosomes. This event speeds up the translation by the protected ribosomes, which is normally slow due to the presence of rare codons in *orf12*, and is predicted to allow the ribosome to catch up to the RNA polymerase and therefore prevent the formation of, or destroy, the terminator structures 5S:6S and 7:8. This allows transcription from the promoter upstream of *orf12* to extended into, and through, *tet*(M) and into the downstream genes (Fig. 6). However, in the absence of tetracycline, the ribosome pauses on the leader peptide of *orf12* due to both the shortage of charged t-RNA molecules and the rare codons within *orf12*. This results in the ribosome lagging behind the RNA polymerase allowing the formation of the predicted strong 5S:6S terminator and/or the weaker 7:8 terminator which is predicted to terminate the majority of transcription.

Orf9 is proposed to repress the transcription from the *orf7* promoter Porf7 (Fig. 6). Increased transcription through *tet*(M) and downstream regions will lead to the production of antisense *orf9* RNA which leads to de-repression of Porf7 and increased transcription of Orf7 and Orf8 from the upstream promoter Porf7. Increased transcription from Porf7 will lead to an increase in the translation of Orf7 and Orf8. Orf7 and Orf8 are predicted to upregulate their own transcription from Porf7, thereby providing an amplification of the environmental signal (tetracycline) sensed upstream of *tet*(M). The increase in transcription from Porf7 will lead to an increase in the translation of downstream genes (*xisTn* and *intTn*) promoting excision of the element from the host replicon. In its circular form, transcription continues into the conjugation module (Figs. 1 and 6) presumably promoting transfer.

The regulatory model of Tn*916* has never been experimentally proven although it is fundamentally important. When considering the regulation of Tn*916;* it is not actually dependent on the presence or absence of tetracycline but on the translation rate, where an increased pool of charged t-RNAs is likely to result in the upregulation of Tn*916* genes. This means that any malfunction in the cell's translational apparatus will cause the translation rate to drop and therefore increase the tRNA concentration, which is expected to be deleterious to the cell. Tn*916* should be able to sense this response to cellular distress and respond by activating its own transcription and movement.

Table 1. Functions of orf24–13 of the Tn916 conjugation module

Coding Region	Closest Homolog	Percentage Identity and Coverage (%)	Accession Number	Experimental Evidence of Function	Reference
orf24	*Streptococcus agalactiae* 2603V/R Tn916 hypothetical protein	100, 97	NP_687949		
orf23	*Streptococcus infantis* ATCC 700779 conjugative transposon protein	99, 99	ZP_08061828		
orf22	*Peptostreptococcus anaerobius* 653–L conjugative transposon protein	99, 99	ZP_06424608		
orf21	*Streptococcus agalactiae* 2603V/R Tn916, FtsK/SpoIIIE family protein	100, 99	NP_687949	FtsK/SpoIIIE family protein; required for DNA segregation during cell division	61
orf20	*Streptococcus pneumoniae* putative conjugative transposon replication initiation factor	100, 99	CBW39427	Endonuclease which cleaves Tn916 at *oriT*	23
orf19	*Streptococcus agalactiae* 2603V/R Tn916 hypothetical protein	100, 99	NP_687949		
orf18	*Ureaplasma urealyticum* serovar 9 str. ATCC 33175 conjugative transposon protein	100, 99	ZP_03079519	Anti-restriction protein responsible for DNA modification immunity (Ard)	25
orf17	*Streptococcus pneumoniae* putative conjugative transposon membrane protein	99, 99	CBW38812		
orf16	*Enterococcus faecalis* TX0309B putative ATP/GTP-binding protein	100, 99	EFU87609		
orf15	*Peptoniphilus duerdenii* ATCC BAA-1640 conjugative transposon membrane protein	100, 95	ZP_07400188		
orf14	*Streptococcus agalactiae* 2603V/R Tn916, NLP/P60 family protein	100, 99	NP_687949	NLP/P60 extracellular lipoprotein	26
orf13	*Streptococcus suis* BM407 membrane protein	99, 99	YP_003028726		

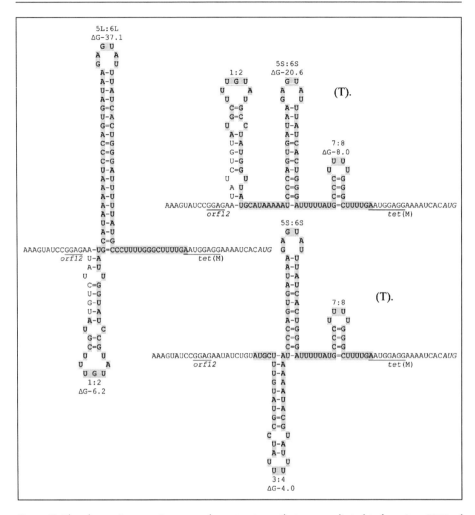

Figure 5. The alternative putative secondary structures that are predicted to form in mRNA of Tn*916*. The stem-loops named 1:2, 3:4 and 5S:6S are mutually exclusive to that of the 5L:6L. The free-energy values are shown in kcal/mol. The gray-shaded area is *orf12*. The Shine-Dalgarno sequences are underlined and labeled. The *tet*(M) start codon is italicised. The structures which include the terminators are labeled (T). Schematic adapted from: Roberts AP, Mullany P. Trends Microbiol 2009; 17:251-8;[4] ©2009 with permission from Elsevier.

Accessory Genes

Most of the Tn*916*/Tn*1545* family of elements possess *tet*(M) which encodes a ribosomal protection protein (RPP) conferring resistance to tetracyclines.[17,28] A number of other accessory genes have also been found among the Tn*916*/Tn*1545* family of transposons (Fig. 2, Table 2). These range from other RPP genes such as *tet*(S) and the efflux gene *tet*(L), a number of macrolide resistance genes such as *erm*(B) and *mef*(E), as well as the mercury resistance gene *mer*(A), the kanamycin resistance gene *aph*A, and the quarternary ammonium compound (QAC) resistance gene *qrg* (Table 2).

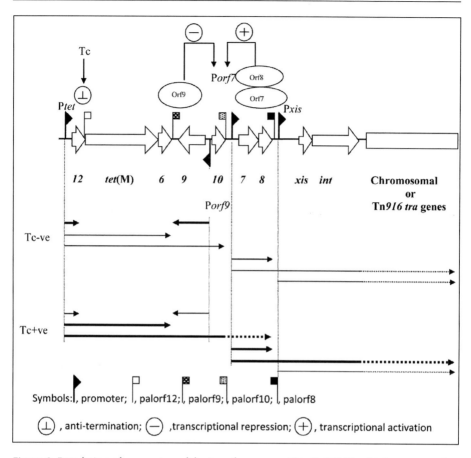

Figure 6. Regulation of expression of the transfer genes within Tn*916* The thick arrows under-neath the figure represent the majority of the transcripts, the thin lines represent lower levels of transcription. The dotted lines represent possible basal level, read-through transcripts. In the absence of tetracycline (Tc), most of the transcripts initiated at P*tet* terminate at pal*orf12*, P*orf9* transcribes *orf9* efficiently, whereas P*orf7* directs a low level of transcription through *orf7* and *orf8*. In these conditions, P*orf7* and P*xis* direct a low level of transcription through the transposition-associated and down stream genes. In the presence of Tc, the transcripts initiated at P*tet* read through pal*orf12*, pal*orf9* and pal*orf10* which leads to a decreased tran-scription of *orf9* and an increased transcription of *orf7* and *orf8*. The resulting overexpression of *orf7* and *orf8* stimulates the activity of P*orf7*, which leads to an increased transcription of the downstream genes. *orf9* could repress the activity of P*orf7*. (Adapted from ref. 60).

Tn916 Can Have Multiple Effects on the Host Genome

The insertion of Tn*916* into a genome can affect the host in various ways. An insertion within a gene can lead to loss or alteration of gene function. Insertion near genes may lead to polar effects, e.g., in some *E. faecalis* strains Tn*916* has inserted upstream of a hemolysin located on plasmid pAD1 which has resulted in its overexpression.[30] Another mechanism of introducing heritable change in host cells is by transporting non Tn*916* DNA (the coupling sequences) into the cell when the element transfers. In a study Tn*916* was used to create insertional mutations in *Desulfitobacterium dehalogenans*, an anaerobic organism capable

Table 2. Accessory genes found among the Tn916 family of transposons

Transposon	Assessory Genes	Function	Location	Reference
CTn1	ABC transporter	unknown substrate	orf13–7	56
Tn1545	aphA	kanamycin resistance	between orf20–19	62
Tn2009	MEGA mef(E)	macrolide resistance	between orf9–6	5
Tn2010	erm(B)	MLS* resistance	between orf21–20	63
Tn2010	MEGA mef(E)	macrolide resistance	between orf9–6	63
Tn2017	MEGA mef(E)	macrolide resistance	between orf9–6	64
Tn2017	Tn917 erm(B)	MLS resistance	between orf9–6	64
Tn3872	Tn917 erm(B)	MLS resistance	between orf9–6	65
Tn5386	spa	subtilisin immunity	in place of tet(M)	34
Tn6000	tet(S)	tetracycline resistance	in place of tet(M)	9
Tn6002	erm(B)	MLS resistance	between orf20–19	6
Tn6003	erm(B)	MLS resistance	between orf20–19	62
Tn6003	MAS erm(B)	MLS resistance	between orf20–19	62
Tn6003	aphA	kanamycin resistance	between orf20–19	62
Tn6009	mer(A)	mercury resistance	upstrem of orf24	7
Tn6079	tet(L)	tetracycline resistance	downstream of tet(M)	54
Tn6079	erm(T)	MLS resistance	upstream of orf5	54
Tn6087	smr	multidrug resistance	orf15	57
Tn916S	tet(S)	tetracycline resistance	in place of tet(M)	66

*MLS; Macrolide, lincosamide and streptogramin

of halorespiration. The transposon mutagenesis generated a relatively high percentage of halorespiration deficient mutants in the strain.[31] This study also revealed that the coupling sequences formed during recombination remained in some of the mutants in the empty target sites following excision of Tn916. Therefore, Tn916 was responsible for the introduction of short fragments of foreign DNA which remained after excision of the element causing deficiency in halorespiration. Another study has shown that Tn916 had replaced six nucleotides of its target site in *Erysipelothrix rhusiopathiae* with six alternative nucleotides, most likely the coupling sequence.[32] In this case the transposon was responsible for the replacement of six chromosomal nucleotides (AAACAA) by a six new nucleotides (GTATTA) as a result of its insertion and subsequent excision.

Interactions with Other Mobile Genetic Elements

When investigating the structure and functions of the Tn*916*/Tn*1545* family of transposons, it quickly becomes apparent that one is not looking at a static picture. These elements are constantly evolving and interacting with other MGEs including transposons, plasmids, insertion sequences and introns.

Group II introns have been found inserted into Tn*916*/Tn*1545*-like elements e.g., Tn*5397*,[33] Tn*5386*,[34] Tn*6000*,[9] and Tn*6084*.[10] All of the above are inserted into various *orfs* of the conjugation module of the host elements.

There are multiple insertions in Tn*916*/Tn*1545*-like elements which contain the *erm*(B) gene conferring resistance to macrolide, lincosamide and streptogramin (MLS) antibiotics. Tn*917*[35,36] is found upstream of the recombination module of the elements disrupting *orf9* in SPnRi3*erm*, Tn*3872* and Tn*2017*. In fact the transposition of Tn*917* has been found to be inducible by the presence of erythromycin,[37] much like Tn*916* transfer is thought to be induced by the presence of tetracycline although the molecular mechanisms of induction are different. The macrolide, aminoglycoside, streptothricin (MAS) element[38] has been found in SPnRi3*erm*, Tn*6003* and Tn*1545* within the conjugation module within *orf20*. Macrolide efflux genetic assembly (MEGA) elements[5,39] are also found among the Tn*916*/Tn*1545* family. MEGA, which includes the *mef*(E) efflux gene, has been found in the regulatory region between *orf6* and *orf9* of Tn*2009*, Tn*2010* and Tn*2017* (Fig. 2).

Multiple copies of Tn*916*/Tn*1545*-like elements have been found in various genomes and mobilisation of other Tn*916*/Tn*1545*-like elements has been shown to occur.[40] A recent study describes the presence of three highly similar elements (Tn*6085*a, Tn*6085*b and Tn*6084*) which are all found in one strain, *E. faecium* C68.[10] Interestingly the presence of three transposons does not significantly increase the organism's resistance to tetracycline. In a previous study by the same group, the presence of two related elements in the same strain (Tn*916* and Tn*5386*) resulted in the deletion of a large 178 kb genomic fragment suggesting interaction between the elements.[41] Further investigations indicated that excision of Tn*5386* was catalyzed by the Tn*916* integrase, IntTn, resulting in the simultaneous excision of both elements and the region between them. Another study investigating the target site of Tn*5397* demonstrated that introduction of Tn*916* to strains already containing Tn*5397*, resulted in its loss in > 95% of cases, presumably due to *trans* acting factors from the other element.[42]

Variations on the Tn*916* Theme

Tn*5397* from Clostridium difficile

Tn*5397* was originally identified in *Clostridium difficile* and has subsequently been found or transferred into *E. faecalis*,[43,44] *B. subtilis*,[45] and an oral *Streptococcus* sp.[46] A Tn*5397*-like element; Tn*1116*, has also been discovered in *Streptococcus pyogenes*.[19]

Instead of the tyrosine integrase and the excisionase genes Tn*5397* encodes a large serine recombinase; TndX which catalyzes recombination (Fig. 2).[42] This protein is related to the TnpX resolvase found in the chloramphenicol resistance encoding *Clostridium perfringens* and *Clostridium difficile* transposons, Tn*4451* and Tn*4453* respectively.[42,47] Copies of Tn*5397* are always flanked by a direct repeat of a GA dinucleotide, during excision endonucleolytic staggered cuts, mediated by TndX, occur at the GA leading to G/C and A/T Crick and Watson base pairing at the joint of the circular form.[42,43] Upon excision the target sequence is also regenerated.

The regulation of Tn*5397* is subtly different to that hypothesized for Tn*916*. While the ORFs *orf7* and *orf8* of Tn*5397* and Tn*916* are homologous and the promoters upstream of *orf7* (Porf7) and *tet*(M) are almost identical, an 88-bp deletion in Tn*5397* effectively removes *orf12* replacing this region with two alternative ORFs; *orf25* and *orf26*.[48] This deletion results in the disruption of the 5S:6S terminator (Fig. 5) which is predicted to be crucial for the regulation of *tet*(M) in

Tn*916*.[28] Despite these differences, we have shown, both at the phenotypic and genotypic level, that the expression of tetracycline resistance is still inducible (Roberts et al., unpublished). An alternative hypothesis for the regulation of Tn*5397* is currently being tested.

The other major difference between Tn*5397* and Tn*916* is that Tn*5397* has a group II intron inserted in the 3′ end of *orf14*. This intron has been shown to splice out of the pre-mRNA,[33] however splicing is not a prerequisite for conjugal transfer as Tn*5397* containing a mutant intron (with a kanamycin resistance gene inserted into the reverse transcriptase) incapable of splicing, could still transfer from *B. subtilis* to *C. difficile*. As the intron has inserted close to the 3′ end of *orf14* the interrupted gene can presumably still produce a functional protein.[33]

Tn*6000 from* Enterococcus casseliflavus

Tn*6000* was originally isolated from a cynomolgus monkey in a study investigating the microbiological effects of amalgam fillings.[49] Tn*6000* encodes a tyrosine integrase; Int6000 which is homologous to Int (42% identical) and Sip (41% identical), the integrases from the bovine staphylococcal pathogenicity islands SaPIbov and SaPIbov2, respectively (Fig. 4).[9,50] It has been shown that the element is flanked by perfect 18 bp direct repeats which are also found in the target site as well as the circular form.[18] This is also the case in SaPIbov2, although the 18bp sequences are different.

Tn*6000* contains some insertions and additions likely derived from diverse sources. Upstream of the conjugation region are a group of five genes of which four are predicted to be involved in restriction/modification and anti-restriction (Fig. 2). These genes are in addition to the Tn*916* anti-restriction gene *orf18* which Tn*6000* also possesses meaning that five of the 29 predicted ORFs are likely to be involved with protecting DNA against restriction enzymes. Next, there is an insertion of a fragment of DNA that shares nucleotide identity and gene order to a region of the virulence-related locus (*vrl*) from *Dichelobacter nodosus*, the causative agent of ovine foot rot.[51] The *vrl* is a 27.1-kb genomic island associated with more virulent strains of *D. nodosus*. It has also been found in *Desulfococcus multivorans*, indicating that it has undergone horizontal gene transfer. The *vrl* is hypothesized to undergo horizontal gene transfer possibly mediated by a bacteriophage such as DinoHI.[52] In Tn*6000*, the genes *vap* and *hel* (Fig. 2) are in the same order as *vrlR* and *vrlS*, a virulence-associated protein and a DEAD helicase of the Super-family 2 from *vrl*. The proteins Vap and Hel are 35% and 36% identical to VrlR and VrlS, respectively. The DEAD-DEAH helicases are involved in ATP-dependent unwinding of nucleic acids and may have a role in the conjugation process of Tn*6000*.

Finally, there is a group II intron present in exactly the same place as the one present in Tn*5397*. These two group II introns are > 99% identical to each other at the nucleotide level. It is therefore likely that the progenitors of one of these elements has previously inhabited the same cell as the other and acquired the intron, either by a retro-homing mechanism or by homologous recombination.

Tn*6079 from* Streptococcus gallolyticus

Tn*6079* was recently isolated from a fecal metagenomic fosmid library of a one month old healthy infant boy.[53] It is a composite transposon (28872 bp) carrying both tetracycline and erythromycin resistance genes (Fig. 2). The sequence and overall structure of Tn*6079* is highly similar to putative Tn*916*-like transposons detected in *S. gallolyticus*-like strains, and flanking sequences from the fosmid insert of Tn*6079* were used to assign the original host of the fragment to species level.[53]

Tn*6079* is located at the 3′ end of a gene predicted to encode protein L33 from the ribosomal 50S subunit. The element contains complete Tn*916*-like conjugation and recombination modules, but in the regulation module only *orf12* and *orf5* is present (Fig. 1 and 2). Regarding accessory genes, in addition to *tet*(M), Tn*6079* carries another tetracycline resistance gene, *tet*(L) (Table 2) predicted to encode an efflux protein and in addition it carries an erythromycin resistance gene, *erm*(T). The *tet*(L) gene is located just downstream of *tet*(M) and is closely linked to plasmid recombination/ mobilization (pre/mob) and replication (rep) genes. Next to this, *erm*(T) is surrounded by IS*1216* transposase sequences. Thus, apparently Tn*6079* has evolved by the integration of different MGEs.

Comparison of sequence and structure of Tn*6079* and corresponding MGEs detected in other *S. gallolyticus* strains[54,55] showed that the element with *erm*(T) and IS*1216* genes most likely was introduced into Tn*6079* by intraspecific genetic exchange.[53] Another accessory gene predicted to encode a cell-surface protein is located just upstream of *orf24*. This gene is highly similar to genes present in the end of CTn*1* from *C. difficile* 630 and Tn*5386* from *E. faecium*.[34,56] Finally, two hypothetical genes with unknown functions are present on both sides of *orf21*.

Tn*6087 from* Streptococcus oralis

Tn*6087* was isolated from *Streptococcus oralis* cultured from pooled saliva collected from healthy volunteers.[57] Its architecture is much like that of Tn*916* with the same functional modules and identity in both the regulation and recombination genes. The Tn*6087 tet*(M) is more similar to that found in an *E. faecalis* Tn*916*-like conjugative transposon (DQ223248) than the one in Tn*916* (U09422). The main differences are found in the conjugation region where a number of the ORFs are truncated and a 3 kb insertion is found within *orf15* (Fig. 2).

The insertion within *orf15* was found to be a composite transposon consisting of a QAC resistance gene (*qrg*) and a gene predicted to encode a hypothetical protein flanked by two nearly identical IS*1216* sequences. The Qrg protein sequence showed some limited identity (46–57%) to known small multidrug resistance (SMR) proteins. SMR proteins are known to increase resistance to QACs, and the MIC to cetyltrimethyl ammonium bromide (CTAB) was found to be higher than expected (64 µg/ml) in the *Streptococcus oralis* isolate. To determine whether the increase in resistance to CTAB was due to the presence of *qrg*, the gene was mutated by allelic replacement with a chloramphenicol resistance gene. Resistance to CTAB was indeed found to be lower (16 µg/ml) in the *qrg* deficient mutant, and was restored when the mutant was complemented with *qrg* on a plasmid (32 µg/ml). To investigate how widespread the novel *qrg* was, the gene was amplified from eight metagenomic DNA samples. One sample extracted from pooled saliva and pooled faecal samples taken from volunteers from four European countries.[58] All samples were found to be positive for the novel *qrg* gene and sequence analysis of the amplicons showed at least 97.81% identity to the Tn*6087 qrg*.

PCR analysis demonstrated that recombination occurred in this region resulting in a range of different molecules including the excision of the entire 3 kb composite transposon and a circular form of the *qrg* region and one of the IS elements was also detected (Fig. 7). As in the case of Tn*6079*, it seems likely that the IS*1216* elements were able to mobilise genes foreign to Tn*916* and insert them into this MGE thereby facilitating their spread.

Tn*6087* also differs from Tn*916* in that it has not been shown to have the ability to transfer by conjugation. Single nucleotide polymorphisms in the Tn*6087* sequence within the conjugation module resulted in a number of truncations within these ORFs: specifically in *orf24, orf20, orf16*, and *orf15*. Furthermore, the presence of the composite transposon within *orf15* was also likely to have an effect on conjugation ability. However, Tn*6087* could be transformed into another *Streptococcus* sp.[57]

Conclusion

The Tn*916*/Tn*1545* family of transposons is diverse and ubiquitous among many bacterial genera. Nearly all members of the family encode tetracycline resistance, but some also have resistance to other antimicrobial agents, such as macrolide antibiotics and some antiseptics. These traits give the host an advantage in certain environments, and importantly in the healthcare setting. Some of the Tn*916*/Tn*1545* family of elements have a composite structure which includes smaller mobile elements within a larger Tn*916*/Tn*1545*-like structure which are also mobilisable. Furthermore, the presence of these elements can have a number of effects on the host genome, from interruptions to genes to the deletions of large parts of the genome. All of the above traits make the Tn*916*/Tn*1545* family of elements important players in bacterial genome evolution. Their ability to modify their host's genotype and phenotype makes further study of these MGEs of high importance.

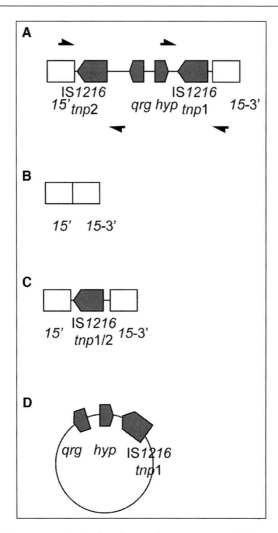

Figure 7. A schematic representing the four forms of the *qrg* gene and IS*1216* insertion found in Tn*6087*. A) The entire 3124 bp insertion within *orf15*; B) the entire insertion excised from *orf15*; C) only a chimeric form of the two IS*1216* sequences remaining; D) and a circular molecule consisting *qrg*, the hypothetical protein and the IS*1216* 1 sequence. The forms described were obtained by sequencing amplicons resulting from reactions using the primers shown in (A).

Questions for the Future

The major gaps in our understanding of Tn*916* and the members of its family lie within the regulation and conjugation modules of the elements. We have only minimal knowledge of the genes involved in conjugation which is one of the most important characteristics of these elements. There is also evidence that some of the genes within the regulation module are affected by the presence of tetracycline in the cell's environment which trigger the transfer of the element. It would also be interesting to investigate how the cells detect the presence of the antibiotic. We have described a high level of plasticity among the members of the Tn*916*/Tn*1545* family, with more being discovered all the time.

Acknowledgments

Research in the authors (PM and APR) laboratories is performed with financial support from the European Union Seventh Framework Programme (FP7/2007–2013) under grant agreement no. 241446 (ANTIRESDEV) and under grant agreement no. 223585 (HYPERDIFF) and the Medical Research Council (grant no. G0601176). AJ is funded by the Malaysian Government.

References

1. Franke AE, Clewell DB. Evidence for a chromosome-borne resistance transposon (Tn916) in Strepto-coccus-faecalis that is capable of conjugal transfer in the absence of a conjugative plasmid. J Bacteriol 1981; 145:494-502. PMID:6257641

2. Toussaint A, Merlin C. Mobile elements as a combination of functional modules. Plasmid 2002; 47:26-35. PMID:11798283 doi:10.1006/plas.2001.1552

3. Osborn AM, Boltner D. When phage, plasmids, and transposons collide: genomic islands, and conjuga-tive- and mobilizable-transposons as a mosaic continuum. Plasmid 2002; 48:202-12. PMID:12460536 doi:10.1016/S0147-619X(02)00117-8

4. Roberts AP, Mullany P. A modular master on the move: the Tn916 family of mobile genetic elements. Trends Microbiol 2009; 17:251-8. PMID:19464182 doi:10.1016/j.tim.2009.03.002

5. Del Grosso M, d'Abusco AS, Iannelli F, et al. Tn2009, a Tn916-like element containing mef(E) in Streptococcus pneumoniae. Antimicrob Agents Ch. Jun 2004;48(6):2037-2042.

6. Warburton PJ, Palmer RM, Munson MA, Wade WG. Demonstration of in vivo transfer of doxycycline re-sistance mediated by a novel transposon. J Antimicrob Chemother 2007; 60:973-80. PMID:17855723 doi:10.1093/jac/dkm331

7. Soge OO, Beck NK, White TM, et al. A novel transposon, Tn6009, composed of a Tn916 element linked with a Staphylococcus aureus mer operon. J Antimicrob Chemother 2008; 62:674-80. PMID:18583328 doi:10.1093/jac/dkn255

8. Mullany P, Pallen M, Wilks M, et al. A Group II intron in a conjugative transposon from the Gram-pos-itive bacterium, Clostridium difficile. Gene 1996; 174:145-50. PMID:8863741 doi:10.1016/0378-1119(96)00511-2

9. Brouwer MSM, Mullany P, Roberts AP. Characterization of the conjugative transposon Tn6000 from Enterococcus casseliflavus 664.1H1 (formerly Enterococcus faecium 664.1H1). FEMS Microbiol Lett 2010; 309:71-6. PMID:20528943

10. Rice LB, Carias LL, Rudin S, et al. Multiple copies of functional, Tet(M)-encoding Tn916-like elements in a clinical Enterococcus faecium isolate. Plasmid 2010; 64:150-5. PMID:20600284 doi:10.1016/j. plasmid.2010.06.003

11. Caparon MG, Scott JR. Excision and insertion of the conjugative transposon Tn916 involves a novel re-combination mechanism. Cell 1989; 59:1027-34. PMID:2557157 doi:10.1016/0092-8674(89)90759-9

12. Salyers AA, Shoemaker NB, Stevens AM, Li LY. Conjugative transposons - an unusual and diverse set of integrated gene-transfer elements. Microbiol Rev 1995; 59:579. PMID:8531886

13. Scott JR, Churchward GG. Conjugative transposition. Annu Rev Microbiol 1995; 49:367-97. PMID:8561465 doi:10.1146/annurev.mi.49.100195.002055

14. Jaworski DD, Flannagan SE, Clewell DB. Analyses of traA, int-Tn, and xis-Tn mutations in the conjugative transposon Tn916 in Enterococcus faecalis. Plasmid 1996; 36:201-8. PMID:9007015 doi:10.1006/plas.1996.0047

15. Rudy CK, Scott JR. Length of the coupling sequence of Tn916. J Bacteriol 1994; 176:3386-8. PMID:8195096

16. O'Keeffe T, Hill C, Ross RP. In situ inversion of the conjugative transposon Tn916 in En-terococcus faecium DPC3675. FEMS Microbiol Lett 1999; 173:265-71. PMID:10220904 doi:10.1111/j.1574-6968.1999.tb13511.x

17. Flannagan SE, Zitzow LA, Su YA, Clewell DB. Nucleotide-sequence of the 18-Kb conjugative trans-poson Tn916 from Enterococcus-faecalis. Plasmid 1994; 32:350-4. PMID:7899523 doi:10.1006/plas.1994.1077

18. Roberts AP, Davis IJ, Seville L, et al. Characterization of the ends and target site of a novel tetracycline resistance-encoding conjugative transposon from Enterococcus faecium 664.1H1. J Bacteriol 2006; 188:4356-61. PMID:16740942 doi:10.1128/JB.00129-06

19. Brenciani A, Bacciaglia A, Vecchi M, et al. Genetic elements carrying erm(B) in Streptococcus pyogenes and association with tet(M) tetracycline resistance gene. Antimicrob Agents Ch. Apr 2007;51(4):1209-1216.

20. Cookson AL, Noel S, Hussein H, et al. Transposition of Tn916 in the four replicons of the Butyrivibrio proteoclasticus B316(T) genome. FEMS Microbiol Lett 2011; 316:144-51. PMID:21204937 doi:10.1111/j.1574-6968.2010.02204.x

21. Wang H, Roberts AP, Mullany P. DNA sequence of the insertional hot spot of Tn916 in the Clostridium difficile genome and discovery of a Tn916-like element in an environmental isolate integrated in the same hot spot. FEMS Microbiol Lett 2000; 192:15-20. PMID:11040422 doi:10.1111/j.1574-6968.2000.tb09352.x

22. Senghas E, Jones JM, Yamamoto M, et al. Genetic organization of the bacterial conjugative transposon Tn916. J Bacteriol 1988; 170:245-9. PMID:2826394

23. Rocco JM, Churchward G. The integrase of the conjugative transposon Tn916 directs strand- and sequence-specific cleavage of the origin of conjugal transfer, oriT, by the endonuclease Orf20. J Bacteriol 2006; 188:2207-13. PMID:16513750 doi:10.1128/JB.188.6.2207-2213.2006

24. Jaworski DD, Clewell DB. A functional origin of transfer (oriT) on the conjugative transposon Tn916. J Bacteriol 1995; 177:6644-51. PMID:7592445

25. Serfiotis-Mitsa D, Roberts GA, Cooper LP, et al. The orf18 Gene Product from conjugative transposon Tn916 is an ArdA antirestriction protein that inhibits Type I DNA restriction-modification systems. J Mol Biol 2008; 383:970-81. PMID:18838147 doi:10.1016/j.jmb.2008.06.005

26. Rigden DJ, Jedrzejas MJ, Galperin MY. Amidase domains from bacterial and phage autolysins define a family of gamma-D,L-glutamate-specific amidohydrolases. Trends Biochem Sci 2003; 28:230-4. PMID:12765833 doi:10.1016/S0968-0004(03)00062-8

27. Clewell DB, Flannagan SE, Jaworski DD. Unconstrained bacterial promiscuity - the Tn916-Tn1545 family of conjugative transposons. Trends Microbiol 1995; 3:229-36. PMID:7648031 doi:10.1016/ S0966-842X(00)88930-1

28. Su YA, Ping H, Clewell DB. Characterization of the tet(M) determinant of Tn916 - evidence for regulation by transcription attenuation. Antimicrob Agents Chemother. 1992; 36:769-78. PMID:1323953

29. Celli J, Trieu-Cuot P. Circularization of Tn916 is required for expression of the transposon-encoded transfer functions: characterization of long tetracycline-inducible transcripts reading through the attachment site. Mol Microbiol 1998; 28:103-17. PMID:9593300 doi:10.1046/j.1365-2958.1998.00778.x

30. Ike Y, Flannagan SE, Clewell DB. Hyperhemolytic phenomena associated with insertions of Tn916 into the hemolysin determinant of Enterococcus faecalis Plasmid Pad1. J Bacteriol 1992; 174:1801-9. PMID:1312528

31. Smidt H, Song DL, van der Oost J, de Vos WM. Random transposition by Tn916 in Desulfitobacterium dehalogenans allows for isolation and characterization of halorespiration-deficient mutants. J Bacteriol 1999; 181:6882-8. PMID:10559152

32. Shimoji Y, Mori Y, Sekizaki T, et al. Construction and vaccine potential of acapsular mutants of Erysipelothrix rhusiopathiae: Use of excision of Tn916 to inactivate a target gene. Infect Immun 1998; 66:3250-4. PMID:9632592

33. Roberts AP, Braun V, von Eichel-Streiber C, Mullany P. Demonstration that the group II intron from the clostridial conjugative transposon Tn5397 undergoes splicing in vivo. J Bacteriol 2001; 183:1296-9. PMID:11157942 doi:10.1128/JB.183.4.1296-1299.2001

34. Rice LB, Carias LL, Marshall SH, et al. Characterization of Tn5386, a Tn916-related mobile element. Plasmid 2007; 58:61-7. PMID:17408741 doi:10.1016/j.plasmid.2007.01.002

35. Tomich PK, An FY, Clewell DB. Transposon (Tn917) in Streptococcus faecalis that exhibits enhanced transposition during induction of drug-resistance. Cold Spring Harb Sym. 1978;43:1217-1221.

36. Shaw JH, Clewell DB. Complete nucleotide-sequence of "macrolide-lincosamide-streptogramin B-resistance transposon Tn917 in Streptococcus faecalis. J Bacteriol 1985; 164:782-96. PMID:2997130

37. Tomich PK, An FY, Clewell DB. Properties of Erythromycin-inducible transposon-Tn917 in Streptococcus faecalis. J Bacteriol 1980; 141:1366-74. PMID:6245068

38. Cochetti I, Tili E, Vecchi M, et al. New Tn916-related elements causing erm(B)-mediated erythromycin resistance in tetracycline-susceptible pneumococci. J Antimicrob Chemother 2007; 60:127-31. PMID:17483548 doi:10.1093/jac/dkm120

39. Maria-Marimon J, Valiente A, Ercibengoa M, et al. Erythromycin resistance and genetic elements carrying macrolide efflux genes in Streptococcus agalactiae. Antimicrob Agents Chemother. 2005; 49:5069-74 PMID:16304174 doi:10.1128/AAC.49.12.5069-5074.2005.

40. Flannagan SE, Clewell DB. Conjugative transfer of Tn916 in Enterococcus faecalis - Transactivation of homologous transposons. J Bacteriol 1991; 173:7136-41. PMID:1657880

41. Rice LB, Carias LL, Marshall S, et al. Tn5386, a novel Tn916-like mobile element in Enterococcus faecium D344R that interacts with Tn916 to yield a large genomic deletion. J Bacteriol 2005; 187:6668-77. PMID:16166528 doi:10.1128/JB.187.19.6668-6677.2005

42. Wang H, Roberts AP, Lyras D, et al. Characterization of the ends and target sites of the novel conjugative transposon Tn5397 from Clostridium difficile: Excision and circularization is mediated by the large resolvase, TndX. J Bacteriol 2000; 182:3775-83. PMID:10850994 doi:10.1128/JB.182.13.3775-3783.2000

43. Jasni AS, Mullany P, Hussain H, Roberts AP. Demonstration of conjugative transposon (Tn5397)-mediated horizontal gene transfer between Clostridium difficile and Enterococcus faecalis. Antimicrob Agents Chemother. 2010; 54:4924-6. PMID:20713671 doi:10.1128/AAC.00496-10

44. Agersø Y, Pedersen AG, Aarestrup FM. Identification of Tn5397-like and Tn916-like transposons and diversity of the tetracycline resistance gene tet(M) in enterococci from humans, pigs and poultry. J Antimicrob Chemother 2006; 57:832-9. PMID:16565159 doi:10.1093/jac/dkl069

45. Mullany P, Wilks M, Puckey L, Tabaqchali S. Gene cloning in Clostridium difficile using Tn916 as a shuttle conjugative transposon. Plasmid 1994; 31:320-3. PMID:8058827 doi:10.1006/plas.1994.1036

46. Roberts AP, Pratten J, Wilson M, Mullany P. Transfer of a conjugative transposon, Tn5397 in a model oral biofilm. FEMS Microbiol Lett 1999; 177:63-6. PMID:10436923 doi:10.1111/j.1574-6968.1999.tb13714.x

47. Lyras D, Adams V, Lucet I, Rood JI. The large resolvase TnpX is the only transposon-encoded protein required for transposition of the Tn4451/3 family of integrative mobilizable elements. Mol Microbiol 2004; 51:1787-800. PMID:15009902 doi:10.1111/j.1365-2958.2003.03950.x

48. Roberts AP, Johanesen PA, Lyras D, Mullany P, Rood JI. Comparison of Tn5397 from Clostridium difficile, Tn916 from Enterococcus faecalis and the CW459tet(M) element from Clostridium perfringens shows that they have similar conjugation regions but different insertion and excision modules. Microbiology 2001; 147:1243-51. PMID:11320127

49. Summers AO, Wireman J, Vimy MJ, et al. Mercury released from dental "silver" fillings provokes an increase in mercury- and antibiotic-resistant bacteria in oral and intestinal floras of primates. Antimicrob Agents Chemother 1993; 37:825-34. PMID:8280208

50. Ubeda C, Tormo MA, Cucarella C, et al. Sip, an integrase protein with excision, circularization and integration activities, defines a new family of mobile Staphylococcus aureus pathogenicity islands. Mol Microbiol 2003; 49:193-210. PMID:12823821 doi:10.1046/j.1365-2958.2003.03577.x

51. Billington SJ, Huggins AS, Johanesen PA, et al. Complete nucleotide sequence of the 27-kilobase virulence related locus (vrl) of Dichelobacter nodosus: Evidence for extrachromosomal origin. Infect Immun 1999; 67:1277-86. PMID:10024571

52. Cheetham BF, Parker D, Bloomfield GA, et al. Isolation of the bacteriophage DinoHI from Dichelobacter nodosus and its interactions with other integrated genetic elements. Open Microbiol J. 2008; 2:1-9. PMID:19088904 doi:10.2174/1874285800802010001

53. de Vries LE, Valles Y, Agerso Y, et al. The gut as reservoir of antibiotic resistance: microbial diversity of tetracycline resistance in mother and infant. PLoS ONE 2011; 6:e21644 doi:10.1371/journal.pone.0021644.

54. Tsai JC, Hsueh PR, Chen HJ, et al. The erm(T) gene is flanked by IS1216V in inducible erythromycin-resistant Streptococcus gallolyticus subsp. pasteurianus. Antimicrob Agents Chemother. 2005; 49:4347-50. PMID:16189118 doi:10.1128/AAC.49.10.4347-4350.2005

55. Rusniok C, Couve E, Da Cunha V, et al. Genome sequence of Streptococcus gallolyticus: Insights into its adaptation to the bovine rumen and its ability to cause Endocarditis. J Bacteriol 2010; 192:2266-76. PMID:20139183 doi:10.1128/JB.01659-09

56. Sebaihia M, Wren BW, Mullany P, et al. The multidrug-resistant human pathogen Clostridium difficile has a highly mobile, mosaic genome. Nat Genet 2006; 38:779-86. PMID:16804543 doi:10.1038/ng1830

57. Ciric L, Mullany P, Roberts AP. Antibiotic and antiseptic resistance genes are linked on a novel mobile genetic element: Tn6087. J Antimicrob Chemother 2011; In press.

58. Seville LA, Patterson AJ, Scott KP, et al. Distribution of tetracycline and erythromycin resistance genes among human oral and fecal metagenomic DNA. Microb Drug Resist 2009; 15:159-66. PMID:19728772 doi:10.1089/mdr.2009.0916

59. Ludwig W, Strunk O, Westram R, et al. ARB: a software environment for sequence data. Nucleic Acids Res 2004; 32:1363-71. PMID:14985472 doi:10.1093/nar/gkh293

60. Celli J, Trieu-Cuot P. Circularization of Tn916 is required for expression of the transposon-encoded transfer functions: characterization of long tetracycline-inducible transcripts reading through the attachment site. Mol Microbiol 1998; 28:103-17. PMID:9593300 doi:10.1046/j.1365-2958.1998.00778.x

61. Wu LJ, Errington J. Bacillus subtilis SpoIIIE protein required for DNA segregation during asymmetric cell division. Science 1994; 264:572-5. PMID:8160014 doi:10.1126/science.8160014

62. Cochetti I, Tili E, Mingoia M, et al. erm(B)-carrying elements in tetracycline-resistant pneumococci and correspondence between Tn1545 and Tn6003. Antimicrob Agents Chemother. 2008; 52:1285-90. PMID:18285489 doi:10.1128/AAC.01457-07

63. Del Grosso M, Camilli R, Iannelli F, et al. The mef(E)-carrying genetic element (MEGA) of Streptococcus pneumoniae: Insertion sites and association with other genetic elements. Antimicrob Agents Chemother. 2006; 50:3361-6. PMID:17005818 doi:10.1128/AAC.00277-06

64. Del Grosso M, Camilli R, Libisch B, et al. New composite genetic element of the Tn916 family with dual macrolide resistance genes in a Streptococcus pneumoniae isolate belonging to clonal Complex 271. Antimicrob Agents Chemother. 2009; 53:1293-4. PMID:19104015 doi:10.1128/AAC.01066-08

65. McDougal LK, Tenover FC, Lee LN, et al. Detection of Tn917-like sequences within a Tn916-like conjugative transposon (Tn3872) in erythromycin-resistant isolates of Streptococcus pneumoniae. Antimicrob Agents Chemother. 1998; 42:2312-8. PMID:9736555

66. Lancaster H, Roberts AP, Bedi R, et al. Characterization of Tn916S, a Tn916-like element containing the tetracycline resistance determinant tet(S). J Bacteriol 2004; 186:4395-8. PMID:15205444 doi:10.1128/JB.186.13.4395-4398.2004

CHAPTER 10

Tn*1549* and Closely Related Elements

Thierry Lambert*

Abstract

T
n*1549* and closely related elements are largely responsible for VanB-type resistance in enterococci. The presence on these elements of three functional modules involved in vancomycin resistance, transposition, and intercellular transfer is indicative of conjugative transposons. Their transposition module is very similar to that of Tn*916* and leads to formation of detectable circular intermediate. However, Tn*1549*-type elements are passively transferred between *Enterococcus* spp. by plasmid or large chromosomal fragments. They have been also detected in anaerobes, and their capacity to transfer to *Enterococcus* was established. They possess a functional mobilization structure composed by an origin of transfer and mobilization proteins, which enable the conjugation process. Detection of *vanB* gene in stool samples in the absence of cultivable vancomycin-resistant enterococci suggests that anaerobes constitute the reservoir for Van-B type resistance and that Tn*1549* is the vehicle involved for its mobilization.

Introduction

Emergence of vancomycin resistance in enterococci was reported in 1986,[1,2] approximately 30 years after the introduction of this antibiotic into clinical practice. Since then, glycopeptide-resistant enterococci represent important nosocomial pathogens.[3] Glycopeptide resistance in enterococci is due to gene operons which lead to synthesis of modified peptidoglycan precursors ending in D-Ala-D-Lac (VanA, VanB, VanD, VanM) or D-Ala-D-Ser (VanC, VanE, VanG, VanL and VanN) combined with the elimination of precursors produced by the ligase of the host ending in D-Ala-D-Ala.[4-6] VanA and VanB are the most frequently encountered types and are responsible for more than 95% of the vancomycin-resistant enterococci (VRE) isolates.[5,7] The *van* operons are under control of two-component regulatory systems directing their transcription. In the case of the *vanB* operon, vancomycin is an inducer but not teicoplanin. The *van* genes origin remains hypothetical, however recent studies indicated that *vanA* might originate from soil microorganisms[8,9] whereas *vanB* might arise from gene transfer from human intestinal microbiota.[10]

Tn*5382* and Tn*1549* are quasi identical elements which were discovered in VanB-type vancomycin resistant *Enterococcus* (VRE) at the end of the 90s. They contained the *vanB* operon conferring vancomycin but not teicoplanin resistance. These elements possess features of conjugative transposon via three functional modules. Compared to Tn*916* the transposition modules are very similar, whereas conjugative modules are completely distinct. Tn*1549* was found located both on chromosome and plasmids. Despite, the presence of detectable circularized form of the transposon, movement of Tn*1549* in enterococci was attributed to passive processes mediated by plasmids or exchange of large chromosomal fragments.[11-14] More recently, these elements were also detected in anaerobes, and their capacity to transfer to *Enterococcus* has brought a different perspective to viewing these elements.[15]

*Département de Microbiologie, UFR de Pharmacie, Université Paris 11, Châtenay-Malabry, France. Email: thierry.lambert@u-psud.fr

Bacterial Integrative Mobile Genetic Elements, edited by Adam P. Roberts and Peter Mullany. ©2013 Landes Bioscience.

Structure of Tn*1549*

The sequence of Tn*1549* from *Enterococcus faecalis* E93/268 was entirely characterized,[12] meanwhile that of Tn*5382* was partially determined.[16] More recently the complete sequence of the Tn*1549*-like element borne by the pheromone responsive plasmid pGM2200 was also reported.[17] In addition, various portions of Tn*1549*-like elements found in distinct host are available in data banks. All these elements are closely related. Tn*1549* is a ca. 34 kb element which contains 32 Orfs and is likely organized such as Tn*916* family of conjugative transposons throughout three functional regions.[12,18] The genetic structure of Tn*1549* is shown in Figure 1. The "conjugation" module at the left end is formed by two gene clusters. The first contains 14 Orfs (*orfs 13* to *orf26*) in the same orientation. Inside this cluster *orf24* and *orf25* are separated by a terminator. In contrast, the second gene cluster carrying the relaxase gene is transcribed in the opposite orientation. The resistance module involves the *vanB* operon, whereas the latter module located at the right end of the element correspond to the transposition function containing *orf7*, *orf8*, *xis* and *int* genes. Imperfect inverted repeats terminate the extremities of the element. The G+C contents of the transposition module and of the conjugation module are significantly different (55.4 and 38%, respectively). This suggests that these modules have different origins, and such as most of Integrative and Conjugative Elements (ICEs), Tn*1549* has evolved by modules exchanges.[18]

The Resistance Module: The *vanB* Operon

The *vanB* operon is composed of the *vanR*$_B$*S*$_B$ gene encoding a two-component regulatory system and the *vanY*$_B$*WH*$_B$*BX*$_B$ resistance genes that are transcribed from promoters P_{RB} and P_{YB} respectively.[19] Heterogeneity in *vanB* gene led to the description of three subtypes van-*B1*, -*B2*, and -*B3* on the basis of differences in the sequence of the *vanB* ligase.[20] These mutations do not affect the vancomycin resistance level.[5] The *vanB2* determinant is part of conjugative transposon Tn*1549* and closely related elements, whereas van-*B1*, and van-*B3* are found in composite transposons.[20,21] Tn*1549* from *E. faecalis* E93/268 is composed of 33,799 nucleotides and differs from its variant in pGM2200 by 94 point mutations (99.7% identity). Curiously, 76 of these mutations are focused on a 7,252-bp fragment which contains the *vanB2* operon. This finding shows that polymorphism is significantly more important in the resistance module than in the transposition and the transfer modules. In addition, insertion sequences have been reported into Tn*1549*, such as, ISEnfa*200* and ISEnfa*150* inserted within the intergenic region upstream from *vanY*$_B$ and *orfC* (located between *vanX*$_B$ and *orf7*) respectively.[22,23]

As mentioned, the *vanB2* operon, which is part of Tn*1549*-like elements is the most prevalent *vanB* subtype responsible for vancomycin resistance in *Enterococcus* all around the world. A particular high prevalence of *vanB* was observed in Australia.[24,25] VanB phenotype is characterized by inducible resistance with various levels to vancomycin (MIC, 4 to 1,024 µg/ml) and susceptibility to teicoplanin.[5] Several studies reported high rates of *vanB* carriage in the absence of cultivable VRE in fecal samples.[24-26] This result was attributed to the presence of commensal anaerobes harbouring Tn*1549*. Culture of various gram positive anaerobes carrying Tn*1549*-like elements from stool samples along control for detection of VRE in Australia and Canada, led to identification of *Clostridium bolteae* MLG080-1 and *Clostridium hathewayi* MLG392, *Clostridium lavalense*, *Clostridium symbiosum* MLG101, *Clostridium* sp CCRI-9842, *Clostridium* sp. MLG055, *Clostridium* sp. MLG480, *Eggerthella lenta* MLG043, and *Ruminococcus* sp. MLG080-3.[10,27-29] *Atopobium minutum* CIP 110250 and *Clostridium clostridioforme* CIP 110249 are two VanB2 anaerobes carrying a Tn*1549* element isolated recently in France from an infected wound and peritoneal fluid, respectively (submitted). The isolation of *vanB*-anaerobes from clinical samples exemplifies the risk of VanB-type resistance to disseminate from the intestinal microbiota, which clearly constitutes a reservoir for antibiotic resistance.

The Transposition Module

This module is formed by *orf7*$_{Tn1549}$, *orf8*$_{Tn1549}$, *int* and *xis*. Corresponding proteins Orf7, Orf8, Int, and Xis show similarity with their homologous from Tn*916* (30%, 28%, 68%, and 86% identity, respectively). Int belongs to the tyrosine recombinase super family which includes the lambda

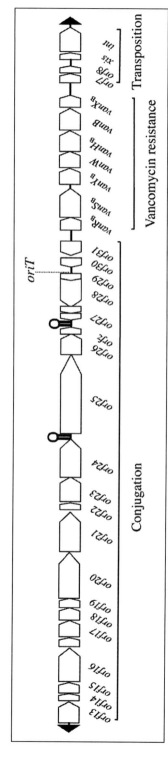

Figure 1. The shematic representation of Tn1549 and of its functional modules. Open arrows represent ORFs and indicate the sense of transcription. Large stem-loops are indicated.

site-specific integrase.[30] The transposition mechanism has been extensively analyzed for Tn916. It was demonstrated that Int$_{Tn916}$ and Xis$_{Tn916}$ cooperate for excision,[30] Int$_{Tn916}$ creates staggered cleavages at the extremities of the transposon (the overlap at the target level are designated the coupling sequences) leading to the formation of a heteroduplex structure during circular intermediate formation (CI) by joining of the extremities.[31,32] The N-terminal domain binds the repeated sequences (DR2) as reported.[33-35] Conserved DR (95.5% identity with DR2 of Tn916) are present in the extremities of Tn1549 and are also involved in binding Int as shown by gel retardation (unpublished).

The transposition of Tn1549 was initially suggested by detection of a circular intermediate, detected by nested PCR.[12] Afterward, transposition event was demonstrated by pulsed-field gel electrophoresis in 48 *Enterococcus faecium* transconjugants obtained by horizontal transfer from *C. symbiosum* MLG101.[15] Sequencing of the flanking regions of Tn1549-like has revealed 29 unique integration events in 26 loci in the *E. faecium* genome, and two hot spots for insertion were identified. Target sequences corresponded to a Trich region separated from an A-rich region by ca. six nucleotides (coupling sequence). Integration of the transposon was associated with the acquisition of 5 (n=18) or 6 (n=7) bp of donor DNA, in agreement with the model of Tn916 transposition which involves a tyrosine recombinase and coupling sequences leading to a non-replicative circular intermediate.[15] One transconjugant had two insertions, a feature also reported for Tn916.[32] Surprisingly, in five transconjugants the transposition of Tn1549 occurred with a 5-bp duplication of target DNA, a phenomenon which remains to be explained. Sequencing of circular intermediates in *A. minutum* CIP 110250 has also revealed coupling sequences of 7-bp (unpublished).

The Conjugation Module

As already mentioned, Tn1549 possesses features of conjugative transposons of the Tn916 family, however, transfer in *Enterococcus* was shown to occur passively as an integral part of variable-size chromosomal fragments or plasmids. Conjugative transfer of Tn1549-like elements from *C. symbiosum* to *Enterococcus* spp has been demonstrated both in vitro and in vivo.[15] However, attempts to retransfer Tn1549 from transconjugants were unsuccessful, except a single event obtained after introduction of Tn916. These results strongly suggest that Tn1549 is a mobilizable transposon. Transfer also occurred from *Clostridium bolteae* MLG080-1 and *Clostridium hathewayi* MLG392 but not from *Atopobium minutum* CIP 110250, *Clostridium clostridioforme* CIP 110249, *Clostridium* sp. MLG055, *Clostridium* sp. MLG480, and *Ruminococcus* sp. MLG080-3 (unpublished).

In fact, Tn1549 has the capacity to be mobilized by heterologous transfer system when plasmids or ICEs are present in the bacterial host.[36] This accounts for the observation that not all the *vanB*-anaerobe strains are able to transfer Tn1549.

Organization of Tn1549 is very similar to that of Tn916 in respect of conservation of the polarity of transcription of the functional modules. This suggests, that similarity can also concern the role of antibiotic in transfer, such as tetracycline known to increase the frequency of transfer of Tn916.[37]

The conjugation module of Tn1549 is part of a 24-kb DNA region unrelated with that of Tn916 (12 kb). It displays homology with sequences of ICEs found in anaerobes such as in genome of *Faecalibacterium prausnitzii* SL3/3, CTn4 of *C. difficile* 630,[38] TnCM1 a conjugative transposon of *Clostridium innocuum* (unpublished). The function of Orfs 16 to 26 is only deduced from sequence comparisons and remains to be established.

Orf16 is a putative ATPase with Walker A and Walker B motifs which displays similarity to TrsK protein of *Staphylococcus aureus* (29% identity) and could play the role of a VirD4-like coupling protein. To promote the conjugative process ATPases are needed to energize DNA and protein substrate transfer.[39] VirD4 proteins are thought to mediate interactions between the DNA-processing formed by relaxosome and mating pair formation of type IV secretion systems (T4SS).[40-43] Orf20 has significant similarity to TrsE of pGO1 (27% identity) a VirB4 homologue. Orf21 is a putative cell wall hydrolase, and Orf23 is a putative bacteriocin. Orf24 is homologous to Topoisomerase III (Topo III). Topo III homologues have been found in broad host range plasmid and conjugative transposons and may have an essential role in conjugative DNA transfer.[44]

Orf25 presents similarity with a portion of the LtrC-like protein involved in conjugation of *Staphylococcus aureus* plasmid pSK41.[45] In contrast, the function of Orf26 remains unknown.

The second gene cluster, including the *orfs* 27 to 31 and *orfz* corresponds to the mobilization functions. This cluster is transcribed in opposite direction to the other genes of Tn*1549*. Orf31 is a putative regulatory protein of the LysR family. Orf30 is a Helix-Turn-Helix (HTE) XRE-superfamily like protein. Orf28 is a relaxase belonging to the 3H family of MOB_p superfamily relaxases. DNA relaxases are key enzymes in the initiation of the conjugative transfer by catalyzing single-strand cleavage of a phosphodiester bond at the nick site (*nic*) in a specific and reversible manner.[46] A minimal 28-bp *oriT* has been delimited in a mobilization assay in *Escherichia coli*. It has also been established that Rlx_{Tn1549} requires the cis acting protein $mobC_{Tn1549}$ (Orf29), whereas Orf27 and OrfZ do not influence mobilization. The *nic* site has been functionally identified, and potential related *oriTs* consisting in a consensus motif CTRTGCTTG'CT, adjacent to an inverted repeat have been detected associated with both putative *mobC* and *rlx* as part of various ICEs from data bank. Interestingly, a plasmid vector containing $oriT_{Tn1549}$ can be mobilized in *E. coli* harbouring the RP4 transfer system provided that *rlx* and *mobC* are delivered in *trans*.[36] This supports the notion that the relaxosome resulting in the cleavage by Rlx interacts with the coupling protein TraG, and therefore the capacity of Tn*1549* to be mobilized by heterologous transfer systems. This is a remarkable feature of this mobilization system, since coupling proteins are normally specific.[47]

Transfer experiments from *C. symbiosum* to *Enterococcus* spp. have revealed that vancomycin is not necessary for transfer to occur, however a role in facilitating transfer remained possible. Unfortunately, the impact of vancomycin cannot be deduced by analysis transfer frequency due to a lack of reproducibility in experiments combined to a very low frequency (ca. 10^{-9}). The detection of transconjugants is also made difficult because *vanB* can confer a low level of resistance to vancomycin and requires induction for phenotypic expression. For these reasons, the effect of vancomycin was evaluated on the transcription of Tn*1549* genes. The mRNA corresponding to *orfs* representative of each functional module was quantified by qRT-PCR in presence or absence of vancomycin (manuscript in preparation). Search for a cotranscription was performed by amplifying trans-gene fragments from the cDNA of Tn*1549* by PCR. Briefly (Fig. 2), (i) a cotranscription was observed for *orf16* to *orf23*, and for *orf7, orf8, xis,* and *int*; (ii) qRT-PCR experiments indicated that in presence of vancomycin the transcription of the *vanB* operon increased enabling transcription

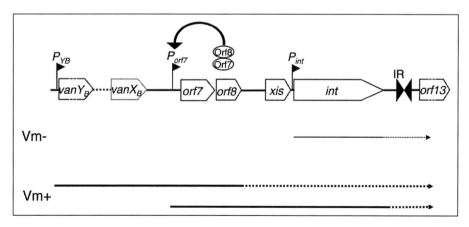

Figure 2. Regulation of the transposition module of Tn*1549*. In the absence of vancomycin the *vanB* operon is not transcribed due to a tight regulation by the two component regulatory system. *orf7, orf8,* and *xis* are down-regulated, in contrast *int* is transcribed by its own promoter. In the presence of vancomycin the *vanB* operon is induced leading to an increased transcription of *orf7* and *orf8*. The resulting overexpression of ORF7 and ORF8 enhance the activity of P*orf7* leading to an increased expression of *orf7, orf8, xis,* and at a lesser measure *int,* and reading through the joined ends of the transposon *orf16* and downstream *orfs*.

from promoter P_{YB} of the $orf7_{Tn1549}$ and $orf8_{Tn1549}$ located just downstream from the terminator of the *vanB* operon; (iii) P_{orf7} directs the transcription of xis_{Tn1549} and int_{Tn1549} and the transfer function by reading through the joined ends of circularized transposon. Synthesis of Orf7$_{Tn1549}$ and Orf8$_{Tn1549}$ activates the promoter P_{orf7} as demonstrated in *E. faecalis* JH2-2 by cloning the *cat* gene as reporter under the control of P_{orf7} and by delivering *orf7* and *orf8* in *trans*; (iv) mRNA quantification indicated that vancomycin increases significantly transcription of *orf7*, *orf8*, and *xis*, and at a lesser extent *int*, which is in part transcribed from its own promoter. In contrast, the transcription of the *orf16-23* operon was weakly affected by vancomycin; (v) real-time PCR indicated that vancomycin increases the quantity of circular intermediate by about ten fold. Therefore, vancomycin influences at least the recombination of Tn*1549* by a regulation by attenuation similar to that reported in Tn*916* which involves tetracycline as inducer and two small regulatory genes (*orf7* and *orf8*) downstream from *tetM*.[48] Excision is the first step in movement of conjugative transposons, the synthesis of circular intermediate is thus crucial in the transfer process. This suggests that role of vancomycin in transfer of Tn*1549* is perhaps underestimated.

Fitness Cost of VanB-Type Resistance in Enterococci

Recently, the biological cost of inducible or constitutive expression of VanB-type resistance and of carriage of Tn*1549* was evaluated in an isogenic set of *E. faecalis*.[49] Tn*1549* carriage does not alter fitness of *E. faecalis*, and noninduced VanB-type resistance is not costly for the host. In contrast, expression of VanB reduces the fitness significantly and constitutive expression which leads to an altered peptidoglycan is very costly. These results are consistent with the necessity of a tight regulation mechanism for the maintenance of resistance when antibiotic pressure is absent. They also account for the worldwide dissemination of VanB resistance carried by Tn*1549*.

Conclusion

Analysis of the world of ICEs suggests that these elements have evolved by exchanges of the functional modules.[50] The *vanB* operon would be recruited by an ancestor transposon in anaerobic bacteria leading to Tn*1549* related elements. These elements are widely spread among human intestinal microbiota and constitute a reservoir of *vanB*. Their ability to transpose via a circular intermediate combined with the presence of a transfer origin enables their transfer by mobilization. In contrast to plasmids, conjugative or mobilizable transposons do not require a replication origin, a feature which enhances their potential to disseminate in various bacteria. Therefore, Tn*1549* and related elements represent a real threat for dissemination of vancomycin resistance to pathogenic microorganisms from a silent reservoir.

References

1. Leclercq R, Derlot E, Duval J, et al. Plasmid-mediated resistance to vancomycin and teicoplanin in Enterococcus faecium. N Engl J Med 1988; 319:157-61. PMID:2968517 doi:10.1056/NEJM198807213190307
2. Uttley AH, Collins C, Naidoo J, et al. Vancomycin-resistant enterococci. Lancet 1988; 1:57-8. PMID:2891921 doi:10.1016/S0140-6736(88)91037-9
3. Murray BE. Vancomycin-resistant enterococcal infections. N Engl J Med 2000; 342:710-21. PMID:10706902 doi:10.1056/NEJM200003093421007
4. Arthur M, Reynolds P, Courvalin P. Glycopeptide resistance in enterococci. Trends Microbiol 1996; 4:401-7. PMID:8899966 doi:10.1016/0966-842X(96)10063-9
5. Courvalin P. Vancomycin resistance in gram-positive cocci. Clin Infect Dis 2006; 42:S25-34. PMID:16323116 doi:10.1086/491711
6. Xu X, Lin C, Yan G, et al. vanM, a new glycopeptide resistance gene cluster found in Enterococcus faecium. Antimicrob Agents Chemother 2010; 54:4643-7. PMID:20733041 doi:10.1128/AAC.01710-09
7. Toye B, Shymanski M, Bobrowska M, et al. Clinical and epidemiological significance of enterococci intrinsically resistant to vancomycin (possessing the vanC genotype). J Clin Microbiol 1997; 35:3166-70. PMID:9399514
8. Guardabassi L, Perichon B, van Heijenoort J, et al. Glycopeptide resistance vanA operons in Paenibacillus strains isolated from soil. Antimicrob Agents Chemother 2005; 49:4227-33. PMID:16189102 doi:10.1128/AAC.49.10.4227-4233.2005

9. Marshall CG, Lessard IA, Park I, et al. Glycopeptide antibiotic resistance genes in glycopeptide-producing organisms. Antimicrob Agents Chemother 1998; 42:2215-20. PMID:9736537

10. Stinear TP, Olden DC, Johnson PD, et al. Enterococcal vanB resistance locus in anaerobic bacteria in human faeces. Lancet 2001; 357:855-6. PMID:11265957 doi:10.1016/S0140-6736(00)04206-9

11. Dahl KH, Røkenes TP, Lundblad EW, et al. Non conjugative transposition of the vanB-containing Tn5382-like element in Enterococcus faecium. Antimicrob Agents Chemother 2003; 47:786-9. PMID:12543693 doi:10.1128/AAC.47.2.786-789.2003

12. Garnier F, Taourit S, Glaser P, et al. Characterization of transposon Tn1549, conferring VanB-type resistance in Enterococcus spp. Microbiology 2000; 146:1481-9. PMID:10846226

13. Rice LB, Carias LL, Donskey CL, et al. Transferable, plasmid-mediated VanB-type glycopeptide resistance in Enterococcus faecium. Antimicrob Agents Chemother 1998; 42:963-4. PMID:9559822

14. Woodford N, Morrison D, Johnson A, et al. Plasmid-mediated vanB glycopeptide resistance in enterococci. Microb Drug Resist 1995; 1:235-40. PMID:9158780 doi:10.1089/mdr.1995.1.235

15. Launay A, Ballard SA, Johnson PD, et al. Transfer of vancomycin resistance transposon Tn1549 from Clostridium symbiosum to Enterococcus spp. in the gut of gnotobiotic mice. Antimicrob Agents Chemother 2006; 50:1054-62. PMID:16495268 doi:10.1128/AAC.50.3.1054-1062.2006

16. Carias LL, Rudin SD, Donskey CJ, et al. Genetic linkage and cotransfer of a novel, vanB-containing transposon (Tn5382) and a low-affinity penicillin-binding protein 5 gene in a clinical vancomycin-resistant Enterococcus faecium isolate. J Bacteriol 1998; 180:4426-34. PMID:9721279

17. Zheng B, Tomita H, Inoue T, et al. Isolation of VanB-type Enterococcus faecalis strains from nosocomial infections: first report of the isolation and identification of the pheromone-responsive plasmids pMG2200, encoding VanB-type vancomycin resistance and Bac41-Type bacteriocin, and pMG2201, encoding erythromycin resistance and cytolysin (Hly/Bac). Antimicrob Agents Chemother 2009; 53:735-47. PMID:19029325 doi:10.1128/AAC.00754-08

18. Burrus V, Pavlovic G, Decaris B, et al. The ICESt1 element of Streptococcus thermophilus belongs to a large family of integrative and conjugative elements that exchange modules and their specificity of integration. Plasmid 2002; 48:77-97. PMID:12383726 doi:10.1016/S0147-619X(02)00102-6

19. Depardieu F, Courvalin P, Kolb A. Binding sites of VanRB and sigma70 RNA polymerase in the vanB vancomycin resistance operon of Enterococcus faecium BM4524. Mol Microbiol 2005; 57:550-64. PMID:15978084 doi:10.1111/j.1365-2958.2005.04706.x

20. Dahl KH, Simonsen GS, Olsvik Ø, et al. Heterogeneity in the vanB gene cluster of genomically diverse clinical strains of vancomycin-resistant enterococci. Antimicrob Agents Chemother 1999; 43:1105-10. PMID:10223921

21. McGregor KF, Nolan C, Young HK, et al. Prevalence of the vanB2 gene cluster in vanB glycopeptide-resistant enterococci in the United Kingdom and the Republic of Ireland and in association with a Tn5382-like element. Antimicrob Agents Chemother 2001; 45:367-8. PMID:11221724 doi:10.1128/AAC.45.1.367-368.2001

22. Dahl KH, Lundblad EW, Røkenes TP, et al. Genetic linkage of the vanB2 gene cluster to Tn5382 in vancomycin-resistant enterococci and characterization of two novel insertion sequences. Microbiology 2000; 146:1469-79. PMID:10846225

23. Lu JJ, Chang TY, Perng CL, et al. The vanB2 gene cluster of the majority of vancomycin-resistant Enterococcus faecium isolates from Taiwan is associated with the pbp5 gene and is carried by Tn5382 containing a novel insertion sequence. Antimicrob Agents Chemother 2005; 49:3937-9. PMID:16127076 doi:10.1128/AAC.49.9.3937-3939.2005

24. Ballard SA, Grabsch EA, Johnson PD, et al. Comparison of three PCR primer sets for identification of vanB gene carriage in feces and correlation with carriage of vancomycin-resistant enterococci: interference by vanB-containing anaerobic bacilli. Antimicrob Agents Chemother 2005; 49:77-81. PMID:15616278 doi:10.1128/AAC.49.1.77-81.2005

25. Graham M, Ballard SA, Grabsch EA, et al. High rates of fecal carriage of nonenterococcal vanB in both children and adults. Antimicrob Agents Chemother 2008; 52:1195-7. PMID:18180361 doi:10.1128/AAC.00531-07

26. Mak A, Miller MA, Chong G, et al. Comparison of PCR and culture for screening of vancomycin-resistant enterococci: highly disparate results for vanA and vanB. J Clin Microbiol 2009; 47:4136-7. PMID:19846635 doi:10.1128/JCM.01547-09

27. Ballard SA, Pertile K, Lim M, et al. Molecular characterisation of vanB elements in naturally occurring gut anaerobes. Antimicrob Agents Chemother 2005; 49:1688-94. PMID:15855482 doi:10.1128/AAC.49.5.1688-1694.2005

28. Domingo MC, Huletsky A, Bernal A, et al. Characterization of a Tn5382-like transposon containing the vanB2 gene cluster in a Clostridium strain isolated from human faeces. J Antimicrob Chemother 2005; 55:466-74. PMID:15731199 doi:10.1093/jac/dki029

29. Domingo MC, Huletsky A, Boissinot M, et al. Clostridium lavalense sp. nov. a glycopeptide-resistant species isolated from human faeces. Int J Syst Evol Microbiol 2009; 59:498-503. PMID:19244429 doi:10.1099/ijs.0.001958-0

30. Poyart-Salmeron C, Trieu-Cuot P, Carlier C, et al. Molecular characterization of two proteins involved in the excision of the conjugative transposon Tn1545: homologies with other site-specific recombinases. EMBO J 1989; 8:2425-33. PMID:2551683

31. Storrs MJ, Poyart-Salmeron C, Trieu-Cuot P, et al. Conjugative transposition of Tn916 requires the excisive and integrative activities of the transposon-encoded integrase Tnt-Tn. J Bacteriol 1991; 173:4347-52. PMID:1648556

32. Scott JR, Churchward GG. Conjugative transposition. Annu Rev Microbiol 1995; 49:367-97. PMID:8561465 doi:10.1146/annurev.mi.49.100195.002055

33. Lu F, Churchward G. Conjugative transposition: Tn916 integrase contains two independent DNA binding domains that recognize different DNA sequences. EMBO J 1994; 13:1541-8. PMID:8156992

34. Jia Y, Churchward G. Interactions of the integrase protein of the conjugative transposon Tn916 with its specific DNA binding sites. J Bacteriol 1999; 181:6114-23. PMID:10498726

35. Hinerfeld D, Churchward G. Specific binding of integrase to the origin of transfer (oriT) of the conjugative transposon Tn916. J Bacteriol 2001; 183:2947-51. PMID:11292817 doi:10.1128/JB.183.9.2947-2951.2001

36. Tsvetkova K, Marvaud JC, Lambert T. Analysis of the mobilization functions of the vancomycin resistance transposon Tn1549, a member of a new family of conjugative elements. J Bacteriol 2010; 192:702-13. PMID:19966009 doi:10.1128/JB.00680-09

37. Celli J, Trieu-Cuot P. Circularization of Tn916 is required for expression of the transposon-encoded transfer functions: characterization of long tetracycline-inducible transcripts reading through the attachment site. Mol Microbiol 1998; 28:103-17. PMID:9593300 doi:10.1046/j.1365-2958.1998.00778.x

38. Sebaihia M, Wren BW, Mullany P, et al. The multidrug-resistant human pathogen Clostridium difficile has a highly mobile, mosaic genome. Nat Genet 2006; 38:779-86. PMID:16804543 doi:10.1038/ng1830

39. Cabezón E, de la Cruz F. TrwB: An F1-ATPase-like molecular motor involved in DNA transport during bacterial conjugaison. Res Microbiol 2006; 157:299-305. PMID:16427770 doi:10.1016/j.resmic.2005.12.002

40. Christie, P.J. Type IV secretion: the Agrobacterium VirB/D4 and related conjugation systems. Biochim Biophys Acta 2004; 1694:219-234.

41. Llosa M, Gomis-Rüth FX, Coll M, de la Cruz F. Bacterial conjugation: a two-step mechanism for DNA transport. Mol Microbiol 2002; 45:1-8. PMID:12100543 doi:10.1046/j.1365-2958.2002.03014.x

42. Moncalián G, Cabezón E, Alkorta I, et al. Characterization of ATP and DNA binding activities of Trw, the coupling protein essential in plasmid R388 conjugation. J Biol Chem 1999; 274:36117-24. PMID:10593894 doi:10.1074/jbc.274.51.36117

43. Schröder G, Lanka E. The mating pair formation system of conjugative plasmids- A versatile secretion machinery for transfer of proteins and DNA. Plasmid 2005; 54:1-25. PMID:15907535 doi:10.1016/j.plasmid.2005.02.001

44. Grohmann E, Muth G, Espinosa M. Conjugative plasmid transfer in gram-positive bacteria. Microbiol Mol Biol Rev 2003; 67:277-301. PMID:12794193 doi:10.1128/MMBR.67.2.277-301.2003

45. Berg T, Firth N, Apisiridej S, et al. Complete nucleotide sequence of pSK41: evolution of staphylococcal conjugative multiresistance plasmids. J Bacteriol 1998; 180:4350-9. PMID:9721269

46. Pansegrau W, Schröder W, Lanka E. Relaxase (TraI) of IncPα plasmid RP4 catalyzes a site-specific cleaving-joining reaction of single-stranded DNA. Proc Natl Acad Sci USA 1993; 90:2925-9. PMID:8385350 doi:10.1073/pnas.90.7.2925

47. Llosa M, Zunzunegui S, de la Cruz F. Conjugative coupling proteins interact with cognate and heterologous VirB10-like proteins while exhibiting specificity for cognate relaxosomes. Proc Natl Acad Sci USA 2003; 100:10465-70. PMID:12925737 doi:10.1073/pnas.1830264100

48. Celli J, Trieu-Cuot P. Circularization of Tn916 is required for expression of the transposon–encoded transfer function: characterization of long tetracycline-inducible transcripts reading through the attachment site. Mol Microbiol 1998; 28:103-17. PMID:9593300 doi:10.1046/j.1365-2958.1998.00778.x

49. Foucault ML, Depardieu F, Courvalin P, et al. Inducible expression eliminates the fitness cost of vancomycin resistance in enterococci. Proc Natl Acad Sci USA 2010; 107:16964-9. PMID:20833818 doi:10.1073/pnas.1006855107

50. Osborn AM, Böltner D. When phage, plasmids, and transposons collide: genomic islands, and conjugative- and mobilizable-transposons as a mosaic continuum. Plasmid 2002; 48:202-12. PMID:12460536 doi:10.1016/S0147-619X(02)00117-8

CHAPTER 11

The Tn4371 ICE Family of Bacterial Mobile Genetic Elements

Rob Van Houdt,[1] Ariane Toussaint,[2] Michael P. Ryan,[3]
J. Tony Pembroke,[4] Max Mergeay*,[1] and Catherine C. Adley[3]

Abstract

The Tn4371 ICE (Integrative Conjugative Element) family refers to a group of mobile genetic elements with four modules containing genes involved in integration (via a tyrosine-based site-specific recombinase), maintenance/stability, accessory genes conferring a special phenotype to the host bacteria and genes involved in conjugational transfer. The latter display similarity with conjugative genes of Ti plasmids and IncP broad host range plasmids. Currently this ICE$_{Tn4371}$ family harbours around 40 elements with sizes ranging from 38 to 101 kb. Elements carrying accessory genes, which are mostly flanked by the conjugative genes $rlxS$ ($virD2$) and $traR$, reside in Beta- and Gammaproteobacteria, while elements in Alphaproteobacteria apparently lack accessory genes and display a more divergent pattern. Accessory genes have very diverse functions including catabolism of xenobiotic compounds, resistance to heavy metals, to antibiotics, chemolithotrophic metabolism, but also some unknown functions. Strains from man-made environments (sewages, industrial wastes and clinical settings) are predominant among the bearers of Tn4371 ICEs next to plant pathogens, which are also well represented. Insights into ~30 characteristic genes (such as those encoding site-specific recombinases and excisionases) as well as their distribution in the described elements and their relationship with other mobile genetic elements is provided.

Introduction

Tn4371 is a 55 kb mobile element, which allows its host to degrade biphenyl and 4-chlorobiphenyl. It was originally isolated after mating between *Cupriavidus oxalaticus* A5 (formerly *Alcaligenes eutrophus* A5) carrying the broad host range conjugative plasmid RP4 and *Cupriavidus metallidurans* (formerly *Ralstonia metallidurans*) CH34. Transconjugants were selected by their resistance to heavy metals, which is a key trait of CH34, and their ability to grow on biphenyl as a sole source of carbon and energy.[1] These transconjugants carried an RP4 plasmid with a 55 kb insert near its tetracycline resistance operon. The element was designated Tn4371 as it was shown to transpose to other locations,[1,2] and subsequently sequenced and studied further.[3] Closely related elements have been found in the genome sequences of a number of bacteria in the Beta- and Gammaproteobacteria classes including in *Cupriavidus metallidurans* CH34 and *Pseudomonas aeruginosa*.[3-6] Table 1 shows a full list of all elements.

[1]Unit of Microbiology, Belgian Nuclear Research Centre (SCK•CEN), Mol, Belgium;
[2]Laboratoire de Bioinformatique des Génomes et des Réseaux (BiGRe), Université Libre de Bruxelles, Brussels, Belgium; [3]Microbiology Laboratory, [4]Molecular Biochemistry Laboratory, Department of Chemical and Environmental Sciences, University of Limerick, Limerick, Ireland.
*Corresponding Author: Max Mergeay—Email: mmergeay@sckcen.be

Bacterial Integrative Mobile Genetic Elements, edited by Adam P. Roberts and Peter Mullany.
©2013 Landes Bioscience.

Table 1. ICE$_{Tn4371}$ elements identified in different Proteobacteria

Strain	Element	Location	Environment	Accessory Genes[a]	Acc. No.	%GC	Size (kb)
Alphaproteobacteria							
Xanthobacter autotrophicus Py2	ICE$_{Tn4371}$6139	NLD	Propene and 1-butene enrichment culture	Insertion Sequences	NC_009720	65.77	39
Gammaproteobacteria							
Azotobacter vinelandii DJ	ICE$_{Tn4371}$6144	USA	Soil	MFS transporters, antibiotic resistance genes	NC_012560	65.25	46
Citrobacter sp 30_2	ΔICE$_{Tn4371}$6153	—	Intestinal biopsy specimen	MFS transporter, RND transporter, CDF, heavy metal P-type ATPase	NZ_GG657370	ND	ND
Congregibacter litoralis KT71	ICE$_{Tn4371}$6035	DEU	Ocean-surface water	RND transporter, bacterial luciferase	NZ_AAOA01000008	59.52	59
Dickeya dadantii 3937	ΔICE$_{Tn4371}$6154	FRA	Saintpaulia plant	Hypothetical	NC_014500	ND	ND
Pectobacterium carotovorum subsp brasiliensis PBR1692	ICE$_{Tn4371}$6145	BRA	Blackleg disease in potatoes	MFS transporters	NZ_ABVX01000011	ND	ND
Pseudomonas aeruginosa 2192	ICE$_{Tn4371}$6041	USA	Cystic fibrosis patient	RND transporter	NZ_CH482384	62.62	48
Pseudomonas aeruginosa 39016	ICE$_{Tn4371}$6146	GBR	Cornea of a patient with ulcerative keratitis	RND transporter, ABC transporter	AEEX01000149	ND	ND
Pseudomonas aeruginosa PA7	ICE$_{Tn4371}$6042	ARG	Clinical wound isolate	Antibiotic resistance genes, ATP-driven potassium transport system	NC_009656	52.38	46
Pseudomonas aeruginosa PACS171b	ICE$_{Tn4371}$6070	USA	Cystic fibrosis patient	Universal stress protein (UspA), sulfate permease, arsenic resistance	EU595746	64.12	42

continued on next page

Table 1. Continued

Strain	Element	Location	Environment	Accessory Genes[a]	Acc. No.	%GC	Size (kb)
Pseudomonas aeruginosa UCBPP-PA14	ICE$_{Tn4371}$6069	USA	Burn patient	RND transporter (czc-like)	NC_008463	65.55	43
Shewanella sp ANA-3	ICE$_{Tn4371}$6068	USA	Arsenate treated wood pier	MFS transporters	NC_008577	59.43	45
Stenotrophomonas maltophilia K279a	ICE$_{Tn4371}$6069	GBR	Blood infection	MFS transporter, putative NAD(P)H dehydrogenase and carboxymuconolactone decarboxylase	NC_010943	62.76	44
Thioalkalivibrio sp HL-EbGR7	ICE$_{Tn4371}$6071	USA	Bioreactor removing sulfide from biogas	Potassium transporter system	NC_011901	64.95	43
Xanthomonas fuscans subsp *aurantifolii* str. ICPB 11122	ICE$_{Tn4371}$6147	ARG	Citrus lemon tree	MFS transporters	NZ_ACPX01000208	65.60	38
Betaproteobacteria							
Acidovorax avenae subsp *citrulli* AAC00–1	ICE$_{Tn4371}$6036	USA	Watermelon	MFS transporters, IS elements	NC_008752	63.12	59
Acidovorax ebreus TPSY	ICE$_{Tn4371}$6066	USA	Soil	P-type ATPase, RND transporter (czc-like)	NC_011992	65.30	44
	ICE$_{Tn4371}$6140			ABC transporter, putative RNA processing, metabolism		63.76	61
Acidovorax sp JS42	ICE$_{Tn4371}$6039	USA	Groundwater	Multidrug efflux pump, IS elements	NC_008782	62.88	63
	ΔICE$_{Tn4371}$6050			Undefined		ND	ND

continued on next page

Table 1. *Continued*

Strain	Element	Location	Environment	Accessory Genes[a]	Acc. No.	%GC	Size (kb)
Acidovorax sp KKS102	ICE$_{Tn4371}$KKS	JPN	Soil	Biphenyl degradation, resistance to arsenic and chromate	AB546270	64.44	55
Alicycliphilus denitrificans BC	ICE$_{Tn4371}$6148	NLD	Benzene-degrading, chlorate-reducing enrichment culture	Heavy metal P-type ATPase, RND transporter (czc-like)	NC_014910	64.25	47
	ΔICE$_{Tn4371}$6149			ABC transporter, putative RNA processing, metabolism		ND	ND
Bordetella petrii DSM12804	ICE$_{Tn4371}$6040	DEU	River sediment	Threonine degradation	NC_010170	63.73	47
	ICE$_{Tn4371}$6141			Heavy metal resistance genes		57.75	84
Burkholderia ambifaria AMMD	ΔICE$_{Tn4371}$6151	USA	Pea plant	ABC transporter-related, MFS transporter	NC_008390	ND	ND
Burkholderia multivorans ATCC 17616	ΔICE$_{Tn4371}$6152	USA	Soil	RND transporter, isoprenyl-cysteine carboxyl methyltransferase, heavy metal P-type ATPase	NC_010084	ND	ND
Burkholderia multivorans CGD2 (and CGD2M)[b]	ICE$_{Tn4371}$6057	USA	Chronic granulomatous disease patient	Oxidoreductase, 2,5-dichloro-2,5-cyclohexadiene-1,4-diol dehydrogenase, carboxymuconolactone decarboxylase	NZ_ACFC01000009 (ACFD01000007)	64.65	49
Burkholderia pseudomallei MSHR346	ICE$_{Tn4371}$6064	AUS	Melioidosis patient	Degradation of aromatic compounds	NC_012695	62.21	49

continued on next page

Table 1. Continued

Betaproteobacteria

Strain	Element	Location	Environment	Accessory Genes[a]	Acc. No.	%GC	Size (kb)
Comamonas testosteroni KF-1	ICE$_{Tn4371}$6038	CHE	Activated sludge	RND transporter (czc-like)	NZ_AAUJ02000001	63.77	52
Cupriavidus metallidurans CH34	ICE$_{Tn4371}$6054	BEL	Heavy metal-contaminated sludge of a settling tank in zinc factory	Hydrogenotrophy, degradation of aromatic compounds	NC_007973	63.18	101
	ICE$_{Tn4371}$6055			Hydrogenotrophy, CO_2 fixation		64.03	97
	ΔICE$_{Tn4371}$6056					ND	ND
Cupriavidus oxalaticus A5	Tn4371	USA	PCB contaminated river sediment	Biphenyl degradation	AJ536756	63.49	55
Delftia acidovorans SPH-1	ICE$_{Tn4371}$6037	DEU	Activated sludge	RND transporter (czc-like)	NC_010002	63.66	58
	ICE$_{Tn4371}$6067			Multiple heavy metal resistance mechanisms		64.94	66
Polaromonas naphthalenivorans CJ2 plasmid pPNAP01	ICE$_{Tn4371}$6065	USA	Coal-tar-waste contaminated site	RND transporter (multidrug resistance), putative lipooligosaccharide metabolism, biphenyl degradation	NC_008757	62.89	69
Ralstonia pickettii 12J	ICE$_{Tn4371}$6033	USA	Copper-contaminated sediment from a lake	Lipid metabolism	CP001068	64.63	54
Ralstonia solanacearum GMI1000	ICE$_{Tn4371}$6142	GUF	Tomato Plant	Undefined function	NC_003295	63.11	44
Ralstonia solanacearum MolK2	ICE$_{Tn4371}$6143	PHL	Banana Tree	Undefined function	CU695238	64.56	40

[a]MFS: Major Facilitator Superfamily; RND: Resistance-Nodulation-Cell Division; ABC: ATP-binding cassette. [b]The ICE$_{Tn4371}$ identified in CGD2 and CGD2M are identical.

Integrative Conjugative Elements (ICEs) carry functional modules involved in their conjugative transfer, chromosomal integration and for control of expression of ICE genes. They establish themselves in their host genome at a unique or limited number of sites via site-specific integration. First discovered in the genomes of various Firmicutes, ICEs have been found in various Alpha, Beta- and Gammaproteobacteria, and Bacteroides species.[7] The first ICE found, Tn*916* from *Streptococcus faecalis*[8] (now *Enterococcus faecalis*), has a broad transfer range.[9,10]

As with other ICEs, those in the Tn*4371* family are characterized by a mosaic structure. They consist of four modules, three conserved ones, including an integration/excision module (IntMod), a stabilization/maintenance module (StaMod) and a Ti-RP4-like transfer module (TraMod), and a variable module with accessory genes (AccMod) (Fig. 1).[4,5] Many of the elements have been found integrated into a conserved host *att*B site with the 5'-TTTTTYAT-3' sequence.[2] The ends of the elements can be detected covalently linked as a transfer intermediate indicating that Tn*4371* transposition most likely involves a site-specific integration/excision process.[2,4] Their size varies from 38 to 101 kb depending on the number of accessory genes present. All together, the conserved genes from IntMod, StaMod and TraMod account for approximately 24 kb (a 1.5 kb IntMod, a 8.5 kb StaMod and a 14 kb TraMod).

Taxonomy of Tn*4371*-Like ICEs and Their Hosts

Elements that carry genes characteristic to Tn*4371* (see Fig. 2) and a main AccMod between *rlxS* and *traR* (see below), have been grouped as ICE$_{Tn4371}$-elements. This large group also includes elements with more divergent tyrosine-based site-specific recombinases (TBSSRs) previously identified by Ryan et al.[4] The nomenclature system designed by Ryan et al.,[4] which was based on an adaptation of the system used for naming transposons described by Roberts et al.,[11] was preserved and continued.

Previous in silico comparative studies identified ICEs closely related to Tn*4371* in the genomes of several Betaproteobacteria. Three elements were identified in *C. metallidurans* CH34,[3,5] two in *Delftia acidovorans* SPH-1[4,5] and *Bordetella petrii* DSM12804,[4,5,12] and a single one in *Acidovorax avenae* subsp *citrulli* AAC00–1, *Acidovorax ebreus* TPSY, *Acidovorax* sp JS42, *Burkholderia pseudomallei* MSHR346, *Comamonas testosteroni* KF-1, *Polaromonas naphthalenivorans* CJ2 plasmid pPNAP01, *Ralstonia pickettii* 12J and *Ralstonia solanacearum* GMI1000.[3,4] Tn*4371*-like elements were also found in Gammaproteobacteria including *Azotobacter vinelandii* DJ, *Congregibacter litoralis* KT71, *Dickeya dadantii* 3937 (formerly *Erwinia chrysanthemi*[3]), *P. aeruginosa* strains 2192, PA7, PACS171b and UCBPP-PA14, *Shewanella* sp ANA-3, *Stenotrophomonas maltophilia* K279a and *Thioalkalivibrio* sp HL-EbGR7.[3,4]

This chapter also includes (or further characterizes) new elements in the Betaproteobacteria such as *Acidovorax* sp KKS102, *Alicycliphilus denitrificans* BC, *Burkholderia ambifaria* AMMD, *Burkholderia multivorans* strains CGD2, CGD2M and ATCC 17616, and *R. solanacearum* MolK2. The Gammaproteobacteria included *Citrobacter* sp 30_2, *Pectobacterium carotovorum* subsp *brasiliensis* PBR1692, *P. aeruginosa* 39016 and *Xanthomonas fuscans* subsp *aurantifolii* str. 11122 (Table 1). New elements were also discovered in the Alphaproteobacteria and will be discussed below.

When a set of characteristic proteins from the IntMod, StaMod and TraMod of the ICE$_{Tn4371}$ elements was collected and used to perform pairwise alignments and calculate protein similarities with the original Tn*4371* (Fig. 2) ParA and TraG emerged as the most conserved proteins followed by proteins of the Trb cluster.

Clustering analysis of all these pairwise alignments produced 3 main groups. A large cluster with multiple subgroups (A, B, C and D) contained most of the elements and within this large cluster, subgroup A (Fig. 2) contained elements from Beta- and Gammaproteobacteria, which belong to a variety of environments, strongly suggesting that ICE$_{Tn4371}$ elements may have a broad range of transfer and replication abilities. Subgroup B contained four from Gamma- and two from Betaproteobacteria, while subgroup C contained seven elements from Betaproteobacteria and one from Gammaproteobacteria. All these strains originate from environmentally polluted or industrial settings.

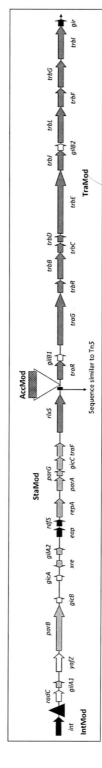

Figure 1. Characteristic genes of the Tn4371-ICE family. Black arrows: Involved in integration and excision, Light gray arrows: Involved in stabilization of the element, Dark gray arrows: Involved in transfer of the element, White arrows: various genes whose functions are not known (not always present in every element), Checkered box: accessory gene module, Black triangle: accessory genes can also be present.

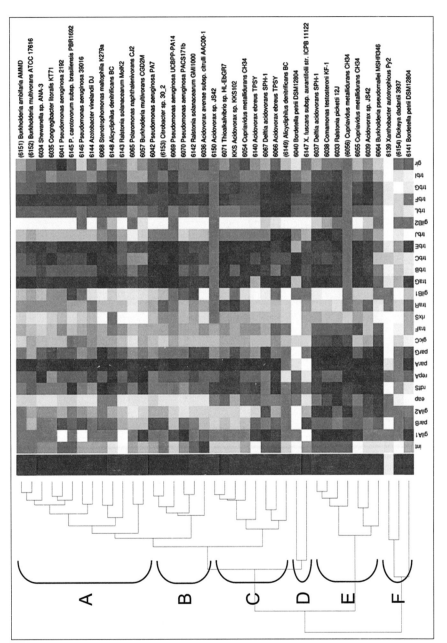

Figure 2. Please see the figure legend on the following page.

Figure 2, viewed on previous page. Hierarchical representation of proteins characteristic of the ICE$_{Tn4371}$ family. Protein similarity with their respective orthologs in the original Tn*4371* are shown with identities ranging from 50 to 100% scaled from light yellow to dark red. Identities below 40% are not colored and absent proteins are marked in gray. The column on the right displays the ICEs (numbers between brackets are partial elements), the column on the left indicates the phylum of the host with Alpha, Beta- and Gammaproteobacteria highlighted red, blue and yellow, respectively.

A second large cluster (group E in Fig. 2) included ICE$_{Tn4371}$6037 from *D. acidovorans* SPH-1, ICE$_{Tn4371}$6038 from *C. testosteroni* KF-1, ICE$_{Tn4371}$6033 from *R. pickettii* 12J, ICE$_{Tn4371}$6039 from *Acidovorax* sp JS42, ICE$_{Tn4371}$6064 from *B. pseudomallei* MSHR346, and ICE$_{Tn4371}$6055 and ΔICE$_{Tn4371}$6056 from *C. metallidurans* CH34, for which the corresponding hosts are Betaproteobacteria isolated from waste waters or again from industrial environments.

The last cluster (group F in Fig. 2) grouped the elements from *Xanthobacter autotrophicus* Py2 (ICE$_{Tn4371}$6139), *D. dadantii* 3937 (ΔICE$_{Tn4371}$6154) and *B. petrii* DSM12804 (ICE$_{Tn4371}$6141) and appeared to be more distant from all the other groups. The element from *X. autotrophicus* Py2 is shown here as a representative of ICE$_{Tn4371}$-like elements that we have observed in a large group of Alphaproteobacteria. These elements do not contain an AccMod between *rlxS* and *traG* nor a *traR* gene, have a shorter *rlxS* gene, and contain a *slt* gene coding for a lytic transglycosylase. The latter is so far specific for ICE$_{Tn4371}$-like elements in Alphaproteobacteria and located between *traF* and *rlxS*. The large group of related ICE$_{Tn4371}$ variants in Alphaproteobacteria is interesting as genes in the StaMod and the TraMod of the ICE*Ml*Sym^{R7A} element from the Alphaproteobacterium *Mesorhizobium loti* R7A are also highly similar to ICE$_{Tn4371}$ (see below) indicating that there may be a wide range of relationships with at least one other family of elements. Other Alphaproteobacteria that carry ICE$_{Tn4371}$-like elements similar to *X. autotrophicus* Py2 (ICE$_{Tn4371}$6139) include *Oligotropha carboxidovorans* OM5, *Rhodopseudomonas palustris* DX1, *Gluconacetobacter diazotrophicus* PAl5, *Chelativorans* sp BNC1 (formerly *Mesorhizobium*), *Azorhizobium caulinodans* ORS571, *Bradyrhizobium japonicum* USDA110, *Bradyrhizobium* sp BTAi1, *Caulobacter* sp K31 and *Parvibaculum lavamentivorans* DS1 among others. Very likely, further studies will show that cluster F may further subdivide in various subgroups.

In various strains that contain two or more ICE$_{Tn4371}$ elements, the corresponding elements belong to different groups: C and E for *D. acidovorans* SPH-1, *Acidovorax* sp JS42 and *C. metallidurans* CH34, A and C for *A. denitrificans* BC, D and F for *B. petrii* DSM12804. This break down of elements into groups could be an example of a phenomenon known as entry exclusion that is seen in some families of ICEs most notably the R391/SXT family found in *Enterobacteriaceae*.[13] These ICEs can be divided into two exclusion groups, called S and R.[14] Bacteria that host an S group ICE are less susceptible to acquire another S group ICE while the acquisition of an R group ICE is not restricted. This exclusion activity is controlled by the TraG protein and a protein termed the entry exclusion protein, Eex (EexR or EexS based on the exclusion group).[15] These are inner-membrane proteins expressed in donor and recipient cells, respectively.[16] An entry exclusion system is also found in IncPα plasmid RP4 (to which the transfer system of ICE$_{Tn4371}$ is related). The *trbK* gene in the Tra2 region of the plasmid has been shown to code for the single protein involved in entry exclusion. This small hydrophilic polypeptide of 69 amino acid residues contains a lipoprotein signature.[17] Four lipoproteins (*gilA1*[*g*enomic *i*sland *l*ipoprotein *A1*], *gilA2, gilB1* and *gilB1*) are found in the ICE$_{Tn4371}$ elements (see below) but none of these proteins show homology to TrbK. Although Figure 2 shows that the GilA2 proteins of group E are highly identical with GilA2 of the original Tn*4371* and as such quite distinct from the other groups, further investigation is required to determine if these groupings are an example of entry exclusion and if so what is the associated mechanism.

In an alternative approach, separate phylogenetic trees were constructed for each of the characteristic proteins using the neighbor-joining method. To explore the underlying phylogenomic information, these (partially) overlapping data sets were used to construct a supertree using Clann

Software.[18] However, bootstrap analysis did not support such a supertree (data not shown). Thus the individual trees generated from ICE_{Tn4371} signature proteins are not congruent, suggesting that different protein families have been subjected to different evolutionary processes.

Integration and Excision (IntMod)

The IntMod makes up the left part of the ICE_{Tn4371} and generally begins with an *int* gene coding for a tyrosine-based site-specific recombinase (TBSSR). In some elements the *int* gene is followed by a few accessory genes (observed in around 20% of the elements). These have an unknown function ($ICE_{Tn4371}6055$ and $ICE_{Tn4371}6056$) or are involved in heavy metal resistance as in *D. acidovorans* SPH1 $ICE_{Tn4371}6067$. These accessory genes will be further described in the section about accessory genes.

Tyrosine-based site-specific recombinases are well known DNA breaking-rejoining enzymes, which are major actors in the comings and goings of mobile genetic elements in bacterial genomes. The 3D structure and molecular mechanisms of action of several enzymes of the family are well documented.[19-24] TBSSRs are usually called "integrases" since they catalyze integration and excision of phages, integron cassettes,[25] ICEs[13] and other genomic islands.[24,26] However some, the XerCD enzymes, resolve chromosome dimers to ensure successful bacterial circular chromosome segregation, plasmid di/multimers or cointegrates generated by some transposons (for a review see ref. 27). TBSSRs specifically bind to their cognate *att* sites and usually need assistance from element-coded directionality proteins called excisionases (Xis) and host-coded nucleoid associated proteins such as IHF or HU for the catalysis of excision.

ICE_{Tn4371} Int Proteins

As shown in Figure 1, a TBSSR gene lies adjacent to the left end/*attL* site of ICE_{Tn4371} elements. Ryan et al.[4] reported that ICE_{Tn4371} TBSSRs and *att* sites are well conserved and more related among them than they are to those encoded by phages and other ICEs. This is further supported by the grouping of most of the ICE_{Tn4371} TBSSR proteins into a single clique upon clustering of all bacterial proteins in the NCBI RefSeq repository (CLSK897396).[28] However, multiple alignments of all TBSSRs encoded by the ICE_{Tn4371} described in Table 1 showed that seven sequences deviate from the very well conserved bulk, both overall and in the probable C-terminal catalytic motif (RHR-Y, aa 353–385, for the bulk; probably RRT-Y, aa 363–397, for five of the diverging ones) (Fig. 3). These seven proteins are from *R. solanacearum* MolK2 ($ICE_{Tn4371}6143$; YP_002253784), *B. petrii* DSM12804 ($ICE_{Tn4371}6141$; YP_001633162), *C. metallidurans* CH34 ($ICE_{Tn4371}6055$; YP_583617), *A. ebreus* TPSY ($ICE_{Tn4371}6140$; YP_002552729), *B. ambifaria* AMMD ($\Delta ICE_{Tn4371}6151$; YP_773911) and *D. dadantii* 3937 ($\Delta ICE_{Tn4371}6154$). The left end of the latter element carries two TBSSRs coding genes, one of which is more related to the other TBSSRs and hence appears as a more likely candidate for being the element's integrase. The variation in IntMod within an ICE family is not unique. Tn5397, a 21 kb ICE from *Clostridium difficile,* shares sequence similarity with Tn916 but bears a serine-based site-specific recombinase (TndX) instead of the TBSSR present in Tn916.[29]

The strong clustering (with almost 100% identity) of five TBSSRs found on elements with different AccMods but residing in hosts isolated from similar environments (wastes, industrial areas) suggests recent horizontal gene transfer and rearrangements. These features and the presence of divergent IntMods associated with a conserved StaMod and TraMod clearly points to the modular structure of ICE_{Tn4371}.

ICE_{Tn4371} Xis Proteins: Excisionases

Phage excisionases are poorly conserved, short and basic,[32] and only few of their homologs in other mobile genetic elements have been characterized (see for instance ref. 33, TorI). Ryan et al.[4] suggested that ORF00035 of Tn4371 (CAD61129) and its orthologs on other ICE_{Tn4371} are excisionases because they show sequence similarity to the RdfS protein of $ICEMlSym^{R7A}$, the symbiosis island from *M. loti* R7A.[34] This RdfS protein (**r**ecombination **d**irectionality **f**actor or excisionase) was shown to be necessary for efficient excision.[34] In addition, ORF00034 of Tn4371

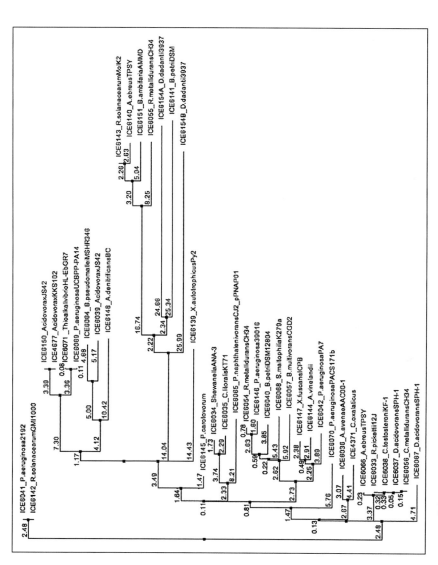

Figure 3. Phylogenetic tree of the Tn*4371* TBSSR and its orthologs. The Neighbor joining tree using percent identity was built with the Jalview interface[30] using a multiple alignment built with the Muscle multiple alignment algorithm[31] through the web service available via the Jalview interface.

located upstream of and in the same orientation as *rdfS* codes for a protein similar to a fused product of *msi172* and *msi171* of ICE*Ml*Sym^{R7A}, which are required for excision and transfer of ICE*Ml*Sym^{R7A}.[35] An ortholog of ORF00034 is found in all ICE$_{Tn4371}$ discovered to date and has been called the *excision associated protein* (*eap*).[35] Expression of *msi172* and *msi171* is regulated by quorum sensing via an *N*-acyl-L-homoserine lactone LuxR/LuxI (TraR/TraI) system, which is not present in ICE$_{Tn4371}$ elements.

Ramsay et al.[35] hypothesized that this regulation by TraR/TraI may be a recent addition to an older regulation of excision such that this event is now coupled with ICE*Ml*Sym^{R7A} transfer in response to autoinducer signaling. Such a coupling between excision and transfer would be beneficial to ICE elements in general. In the ICE$_{Tn4371}$ family there is yet no experimental evidence for such a coupling.

One possibility would be that the last gene before *attR*, which has been designated *gir* (*genomic island right*), is also required for excision. This conserved gene codes for a protein of unknown function and has occasionally been annotated as a transfer gene because it appears to belong to the putative *trb* operon. However, the Gir protein has about 80 residues and a calculated isoelectric point around eight, which are properties consistent with those expected for an excisionase or a nucleoid associated protein.

The Attachment Site (att)

When the Tn*4371* DNA sequence is compared with complete bacterial genomic sequences, the left end comes out very similar in all genomes containing the ICE$_{Tn4371}$ elements described in Table 1 except six, which are those that have different TBSSRs. While most alignments cover at least 1347 bp from the very first Tn*4371* nucleotide at the left end (i.e., the left end and most of the *int* gene, bases 181..1389), five start somewhat further (though still before the *int* start codon), consistent with the slightly more distant position of their cognate enzyme in the tree of the ICE$_{Tn4371}$ TBSSRs. For the six sequences that have the most diverging integrases, there is no significant automatic alignment with Tn*4371* left end. In addition when the putative *C. metallidurans* CH34 ICE$_{Tn4371}$6055 left end and *int* gene (bp 1,585,250–1,586,550) are used as query for a Blastn comparative search, only one among the five remaining elements (*A. ebreus* TPSY ICE$_{Tn4371}$6140; bp 1,324,867–1,326,166) shows up within the best hits.

Upon integration, the *attTn* site is normally split between the left and right ends of the integrated ICEs. Nucleotide alignments between Tn*4371* right end and other ICEs in Table 1 are less straightforward. They usually cover the entire *trb* region and only four ICE$_{Tn4371}$ align up to the end of the Tn*4371* right end (nucleotide 54657).

Many of the elements have been found to integrate into a conserved host *att*B site with the 5′-TTTTTYAT-3′ sequence,[2] however, the deviating 5′-TTTTTYGT-3′ sequence has been occasionally observed like for ICE$_{Tn4371}$6042, ICE$_{Tn4371}$6054 and ICE$_{Tn4371}$6068.

Transcriptional Units

At least ten *int* genes (from ICE$_{Tn4371}$6033, 6036, 6037, 6038, 6054, 6056, 6066, 6069, 6071 and 6142) appear to form a two-gene transcriptional unit (4bp overlap) with a conserved gene coding for a secreted protein of unknown function. A similar coupling (TBSSR coupled to a putative repressor gene) has already been described in plasmid pMOL28 and pMOL30 from *C. metallidurans* CH34 in which these units flank a genomic island involved in chromate resistance and a *czc*-like cluster involved in the resistance to Cd^{2+}, Zn^{2+} and Co^{2+}, respectively. The latter is fully conserved in *R. pickettii* 12J.[5,6]

Stability and Maintenance Genes (StaMod)

Downstream from the IntMod, a gene coding for the GilA1 lipoprotein and the genes *yafZ* and *parB* are present in most elements. YafZ (often annotated as F-plasmid gene 32-like protein) and ParB are related to proteins encoded by genes located near the transfer origin of *Escherichia coli* F plasmid. The function of YafZ is unknown, while ParB is similar to ParB-like nucleases initially identified as crucial for the proper partitioning of plasmid DNA during cell division in the absence of selection pressure.[36-38] Additional plasmid-related genes are located downstream of a cluster containing *gilA2* (coding for a lipoprotein and similar to *gilA1*) and the genes *eap* and *rdfS* involved in excision. These plasmid-related genes code for a RepA (replication) protein similar to various plasmid encoded replication proteins,[3] the ParA partition protein of the type Ib family[39] and an associated ParG protein. Rep and Par proteins have been proposed to act as a stabilization system for the maintenance of mobile elements in bacterial genomes,[38,40] similar to the toxin-anti-toxin system (*mosAT*) coded by ORFs *s052* and *s053* of the SXT-ICE.[41] *P. aeruginosa* ICE PAPI-1 contains an ortholog of the plasmid and chromosome partitioning system *soj* (*parA*) and deletion of *soj* from PAPI-1 resulted in the complete loss of the PAPI-1 pathogenicity island. The mechanism by which the Soj protein promotes PAPI-1 maintenance remains to be elucidated.[42] Genes similar to *soj* have also been found in ICE*Hin1056* (*Hemophilus influenzae*),[43] ICEA (*Mycoplasma agalactiae* strain 5632)[44] and (although with a low identity) in the major genomic island CMGI-1 of *C. metallidurans* CH34. The *parA* and *parG* genes are part of a four-gene operon with 4 bp overlaps (suggesting transcriptional coupling) that also includes a gene of unknown function (*gicC* for *g*enomic *i*sland *c*onserved; see also below) and, rather unexpectedly, *traF* involved in conjugational transfer. Syntenic annotation (via the MaGe[45] annotation platform from Genoscope) shows that this structure is maintained in almost all ICE_{Tn4371} including those of the Alphaproteobacteria. This four-gene operon will be further discussed in the TraMod section as well as the *rlxS* (*virD2*) gene coding for the relaxase protein that is essential in the conjugation process.

A rather large region in the middle of the StaMod between *parB* and *gilA2* appears to be very variable from element to element with a variety of genes coding for small hypothetical proteins. Nevertheless, as shown in Figure 1 three small genes are apparently conserved in this region, namely *gicB* (RO00015), *xre* (RO00055) and *gicA* (RO00018).[3] The *gicB* gene is located immediately downstream of *parB* and codes for a protein that is conserved in a large part of the ICEs reported in Table 1. The *xre* gene codes for a well recognizable transcriptional regulator.[4] The *gicA* gene is remarkably conserved with very high identity in all the ICE_{Tn4371} including the related ICEs found in Alphaproteobacteria. The ubiquity and high sequence conservation of this ortholog inside this family suggests it codes for an essential function in ICE_{Tn4371} and ICE_{Tn4371}-like elements.

The Accessory Genes (AccMod)

The great majority of accessory genes of the ICE_{Tn4371} elements are carried by the diverse AccMods but, more rarely, some are also associated with IntMods (and exceptionally to a StaMod like a gene coding for a cation diffusion factor in *D. acidovorans* SPH1). Accessory genes associated with IntMods include putative DNA helicases and nucleases, proteins with β-lactamase domains, putative reductases, insertion sequences, putative ubiquitin-activating enzymes, putative transcriptional regulators and many different hypothetical proteins whose functions are unknown. More defined systems are discussed below. The main AccMods are systematically located between two ORFs coding for the relaxase protein RlxS and a transcriptional regulator TraR, respectively. The *traF* gene, which codes for a protease, always lies upstream of the *rlxS* gene. The *traR* gene is followed by *traG* (*virD4*) coding for the conjugative coupling protein. The sequence between the *rlxS* gene and the first gene of the variable region, in all elements, is similar to the sequence of a region of Tn5 [U00004, 3787–4143] between the gene for streptomycin phosphotransferase and the transposase. This indicates that the diversity in this region of the element may be due to one or a number of Tn5-mediated insertion events.

The accessory genes vary from element to element and code for various traits such as degradative or metabolic properties, resistance mechanisms to either antibiotics or heavy metals, all of which are advantageous under certain environmental conditions. Yet some functions are hard to define while other AccMods are clearly a mosaic of different functions. Some gene groups are flanked by IS or/and carried by transposons. The main AccMod categories are often clearly recognizable and will be described below.

ICE_{Tn4371} Elements Coding for Resistance to Antibiotics

ICE_{Tn4371}6042 from *P. aeruginosa* PA7 (a clinical wound isolate) carries three genes involved in multiple antibiotic resistance genes (aminoglycoside 3'-phosphotransferase, streptomycin 3'-phosphotransferase, bleomycin resistance). Major Facilitator Superfamily (MFS) transporters and putative determinants for resistance to antibiotics are present in ICE_{Tn4371}6144 from *A. vinelandii* DJ (chloramphenicol and β-lactam) and ICE_{Tn4371}6075 from *Shewanella* sp ANA-3 (chloramphenicol). ICE_{Tn4371}6035 from *C. litoralis* KT71 carries a Resistance-Nodulation-Cell Division (RND) transporter (subfamily HAE5-RND) related to those involved in acriflavine resistance. The three component RND transporters are made of a cytoplasmic membrane export system, a membrane fusion protein, and an outer membrane factor and extrude substrates via an H^+ antiport mechanism.[46] The RND transporter of ICE_{Tn4371}6041 from *P. aeruginosa* 2192 is almost identical (99% protein identity) to that of ICE_{Tn4371}6035 from *C. litoralis* KT71. ICE_{Tn4371}6069 from *S. maltophilia* K279a carries a putative MFS transporter related to the *emrAB* system from *E. coli*,[47] which usually functions as a specific exporter for certain classes of antimicrobial agents.

ICE_{Tn4371} Elements Coding Various Other Transporters

Next to the genes involved in antibiotic resistance, ICE_{Tn4371}6042 from *P. aeruginosa* PA7 carries a group of six genes coding for a KDP ATP-driven potassium transport system in which the KdpFABC complex acts as a high affinity K^+ uptake system. In *E. coli*, the complex is synthesized when the constitutively expressed low affinity K^+ uptake systems Trk and Kup no longer meet the cell's demand for potassium when external K^+ becomes limiting.[48,49] ICE_{Tn4371}6071 from *Thioalkalivibrio* sp HL-EbGR7 also contains a putative *kdpFABC* system, which is very similar (88% aa identity) to that of ICE_{Tn4371}6042 from *P. aeruginosa* PA7. ICE_{Tn4371}6039 from *Acidovorax* sp JS42 contains genes coding for a multidrug efflux pump, which appears inactivated by one of the many insertion sequences carried by this element. The partial element ΔICE_{Tn4371}6151 element from *Burkholderia amfibaria* AMMD codes for an ATP-binding cassette (ABC) transporter-like protein and a MFS transporter. The partial element ΔICE_{Tn4371}6153 in *Citrobacter* sp 30_2 contains multiple MFS transporters, a RND transporter (HAE3-RND family), a Cation Diffusion Facilitator (CDF) transporter and a P-type ATPase. ICE_{Tn4371}6145 of *P. carotovorum* subsp *brasiliensis* PBR1692 contains genes coding for a MFS transporter and an excinuclease ABC. The accessory genes of ICE_{Tn4371}6147 from *Xanthomonas fuscans* subsp *aurantifolii* str. ICPB 11122 code for multiple MFS transporters. ICE_{Tn4371}6146 from *P. aeruginosa* 39016 contains a RND transporter (uncharacterized family) and an ABC-type transporter. ICE_{Tn4371}6070 from *P. aeruginosa* PACS171b contains genes coding for a sulfate permease and a universal stress protein UspA. These genes share respectively 69 and 90% protein identity with those on transposon Tn*6050* from *C. metallidurans* CH34.[5]

ICE_{Tn4371} Elements Encoding for Metal Resistance

ICE_{Tn4371}6038 from *C. testosteroni* KF-1 and ICE_{Tn4371}6037 from *D. acidovorans* SPH-1 are almost identical (98% DNA similarity) and accessory genes code for a Czc-like RND transporter, which probably belongs to the HME3b- or HME4-RND family.[50] The second ICE_{Tn4371} element in *D. acidovorans* SPH-1 (ICE_{Tn4371}6067) carries genes involved in copper resistance (*copSR copAB copGOF copCD copK*) similar to the partial element ΔICE_{Tn4371}6152 from *B. multivorans* ATCC 17616 except the putative isoprenyl cysteine carboxyl methyltransferase. In addition, multiple heavy metal resistance clusters are located in the IntMod including resistance to silver/copper (HME4-RND family transporter), lead/cadmium (P-type ATPase and phosphatase), mercury (reductase) and arsenic (reductase and efflux). A CDF family transporter in the StaMod completes this diversified arsenal of metal resistance genes carried by ICE_{Tn4371}6067, which is probably the best example of an ICE_{Tn4371} specialized in the multiple resistances to heavy metals. These various gene clusters in ICE_{Tn4371}6067 contain the basic resistance genes without the diversity of satellite genes that is found in *C. metallidurans* CH34. ICE_{Tn4371}6148 from *Alicycliphilus denitrificans* BC contains a heavy metal P-type ATPase family and a *czc*-like RND transporter (family HME1-RND). ICE_{Tn4371}6141 from *B. petrii* DSM12804 contains multiple heavy metal resistance mechanisms including a *silP*-like gene coding for a P-type ATPase and a RND transporter putatively involved in

Ag$^+$/Cu$^+$ resistance (family HME4-RND), a cluster involved in copper resistance (*copDCBARS*) and a *czc*-like RND transporter (family unclassified). In addition, the element carries accessory genes coding for arsenic resistance in the IntMod near the site-specific recombinase, which is also observed for ICE$_{Tn4371}$6057, ICE$_{Tn4371}$6067, ICE$_{Tn4371}$6070 and ΔICE$_{Tn4371}$6152.

ICE$_{Tn4371}$6066 from *A. ebreus* TPSY contains two magnesium transporters (MgtC and P-type ATPase family) and a *czc*-like RND transporter (family HME1-RND). ICE$_{Tn4371}$6069 from *P. aeruginosa* UCBPP-PA14 contains a *czc*-like RND transporter (family HME1-RND).

ICE$_{Tn4371}$ Elements Coding for Catabolic Gene Products

ICE$_{Tn4371}$6065 (on plasmid pPNAP01 from *P. naphthalenivorans* CJ2) and ICE$_{Tn4371}$KKS of *Acidovorax* sp KKS102 contain genes involved in biphenyl degradation similar to those found in the original *Tn4371*.[4] However, for ICE$_{Tn4371}$6065 these genes are located in the IntMod while the AccMod contains genes involved in lipooligosaccharide metabolism, which could putatively have a role in tolerance toward desiccation,[51] and multidrug resistance (family HAE1-RND). ICE$_{Tn4371}$6057 from *B. multivorans* CGD2, which is 100% identical to that of *B. multivorans* CGD2M, ICE$_{Tn4371}$6064 from *B. pseudomallei* MSHR346 and ICE$_{Tn4371}$6069 from *S. maltophilia* K279a are putatively involved in the degradation of aromatic compounds.[4] ICE$_{Tn4371}$6033 from *R. pickettii* 12J and ICE$_{Tn4371}$6040 from *B. petrii* DSM12804 code for proteins putatively involved in lipid metabolism and threonine degradation, respectively. *C. metallidurans* CH34 carries three ICE$_{Tn4371}$ elements with different accessory genes. ICE$_{Tn4371}$6054 carries genes involved in the degradation of aromatic compounds and a second cluster involved in hydrogenotrophy flanked by two copies of the same IS element (IS*Rme5*). The latter putatively mediated the integration of this cluster into ICE$_{Tn4371}$6054.[52] A second element (ICE$_{Tn4371}$6055) contains genes involved in hydrogenotrophy and fixation of carbon dioxide through the Calvin-Benson-Bassham cycle.[5] The complete AccMod is flanked by two copies of IS*1071* and loss of autotrophic growth occurs through IS*1071*-mediated excision.[52]

ICE$_{Tn4371}$ Elements Coding for Multiple Functions and with Mosaic Structure

Numerous ICE$_{Tn4371}$ elements described above carry genes that code for multiple functions such as ICE$_{Tn4371}$6042 (antibiotic resistance and K$^+$ transport), ICE$_{Tn4371}$6054 (hydrogenotrophy and aromatic compound degradation), ICE$_{Tn4371}$6055 (hydrogenotrophy and carbon dioxide fixation), ICE$_{Tn4371}$6065 (multidrug resistance, biphenyl degradation and lipooligosaccharide metabolism) and ICE$_{Tn4371}$6069 (transporter, aromatic compound degradation). ICE$_{Tn4371}$6036 from *A. avenae* subsp *citrulli* AAC00–1 contains multiple IS elements, a MFS transporter and metabolic genes with unknown function. All genes except one just upstream *traR*, which encodes a Small Multidrug Resistance (SMR) family transporter, are located between two identical IS elements (IS*3* family). Next to genes putatively involved in antibiotic resistance ICE$_{Tn4371}$6035 from *C. litoralis* KT71 also carries genes encoding a bacterial luciferase and a flavin reductase. ICE$_{Tn4371}$6140 from *A. ebreus* TPSY contains genes involved in metabolism (putative propanoate and butanoate metabolism) and a cluster encoding an ATP-Binding Cassette (ABC) transporter, an outer membrane protein and putative RNA modifying enzymes. The region *rlxS*-AccMod-TraMod of this element is highly similar (> 90% nucleotide identity) to the corresponding part of a partial element ΔICE$_{Tn4371}$6149 from *A. denitrificans* BC.

ICE$_{Tn4371}$ Elements Coding for Unknown or Undefined Functions

R. solanacearum GMI1000 ICE$_{Tn4371}$6142 and MolK2 ICE$_{Tn4371}$6143[3] have AccMods with less defined genes. For the partial element ΔICE$_{Tn4371}$6150 from *Acidovorax* sp JS42 it is more difficult to describe the AccMod region as this element has no transfer module (to delineate the accessory genes). As mentioned above, numerous cases of ICE$_{Tn4371}$-like elements have been observed in Alphaproteobacteria and the element ICE$_{Tn4371}$6139 in *Xanthobacter autotrophicus* Py2 has been shown as a representative. Although it does not contain an AccMod between *rlxS* and *traG*, it does contain numerous IS elements between *int* and TraMod. The spectrum of accessory genes in these Alphaprotobacterial elements and their location in the elements will need further study.

Transfer Genes and TraMod

Transfer genes that make up the transfer module are organized into two groups separated by the AccMod. The first group is composed of two genes, one coding for the TraF protease acting upon the pilus assembly protein TrbC,[53] and the other coding for a RlxS (VirD2) relaxase.[54,55] In fact, the *traF* gene likely belongs to a four-gene transcriptional unit *parAparGgicCtraF* with four base pairs overlaps. This organization, which links key functions involved in partition/replication and conjugational transfer, is conserved in almost all ICE$_{Tn4371}$ including the Tn*4371*-like elements of the Alphaproteobacteria. It is not observed in the IncP/Ti plasmids and hence appears as a characteristic feature of ICE$_{Tn4371}$. The ICE$_{Tn4371}$ relaxases are similar to the RlxS (25%) of ICE*Ml*Sym^{R7A} (Fig. 4) where this protein has been shown to be necessary for transfer and maintenance.[34]

The second part of the gene cluster involved in conjugational transfer is located downstream of the AccMod. This cluster begins with a gene coding for a putative transcriptional regulator protein TraR and a homolog of the coupling protein TraG. The latter, which is a putative DNA binding protein responsible for DNA transfer during conjugation,[56] is similar to those in IncP plasmids. IncP plasmids are the best-studied broad host range plasmids, and thought to be among the most promiscuous of all plasmids known to-date. IncP plasmids are the subject of considerable research as they can spread across taxonomic barriers and therefore contribute to the rapid adaptation of various bacterial populations in natural and clinical environments.[57] IncP plasmids, which replicate in the major branches of the Proteobacteria (Alpha, Beta and Gamma), are also known to have a transfer/mobilisation range that covers the Deltaproteobacteria[58] and even extends far outside the Proteobacteria to the Actinobacteria,[59] cyanobacteria and even yeasts.[60] Also for Ti plasmids, which have a narrow replicative host range, the transfer range may be broader than expected although extensive studies remain to be done.[61,62] IncP plasmids carry accessory genes involved in mercury resistance and degradation of xenobiotic organic compounds as well as antibiotic resistance genes and also share with many other broad host range plasmids the property to capture genes (also called "retrotransfer").[63,64] Putatively, these gene dissemination features could be conserved, adapted and used by the ICE$_{Tn4371}$ (maybe, for example, to capture AccMods). Interestingly, the gene order of two TraMod parts is also suggestive of an insertion of the AccMod into a primordial transfer module. The presence of transfer genes similar to plasmid genes exists in other ICE families. In the R391/SXT type of ICEs, the transfer genes (and a number of other R391/SXT core genes) are highly similar to those in IncA/C conjugative plasmids.[13]

The *traG* gene is followed by a group of genes coding for proteins (TrbBCDEJLFGI) with similarity to the mating-pair formation (mpf) apparatus of IncP and Ti plasmids (except for genes *trbK* and *trbH*, which are missing on the ICE) that are also related to the Type IV secretion system (Table 2).[65] Together TraG and the TrbBCDEJLFGI proteins most likely mediate the transfer of ICE$_{Tn4371}$ to recipient cells.[66,67] In all the ICE$_{Tn4371}$ elements the gene order is *trbBCDEJLFGI*, similar to the organization in ICE*Ml*Sym^{R7A} (Fig. 4).[43] Yet between *traG* and *trbB* lies a gene showing a strong identity with genes encoding repressors of the *copG/metJ/arc/mnt* family, which regulate the copy number of plasmids and phages and for which the *trbR* appellation is proposed. This *copG*-like gene very often overlaps by four base pairs with its neighbors *traG* and *trbB* including in Tn*4371*-like elements from Alphaproteobacteria. This may reflect a functional coupling between the partition/replication regulation and conjugational transfer as it may also be hypothesized with the *parAG gicC traF* operon in the StaMod.

A comparison of the Tra2 region of the IncP plasmid RP4, and the mating-pair formation apparatus of both *R. pickettii* 12J ICE$_{Tn4371}$6033 and ICE*Ml*Sym^{R7A} can be seen in Figure 4. Protein similarities varied between 21–65% for ICE$_{Tn4371}$6033 and ICE*Ml*Sym^{R7A} and ICE$_{Tn4371}$6033 and RP4 protein similarity varied between 11–38%. A similar study was performed with the VirB system from the Ti plasmid pNGR234a (CP000874) and ICE$_{Tn4371}$6033 however little homology was found.

Additional genes lacking in Ti and IncP plasmids can be found inserted between the *traR* and *traG* genes and the *trbJ* and *trbL* genes. These genes code for hypothetical lipoproteins, which are now annotated as *gilB1* and *gilB2* and that share some identity. These genes are found in most of the elements (Fig. 2) as well as the last gene *gir*, tentatively considered as involved in the excision process (see above).

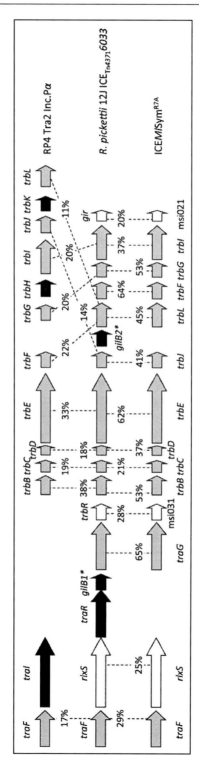

Figure 4. Comparison of the transfer region of RP4 Tra2 region, ICEMlSymR7A and ICE$_{Tn4371}$6033. Black: Only present in that element, Gray: Present in all elements, White: Present in two elements, *not present in ICE$_{Tn4371}$6033 but present in most other ICE$_{Tn4371}$.

Table 2. Similarities/homologies between the transfer region of Tn4371 and other Type IV secretion systems

Tn4371-Like	Ti Plasmid	IncPα	Function
TraF	virD	TraF	Cyclase
RlxS	virD2	TraI	Relaxase
TraR*			
GilB1			Lipoprotein
TraG	VirD4	TraG	Transport
TrbR			
TrbB	VirB11	TrbB	Secretion
TrbC	VirB2	TrbC	Pilin
TrbD	VirB3	TrbD	Pore
TrbE	VirB4	TrbE	Secretion
TrbJ	VirB5	TrbJ	Pore
GilB2			Lipoprotein
		TrbK	Lipoprotein
TrbL	VirB6	TrbL	Pore
	VirB7	TrbH	Lipoprotein
TrbF	VirB8	TrbF	Pore
TrbG	VirB9	TrbG	Secretion
TrbI	VirB10	TrbI	Pore

*previously called BphR

Conclusion

Tn4371-related integrative and conjugative elements are found in a wide range of Alpha-, Beta- and Gammaproteobacteria from both clinical and environmental origin. These types of bacteria are known for their large metabolic repertoires and the ICE$_{Tn4371}$ elements appear to be a source of adaptive functions. The recent attention to environmental microbiology, the incidence of chemical pollution and increase of nosocomial infections has contributed to reveal the role of this otherwise elusive family of mobile genetic elements in microbial evolution and adaptation to anthropogenic conditions.

Although ICE$_{Tn4371}$-related elements are present in Alpha-, Beta- and Gammaproteobacteria two major groups could be discerned. One group specific for the Beta- and Gammaproteobacteria, which characteristically carries its accessory genes between the relaxase (*rlxS*) and a transcriptional regulator (*traR*), while the second group is specific for the Alphaproteobacteria and the presence of accessory genes is scarce. These groups look quite exclusive as it would be expected from a taxonomic barrier. Still, the order of the transfer (*trb*) genes of both ICE$_{Tn4371}$ groups is the same and similar to that of the Ti plasmid (virulence plasmid of Alphaproteobacterium *Agrobacterium*) (Table 2). In the Ti plasmid and ICE$_{Tn4371}$ elements the *trbJ*, *trbK* and *trbL* genes lie further upstream of the *trb* operon, while these are the last genes of the IncP *trb* operon. This obvious translocation may suggest that the ancestral ICE$_{Tn4371}$ was carried by a host most closely related to Alphaproteobacteria, which branched earlier in the tree of Proteobacteria than the Beta- and

Gammaproteobacteria. Phylogenetic studies of the Type IV secretion/transfer systems may help to explore if these four types of elements, viz. ICE_{Tn4371} elements in Beta- and Gammaproteobacteria, ICEs in Alphaproteobacteria, Ti plasmids involved in plant-bacteria interactions, and broad host range IncP plasmids, indeed share a common evolutionary path, and to explore their relationship with integration vs. autonomy of mobile genetic elements.

The gene core of ICE_{Tn4371}, which is composed of approximately 31 genes, harbours an *int* gene coding for a TBSSR, 13 genes in the StaMod and 17 genes involved in the TraMod. Among this group at least eight conserved genes are so far specific for ICE_{Tn4371} elements including four coding for putative lipoproteins (*gilA1* and *gilA2* in StaMod, *gilB1* and *gilB2* in TraMod), three coding for hypothetical proteins (*gicA, gicB* downstream *parB*, and *gicC* in *parAGgicCtraF* operon), and *gir* putatively involved in excision. Their association with the different modules and key functions in these modules suggest a unique role in the biology of ICE_{Tn4371} elements and justifies future genetic and functional studies.

The accessory gene module, which can be quite extensive ranging up to 70 kb, is always inserted into the TraMod (even if accessory genes are regularly found in the IntMod). Hereby it separates the relaxase (*rlxS*) from the transfer (*trb*) system and such a placement could be favorable for on the one hand maintaining the connection between the IntMod-StaMod and TraMod, and on the other be in control of the maximum accessory module insert size. In addition, the transfer function is transcriptionally coupled to the stabilization functions in a four-gene operon (*parAGgicCtraF*) and together with the distribution of the ICE_{Tn4371}-specific conserved genes in both the StaMod and TraMod could act as additional regulatory aspects that allow the acquisition of an extensive baggage of accessory genes without the loss of its integration and transfer functions.

Acknowledgments

This work was supported by the European Space Agency (ESA-PRODEX) and the Belgian Science Policy (Belspo) through the COMICS project (C90356). Thanks to Pieter Monsieurs for constructing the heat map.

References

1. Springael D, Kreps S, Mergeay M. Identification of a catabolic transposon, Tn4371, carrying biphenyl and 4-chlorobiphenyl degradation genes in Alcaligenes eutrophus A5. J Bacteriol 1993; 175:1674-81. PMID:8383664
2. Merlin C, Springael D, Toussaint A. Tn4371: A modular structure encoding a phage-like integrase, a Pseudomonas-like catabolic pathway, and RP4/Ti-like transfer functions. Plasmid 1999; 41:40-54. PMID:9887305 doi:10.1006/plas.1998.1375
3. Toussaint A, Merlin C, Monchy S, et al. The biphenyl- and 4-chlorobiphenyl-catabolic transposon Tn4371, a member of a new family of genomic islands related to IncP and Ti plasmids. Appl Environ Microbiol 2003; 69:4837-45. PMID:12902278 doi:10.1128/AEM.69.8.4837-4845.2003
4. Ryan MP, Pembroke JT, Adley CC. Novel Tn4371-ICE like element in Ralstonia pickettii and genome mining for comparative elements. BMC Microbiol 2009; 9:242. PMID:19941653 doi:10.1186/1471-2180-9-242
5. Van Houdt R, Monchy S, Leys N, Mergeay M. New mobile genetic elements in Cupriavidus metallidurans CH34, their possible roles and occurrence in other bacteria. Antonie van Leeuwenhoek 2009; 96:205-26. PMID:19390985 doi:10.1007/s10482-009-9345-4
6. Janssen PJ, Van Houdt R, Moors H, et al. The complete genome sequence of Cupriavidus metallidurans strain CH34, a master survivalist in harsh and anthropogenic environments. PLoS ONE 2010; 5:e10433. PMID:20463976 doi:10.1371/journal.pone.0010433
7. Burrus V, Waldor MK. Shaping bacterial genomes with integrative and conjugative elements. Res Microbiol 2004; 155:376-86. PMID:15207870 doi:10.1016/j.resmic.2004.01.012
8. Franke AE, Clewell DB. Evidence for a chromosome-borne resistance transposon (Tn916) in Streptococcus faecalis that is capable of "conjugal" transfer in the absence of a conjugative plasmid. J Bacteriol 1981; 145:494-502. PMID:6257641
9. Sen S, Oriel P. Transfer of transposon Tn916 from Bacillus subtilis to Thermus aquaticus. FEMS Microbiol Lett 1990; 55:131-4. PMID:2158473 doi:10.1111/j.1574-6968.1990.tb13849.x

10. Poyart C, Celli J, Trieu-Cuot P. Conjugative transposition of Tn916-related elements from Enterococcus faecalis to Escherichia coli and Pseudomonas fluorescens. Antimicrob Agents Chemother 1995; 39:500-6. PMID:7726521

11. Roberts AP, Chandler M, Courvalin P, et al. Revised nomenclature for transposable genetic elements. Plasmid 2008; 60:167-73. PMID:18778731 doi:10.1016/j.plasmid.2008.08.001

12. Lechner M, Schmitt K, Bauer S, et al. Genomic island excisions in Bordetella petrii. BMC Microbiol 2009; 9:141. PMID:19615092 doi:10.1186/1471-2180-9-141

13. Wozniak RA, Waldor MK. Integrative and conjugative elements: mosaic mobile genetic elements enabling dynamic lateral gene flow. Nat Rev Microbiol 2010; 8:552-63. PMID:20601965 doi:10.1038/nrmicro2382

14. Marrero J, Waldor MK. The SXT/R391 family of integrative conjugative elements is composed of two exclusion groups. J Bacteriol 2007; 189:3302-5. PMID:17307849 doi:10.1128/JB.01902-06

15. Marrero J, Waldor MK. Determinants of entry exclusion within Eex and TraG are cytoplasmic. J Bacteriol 2007; 189:6469-73. PMID:17573467 doi:10.1128/JB.00522-07

16. Marrero J, Waldor MK. Interactions between inner membrane proteins in donor and recipient cells limit conjugal DNA transfer. Dev Cell 2005; 8:963-70. PMID:15935784 doi:10.1016/j.devcel.2005.05.004

17. Haase J, Kalkum M, Lanka E. TrbK, a small cytoplasmic membrane lipoprotein, functions in entry exclusion of the IncP alpha plasmid RP4. J Bacteriol 1996; 178:6720-9. PMID:8955288

18. Creevey CJ, McInerney JO. Clann: investigating phylogenetic information through supertree analyses. Bioinformatics 2005; 21:390-2. PMID:15374874 doi:10.1093/bioinformatics/bti020

19. Jayaram M, Grainge I, Tribble G. Site-Specific recombination by the Flp protein of Saccharomyces cerevisiae. In: Craig NL, Cragie R, Gellert M, Lambowitz AM, eds. Mobile DNA II. Washington, DC: ASM Press; 2002:192-218.

20. Van Duyne GD. A structural view of tyrosine recombinase site-specific recombination. In: Craig NL, Cragie R, Gellert M, Lambowitz AM, eds. Mobile DNA II. Washington, DC: ASM Press; 2002:93-117.

21. Rice PA. Theme and variation in tyrosine recombinases: structure of a Flp-DNA complex. In: Craig NL, Cragie R, Gellert M, Lambowitz AM, eds. Mobile DNA II. Washington, DC: ASM Press; 2002:219-229.

22. Bouvier M, Demarre G, Mazel D. Integron cassette insertion: a recombination process involving a folded single strand substrate. EMBO J 2005; 24:4356-67. PMID:16341091 doi:10.1038/sj.emboj.7600898

23. Dubois V, Debreyer C, Litvak S, et al. A new in vitro strand transfer assay for monitoring bacterial class 1 integron recombinase IntI1 activity. PLoS ONE 2007; 2:e1315. PMID:18091989 doi:10.1371/journal.pone.0001315

24. Dubois V, Debreyer C, Quentin C, Parissi V. In vitro recombination catalyzed by bacterial class 1 integron integrase IntI1 involves cooperative binding and specific oligomeric intermediates. PLoS ONE 2009; 4:e5228. PMID:19381299 doi:10.1371/journal.pone.0005228

25. Cambray G, Guerout AM, Mazel D. Integrons. Annu Rev Genet 2010; 44:141-66. PMID:20707672 doi:10.1146/annurev-genet-102209-163504

26. Boyd EF, Almagro-Moreno S, Parent MA. Genomic islands are dynamic, ancient integrative elements in bacterial evolution. Trends Microbiol 2009; 17:47-53. PMID:19162481 doi:10.1016/j.tim.2008.11.003

27. Hallet B, Vanhoof V, Cornet F. DNA site-specific resolution systems. In: Funnell B, Phillips G, eds. Plasmid Biology. Washington, DC: ASM Press; 2004:145-179.

28. Klimke W, Agarwala R, Badretdin A, et al. The National Center for Biotechnology Information's Protein Clusters Database. Nucleic Acids Res 2009; 37:D216-23. PMID:18940865 doi:10.1093/nar/gkn734

29. Wang H, Mullany P. The large resolvase TndX is required and sufficient for integration and excision of derivatives of the novel conjugative transposon Tn5397. J Bacteriol 2000; 182:6577-83. PMID:11073898 doi:10.1128/JB.182.23.6577-6583.2000

30. Clamp M, Cuff J, Searle SM, Barton GJ. The Jalview Java alignment editor. Bioinformatics 2004; 20:426-7. PMID:14960472 doi:10.1093/bioinformatics/btg430

31. Edgar RC. MUSCLE: multiple sequence alignment with high accuracy and high throughput. Nucleic Acids Res 2004; 32:1792-7. PMID:15034147 doi:10.1093/nar/gkh340

32. Lewis JA, Hatfull GF. Control of directionality in integrase-mediated recombination: examination of recombination directionality factors (RDFs) including Xis and Cox proteins. Nucleic Acids Res 2001; 29:2205-16. PMID:11376138 doi:10.1093/nar/29.11.2205

33. ElAntak L, Ansaldi M, Guerlesquin F, et al. Structural and genetic analyses reveal a key role in prophage excision for the TorI response regulator inhibitor. J Biol Chem 2005; 280:36802-8. PMID:16079126 doi:10.1074/jbc.M507409200

34. Ramsay JP, Sullivan JT, Stuart GS, et al. Excision and transfer of the Mesorhizobium loti R7A symbiosis island requires an integrase IntS, a novel recombination directionality factor RdfS, and a putative relaxase RlxS. Mol Microbiol 2006; 62:723-34. PMID:17076666 doi:10.1111/j.1365-2958.2006.05396.x

35. Ramsay JP, Sullivan JT, Jambari N, et al. A LuxRI-family regulatory system controls excision and transfer of the Mesorhizobium loti strain R7A symbiosis island by activating expression of two conserved hypothetical genes. Mol Microbiol 2009; 73:1141-55. PMID:19682258 doi:10.1111/j.1365-2958.2009.06843.x

36. Austin S, Ziese M, Sternberg N. A novel role for site-specific recombination in maintenance of bacterial replicons. Cell 1981; 25:729-36. PMID:7026049 doi:10.1016/0092-8674(81)90180-X

37. Gerlitz M, Hrabak O, Schwab H. Partitioning of broad-host-range plasmid RP4 is a complex system involving site-specific recombination. J Bacteriol 1990; 172:6194-203. PMID:2172207

38. Bignell C, Thomas CM. The bacterial ParA-ParB partitioning proteins. J Biotechnol 2001; 91:1-34. PMID:11522360 doi:10.1016/S0168-1656(01)00293-0

39. Gerdes K, Moller-Jensen J, Bugge Jensen R. Plasmid and chromosome partitioning: surprises from phylogeny. Mol Microbiol 2000; 37:455-66. PMID:10931339 doi:10.1046/j.1365-2958.2000.01975.x

40. Burrus V, Pavlovic G, Decaris B, Guedon G. Conjugative transposons: the tip of the iceberg. Mol Microbiol 2002; 46:601-10. PMID:12410819 doi:10.1046/j.1365-2958.2002.03191.x

41. Wozniak RA, Waldor MK. A toxin-antitoxin system promotes the maintenance of an integrative conjugative element. PLoS Genet 2009; 5:e1000439. PMID:19325886 doi:10.1371/journal.pgen.1000439

42. Qiu X, Gurkar AU, Lory S. Interstrain transfer of the large pathogenicity island (PAPI-1) of Pseudomonas aeruginosa. Proc Natl Acad Sci USA 2006; 103:19830-5. PMID:17179047 doi:10.1073/pnas.0606810104

43. Sullivan JT, Trzebiatowski JR, Cruickshank RW, et al. Comparative sequence analysis of the symbiosis island of Mesorhizobium loti strain R7A. J Bacteriol 2002; 184:3086-95. PMID:12003951 doi:10.1128/JB.184.11.3086-3095.2002

44. Marenda M, Barbe V, Gourgues G, et al. A new integrative conjugative element occurs in Mycoplasma agalactiae as chromosomal and free circular forms. J Bacteriol 2006; 188:4137-41. PMID:16707706 doi:10.1128/JB.00114-06

45. Vallenet D, Labarre L, Rouy Z, et al. MaGe: a microbial genome annotation system supported by synteny results. Nucleic Acids Res 2006; 34:53-65. PMID:16407324 doi:10.1093/nar/gkj406

46. Tseng TT, Gratwick KS, Kollman J, et al. The RND permease superfamily: an ancient, ubiquitous and diverse family that includes human disease and development proteins. J Mol Microbiol Biotechnol 1999; 1:107-25. PMID:10941792

47. Furukawa H, Tsay JT, Jackowski S, et al. Thiolactomycin resistance in Escherichia coli is associated with the multidrug resistance efflux pump encoded by emrAB. J Bacteriol 1993; 175:3723-9. PMID:8509326

48. Altendorf K, Gassel M, Puppe W, et al. Structure and function of the Kdp-ATPase of Escherichia coli. Acta Physiol Scand Suppl 1998; 643:137-46. PMID:9789555

49. Altendorf K, Siebers A, Epstein W. The KDP ATPase of Escherichia coli. Ann N Y Acad Sci 1992; 671:228-43. PMID:1288322 doi:10.1111/j.1749-6632.1992.tb43799.x

50. Nies DH. Efflux-mediated heavy metal resistance in prokaryotes. FEMS Microbiol Rev 2003; 27:313-39. PMID:12829273 doi:10.1016/S0168-6445(03)00048-2

51. Hundertmark M, Hincha DK. LEA (late embryogenesis abundant) proteins and their encoding genes in Arabidopsis thaliana. BMC Genomics 2008; 9:118. PMID:18318901 doi:10.1186/1471-2164-9-118

52. Mijnendonckx K, Provoost A, Monsieurs P, et al. Insertion sequence elements in Cupriavidus metallidurans CH34: Distribution and role in adaptation. Plasmid 2011; 65:193-203. PMID:21185859 DOI:doi:10.1016/j.plasmid.2010.12.006

53. Haase J, Lanka E. A specific protease encoded by the conjugative DNA transfer systems of IncP and Ti plasmids is essential for pilus synthesis. J Bacteriol 1997; 179:5728-35. PMID:9294428

54. Byrd DR, Matson SW. Nicking by transesterification: the reaction catalysed by a relaxase. Mol Microbiol 1997; 25:1011-22. PMID:9350859 doi:10.1046/j.1365-2958.1997.5241885.x

55. Lanka E, Wilkins BM. DNA processing reactions in bacterial conjugation. Annu Rev Biochem 1995; 64:141-69. PMID:7574478 doi:10.1146/annurev.bi.64.070195.001041

56. Gomis-Rüth FX, de la Cruz F, Coll M. Structure and role of coupling proteins in conjugal DNA transfer. Res Microbiol 2002; 153:199-204. PMID:12066890 doi:10.1016/S0923-2508(02)01313-X

57. Thomas CM. The Horizontal Gene Pool—Bacterial Plasmids and Gene Spread. Amsterdam: Harwood Academic Publishers; 2000.

58. Powell B, Mergeay M, Christofi N. Transfer of broad host-range plasmids to sulphate-reducing bacteria. FEMS Microbiol Lett 1989; 59:269-73 doi:10.1111/j.1574-6968.1989.tb03123.x.

59. Gormley EP, Davies J. Transfer of plasmid RSF1010 by conjugation from Escherichia coli to Streptomyces lividans and Mycobacterium smegmatis. J Bacteriol 1991; 173:6705-8. PMID:1657866

60. Heinemann JA, Sprague GF Jr. Bacterial conjugative plasmids mobilize DNA transfer between bacteria and yeast. Nature 1989; 340:205-9. PMID:2666856 doi:10.1038/340205a0

61. Sprinzl M, Geider K. Transfer of the Ti plasmid from Agrobacterium tumefaciens into Escherichia coli cells. J Gen Microbiol 1988; 134:413-24. PMID:3049936
62. Farrand SK. Conjugal transfer of Agrobacterium plasmids. In: Clewell DB, ed. Bacterial Conjugation. New York: Plenum Press; 1993:255-291.
63. Szpirer C, Top E, Couturier M, Mergeay M. Retrotransfer or gene capture: a feature of conjugative plasmids, with ecological and evolutionary significance. Microbiology 1999; 145:3321-9. PMID:10627031
64. Haines AS, Akhtar P, Stephens ER, et al. Plasmids from freshwater environments capable of IncQ retrotransfer are diverse and include pQKH54, a new IncP-1 subgroup archetype. Microbiology 2006; 152:2689-701. PMID:16946264 doi:10.1099/mic.0.28941-0
65. Li PL, Everhart DM, Farrand SK. Genetic and sequence analysis of the pTiC58 trb locus, encoding a mating-pair formation system related to members of the type IV secretion family. J Bacteriol 1998; 180:6164-72. PMID:9829924
66. Lawley TD, Klimke WA, Gubbins MJ, Frost LS. F factor conjugation is a true type IV secretion system. FEMS Microbiol Lett 2003; 224:1-15. PMID:12855161 doi:10.1016/S0378-1097(03)00430-0
67. Schröder G, Lanka E. The mating pair formation system of conjugative plasmids-A versatile secretion machinery for transfer of proteins and DNA. Plasmid 2005; 54:1-25. PMID:15907535 doi:10.1016/j.plasmid.2005.02.001

CHAPTER 12

The Integrative and Conjugative Element ICE*Bs1* of *Bacillus subtilis*

Melanie B. Berkmen,*,[1] Stephanie J. Laurer,[1] Bridget K. Giarusso,[1] and Rodrigo Romero[2]

Abstract

ICE*Bs1* is an integrative and conjugative element (ICE) integrated in the *trnS-leu2* gene in *Bacillus subtilis*. Related to Tn*916* and ICE*St1*, ICE*Bs1* contains 25 putative genes and is a new model system for understanding Gram-positive ICEs. Here, we review the molecular mechanisms of how ICE*Bs1* gene expression and mating are induced by the global DNA damage response or when cells are crowded by potential recipients lacking ICE*Bs1*. Induction is achieved by proteolysis of the phage-like repressor ImmR by the anti-repressor ImmA protease. Following induction, ICE*Bs1* can excise from the chromosome to form a double-stranded DNA circle. Excision requires the ICE*Bs1*-encoded integrase Int and excisionase Xis. The circular intermediate is nicked by the relaxase NicK and undergoes rolling circle-like replication. Presumably a single strand of DNA is transferred through the mating machinery, composed of the VirB4-like ConE ATPase along with several other ICE*Bs1*-encoded proteins not yet identified. Transfer of ICE*Bs1* from a donor to a recipient can occur at either the cell poles or along the lateral surfaces of either cell. Transconjugants often become donors, and ICE*Bs1* spreads particularly efficiently within chains of connected cells. After transfer, Int catalyzes integration of ICE*Bs1* into the preferred 17 bp *attB* site in *trnS-leu2*, or at *attB*-like sites elsewhere on the chromosome in strains lacking *attB*. In transconjugants, immunity mechanisms limit the acquisition of a second copy of ICE*Bs1*. We conclude with a discussion of some of the most fascinating outstanding questions concerning ICE*Bs1*.

Introduction to ICE*Bs1*

ICE*Bs1* is an ~21 kb mobile genetic element found integrated in the chromosome in many *Bacillus subtilis* isolates (Fig. 1A).[1,3,4] ICE*Bs1*'s mating efficiency is among the highest observed for a conjugative element.[3] As many as five transconjugants can be formed per donor during a 3 h mating experiment. The high mating frequency is possible because ICE*Bs1* can be induced in > 95% of cells under some conditions.[3,5] Furthermore, this high mating efficiency might be achieved in part because transconjugants can serve as donors soon after they receive ICE*Bs1* and because a single donor can mate with multiple recipients.[6] While the host range has not been extensively studied, ICE*Bs1* mates into *B. subtilis*, *Bacillus anthracis*, *Bacillus licheniformis*, and *Listeria monocytogenes*.[3]

The recent *B. subtilis* genome annotation[7] has revealed that ICE*Bs1* contains 25 putative genes (Table 1), one more than originally reported.[3] ICE*Bs1* was first identified by comparative sequence analysis in 2002.[1] ICE*Bs1* contains several genes related to those found in other ICEs such as Tn*916* in *Enterococcus faecalis*, ICE*St1* in *Streptococcus thermophilus*, ICE*Lm1* in *L. monocytogenes*, ICE*Sma1*

[1]Department of Chemistry and Biochemistry, [2]Department of Biology, Suffolk University, Boston, Massachusetts, USA.
*Corresponding Author: Melanie B. Berkmen—Email: mberkmen@suffolk.edu

Bacterial Integrative Mobile Genetic Elements, edited by Adam P. Roberts and Peter Mullany.
©2013 Landes Bioscience.

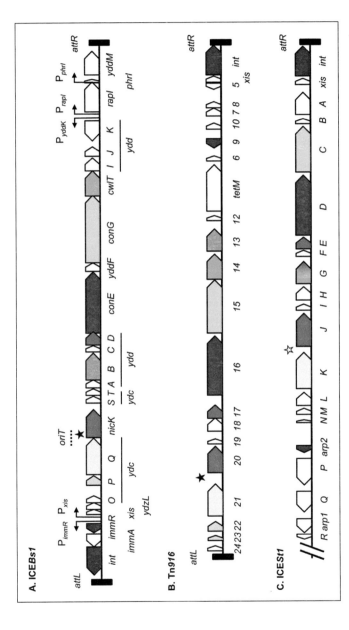

Figure 1. Comparison of the genetic structures of (A) ICE*Bs1*, (B), Tn*916*, and (C) the right portion of ICE*St1* as described previously.[1] Each ORF is represented by an arrowed box oriented in the direction of its transcription. Gene names are indicated underneath. ORF names starting with "orf" have been abbreviated with the corresponding number or letter. Black boxes denote the ends of the element (*attL* and *attR*). Experimentally verified *oriT*s of ICE*Bs1* and Tn*916* are labeled with black stars, while the putative origin of transfer *oriT* of ICE*St1* is labeled with an open star.[1,2] Significant sequence similarities between ORF products between ICEs is indicated by arrowed boxes of the same color. ICE*St1*'s *orfC* is shaded green to blue because it contains similarity to both *orf14* and *orf13* of Tn*916* and *cwlT* of ICE*Bs1*. Ints of the three ICEs are only distantly related; they belong to three different subfamilies of tyrosine recombinases. A) ICE*Bs1* ORF names starting with *ydz*, *ydc*, and *ydd* indicate genes with unknown function. The positions of promoters P*immR*, P*xis*, P*rapI*, and P*phrI* as well as the putative promoter P*yddK* are indicated above with thin arrows above the genes pointing in the direction of transcription. The ~0.8 kb *oriT* region is indicated above with a dotted line.

Table 1. *ICEBs1 contains 25 ORFs*

ORF (former name)	Protein Length (aa)[a]	Predicted Transmembrane Helices	Demonstrated Function	Number of Homologs[e]	Conserved Domains (CD)[f], Gene Ontology (GO)[g], and Predicted Functions
int (*ydcL*)	368		Integrase (recombinase) required for excision and integration[5]	135	CD: φLC3 phage and phage-related integrases, site-specific recombinases, and DNA breaking-rejoining enzymes with C-terminal catalytic domain
immA (*ydcM*)	169		Anti-repressor protease that cleaves ImmR[8,9]	127	CD: Domain of unknown function (DUF955) found in bacteria and viruses that contains a conserved H-E-X-X-H motif suggestive of a catalytic active site
immR (*ydcN*)	127		ICE*Bs1* transcriptional regulator/ immunity repressor[9,10]	111	CD: Helix-turn-helix XRE-family like prokaryotic DNA binding proteins belonging to the xenobiotic response element family of transcriptional regulators
xis (*sacV*)	64		Excisionase (recombination directionality factor)[5]	104	GO:0006310 – DNA recombination GO:0015074 – DNA integration GO:0003677 – DNA binding
ydzL	86			14	GO:0045449 - Regulation of transcription GO:0003677 – DNA binding
ydcO	86			7	
ydcP	126			197	CD: Domain of unknown function (DUF961) found in bacteria
ydcQ	480	2[c,d]		172	Putative coupling protein; FtsK/SpoIIIE superfamily of ATPases involved in the secretion of DNA and proteins across membranes[11,12]
nicK (*ydcR*)	352		Relaxase (nickase) required for nicking and replication[2]	194	CD: Putative phage replication protein RstA family involved in DNA replication, recombination, and repair
ydcS	89			10	

continued on next page

Table 1. Continued

ORF (former name)	Protein Length (aa)[a]	Predicted Transmembrane Helices	Demonstrated Function	Number of Homologs[e]	Conserved Domains (CD)[f], Gene Ontology (GO)[g], and Predicted Functions
ydcT	88			9	
yddA	102			43	
yddB	354	1[c,d]		193	
yddC	82	2[c,d]		26	
yddD	174	2[c] or 3[d]		81	
conE (yddE)	797		Mating machinery ATPase[13]	146	VirB4-like ATPase; FtsK/SpoIIIE superfamily of ATPases involved in the secretion of DNA and proteins across membranes[12]
yddF	108			32	CD: Domain of unknown function (DUF1874) found in some viral and bacterial proteins
conG (yddG)	815	6[d] or 7[c]	Required for mating[6]	178	VirB6-like membrane protein;[11] CD: Domain of unknown function (DUF874) found in *Helicobacter pylori* proteins
cwlT (yddH)	329	1[c,d]	Two-domain cell wall hydrolase (N-terminal muraminidase and C-terminal endopeptidase)[14]	194	CD: N-terminal Lytic Transglycosylase (LT) and Goose Egg White Lysozyme (GEWL) domain. C-terminal NlpC/P60 domain of unknown function found in several lipoproteins
yddI	168	1[c,d]		71	
yddJ	126			106	Putative lipoprotein[h]
yddK	266			102	CD: The Toll/interleukin-1 receptor (TIR) homology domain mediates protein-protein interactions between Toll-like receptors (TLRs) and signal-transduction components

continued on next page

Table 1. Continued

ORF (former name)[a]	Protein Length (aa)[a]	Predicted Transmembrane Helices	Demonstrated Function	Number of Homologs[e]	Conserved Domains (CD)[f], Gene Ontology (GO)[g], and Predicted Functions
rapI	391		Cell-cell signaling regulator of ICE*Bs1* gene expression; increases ImmA-dependent cleavage of ImmR[3,9,10]	23	CD: Three tetratricopeptide repeat (TPR) domains that are found in a wide variety of organisms and are involved in a variety of functions including protein-protein interactions
phrI	5[b] (39)		Cell-cell signaling peptide that antagonizes RapI[3,10]	2	
yddM	313	2[c,d]	Not required for mating[2]	16	Putative helicase[i]

[a] Protein lengths obtained from: (http://bsubcyc.org/); [b] PhrI preprotein is 39 amino acids long, but processed secreted peptide is 5 amino acids long; [c] Number of predicted transmembrane helices obtained from TMHMM 2.0 algorithm[15] accessed from: (http://www.cbs.dtu.dk/services/TMHMM/). Topologies have not been experimentally verified. Protein sequences obtained from: (http://bsubcyc.org/); [d] Number of predicted transmembrane helices obtained from Phobius[16] accessed from: (http://phobius.sbc.su.se/). Topologies have not been experimentally verified. Protein sequences obtained from: (http://bsubcyc.org/); [e] Protein sequences obtained from: (http://bsubcyc.org/). Taxonomy report listing the number of organisms containing a homologous gene was obtained March 2011 from http://blast. ncbi.nlm.nih.gov/; [f] Conserved domains (CD) predictions[17] accessed January, 2011 from NCBI Protein: (http://www.ncbi.nlm.nih.gov/protein). Only the conserved domain with the lowest E-value is reported; [g] Gene ontology (GO) predictions are reported for those proteins without conserved domains. Gene ontologies were accessed January 2011 from: (http://bsubcyc.org/). GOA Gene Ontology annotation was based on Swiss-Prot keyword mapping; [h] Prediction obtained from DO-LOP database of predicted lipoproteins:[18] (http://www.mrc-lmb.cam.ac.uk/genomes/dolop/); [i] Prediction obtained from Genolist database: (http://genodb.pasteur. fr/cgi-bin/WebObjects/GenoList) accessed January 2011.[19]

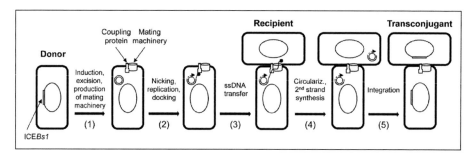

Figure 2. Model for mating of ICE*Bs1*. 1) Upon induction, most ICE*Bs1* genes become highly expressed, leading to the production and assembly of the mating machinery (DNA translocation channel) at the cell membrane. The ICE*Bs1* DNA element is excised from the chromosome, forming a double-stranded circular intermediate. 2) ICE*Bs1* DNA is nicked by the relaxase NicK (shown as a small black circle), which becomes covalently attached to the 5′ end of the cut DNA. The NicK-DNA complex is targeted to the DNA translocation channel via interactions with the coupling protein. 3) Single-stranded ICE*Bs1* DNA, with the attached NicK protein, is transferred into the recipient cell through the mating machinery. Transfer of ICE*Bs1* from a donor to a recipient can occur at the cell pole as shown, or along the lateral surface of either cell. 4) Upon entry into a recipient, the single strand of ICE*Bs1* DNA circularizes prior to second strand synthesis. 5) ICE*Bs1* integrates into the chromosome in both the original donor cell and the newly formed transconjugant.

and ICE*Sma2* in *Staphylococcus aureus*, Tn*1549* in *Enterococcus* sp, and Tn*5397* in *Clostridium difficile* (Fig. 1).[1,20,21] While many of these related elements may confer resistance to vancomycin, tetracycline or cadmium, it is unknown whether any of ICE*Bs1*'s genes provide a growth benefit to the host cell.[22] Like many ICEs, ICE*Bs1*'s genetic structure is modular (Fig. 1A; Table 1).[23] ICE*Bs1* contains blocks of conjugation genes required for DNA transfer, plasmid-like genes involved in replication, and bacteriophage-like genes dedicated to integration, excision, and regulation.[1-3,5,8,13,24]

Work in the Grossman laboratory has led to a model for ICE*Bs1* mating (Fig. 2).[2,3,5,6,8-10,13,24] ICE*Bs1* gene expression and mating are induced under conditions of DNA damage or high density of potential mating partners that lack the element. Upon induction, the element excises from the chromosome, forming a double-stranded circular DNA. The relaxase nicks the circular intermediate, becoming covalently attached to the 5′ end of the cut DNA. ICE*Bs1* undergoes plasmid-like autonomous replication. The relaxase-DNA nucleoprotein complex docks with a coupling protein at the membrane and a single strand of ICE*Bs1* is transferred to the recipient through the mating machinery (DNA translocation channel). Transfer of ICE*Bs1* from a donor to a recipient can occur at the cell pole or along the lateral surface of either cell. Upon entry into the recipient, the single strand of ICE*Bs1* DNA circularizes prior to second strand synthesis. Transconjugants can rapidly become donors, and subsequent spreading of ICE*Bs1* is especially efficient through chains of cells. To maintain genetic stability, ICE*Bs1* integrates onto the chromosome in original donor cell and newly formed transconjugants. This chapter describes these steps in detail and concludes with a description of outstanding questions about ICE*Bs1*.

Regulation of Induction of ICE*Bs1*

ICE*Bs1* gene expression is regulated such that the element exists in two states. In uninduced cells, ICE*Bs1* resides integrated on the chromosome and expression of most ICE*Bs1* genes is repressed.[3,10] Upon induction, almost all of ICE*Bs1*'s 25 genes become more highly expressed, which leads to excision of the element and production of the mating machinery. Induction of ICE*Bs1* is largely controlled through regulation of the P*xis* promoter. P*xis* drives expression of *xis* (encoding excisionase), likely along with the 17 genes downstream (*ydzL-yddJ*; Fig. 1A).

When ICE*Bs1* is uninduced and integrated on the chromosome, P*xis* is repressed by the ImmR protein (Fig. 3).[10] The first 61 residues of ImmR contain the presumed DNA binding domain,

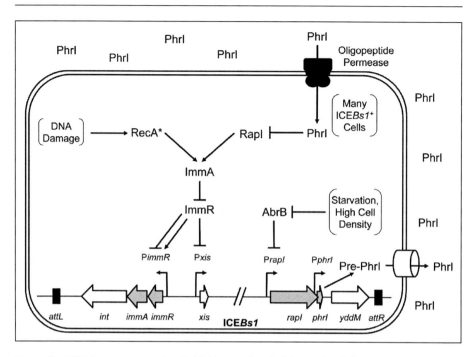

Figure 3. ICE*Bs1* gene expression is highly regulated. Schematic of the genetic structure of ICE*Bs1* is as described in Figure 1 legend, except that genes involved in regulation are shown as gray arrowed boxes. ImmR both activates and represses its own promoter, creating an autoregulatory feedback loop. In uninduced cells, ImmR represses P*xis*, preventing expression of mating genes. Upon induction, the ImmA protease cleaves ImmR, relieving repression of P*xis*. P*xis* and mating are induced under at least two different conditions: (1) the global DNA damage or (2) when cells are surrounded by recipient cells lacking ICE*Bs1*. Upon DNA damage, expression of ICE*Bs1* is induced by a *recA*-dependent process, which stimulates the ImmA-dependent cleavage of ImmR. When ICE*Bs1* is induced by cell crowding, high levels of RapI activate P*xis* by stimulating ImmA-dependent cleavage of ImmR. RapI, in turn, is responsive to PhrI levels in the cell. ICE*Bs1*⁺ cells produce a 39 amino acid Pre-PhrI peptide which is processed and secreted. The mature PhrI pentapeptide is imported back into the cell via oligopeptide permease. High levels of PhrI inhibit RapI, ensuring that mating genes are turned off in the presence of cells containing ICE*Bs1*. Under conditions of low cell density and high nutrients, the transition-state regulator AbrB represses expression of *rapI* and *phrI*.

a conserved helix-turn-helix domain found in several other bacteriophage-like repressors. DNA footprinting shows that purified ImmR binds to several sites in the DNA region between P*xis* and P*immR*, the promoter that drives expression of *immR* along with the two downstream genes *immA* and *int* (Fig. 1A). ImmR probably binds as a dimer to each site, given that ImmR displays self-interaction in yeast two-hybrid assays and that the ImmR binding site consists of an imperfect inverted repeat.[9,10] In addition to repressing P*xis*, ImmR both activates and represses its own promoter, P*immR*.[10] Autoregulation presumably provides high enough levels of ImmR to repress expression of excisionase and other mating genes but low enough to respond efficiently to inducing conditions.

ICE*Bs1* gene expression is induced by derepression of P*xis* through destruction of ImmR by the ICE*Bs1*-encoded anti-repressor ImmA.[9] ImmA is a protease that site-specifically cleaves ImmR between residues F95 and M96. An intact HEXXH motif, a signature of many zinc-dependent metalloproteases, is required for ImmA's activity. ImmR and ImmA homolog pairs are found in over 35 known mobile genetic elements and over 100 putative mobile genetic elements, indicating that this type of regulation is likely conserved.[9,25] For example, ImmR and ImmA homologs are present in the

B. subtilis φ105 and PBSX phages, the *Bacillus thuringiensis* MZTP02 phage, the *L. monocytogenes* A118 phage, and the *Streptococcus pneumoniae* MM1 phage.

There are two conditions known to induce ICE*Bs1*: (1) crowding of donor cells by potential mating partners that lack ICE*Bs1* or (2) the global DNA damage response (SOS) (Fig. 3).[3] In the first case, induction of ICE*Bs1* involves ImmR, ImmA, along with the ICE*Bs1*-encoded cell sensory protein RapI and secreted peptide PhrI.[3,8-10] *rapI* and *phrI* expression increase under conditions of high cell density and starvation.[3] High levels of RapI relieve repression of P*xis* by stimulating ImmA-dependent cleavage of ImmR.[9,10] In turn, high levels of PhrI inhibit RapI.[3] ICE*Bs1*-containing cells produce the pre-PhrI peptide which is then secreted and processed to form the active PhrI pentapeptide. If the donor cell is surrounded by a large number of cells containing ICE*Bs1*, PhrI peptide accumulates extracellularly. PhrI is imported back into the cell through the oligopeptide permease Opp (a.k.a. Spo0K). Once inside, PhrI antagonizes RapI through an unknown mechanism. Altogether, these events ensure that mating is reduced when donors are surrounded by large numbers of cells that already contain the element. This type of regulation may be common, as *rapI-phrI* pairs are found on several phages and conjugative plasmids.

An extra layer of regulation prevents induction of mating when cells are growing rapidly during exponential phase. Under conditions of low cell density and high nutrient availability, expression of *rapI*, along with *phrI*, is repressed directly or indirectly by the transition-state regulator AbrB (Fig. 3).[3] Thus, RapI levels are kept low during exponential growth, even when PhrI levels would be expected to be low. As the cell density rises and nutrients become depleted, *rapI* expression is derepressed which can lead to derepression of P*xis*, so long as the concentration of PhrI is low.

A second condition known to induce ICE*Bs1* is the SOS response, caused by DNA-damaging agents such as mitomycin C (Fig. 3).[3,9,10] Induction by DNA damage is *recA*-dependent but *rapI*- and *phrI*-independent. Presumably regulation by the DNA damage response allows ICE*Bs1* to escape from a host cell in distress.

Induction of ICE*Bs1* by either RapI or the SOS response requires the ImmA-dependent cleavage of ImmR (Fig. 3). Overproduction of RapI does not increase the amount of ImmA in cells.[8] Through an unknown mechanism, RapI likely increases the specific activity of ImmA, leading to proteolysis of ImmR and derepression of P*xis*. Further evidence for this mechanism comes from the analysis of hyperactive *immA^b* mutants that cause partial induction of P*xis* in the absence of inducing signals (such as high levels of RapI). Some *immA^b* mutants produce variant proteins with higher specific activity than wild-type ImmA. Interestingly, these mutations lie near ImmA's predicted zinc-binding motif. The SOS response might also stimulate cleavage of ImmR by increasing the activity of ImmA, but the molecular details have yet to be elucidated.

DNA Processing Events Prior to Transfer

Once P*xis* is induced, *xis* (encoding the excisionase) and the 17 genes downstream become highly expressed.[3] Some of the expressed gene products mediate a series of DNA processing events, including excision, nicking, and replication. The nicked circular element is presumably the substrate for transfer through the mating machinery.

Excision

ICE*Bs1* normally resides integrated in the chromosome at the 3′ end of the *trnS-leu2* (tRNA-Leu2) gene, flanked by the left and right attachment sites, *attL* and *attR*, respectively (Fig. 1A).[3] Excision creates two circular double-stranded DNA products: (1) the circular ICE*Bs1* intermediate with the single attachment site *att*ICE*Bs1* and (2) the repaired bacterial chromosome with attachment site *attB*. Excision occurs through precise recombination between sequences in *attL* and *attR*.[5] As expected, excision requires the flanking *att* sequences, as ICE*Bs1* fails to excise in a strain deleted for *attR*.

Similar to excision of phage lambda, ICE*Bs1* excision requires an integrase (recombinase) and an excisionase (recombination directionality factor), both encoded by the element. The ICE*Bs1* integrase Int belongs to the tyrosine recombinase family of integrases.[3,5] Int is homologous to integrases from Tn*916*, ICE*St1*, ICE*Lm1*, and *Escherichia coli* phage lambda.[1] Like lambda integrase,[26]

Int mediates site-specific recombination.[5] Excision also requires the excisionase Xis.[5] Like other excisionases, Xis is small, highly charged, not well conserved, and is not required for integration. Int and Xis are the only ICE*Bs1*-encoded proteins required for excision. Both presumably bind to several sites within the attachment sites to mediate excision. *int* is constitutively expressed, but *xis* is only conditionally expressed.[3,10] Thus, whether ICE*Bs1* excises or integrates is determined by regulation of the P*xis* promoter.

The frequency of excision has been measured under various conditions.[3,5] Induction by DNA damage or high levels of RapI can lead to very efficient excision (ranging from 3 to > 95% of all cells), which likely contributes to the high mating efficiencies observed. On the other hand, in the absence of inducing conditions, spontaneous excision of the element is rare (~0.005% of all cells). Interestingly, spontaneous excision of ICE*Bs1* increases approximately four-fold in a strain deleted for *rok*.[27] Rok is a small DNA-binding protein that preferentially binds AT-rich DNA, including both the left and right regions of ICE*Bs1*. While the mechanism is unknown, Rok may inhibit ICE*Bs1* excision through repressing P*xis* and/or through inhibiting recombination between *attL* and *attR*.

Nicking

Similar to conjugative plasmids, ICE*Bs1* is nicked at a *nic* site in its origin of transfer, *oriT*, prior to mating.[2] *oriT* was identified by determining which sequences in ICE*Bs1* were required in cis for DNA transfer. *oriT* encompasses all or a portion of a ~0.8 kb DNA fragment that overlaps part of *ydcQ* and *nicK* (Fig. 1A). A *nic* site was identified in the top strand in the 10 bp intervening sequence between two inverted repeats (ACCCCCCCCACGCTAACAGGGGGGT; repeats underlined) in *nicK*. Likely only the top strand is nicked and transferred, as nicking was not detected in the bottom strand. Mutations within the inverted repeats greatly diminish ICE*Bs1* mating. Together, these results indicate that nicking is likely critical for producing the single-stranded DNA substrate for transfer. Nicking requires induction of ICE*Bs1* gene expression, but not excision or circularization of the element.

ICE*Bs1*'s relaxase (nickase) is NicK, which is homologous to the Tn916 relaxase Orf20 and to relaxases required for rolling-circle replication of pT181-like plasmids.[2] *nicK* is the only gene in ICE*Bs1* required for nicking of *oriT* in vivo. While *oriT* is required in cis, *nicK* can function in trans. Mating is undetectable in strains deleted for *nicK*, consistent with the notion that nicking of ICE*Bs1* is required for mating.

A model for nicking and transfer of ICE*Bs1* has been proposed based on work on NicK[2] and analogous plasmid relaxases[28] (Fig. 2). Following induction, ICE*Bs1* excises and forms a double-stranded circular intermediate. After making a single-stranded cut in the top strand of *oriT* of ICE*Bs1*, NicK presumably becomes covalently attached to the 5' end of the cut DNA via a phosphotyrosyl linkage. The NicK-DNA complex would be targeted to the DNA translocation channel via interactions with the putative coupling protein YdcQ at the membrane (see below). Presumably, single-stranded ICE*Bs1* DNA, with the attached NicK protein, is transferred 5' to 3' into the recipient cell. DNA transfer begins by moving the 3' end of *nicK* and finishes with moving *ydcQ* and the 5' end of *nicK* upstream of the *nic* site (Fig. 1A). The single strand of ICE*Bs1* DNA then circularizes and second strand synthesis is required before integration of the element into the chromosome (Fig. 2).

Replication

Under normal conditions, ICE*Bs1* is stably integrated in a tRNA gene, and therefore passively replicated along with the rest of the genome. Generally it has been assumed that ICEs do not replicate autonomously.[22] However, it was recently shown that ICE*Bs1* undergoes plasmid-like autonomous replication upon induction.[24] Rolling circle-type replication initiates from *oriT* and proceeds unidirectionally toward the right (Fig. 1A). The region of ICE*Bs1* to the left of *oriT* is only replicated after excision and circularization. Interestingly, replication may precede excision in some cells, as the copy numbers of chromosomal genes flanking the right end (*attR*) of ICE*Bs1* also increase slightly. Thus, replication initiation does not require excision of the element.

Replication of ICE*Bs1* requires several different replication proteins.[24] Replication requires the ICE*Bs1*-encoded relaxase NicK. A NicK-*oriT* nucleoprotein complex likely recruits the host replication machinery, similar to how some plasmids initiate rolling circle replication. ICE*Bs1* replication requires both the catalytic subunit (PolC) and the β-clamp (DnaN) of the chromosomal replicative DNA polymerase. Like many rolling circle replicating plasmids, replication of ICE*Bs1* requires the chromosomally-encoded helicase PcrA, but not the replicative helicase DnaC or its loader DnaB. Consistent with these results, PolC, PcrA, and single-strand DNA binding protein (Ssb) associate with replicating ICE*Bs1* but DnaC does not.

The location of ICE*Bs1* replication in cells has also been studied. Replication of the *B. subtilis* chromosome occurs near midcell, as fluorescently-tagged replication components are generally observed in the middle region of the cell and only rarely (~5%) observed in the polar quarters.[29-32] Upon induction of ICE*Bs1*, about a third of all replication foci are observed near the cell poles, which is where excised ICE*Bs1* DNA is commonly observed.[13] The replication machinery is likely recruited to the location of the element, similar to the case with plasmid replication in *B. subtilis*.[33]

Replication of ICE*Bs1* might serve a number of purposes. Autonomous replication certainly contributes to greater genetic stability of ICE*Bs1*.[24] ICE*Bs1* re-integrates into the chromosome only several hours after induction. Nevertheless, ICE*Bs1* is maintained in > 95% of donor cells. In contrast, ICE*Bs1* is lost with very high frequency during that timeframe in donor cells that contain mutations that prevent its replication (e.g., *nicK* cells). Replication may also allow a single donor cell to transfer ICE*Bs1* to multiple cells rapidly, consistent with the high mating frequencies observed under certain conditions.[3] Given these advantages, replication may be a general feature common to many ICEs.[24,34-36]

While replication might contribute to increased stability and high mating frequencies, replication of ICE*Bs1* DNA in the donor cell is not required for mating.[24] This result was found by measuring mating frequencies in conditional mutants of essential replication genes. Interestingly, the helicase PcrA is required for mating. Thus, although DNA transfer does not require replication, PcrA is likely needed to unwind the double-stranded DNA to generate a single-stranded substrate for transfer through the mating machinery.

It is interesting to note that NicK is required for nicking of ICE*Bs1* for both conjugation and autonomous replication.[2,24] In contrast, conjugative and mobilizable plasmids that replicate by a rolling circle mechanism generally encode two relaxases, one for mating and the other dedicated to replication. Conjugative relaxases and rolling circle replication relaxases are structurally related, albeit the relative positions of secondary structural elements in conjugative relaxases are switched as compared with those in replication relaxases by a linear permutation event.[37] Thus far, the one exception is NicK, which belongs to the family of relaxases that specialize in replication, yet facilitates both replication and conjugation.[2,24] NicK may be a replication relaxase that subsequently evolved a second conjugation function.

DNA Transfer through the ICE*Bs1* Mating Machinery

For most well-characterized conjugation systems, a so-called "coupling protein" delivers the relaxase-DNA nucleoprotein complex to the DNA translocation channel at the membrane.[38-40] YdcQ may serve as the ICE*Bs1* coupling protein as it is phylogenetically related to other characterized coupling proteins such as PcfC of *E. faecalis* pCF10 and TcpA of *Clostridium perfringens* pCW3.[11,41] Coupling proteins contain conserved nucleotide binding motifs and are members of the HerA/FtsK superfamily of ATPases.[12] Most characterized members of this diverse superfamily (including FtsK, SpoIIIE, VirB4, and VirD4) are involved in pumping DNA and/or proteins across membranes. Interestingly, YdcQ clusters in the FtsK clade (and not the VirD4 coupling protein clade) of the superfamily, with characterized clade members FtsK and SpoIIIE acting as double-stranded DNA pumps. Like many other coupling proteins, YdcQ is predicted to have two N-terminal transmembrane segments, which may facilitate its interaction with the mating machinery at the membrane (Table 1).[11]

During conjugation, negatively charged DNA is able to cross the hydrophobic membrane by passage through a DNA translocation channel. Little is known about the mating machinery of any Gram-positive element, including ICE*Bs1*.[11,42] The DNA translocation channels of several Gram-negative bacteria have been studied in much greater detail.[11,39,40,43] Gram-negative translocation channels are composed of numerous proteins that span the layers of the cell envelope. They generally employ one or two ATPases which use the energy of ATP hydrolysis to drive DNA transport and/ or assembly of the mating machinery. One of the best studied systems is that of the pTi plasmid of *Agrobacterium tumefaciens*. This translocation channel is composed of 11 mating-pair formation (Mpf) proteins (VirB1-VirB11), which includes the ATPases VirB4 and VirB11.

Gram-positive conjugative elements generally encode homologs of only 2 of the 10 or more Mpf components conserved in Gram-negative elements.[11,12] Using the nomenclature from *A. tumefaciens*, most Gram-positive conjugative systems appear to encode homologs of two Mpf proteins: (1) a conserved ATPase (VirB4) and (2) a polytopic membrane subunit (VirB6). Furthermore, while Gram-negative mating elements generally use pili or surface filaments to mediate attachment to potential recipients, Gram-positives often employ surface adhesins.[11] These observations, along with the obvious differences in cell envelope structure, indicate that the mating machineries between Gram-negative and Gram-positive bacteria are likely to feature many significant differences.

ConE is the conserved ATPase component (VirB4 homolog) of the ICEBs1 DNA translocation channel.[13] ConE is required for mating of ICEBs1, but not required for regulation or DNA processing events.[2,3,5,13,24] Like YdcQ, ConE and other VirB4-like proteins are members of the HerA/FtsK superfamily of ATPases (Table 1).[12] ConE is related to mating proteins Orf16 from Tn916, PrgJ from pCF10, and TcpF from pCW3.[1,11,44] Mutations in the Walker A or B box of ConE eliminates detectable conjugative DNA transfer, indicating that both ATP binding and hydrolysis are required for ConE function.[13] Given its homology to VirB4, ConE is predicted to interact with several components of the ICEBs1 mating machinery, energizing its assembly and/or DNA translocation.

A fusion of ConE to green fluorescent protein (ConE-GFP) localizes to the donor's cell membrane, predominantly at the cell poles (Fig. 4A).[13] One or more ICEBs1 products likely interact with ConE to target it to the membrane, as ConE-GFP is dispersed throughout the cytoplasm in a strain lacking ICEBs1 (Fig. 4B). Consistent with this notion, while ConE was previously identified as one of many proteins found in submembrane fractions of *B. subtilis*,[45] ConE lacks predicted transmembrane domains.[11,13] Interestingly, a ConE Walker A mutant that is unable to support mating localizes similarly to the wild-type protein.[13] Thus, proper positioning of ConE is not sufficient for mating, and neither ATP binding nor hydrolysis are required for subcellular positioning of ConE.

ConG is the putative polytopic membrane protein (VirB6-like protein) of the ICEBs1 DNA translocation channel.[11] VirB6-like proteins are thought to be major components of the mating machinery, as they interact with several Mpf proteins and directly contact the conjugative DNA during secretion.[11,43] VirB6-like proteins generally share very low overall sequence similarity, but contain five or more transmembrane segments.[11] Consistent with this notion, ConG is required for mating and is predicted to contain seven or eight transmembrane segments (Table 1).[6,11] ConG is related to other Gram-positive conjugative proteins such as Orf15 from Tn916, OrfC from ICESt1, and TcpH from pCW3.[1,11,44] Interestingly, the ConG-like TcpH and ConE-like TcpF proteins have been shown to localize to the membrane, predominantly at the cell poles.[46]

The putative integral membrane protein CwlT may also be a component of the ICEBs1 mating machinery. CwlT is a two-domain cell wall hydrolase and is homologous to Orf14 from Tn916 and Orf14 from *C. perfringens* ICE CW459.[1,14] The N terminus has a predicted lipoprotein signal peptide with a possible cleavage site after the 22nd residue.[14] Although bioinformatics predicts that the N terminus is a lytic transglycosylase, purified N-terminal domain displays *N*-acetylmuramidase activity, defining a new class of cell wall hydrolases. E87 and D94 are probable active site residues as indicated by their conservation and in vitro analysis of mutant proteins. Based on homology and modeling, the glutamate residue catalyzes the hydrolysis of the linkage between N-acetyl muramic acid and N-acetylglucosamine. Purified C-terminal domain of CwlT acts as a DL-endopeptidase, cleaving the linkage of D-γ-glutamyl-*meso*-diaminopimelic acid in peptidoglycan. CwlT's C terminus is predicted to function as a cysteine protease, making C237 and H288 the predicted catalytic residues.[47]

Several uncharacterized ICEBs1 genes likely encode critical subunits of the DNA translocation channel (Table 1). *yddB*, *yddC*, and *yddD* are excellent candidates; these genes encode putative membrane proteins with sizes and topologies similar to proteins encoded on other Gram-positive conjugative elements (Table 1; Fig. 1).[1,11] YddD is particularly interesting as it was also identified in

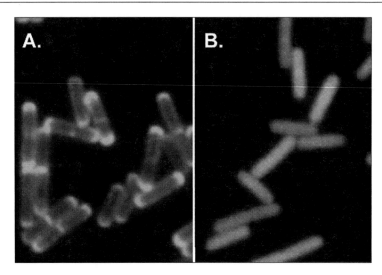

Figure 4. ConE-GFP localizes to the membrane, predominantly at the cell poles. GFP fluorescence is shown. Cells were grown and visualized as previously described.[13] A) ConE-GFP in ICE*Bs1*+ cells. B) ConE-GFP in cells lacking ICE*Bs1*. One or more ICE*Bs1* products likely interact with ConE to target it to the membrane, as ConE-GFP is dispersed throughout the cytoplasm in a strain lacking ICE*Bs1*.

the same membrane subfractions as ConE.[45] Furthermore, *yddD* is encoded immediately upstream of *conE*; typically, small integral membrane proteins are encoded upstream of *virB4*-like genes and may be involved in bringing the VirB4-like subunit to the membrane.[11] Nevertheless, YddD is not sufficient for bringing ConE to the membrane since ConE-GFP is dispersed in the cytoplasm in strains expressing *yddD*, but lacking other ICE*Bs1* genes.[13] Although not as well-conserved, we hypothesize that the putative integral membrane protein YddI and putative lipoprotein YddJ may also be components of the DNA translocation channel. While YddM is also predicted to be an integral membrane protein, it is not required for mating under laboratory conditions.[3] Although not conserved in other ICEs,[11] YddF could conceivably be a component of the DNA translocation channel, as two archaeal homologs have been shown to bind DNA in vitro.[48,49]

During mating, many bacteria produce aggregation substances or elaborate conjugative pili, surface filaments, or adhesins to promote attachment to potential recipients.[11] Thus far, there is no evidence to suggest that ICE*Bs1* employs such specialized means to promote the formation of stable attachment junctions with recipients. While ICE*Bs1* does encode some proteins that are predicted to be surface displayed (e.g., the putative lipoprotein YddJ and cell wall hydrolase CwlT), it does not encode any recognizable pilins or adhesins (Table 1).

Cell Biology of Mating of ICE*Bs1*

Live fluorescence microscopy has been used to visualize the orientations of donors and recipients in ICE*Bs1* mating pairs in real time.[6] Donors and recipients appear to be in close contact during successful mating events, indicating that mating likely cannot occur from a distance through an extended conjugative pilus. Given that the mating machinery component ConE is most heavily concentrated at the cell poles, it had been hypothesized that transfer of ICE*Bs1* preferentially occurs from a donor pole rather than along the lateral sides of the donor cell.[13,50] However, transfer of ICE*Bs1* from a donor to a recipient can occur along the poles or the lateral sides of either cell.[6]

Additional analysis revealed that transconjugants frequently became donors soon after transfer of ICE*Bs1*.[6] The ability of a transconjugant to rapidly become a donor is likely due to the kinetics of repression of P*xis*, which is determined by the levels of ImmR repressor. ImmR both activates and represses its own expression, thus forming a homeostatic autoregulatory loop. When transconjugants initially form, no ImmR is present in the cell which allows for transcription from P*xis* and expression of the mating proteins.

When ICE*Bs1* transfers into a cell in a chain of connected cells, the element spreads rapidly from cell to cell within the chain.[6] Spreading of ICE*Bs1* through cell chains requires the conjugation machinery and thus is not due to replication or segregation of ICE*Bs1* during cell division. The high efficiency of intrachain conjugation is likely attributable to the stable, close contact between cells in chains. The enrichment of conjugation machinery at the cell poles may also contribute to the speed of intrachain transfer events.[13] Many types of bacteria grow in chains and contain conjugative elements, and bacterial cell chains are especially prevalent in biofilms.[51] Accelerated sequential conjugation through cell chains may be a general mechanism that bacteria employ to rapidly spread conjugative elements in microbial communities.

Integration of ICE*Bs1* on the Chromosome

After transfer through the mating machinery, integration of ICE*Bs1* DNA into the recipient's genome allows for its stable inheritance. Since only a single strand of ICE*Bs1* DNA is likely transferred, circularization and complementary strand synthesis in the recipient would be required prior to integration (Fig. 2).[2]

ICE*Bs1*'s preferred integration site in the *B. subtilis* chromosome is at the 3′ end of the *trnS-leu2* tRNA gene.[5] Integration requires the ICE*Bs1*-encoded integrase Int. Int catalyzes site-specific recombination between the attachment site *att*ICE*Bs1* in the element and the attachment site *attB* in *trnS-leu2*. Functional Int protein is not transferred through the mating machinery, as had been proposed for Tn*916*,[52] since the *int* gene is required in the recipient for integration to occur.[5]

ICE*Bs1* integrates into the chromosome at alternative attachment sites when mated into *B. subtilis* strains deleted for the primary attachment site *attB* in *trnS-leu2*.[5] All seven secondary ICE*Bs1* integration sites analyzed were similar in sequence to a 17 bp sequence in *attB* and *att*ICE*Bs1*. Composed of 5 bp inverted repeats separated by a 7 bp spacer (C̲T̲A̲G̲G̲TTGAGGGC̲C̲T̲A̲G̲; repeats underlined), this 17 bp sequence may constitute the minimal *attB* sequence that can mediate integration. A bioinformatic search of the *B. subtilis* genome for sequences similar to the 17 bp sequence indicates that ICE*Bs1* may be able to integrate into hundreds of alternative attachment sites in the *B. subtilis* genome.

ICE*Bs1* can also integrate into the chromosome in the absence of *int*, albeit at a reduced efficiency. Integration is RecA-dependent.[5] In this case, a region of DNA sequence identity between ICE*Bs1* and the chromosome is necessary for homologous recombination.

Immunity

Like many mobile genetic elements, ICE*Bs1* has mechanisms to reduce the likelihood of acquisition of additional copies of the same or similar element. ICE*Bs1* is transferred 50- to 500-fold less efficiently to recipients that contain ICE*Bs1* than to recipients lacking the element.[3,10] Expression of ImmR alone in the recipient confers immunity to a similar extent as the presence of ICE*Bs1* in its entirety.[10] Overproduction of Int in the recipient bypasses ImmR-mediated immunity, indicating that immunity depends on ImmR inhibiting the expression and/or activity of the integrase. In one model, ImmR binding to the *immR* promoter of the incoming ICE*Bs1* DNA prevents expression of *int* which is required for integration. Other evidence indicates that there is an additional, as yet uncharacterized, mechanism that contributes to immunity, independent of the effects of ImmR on Int.

Conclusion and Future Directions

Despite its discovery less than 10 y ago, ICE*Bs1* is quickly becoming an important paradigm for understanding ICEs in Gram-positive bacteria. So far, the roles of 10 of ICE*Bs1*'s 25 putative genes have been at least partially elucidated experimentally (Table 1). We now have a preliminary understanding of the regulation of ICE*Bs1* gene expression by cell-cell signaling and the DNA damage response (Fig. 3). We also have a model for the events that follow induction, including excision, nicking, rolling circle replication, transfer, and integration (Fig. 2). The rapid progress is at least partly attributable to the fact that ICE*Bs1* can be induced in > 95% of cells in a population simply by overproducing the activator RapI.[3,5] Without this high level of induction, some of the critical discoveries, such as the detection of autonomous replication, may not have been possible.

Despite our rapid progress, many fundamental questions remain. For example, how does RapI stimulate ImmA's proteolytic activity? How does the imported PhrI pentapeptide antagonize RapI? How does DNA damage lead to increased proteolysis of ImmR? Is YdcQ the ICE*Bs1* coupling protein? Furthermore, one of the largest gaps in our knowledge regarding ICE*Bs1* (and any Gram-positive conjugative element) concerns the structure and function of the DNA translocation channel. Once the critical components of the channel are identified, numerous questions remain regarding how they interact with each other and translocate DNA. Given ICE*Bs1*'s small size, homology to other elements, and ease of study, the ICE*Bs1* DNA translocation channel will likely become an important model for our understanding of the mating machinery of Gram-positive bacteria in the future.

It will be interesting to determine whether ICE*Bs1* provides a selective advantage to its host. ICE*Bs1* lacks antibiotic resistance or otherwise recognizably beneficial genes found on many of its relatives. One selective value of ICE*Bs1* may lie in the general ability of mobile elements to provide mechanisms for conferring genome plasticity and greater rates of innovation.[53-55] Due to their mosaic structures, ICEs have a propensity to acquire and rearrange genes, often through inter-ICE recombination,[56,57] which may be advantageous to the host.

Since ICE*Bs1* can integrate into the chromosome through both site-specific (Int-dependent) and homologous recombination (RecA-dependent) mechanisms,[5] scientists may find that ICE*Bs1* is a useful genetic tool for the study of other bacteria. In that regard, it will be important to explore ICE*Bs1*'s host range. ICE*Bs1* may be able to mate into *S. aureus, Staphylococcus epidermidis, Listeria innocua, B. halodurans*, and *B. cereus* since they contain close matches to ICE*Bs1*'s *attB* site.[3] Given the vast number of unanswered questions regarding ICE*Bs1*, we expect countless exciting surprises.

Acknowledgments

We are grateful to Alan D. Grossman, Catherine A. Lee, Bijou Bose, Jennifer M. Auchtung, and Tyler A. DeWitt for helpful comments on the manuscript. We thank Ingrid Keseler of SRI International and BsubCyc *B. subtilis* database website and Vincent Burrus for their helpful advice. We also thank the Department of Chemistry and Biochemistry and the Dean of the College of Arts and Sciences of Suffolk University for generously granting a reduction in teaching load to MBB to allow time for writing this chapter.

References

1. Burrus V, Pavlovic G, Decaris B, et al. The ICESt1 element of Streptococcus thermophilus belongs to a large family of integrative and conjugative elements that exchange modules and change their specificity of integration. Plasmid 2002; 48:77-97. PMID:12383726 doi:10.1016/S0147-619X(02)00102-6

2. Lee CA, Grossman AD. Identification of the origin of transfer (oriT) and DNA relaxase required for conjugation of the integrative and conjugative element ICEBs1 of Bacillus subtilis. J Bacteriol 2007; 189:7254-61. PMID:17693500 doi:10.1128/JB.00932-07

3. Auchtung JM, Lee CA, Monson RE, et al. Regulation of a Bacillus subtilis mobile genetic element by intercellular signaling and the global DNA damage response. Proc Natl Acad Sci USA 2005; 102:12554-9. PMID:16105942 doi:10.1073/pnas.0505835102

4. Earl AM, Losick R, Kolter R. Bacillus subtilis genome diversity. J Bacteriol 2007; 189:1163-70. PMID:17114265 doi:10.1128/JB.01343-06

5. Lee CA, Auchtung JM, Monson RE, et al. Identification and characterization of int (integrase), xis (excisionase) and chromosomal attachment sites of the integrative and conjugative element ICEBs1 of Bacillus subtilis. Mol Microbiol 2007; 66:1356-69. PMID:18005101

6. Babic A, Berkmen MB, Lee CA, et al. Efficient gene transfer in bacterial cell chains. mBio 2011;2(2). PMID:21406598 doi:10.1128/mBio.00027-11.

7. Barbe V, Cruveiller S, Kunst F, et al. From a consortium sequence to a unified sequence: the Bacillus subtilis 168 reference genome a decade later. Microbiology 2009; 155:1758-75. PMID:19383706 doi:10.1099/mic.0.027839-0

8. Bose B, Grossman AD. Regulation of horizontal gene transfer in Bacillus subtilis by activation of a conserved site-specific protease. J Bacteriol 2011; 193:22-9. PMID:21036995 doi:10.1128/JB.01143-10

9. Bose B, Auchtung JM, Lee CA, et al. A conserved anti-repressor controls horizontal gene transfer by proteolysis. Mol Microbiol 2008; 70:570-82. PMID:18761623 doi:10.1111/j.1365-2958.2008.06414.x

10. Auchtung JM, Lee CA, Garrison KL, et al. Identification and characterization of the immunity repressor (ImmR) that controls the mobile genetic element ICEBs1 of Bacillus subtilis. Mol Microbiol 2007; 64:1515-28. PMID:17511812 doi:10.1111/j.1365-2958.2007.05748.x

11. Alvarez-Martinez CE, Christie PJ. Biological diversity of prokaryotic type IV secretion systems. Microbiol Mol Biol Rev 2009; 73:775-808. PMID:19946141 doi:10.1128/MMBR.00023-09

12. Iyer LM, Makarova KS, Koonin EV, et al. Comparative genomics of the FtsK-HerA superfamily of pumping ATPases: implications for the origins of chromosome segregation, cell division and viral capsid packaging. Nucleic Acids Res 2004; 32:5260-79. PMID:15466593 doi:10.1093/nar/gkh828

13. Berkmen MB, Lee CA, Loveday EK, et al. Polar positioning of a conjugation protein from the integrative and conjugative element ICEBs1 of Bacillus subtilis. J Bacteriol 2010; 192:38-45. PMID:19734305 doi:10.1128/JB.00860-09

14. Fukushima T, Kitajima T, Yamaguchi H, et al. Identification and characterization of novel cell wall hydrolase CwlT: a two-domain autolysin exhibiting n-acetylmuramidase and DL-endopeptidase activities. J Biol Chem 2008; 283:11117-25. PMID:18305117 doi:10.1074/jbc.M706626200

15. Krogh A, Larsson B, von Heijne G, et al. Predicting transmembrane protein topology with a hidden Markov model: application to complete genomes. J Mol Biol 2001; 305:567-80. PMID:11152613 doi:10.1006/jmbi.2000.4315

16. Käll L, Krogh A, Sonnhammer EL. A combined transmembrane topology and signal peptide prediction method. J Mol Biol 2004; 338:1027-36. PMID:15111065 doi:10.1016/j.jmb.2004.03.016

17. Marchler-Bauer A, Lu S, Anderson JB, et al. CDD: a Conserved Domain Database for the functional annotation of proteins. Nucleic Acids Res 2011; 39:D225-9. PMID:21109532 doi:10.1093/nar/gkq1189

18. Babu MM, Priya ML, Selvan AT, et al. A database of bacterial lipoproteins (DOLOP) with functional assignments to predicted lipoproteins. J Bacteriol 2006; 188:2761-73. PMID:16585737 doi:10.1128/JB.188.8.2761-2773.2006

19. Lechat P, Hummel L, Rousseau S, et al. GenoList: an integrated environment for comparative analysis of microbial genomes. Nucleic Acids Res 2008; 36:D469-74. PMID:18032431 doi:10.1093/nar/gkm1042

20. Churchward G. Back to the future: the new ICE age. Mol Microbiol 2008; 70:554-6. PMID:18761693 doi:10.1111/j.1365-2958.2008.06415.x

21. Schijffelen MJ, Boel CH, van Strijp JA, et al. Whole genome analysis of a livestock-associated methicillin-resistant Staphylococcus aureus ST398 isolate from a case of human endocarditis. BMC Genomics 2010; 11:376. PMID:20546576 doi:10.1186/1471-2164-11-376

22. Burrus V, Waldor MK. Shaping bacterial genomes with integrative and conjugative elements. Res Microbiol 2004; 155:376-86. PMID:15207870 doi:10.1016/j.resmic.2004.01.012

23. Wozniak RA, Waldor MK. Integrative and conjugative elements: mosaic mobile genetic elements enabling dynamic lateral gene flow. Nat Rev Microbiol 2010; 8:552-63. PMID:20601965 doi:10.1038/nrmicro2382

24. Lee CA, Babic A, Grossman AD. Autonomous plasmid-like replication of a conjugative transposon. Mol Microbiol 2010; 75:268-79. PMID:19943900 doi:10.1111/j.1365-2958.2009.06985.x

25. Bose B. Regulation of the mobile genetic element ICEBs1 by a conserved repressor and anti-repressor [Ph.D. thesis]. Cambridge: Department of Biology, Massachusetts Institute of Technology; 2010.

26. Weisberg RA, Landy A. Site-specific recombination in phage lambda. In: Hendrix RW, Roberts JW, Stahl FW, Weisberg RA, eds. Lambda II. Cold Spring Harbor, N.Y.: Cold Spring Harbor Laboratory; 1983.

27. Smits WK, Grossman AD. The transcriptional regulator Rok binds A+T-rich DNA and is involved in repression of a mobile genetic element in Bacillus subtilis. PLoS Genet 2010; 6:e1001207. PMID:21085634 doi:10.1371/journal.pgen.1001207

28. de la Cruz F, Frost LS, Meyer RJ, et al. Conjugative DNA metabolism in Gram-negative bacteria. FEMS Microbiol Rev 2010; 34:18-40. PMID:19919603 doi:10.1111/j.1574-6976.2009.00195.x

29. Lemon KP, Grossman AD. Localization of bacterial DNA polymerase: evidence for a factory model of replication. Science 1998; 282:1516-9. PMID:9822387 doi:10.1126/science.282.5393.1516

30. Berkmen MB, Grossman AD. Spatial and temporal organization of the Bacillus subtilis replication cycle. Mol Microbiol 2006; 62:57-71. PMID:16942601 doi:10.1111/j.1365-2958.2006.05356.x

31. Migocki MD, Lewis PJ, Wake RG, et al. The midcell replication factory in Bacillus subtilis is highly mobile: implications for coordinating chromosome replication with other cell cycle events. Mol Microbiol 2004; 54:452-63. PMID:15469516 doi:10.1111/j.1365-2958.2004.04267.x

32. Meile JC, Wu LJ, Ehrlich SD, et al. Systematic localisation of proteins fused to the green fluorescent protein in Bacillus subtilis: identification of new proteins at the DNA replication factory. Proteomics 2006; 6:2135-46. PMID:16479537 doi:10.1002/pmic.200500512

33. Wang JD, Rokop ME, Barker MM, et al. Multicopy plasmids affect replisome positioning in Bacillus subtilis. J Bacteriol 2004; 186:7084-90. PMID:15489419 doi:10.1128/JB.186.21.7084-7090.2004
34. Grohmann E. Autonomous plasmid-like replication of Bacillus ICEBs1: a general feature of integrative conjugative elements? Mol Microbiol 2010; 75:261-3. PMID:19943906 doi:10.1111/j.1365-2958.2009.06978.x
35. Wang J, Wang GR, Shoemaker NB, et al. Production of two proteins encoded by the Bacteroides mobilizable transposon NBU1 correlates with time-dependent accumulation of the excised NBu1 circular form. J Bacteriol 2001; 183:6335-43. PMID:11591678 doi:10.1128/JB.183.21.6335-6343.2001
36. Ramsay JP, Sullivan JT, Stuart GS, et al. Excision and transfer of the Mesorhizobium loti R7A symbiosis island requires an integrase IntS, a novel recombination directionality factor RdfS, and a putative relaxase RlxS. Mol Microbiol 2006; 62:723-34. PMID:17076666 doi:10.1111/j.1365-2958.2006.05396.x
37. Smillie C, Garcillan-Barcia MP, Francia MV, et al. Mobility of plasmids. Microbiol Mol Biol Rev 2010; 74:434-52. PMID:20805406 doi:10.1128/MMBR.00020-10
38. Llosa M, Gomis-Ruth FX, Coll M, et al. Bacterial conjugation: a two-step mechanism for DNA transport. Mol Microbiol 2002; 45:1-8. PMID:12100543 doi:10.1046/j.1365-2958.2002.03014.x
39. Schröder G, Lanka E. The mating pair formation system of conjugative plasmids-A versatile secretion machinery for transfer of proteins and DNA. Plasmid 2005; 54:1-25. PMID:15907535 doi:10.1016/j.plasmid.2005.02.001
40. Burton B, Dubnau D. Membrane-associated DNA transport machines. Cold Spring Harb Perspect Biol 2010; 2:a000406. PMID:20573715 doi:10.1101/cshperspect.a000406
41. Parsons JA, Bannam TL, Devenish RJ, et al. TcpA, an FtsK/SpoIIIE homolog, is essential for transfer of the conjugative plasmid pCW3 in Clostridium perfringens. J Bacteriol 2007; 189:7782-90. PMID:17720795 doi:10.1128/JB.00783-07
42. Grohmann E, Muth G, Espinosa M. Conjugative plasmid transfer in gram-positive bacteria. Microbiol Mol Biol Rev 2003; 67:277-301. PMID:12794193 doi:10.1128/MMBR.67.2.277-301.2003
43. Waksman G, Fronzes R. Molecular architecture of bacterial type IV secretion systems. Trends Biochem Sci 2010; 35:691-8. PMID:20621482 doi:10.1016/j.tibs.2010.06.002
44. Bannam TL, Teng WL, Bulach D, et al. Functional identification of conjugation and replication regions of the tetracycline resistance plasmid pCW3 from Clostridium perfringens. J Bacteriol 2006; 188:4942-51. PMID:16788202 doi:10.1128/JB.00298-06
45. Bunai K, Nozaki M, Hamano M, et al. Proteomic analysis of acrylamide gel separated proteins immobilized on polyvinylidene difluoride membranes following proteolytic digestion in the presence of 80% acetonitrile. Proteomics 2003; 3:1738-49. PMID:12973734 doi:10.1002/pmic.200300529
46. Teng WL, Bannam TL, Parsons JA, et al. Functional characterization and localization of the TcpH conjugation protein from Clostridium perfringens. J Bacteriol 2008; 190:5075-86. PMID:18487333 doi:10.1128/JB.00386-08
47. Anantharaman V, Aravind L. Evolutionary history, structural features and biochemical diversity of the NlpC/P60 superfamily of enzymes. Genome Biol 2003; 4:R11. PMID:12620121 doi:10.1186/gb-2003-4-2-r11
48. Larson ET, Eilers BJ, Reiter D, et al. A new DNA binding protein highly conserved in diverse crenarchaeal viruses. Virology 2007; 363:387-96. PMID:17336360 doi:10.1016/j.virol.2007.01.027
49. Keller J, Leulliot N, Cambillau C, et al. Crystal structure of AFV3-109, a highly conserved protein from crenarchaeal viruses. Virol J 2007; 4:12. PMID:17241456 doi:10.1186/1743-422X-4-12
50. Grohmann E. Conjugative transfer of the integrative and conjugative element ICEBs1 from Bacillus subtilis likely initiates at the donor cell pole. J Bacteriol 2010; 192:23-5. PMID:19854907 doi:10.1128/JB.01305-09
51. Davey ME, O'Toole GA. Microbial biofilms: from ecology to molecular genetics. Microbiol Mol Biol Rev 2000; 64:847-67. PMID:11104821 doi:10.1128/MMBR.64.4.847-867.2000
52. Bringel F, Van Alstine GL, Scott JR. Conjugative transposition of Tn916: the transposon int gene is required only in the donor. J Bacteriol 1992; 174:4036-41. PMID:1317846
53. Bennett PM. Genome plasticity: insertion sequence elements, transposons and integrons, and DNA rearrangement. Methods Mol Biol 2004; 266:71-113. PMID:15148416
54. Frost LS, Leplae R, Summers AO, et al. Mobile genetic elements: the agents of open source evolution. Nat Rev Microbiol 2005; 3:722-32. PMID:16138100 doi:10.1038/nrmicro1235
55. Gogarten JP, Townsend JP. Horizontal gene transfer, genome innovation and evolution. Nat Rev Microbiol 2005; 3:679-87. PMID:16138096 doi:10.1038/nrmicro1204
56. Wozniak RA, Fouts DE, Spagnoletti M, et al. Comparative ICE genomics: insights into the evolution of the SXT/R391 family of ICEs. PLoS Genet 2009; 5:e1000786. PMID:20041216 doi:10.1371/journal.pgen.1000786
57. Garriss G, Waldor MK, Burrus V. Mobile antibiotic resistance encoding elements promote their own diversity. PLoS Genet 2009; 5:e1000775. PMID:20019796 doi:10.1371/journal.pgen.1000775

CHAPTER 13

Integrating Conjugative Elements of the SXT/R391 Family

Geneviève Garriss and Vincent Burrus*

Abstract

Integrating conjugative elements (ICEs) are bacterial mobile genetic elements that are found integrated in the chromosome of their host and transfer via conjugation. The SXT/R391 family of ICEs is widespread in γ-proteobacteria related to *Vibrio cholerae* and contributes to the massive dissemination of antibiotic resistance genes. Besides promoting their own transfer to new host strains these ICEs have been shown to catalyze the mobilization of otherwise non-mobilizable genetic elements such as genomic islands and virulence plasmids. In this chapter we present a comprehensive review on the main functions of this family of mobile elements, focusing on the major advancements that have recently been published.

Introduction

Integrating and conjugative elements (ICEs) are self-transmissible bacterial mobile elements that play a major role in gene exchange in bacterial populations. These elements are horizontally transferred via conjugation by a process similar to that used by many conjugative plasmids. However, unlike plasmids, ICEs usually do not have a circular replicative form and are rather found integrated into the chromosome. An obvious consequence of this feature is that ICEs are also transmitted vertically with the host chromosome when the cell divides. When subjected to certain environmental conditions these elements excise from their host's chromosome, usually by site-specific recombination, and form a circular intermediate that is the substrate for their conjugative transfer. Upon entry into the recipient cell they recircularize and reintegrate into the chromosome.[1,2]

A large number of ICEs studied to date were discovered in *Vibrio cholerae*, the causative agent of cholera, and related species. Cholera is a severe epidemic infectious disease that is contracted by the ingestion of water or food contaminated with *V. cholerae*. This disease is widespread in regions with limited access to clean water and where poor sanitation allows for easy dissemination of the bacterium in drinking water sources. Cholera is characterized by a profuse diarrhea that rapidly induces massive fluid loss, causing the severe dehydration of the patient and leading to death in less than 24 hours. While rehydration of cholera patients remains the most effective treatment, antibiotics have been widely misused as prophylactic and therapeutic agents, allowing the emergence of multiresistant strains.[3]

Cholera epidemics have succeeded each other for the last two centuries and until the emergence in the early 1990s of the new O139 serogroup of *V. cholerae*, all cholera outbreaks were attributable to the O1 serogroup.[3] Analysis of the antibiotic resistance profile of the new O139 serogroup led to the discovery of SXT, an ICE conferring resistance to sulfamethoxazole, trimethoprim, chloramphenicol and streptomycin,[4] which marked the beginning of two decades

*Département de Biologie, Faculté des Sciences, Université de Sherbrooke, Sherbrooke, Québec, Canada.
Corresponding Author: Vincent Burrus—Email: vincent.burrus@usherbrooke.ca

Bacterial Integrative Mobile Genetic Elements, edited by Adam P. Roberts and Peter Mullany.
©2013 Landes Bioscience.

of extensive study of this new type of mobile genetic element (MGE). Besides opening the way to the discovery of many more ICEs, the discovery of SXT also led to the reanalysis of many known MGEs that had previously been misclassified as conjugative plasmids or conjugative transposons. One of the consequences of this reanalysis was the reclassification of all "R factors" of the IncJ incompatibility group,[5-7] such as R391 from *Providencia rettgeri* (formerly known as *Proteus rettgeri*), as ICEs related to SXT and to their inclusion in a new family of MGEs, the SXT/R391 family (Table 1).[8-10] These elements contain a conserved set of genes that mediate the key functions of integration, excision, conjugative transfer and regulation. SXT/R391 ICEs are found integrated into the 5' end of *prfC*, the gene encoding peptide chain release factor 3 (RF3).[11] Their integration into the chromosome is the result of the site-specific recombination, catalyzed by the integrase Int, of two identical 17-bp sequences found on the extrachromosomal circular ICE molecule (*attP*) and on the chromosome (*attB*). Integration of SXT provides *prfC* with a new 5' non-coding sequence, allowing the expression of a functional RF3. The reverse reaction which leads to the excision of the element from the chromosome is mediated by Int, assisted by Xis, a recombination directionality factor, and involves the site-specific recombination of the *attL* and *attR* sites found on each side of the integrated element.[12] Once excised from the chromosome, it is thought that the circular ICE molecule is cleaved at its origin of transfer (*oriT*) and transferred as a single-stranded molecule into the recipient cell through a mating channel encoded by the ICE itself. The single-stranded molecules generated in the donor and recipient cells are duplicated and subsequently integrate into both chromosomes. The concerted expression of the genes required for the integration/excision and transfer functions relies on the action of three ICE-encoded regulators: *setR*, *setC* and *setD*.[13,14] SetC and SetD are transcriptional activators that stimulate the expression of *int* and conjugation-associated operons. Transcription of *setCD* is itself repressed by SetR, a repressor similar to bacteriophage lambda CI. As is the case for lambda, the repression by SetR is alleviated by the induction of the SOS response, thus promoting the transfer of SXT/R391 ICEs in response to DNA damaging agents such as SOS response-inducing antibiotics.[13]

The SXT/R391 family now contains more than 50 elements that have been discovered in clinical and environmental isolates of various bacterial species from diverse regions of the globe over the past 40 years (Table 1).[10,15] Newly identified elements are classified as part of this family based on the conservation of a scaffold of functional genes, the presence of a conserved integrase gene and by their site-specific integration into the 5' end of *prfC* (see Burrus[10] and references therein for a complete review). Besides being major vectors for the dissemination of antibiotic resistances genes, ICEs of the SXT/R391 family are increasingly recognized for their important contribution in the plasticity of their hosts' genome and in their adaptation to the environment. The present chapter will focus on the recent advancements made in the study of SXT/R391 ICEs.

SXT/R391 ICEs Share a Common Backbone

Minimal Functional Core

Comparison of the complete DNA sequences of 13 SXT/R391 ICEs revealed that these elements share a conserved backbone of genes that mediate their essential functions of integration/excision, conjugative transfer and regulation (Fig. 1A).[15] The conserved core of SXT/R391 ICEs comprises 52 genes and contains approximately as many genes of unknown function as genes of known function. Of this conversed backbone, only 25 genes spanning 29.7 kb are required for the minimal functional element.[15] The core of SXT/R391 ICEs is organized in three discrete functional modules that govern integration/excision, conjugative transfer (DNA processing and mating pair formation) and regulation.

The integration and excision processes are mediated by a P4-type site-specific tyrosine recombinase, Int, and a recombination directionality factor, Xis, that are located in the 5' end of the elements. Int is sufficient to mediate integration into to the host's chromosome, but Xis is required for efficient excision.[12] Right of the integration/excision module is found the DNA processing module, separated in two discrete loci by *rumBA* (involved in UV damage repair)[16] and ORFs *s024*, *s025* and *s026* (unknown function).

Table 1. Most relevant SXT/R391 ICE family members

ICE Name	Host[a]	Origin		Type[b]	Drug(s) to Which Resistance is Conferred[c]	eex Allele[d]	Refs.
		Location	Year				
SXT[MO10]	*Vibrio cholerae* O139 MO10	India	1992	C	Su Tm Cm Sm	S	4,15,28,41
ICEVchInd4	*Vibrio cholerae* O139 AS207	India	1997	C	Su Cm Sm	S	15,28,41
ICEVchBan7	*Vibrio cholerae* O139 2125/98	Bangladesh	1998	C	none	S	28,41
ICEVchHKo1[f]	*Vibrio cholerae* O139	Hong Kong	1993	C	none	na	28
ICEVchSL1	*Vibrio cholerae* O139 E712	Sri Lanka	1994	E	Su Tm Cm Sm	S5	28,41
ICEVchBan1	*Vibrio cholerae* O1	Bangladesh	1998	C	Su Tm Cm Sm	S[h]	28
ICEVchBan2	*Vibrio cholerae* O1	Bangladesh	2005	C	Su Tm Tc	R1	41
ICEVchBan3	*Vibrio cholerae* O1	Bangladesh	2005	C	Su Tm Cm	R1	41
ICEVchBan4	*Vibrio cholerae* O1	Bangladesh	1998	C	Su Tm	R1	41
ICEVchBan5	*Vibrio cholerae* O1	Bangladesh	1998	C	Su Tm Cm Sm	R1	15,41
ICEVchInd1	*Vibrio cholerae* O1	India	1994	C	Su Tm Cm Sm	R1	28,41
ICEVchInd2	*Vibrio cholerae* O1	India	1994	C	Su Tm	R1	41
pJY1[e,f]	*Vibrio cholerae* O1	Philippines	1973	C	Su Cm Sm	na	53
ICEVchLao1[f]	*Vibrio cholerae* O1	Lao PDR	1998	C	Su Tm Cm Sm Tc	na	54,55
ICEVchVie0[f]	*Vibrio cholerae* O1	Vietnam	1990	C	none	na	56
ICEVchVie1[f]	*Vibrio cholerae* O1	Vietnam	2000	C	Su Cm Sm Tc	na	57

continued on next page

Table 1. Continued

| ICE Name | Host[a] | Origin | | Type[b] | Drug(s) to Which Resistance is Conferred[c] | eex Allele[d] | Refs. |
		Location	Year				
ICEVchAlg1[f]	*Vibrio cholerae* O1	Algeria	1994	C	Su Tm Cm Sm Tc	na	58
ICEVchInd5	*Vibrio cholerae* O1	India	1994–2005	C	Su Tm Cm Sm	R1[g]	15,50
ICEVchBan9	*Vibrio cholerae* O1	Bangladesh	1994	C	Cm Tc Sm Su Tm	R1[g]	15,59
ICEVchMoz10	*Vibrio cholerae* O1 B33	Mozambique	2004	C	Cm Tc Sm Su Tm	R1[g]	15,60
ICEVchBan10	*Vibrio cholerae* O1 CIRS101	Bangladesh	2010	C	Su Tm Cm Sm	R1[g]	50
ICEVchMoz3[f]	*Vibrio cholerae* non-O1	Mozambique	2002	E	Pn Sp β-lactams	na	61
ICEVchMex1	*Vibrio cholerae* non-O1/O139	Mexico	2001	E	none	R3	15,38
ICEVchInd6[f]	*Vibrio cholerae* O2	India	1998	E	na	na	62
ICEVchInd7[f]	*Vibrio cholerae* O130	India	1998	E	na	na	62
ICEVchInd8[f]	*Vibrio cholerae* O144	India	1998	E	na	na	62
ICEVchInd9[f]	*Vibrio cholerae* O150	India	1998	E	na	na	62
ICEVchInd10[f]	*Vibrio cholerae* O151	India	1998	E	na	na	62
ICEVflInd1	*Vibrio fluvialis* H-08942	India	2002	C	Su Tm Cm Sm	S1	15,41,63
R391[e]	*Providencia rettgeri*	South Africa	1967	C	Kn Hg	R	5,36,41
R748[e,f]	*Providencia* spp	South Africa	1974	na	Kn Hg	na	6
ICEPalBan1	*Providencia alcalifaciens*	Bangladesh	1999	C	Su Tm Cm Sm	R2	15,28,41

continued on next page

Table 1. Continued

ICE Name	Host[a]	Origin		Type[b]	Drug(s) to Which Resistance is Conferred[c]	eex Allele[d]	Refs.
		Location	Year				
R997[e]	*Proteus mirabilis*	India	1977	C	Su Sm Ap	S2	41,64
ICE*Pmi*Jpn1	*Proteus mirabilis* TUM4660	Japan	2010	C	Extended spectrum cephalosporins	S6[g]	65
ICE*Pmi*USA1	*Proteus mirabilis*	United States	1986	C	none	S7[g]	15,66
R705[e,f]	*Proteus vulgaris*	South Africa	1975	na	Kn Hg	na	7
ICE*Pda*Spa1	*Photobacterium damselae* subsp *piscicidae*	Spain	2001	F	Tc	S3	15,37,41,67
ICE*Spu*PO1	*Shewanella putrefaciens* W3–18–1	Pacific Ocean marine sediment	na	E	Putative multi-drug efflux	S4	15,28,41,68
pMERPH[e]	*Shewanella putrefaciens*	United Kingdom river sediment	1990	E	Hg	R4	41,69

[a] When relevant, the strain is indicated; [b] Type of sample: C, clinical isolate; E, environmental isolate; F, isolate from diseased fish; [c] Ap, ampicillin; Cm, chloramphenicol; Kn, kanamycin; Pn, penicillin; Sm, streptomycin; Su, sulfamethoxazole; Tc, tetracycline; Tm, trimethoprim; Hg, mercury; [d] Variants of entry exclusion determinants S and R determine the S and R exclusion groups, respectively. ; [e] Members of the now obsolete Inc*I* element family. Other members are R392 and R397, which are likely clones of R391, R706, which is likely a clone of R705, and R749, likely clone of R748. ; [f] Members of the SXT/R391 family for which no complete sequence was available. ; [g] Corrected eex allele determined by our analysis. ; [h] Corrected eex allele from Marrero and Waldor.[41] na: data not available.

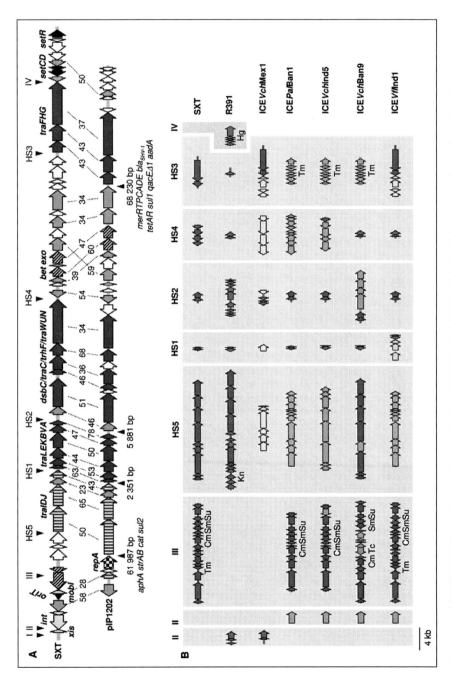

Figure 1. Structure of SXT/R391 ICEs and comparison with plasmids of the IncA/C family. The figure legend is continued on the next page.

Figure 1, continued from previous page. A) Comparison of the SXT/R391 and IncA/C backbones. Alignment of the backbone of 2 prototypical members of each family: SXT and pIP1202 (*Yersinia pestis*). Color code goes as follows: integration/excision, light orange; DNA processing, brown and white vertical stripes; mating pair formation, brown; DNA recombination and repair, hatched; replication, checkered; regulation, gray; entry exclusion, light blue; homologous genes of unknown function, black; genes lacking counterpart in either element, white. The % identity between the orthologous proteins are indicated by numbers in the middle of the two backbones. Downward arrowheads indicate the positions of the hotspots (HS1-HS5) or other variable regions (I-IV) in SXT/R391 ICEs. Upward arrowheads indicate the position of the variable DNA and the resistances markers found in pIP1202. *aph*, kanamycin; *aad7*, spectinomycin; *strAB*, streptomycin; *sul1* and *sul2*, sulfonamides; *cat*, chloramphenicol; *bla*$_{SHV-1}$, β-lactams; *tetAR*, tetracycline; *qacEΔ1*, quaternary ammonium compounds; *merRTPCADE*, mercury ions. B) Variable DNA found in the hotspots (HS1-HS5) and in the variable regions not considered as true hotspots (I-IV) of 7 sequenced ICEs of the SXT/R391 family. The ORFs are depicted in colors that reflect the ICE in which they were first described: SXT, blue; R391, red; ICE*Vch*Mex1, yellow; ICE*Pal*Ban1, orange; ICE*Vch*Ind5, turquoise; ICE*Vch*Ban9, pink ; ICE*Vfl*Ind1, light green. Cm, chloramphenicol; Hg, mercury; Kn, kanamycin; Sm, streptomycin; Su, sulfamethoxazole; Tc, tetracycline; Tm, trimethoprim. A color version of this figure is available online at http://www.landesbioscience.com/curie.

The DNA processing functions are split into two clusters. The first cluster contains the origin of transfer (*oriT*), of all SXT/R391 ICEs, which is the substrate for the putative conjugative relaxase, TraI.[17] The 299-bp sequence encoding the minimal *oriT* of all SXT/R391 ICEs is located in the intergenic region found between *s003* and *rumB* and contains five imperfect direct and inverted repeats, a common feature of *oriT* regions found in conjugative plasmids. These repeats are presumably involved in the specific recognition of this locus by the relaxase, allowing cleavage of one strand of the circular molecule to take place. An additional gene, termed *mobI*, was identified downstream of *oriT* and was found to be absolutely required for conjugative transfer of SXT. MobI is presumably involved in the recognition of *oriT* since its absence does not affect SXT-mediated mobilization of plasmid CloDF13, suggesting that MobI is not a part of the mating pore.[17] The second cluster of DNA processing functions is comprised of *traI*, the putative relaxase, *traD*, a putative coupling factor, and *traJ*, a gene encoding a protein sharing weak homology with TraJ from plasmid F.[15]

The type IV secretion system enabling the translocation of the ICE DNA into the recipient cell is encoded by three gene clusters: (i) *traLEKBVA*, (ii) *s054/traC/trhF/traWUN* and (iii) *traFHG*, that include the genes required for the formation of the pilus, the conjugative machinery and the mating pair formation and stabilization.[15] Another gene, *s063*, is found between *traN* and *traF* in the minimal backbone. Although no precise function has yet been attributed to this gene, its deletion results in a ~100-fold decrease in the frequency of transfer of SXT and is thus considered as part of the minimal functional core.[15]

The locus encoding the regulation genes is located at the 3' end of the conserved backbone and is comprised of a repressor, *setR*, and two transcriptional activators, *setC* and *setD* (Fig. 1A). Four genes (*s082, s083, s084* and *s086*) separate *setR* from *setCD* and although they appear to be part of the operon encoding *setCD*, their function remains unknown. Like the λ cI repressor, SetR contains both a helix-turn-helix DNA-binding motif and a protease motif.[13] SetR binds to an intergenic region between the divergently transcribed *s086* and *setR* genes,[14] and represses expression of *setCD* from a promoter located upstream of *s086*. A marked increase in SXT transfer is observed when donor cells are grown in the presence of DNA-damaging agents that induce the SOS response such as mitomycin C or ciprofloxacin, a widely used fluoroquinolone antibiotic.[13] After DNA damage has stimulated the coprotease activity of RecA, activated RecA appears to promote the autoproteolysis and inactivation of SetR.[13] Depletion of the intracellular pool of SetR alleviates repression of *setC* and *setD* expression, increasing the expression of *int*, *xis*, the DNA processing genes and the three *tra* operons described above, presumably leading to the simultaneous expression of all genes required for efficient transfer of SXT to a new host.[12,18,19]

The SXT/R391 Transfer Genes are Related to the IncA/C Plasmid Backbone

Interestingly, all of the predicted transfer genes of SXT/R391 ICEs are also found in plasmids of the IncA/C group. These multidrug resistant conjugative plasmids are widely distributed in *Salmonella*[20] and have been recently identified in *Yersinia pestis*,[21] the causative agent of bubonic plague, and in aquatic γ-proteobacteria such as *Vibrio cholerae*.[22-24] Besides being their closest relatives, the transfer genes found in IncA/C plasmids display perfect synteny with the four gene clusters encoding the conjugative machinery of SXT/R391 ICEs (Fig. 1A). However, plasmids of the IncA/C group lack the regulation module comprised of *setR* and *setCD*, suggesting that their conjugative transfer is regulated differently. As expected, these conjugative plasmids do not encode integration and excision functions (*int* and *xis*) allowing the maintenance of ICEs integrated into the chromosome. Instead, IncA/C plasmids carry a RepA-like replicon.

Besides the known transfer genes, 10 genes of unknown function found in the backbone of SXT/R391 ICEs are also found in IncA/C plasmids, suggesting that they might somehow play an important role for these two distinct families of mobile elements.[15] Furthermore, all members of both families contain homologs of the bacteriophage lambda recombination genes *bet* and *exo* (please see section on plasticity and evolution below).[25]

As for SXT/R391 ICEs, plasmids of the IncA/C group also display a number of variable regions scattered in their conserved backbone. Interestingly, all of the variable regions are found in the same relative positions of the scaffold of conserved genes as in SXT/R391 ICEs: upstream of *traI* and *traL*, downstream of *traA* and upstream of *traF*.[15] However, the content of the variable regions found in IncA/C plasmids are utterly different from the ones found in SXT/R391 ICEs, suggesting that they took divergent evolutionary paths prior to the acquisition of these variable segments of DNA. Interestingly, it has been recently demonstrated that ISCR2 elements, that are frequently associated with antibiotic resistance genes in SXT/R391 ICEs, as well as other ISCR elements are major contributors to the construction of IncA/C plasmids,[26] suggesting that these two families of conjugative elements might share even deeper evolutionary linkage.

Variable Regions and Diversity of Encoded Functions

All members of the SXT/R391 family that have been sequenced to date share the same structure, comprising of the scaffold of conserved genes and of five loci in which are found different DNA segments of variable length (Fig. 1B). These five loci, termed HS1 to HS5 for "hotspot", are all found in intergenic regions of the backbone where their acquisition appears to have left intact all core functions. Additionally, a few ICEs contain variable DNA inserted outside of these five hotspots (named Variable Region I to IV) as is the case for SXT and other ICEs of this family that contain an ISCR2 element inserted in *rumB* (Fig. 1B). ISCR2 elements are IS91-like transposable elements that are frequently associated with resistance to sulfonamides.[27]

Interestingly, the sequences found in a given hotspot are rarely restricted to one element, and are usually found in at least another ICE of the family, although in most cases the other hotspots of these elements will be completely different. For example, ICE*Vch*Mex1 and ICE*Vfl*Ind1 contain the same variable DNA in HS3 (Fig. 1B) but the content of their other hotspots is completely different. As another example, ICE*Pal*Ban1, ICE*Vch*Ind5 and ICE*Vch*Ban9 contain the same class IV integron conferring resistance to trimethoprim in HS3.[28] However, comparison of the contents of the other hotspots of these three elements shows that they are mostly different (Fig. 1B). (See Wozniak et al.[15] for a complete overview of the variable DNA found in SXT/R391 ICEs.)

While a large number of the variable genes do not have any known homolog, several genes encode functions that likely increase the fitness of their host strain. Most ICEs of the SXT/R391 family contain genes conferring resistance to multiple antibiotics, mainly found as part of associated mobile elements such as ISCR2 elements (e.g., SXT, ICE*Pda*Spa1, and ICE*Vch*Ban9) or class IV integrons (e.g., ICE*Vch*Ind5 and ICE*Vch*Ban9). Conversely and as will be further discussed below, some ICEs of this family also contain genes that offer other possible adaptative advantages to the strains that harbor them or to the ICEs themselves. At least four ICEs of the SXT/R391 family (SXT, ICE*Vch*Mex1, ICE*Vfl*Ind1 and ICE*Vch*Moz3) contain in HS3 genes that are involved in the biosynthesis of bis-(3'-5')-cyclic dimeric guanosine monophosphate (c-di-GMP), a second messenger molecule of primary importance for *V. cholerae*'s lifestyle.[29] Moreover, most members of the family also contain *mosAT*, a toxin-antitoxin system that prevents their loss from the host cell.[30] Besides these functions, there are large number of variable genes that encode putative DNA modification, recombination and repair functions, such as endonucleases, restriction/modification systems and helicases (for an exhaustive listing of ICE-encoded variable function please see Wozniak et al.[15]). These genes might offer a greater adaptative advantage for the host by limiting the entry of foreign DNA and offering a selective advantage for the element itself by helping it preserve its integrity during conjugative transfer.

Modulation of c-di-GMP Signaling Pathways by SXT/R391 ICEs

ICE*Vch*Mex1, an ICE derived from an environmental non-O1 non-O139 isolate of *V. cholerae* from Mexico, and at least two other ICEs of this family (ICE*Vfl*Ind1 and ICE*Vch*Moz3) contain two genes, *dgcK* and *dgcL*, that encode functional diguanylate cyclases (DGCs), enzymes that catalyze the biosynthesis of c-di-GMP. *DgcL* is also found in SXT.

V. cholerae is an inhabitant of the aquatic environment in which it can exist as a free-living planktonic form, colonize various environmental hosts such as phytoplankton, copepods and the egg masses of Chironomid insects and form biofilms on abiotic or chitinous surfaces. The transition between these various lifestyles requires *V. cholerae* to switch efficiently between sessile and motile forms.

As is the case for many other bacteria, *V. cholerae*'s ability to switch between these two behaviors is dictated by the intracellular levels of c-di-GMP that are controlled by the antagonist activities of DCGs and phosphodiesterases (PDEs), respectively producing and degrading the molecule.[31,32] DGCs' enzymatic activity is provided by a GGDEF domain that catalyzes the formation of c-di-GMP from two GTP molecules. In many bacteria increased cellular levels of c-di-GMP are associated with an increased production of adhesins and biofilm matrix components, such as exopolysaccharides, and with the reduced expression of genes involved in motility and virulence. Inversely, PDEs linearize c-di-GMP into pGpG, which can later be degraded into two GMP molecules. Decreased intracellular concentrations of c-di-GMP inhibit biofilm formation and favor expression of genes involved in motility and virulence.

Both DgcK and DgcL have a GGDEF domain and have been shown in vitro and in vivo to produce c-di-GMP from GTP.[29] DgcK and DgcL also contain putative domains that might act as signaling modules through the binding various ligands or by sensing changes in environmental conditions (e.g., light, oxygen, redox potential).[29] However, these conditions are still unknown.

When overexpressed in *V. cholerae* DgcK and DgcL markedly trigger biofilm formation and cause the appearance of the rugose phenotype normally associated with an increased production of exopolysaccharides and biofilm matrix components.[29] As expected, their overexpression also significantly reduces the motility of *V. cholerae* on soft agar plates.[29] Inversely, deletion of *dgcL* or of *dgcKL* significantly increases its motility. However, deletion of *dgcK* alone does not impact motility of *V. cholerae*, suggesting that this gene might not be expressed in the conditions tested. Additionnaly, in contradiction with the clear phenotypes observed in overexpression conditions, deletion of either *dgcK*, *dgcL* or both genes does not significantly increase the ability of *V. cholerae* to form biofilm. Yet, this observation appears to be typical of most individual deletions of c-di-GMP signaling genes in *V. cholerae*.[33] A plausible explanation to this phenomenon might reside in the complexity of c-di-GMP signaling pathways, where many environmental conditions such as changes in light, oxygen and nutrient availability can directly affect the function of c-di-GMP signaling enzymes or the transcription of their encoding genes (please see Hengge[31] for a complete review). Moreover, although it is known that most genes of the backbone of SXT/R391 ICEs are under the control of the two main ICE-encoded transcriptional activators SetC and SetD, that ultimately respond to SOS response inducing conditions, it is yet to be determined if SetCD also impact the expression of variable genes, such as *dgcK* and *dgcL*.

The lack of impact of the deletions of *dgcL* and *dgcK* on biofilm formation and motility of *V. cholerae* also leaves an open question for the reason for their presence in SXT/R391 ICEs. On one hand, it might be that under certain conditions that have yet to be determined DgcK and DgcL positively impact the survival of their host in the environment by enhancing its capacity to form biofilms. Moreover, DgcK and DgcL might have a greater impact on non *V. cholerae* hosts of SXT/R391 ICEs that do not contain as many c-di-GMP turn-over proteins. Alternatively, DgcK and DgcL might confer an advantage to the ICE itself, by favoring the cell-to-cell contacts needed for conjugation via increased biofilm formation.

Genes encoding predicted c-di-GMP regulatory proteins are to date found in more than 180 MGEs such as plasmids, ICEs and bacteriophages.[29] However, the advantages that they confer to their bacterial hosts or to the MGEs themselves remain to be determined. Interestingly, besides *dgcK* and *dgcL*, *orf53* of R391 encodes a putative protein predicted to contain a GGDEF domain. However, comparison of its predicted sequence with known functional DGCs suggests that Orf53 is probably not catalytically functional. As virulence has been shown to be regulated in part by intracellular c-di-GMP levels in several pathogens such as *V. cholerae*, the further investigation of the role of these genes will provide useful information to the understanding of the impact of these MGEs on the pathogenesis of the strains containing them.

Toxin/Anti-Toxin System

SXT and a few other members of the SXT/R391 family have been reported to carry two toxin/antitoxin (TA) systems encoded by genes *s044/s045* and *mosA/mosT*.[30,34] TA systems, sometimes referred to as postsegregational cell killing and addiction systems, are composed of a stable toxin and a labile antitoxin that generally function to inhibit the loss of extrachromosomal elements (for a comprehensive review please see Hayes[35]). If the element carrying the TA gene pair is lost due to erroneous replication or defect in maintenance, the unstable antitoxin is rapidly eliminated from the complex and cannot be resynthesized due to the absence of the element. The stable toxin is then released and interacts with host targets, inhibiting its growth or inducing cell death.

When they are present in their host cell in their integrated form, SXT/R391 ICEs are stably maintained and replicated along with the host chromosome. Upon certain conditions that trigger their transfer, they excise from the host chromosome and transiently exist as a double-stranded non-replicative extrachromosomal molecule.[2] However, excision events do not necessarily lead to transfer events, as is shown by the 100-fold higher frequency of excision than of exconjugant formation,[12,15] thus creating a constant balance between integrated and excised forms of SXT/R391 ICEs in a given cell. This phenomenon poses a permanent threat to the elements' stability, as they become vulnerable to loss if the cell divides while the non-replicative ICE is in its extrachromosomal form. Interestingly, while there are 3.7 detectable excision events per 100 cells containing SXT, only 3.2 loss events can be detected per 10^6 cells, revealing that most excision events do not lead to SXT loss, a phenomenon that could be explained by the presence of a TA system.[12,30] Two independent studies have described such systems in SXT/R391 ICEs.

Dziewit and colleagues[34] have shown that genes *s044/s045* found in HS1 of SXT and ICE*Pal*Ban1 (Table 1) encode a functional TA pair. The toxin encoded by *s045* causes a 100-fold reduction in cell viability when expressed in *E. coli* and its effect is reversed by the expression of *s044*. Moreover, when cloned into an unstable vector, the *s044/s045* system enhances plasmid retention by about 30% after 30 generations under non selective conditions.[34] However, the absence of *s044/s045* from SXT was reported to have no negative impact on its stability.[30]

On the other hand, Wozniak and Waldor[30] recently showed that the stability of SXT and of several other members of the family in the host cell is enhanced by genes *mosA* and *mosT* (for *m*aintenance *o*f *S*XT, *A*ntitoxin and *T*oxin). The two genes are found in HS2 and act to prevent loss of SXT. *mosA* and *mosT* appear to be an operon that is maintained repressed due to autorepression by MosA when SXT is integrated in the chromosome. However, factors that promote SXT excision, such as Xis (the excisionase) and SetCD (the two main activators of transfer genes), derepress *mosAT*, leading to its expression at the moment when SXT is excised from the chromosome and most vulnerable to loss. *mosT* encodes a toxin that, when overexpressed, severely inhibits growth of *E. coli* and *V. cholerae*, but its toxic effect is neutralized by the presence of *mosA*. By analogy to other TA systems, MosA would be less stable than MosT, leading to killing of cells that did not inherit SXT if cell division occurred while the element was in its extrachromosomal form or by inhibiting growth of cells until the element has reintegrated.

Interestingly, many members of the SXT/R391 family, such as R391,[5,36] ICE*Pda*Spa1[37] and ICE*Vch*Mex1,[38] do not possess *mosAT*. However, preliminary studies suggest that R391 is only slightly less stable than SXT, which suggests that members of the family lacking *mosAT* might contain other genes promoting their maintenance.[30] For instance, both R391 and ICE*Vch*Mex1 contain genes distantly related to the HipAB toxin-antitoxin system from *E. coli* in HS2.[15]

Entry Exclusion: Two Distinct Groups Coexist in SXT/R391 ICES

Similar to F and other conjugative plasmids, SXT/R391 family members carry genes for an entry exclusion system mediated by two inner membrane proteins, TraG and Eex, acting respectively in the donor and in the recipient cell.[39-41] Entry exclusion systems function to specifically reduce redundant conjugative transfers between donor cells that carry the same element. Marrero and Waldor[39] showed that even though SXT and R391 have nearly identical conjugative transfer genes, these ICEs do not exclude each other: cells harbouring SXT limit the acquisition of SXT but not of R391 and vice versa.

Eex variants EexS and EexR, and TraG variants TraG$_S$ and TraG$_R$, respectively encoded by SXT and R391, mediate element-specific exclusion activity. *traG* and *eex* genes of 21 ICEs belonging to the SXT/R391 family have recently been characterized both genetically and functionally.[41] Unexpectedly, both genes segregate unambiguously into two exclusion groups, called S and R (Table 1). Consistent with earlier findings for SXT and R391,[39] ICEs harboring an exclusion system of the S group exclude acquisition of ICEs belonging to the same exclusion group but not the ones belonging to the R group and vice versa.[41] Eex proteins from both S and R groups are highly similar (87% identity) in their first 86 N-terminal residues but their remaining 56 carboxyl-terminal residues are more divergent (41% identity) and determine their exclusion specificity.[39] Subtle differences in the carboxyl-terminal residues of EexS-like and EexR-like proteins allow for their classification into variant groups (Table 1), but are not expected to influence exclusion specificity.[41] However, the exclusion activity of TraG and Eex are not perfect since they reduce the transfer of SXT to an SXT$^+$ cell by 10–100 times,[41] a phenomenon which could explain the formation of SXT tandem arrays from multiple acquisition of SXT by a unique recipient cell during conjugative transfer.[42]

In each ICE, the *traG* gene belonging to one of the two exclusion groups pairs with the *eex* gene of the same exclusion group, likely reflecting the functional interaction between TraG and Eex that mediates exclusion. Unlike Eex alleles the two alleles of the 1189-amino acid TraG proteins are highly similar (98% identity). The exclusion specificity dictated by TraG$_S$ and TraG$_R$ relies on three specific amino acid residues found at positions 606–608 (Pro-Gly-Glu in TraG$_S$ and Thr-Asp-Asp in TraG$_R$).[39] Curiously, Marrero and Waldor[40] found that the exclusion specificity residues of both Eex and TraG are cytoplasmic. Although it is not yet fully understood how cytoplasmic regions of inner membrane proteins in mating cells can come in contact, several plausible scenarios arise. In one case, the entire Eex or TraG protein could translocate through the mating pore between the donor and recipient cells. In a second case, one of the proteins might undergo proteolysis, yielding a fragment that could move from one cell to the other. Finally, TraG or Eex could be subjected to a conformational change that would allow for their cytoplasmic region to flip inside the mating channel.

Plasticity and Evolution of SXT/R391 ICEs

When the sequences of SXT[19] and R391[36] were initially reported, it appeared that the variable regions interspersed in the common backbone of members of this family were specific of each element. However, as more ICEs of this family were sequenced, the closer analysis of their variable regions suggested that while some of these are element-specific, a large number of them are shared by at least two elements (e.g., see ICE*Vchi*Mex1 and ICE*Vfl*Ind1 in Fig. 1B). As discussed above, SXT/R391 ICEs sometimes do not exclude each other, allowing the coexistence of two similar ICEs in the same bacterial cell. This, along with their shared unique integration site, enables the formation of chromosomal tandem arrays of similar yet non identical ICEs.[42] For example, R391 and SXT can reside together in the same host, providing an ample substrate for inter-ICE recombination due to the presence in their scaffold of 47 kb of highly conserved DNA.

The propensity of SXT/R391 ICEs to form tandem arrays has recently been shown to allow the formation of hybrid ICEs by recombination between the co-existing parental ICEs.[25] Hybrid ICE formation seems to be a rather frequent phenomenon since in laboratory conditions roughly 7% of exconjugant colonies bear a hybrid ICE when the donor cell used for the mating assay contained a heterogeneous tandem array.[25] Interestingly, neither excision of the ICEs from the host chromosome nor their conjugative transfer is required for hybrid ICE formation. As expected due to the extensive homology between SXT/R391 ICEs' sequences it was shown that RecA plays a major role in hybrid ICE formation. However, its role seems to be in great part restricted to the donor cell, as its absence from the recipient cell does not significantly impair hybrid ICE formation.[25] This, along with the fact that conjugation is not required for hybrid ICE formation reveals that most hybrid ICEs are formed prior to their transfer into a new host cell.

Although it was found that RecA can mediate the formation of the majority of hybrid ICEs, an ICE-encoded recombination system, comprised of genes *s065* and *s066* (referred to as *bet* and *exo* in the present chapter), also catalyzes hybrid ICE formation. The proteins encoded by these genes respectively share 71% and 38% similarity with the bacteriophage lambda-encoded single-strand annealing protein Bet and 5′-3′ dsDNA-dependant exonuclease Exo. Most interestingly, the single-strand annealing protein encoded by *bet*[25,43] and the putative exonuclease encoded by *exo* can catalyze hybrid ICE formation in a RecA-independent fashion.[25] Although the precise mechanism by which this recombination system promotes hybrid ICE formation is not clear, its similarity with phage-encoded recombination systems renders plausible the theory that it works following a similar pathway. Lambda Exo is thought to use double-stranded DNA ends as a principal substrate, digesting their 5′ extremity and providing a suitable 3′ single-strand substrate for recombination by lambda Bet.[44] However, contrarily to bacteriophage lambda, ICEs do not enter the recipient cell in the form of a double-stranded DNA molecule. Yet, double-stranded DNA breaks could occur in the chromosomally integrated ICE or in the excised circular form prior or post-transfer to a recipient cell. Additionally, DNA damages promoted by agents such as UV light and certain antibiotics (e.g., the fluoroquinolone ciprofloxacin), which are known to trigger the conjugative transfer of SXT/R391 by the induction of the SOS response, could also provide suitable recombination substrates in the form of double-stranded DNA breaks.

Interestingly, the fraction of hybrid ICEs formed by only RecA (37%) and by only Bet-Exo (11%) does not account for the total of hybrid ICE formed.[25] Consequently and although hybrid ICEs may be formed by those two independent recombination pathways it appears that both act synergistically to promote hybrid ICE formation. Such cooperation between RecA and lambda Bet-Exo has been previously proposed, as Bet cannot catalyze invasion of unbroken duplex DNA molecules.[45] Given the similarity between lambda Bet and SXT/R391 Bet it is likely that the latter might also take advantage of RecA's ability to invade duplex DNA molecules. Furthermore, single-stranded DNA ends generated by Exo could possibly be used by RecA to catalyze recombination. Nonetheless, solid experimental data proving the interaction between RecA and the ICE-encoded Bet-Exo recombination system are still needed.

The conservation of this recombination system in the backbones of SXT/R391 ICEs and of the related IncA/C conjugative plasmids (Fig. 1A) suggests that recombination plays a key role in the evolution of these conjugative elements. Besides recombining SXT/R391 ICEs, the Bet-Exo recombination systems might also further participate in the plasticity of SXT/R391 ICEs and IncA/C plasmids by promoting the acquisition of the exogenous variable DNA found in their hotspots. Beyond their contribution to element plasticity, Bet and Exo might also play a role in DNA repair, especially for SXT/R391 ICEs since they transfer in conditions affecting DNA integrity.

Mobilization of Plasmids, Mobilizable Genomic Islands and Chromosomal DNA

Mobilization of Plasmids

ICEs of the SXT/R391 family have been known to promote the transfer of mobilizable but non conjugative plasmids such as pCLODF13 by providing them with a conjugative pore.[46] More recently, Osorio and colleagues[37] demonstrated that ICE*Pda*Spa1, an ICE of the SXT/R391 family derived from the fish pathogen *Photobacterium damselae* subsp *piscicidae*, promotes the mobilization of the co-resident virulence plasmid pPHDP10. However, the mechanism by which ICE*Pda*Spa1 promotes the transfer of pPHDP10 is entirely different from SXT-mediated pCLODF13 mobilization since this plasmid does not carry an *oriT* nor any mobilization genes. In fact, Osorio et al.[37] have shown that the mobilization of pPHDP10 happens through the formation of a cointegrate between the plasmid and ICE*Pda*Spa1. Moreover, this phenomenon is not site-specific, as pPHDP10 was found integrated in four different loci of the ICE and does not require RecA-mediated recombination. The mechanism by which pPDHP10 integrates into ICE*Pda*Spa1 is not known, but the large number of tranposases genes found in the plasmid might explain this phenomenon. The integration of pPHDP10 into ICE*Pda*Spa1 allows for the cointegrate to be transferred by conjugation to a new host where it integrates into the chromosome. Remarkably, pPDHP10 remains stably integrated into ICE*Pda*Spa1 and the cointegrate is subsequently transferred at a frequency comparable to that of ICE*Pda*Spa1.[37]

Transfer of Mobilizable Genomic Islands through Cross-Recognition of a Related *oriT*

It has recently been demonstrated that SXT/R391 ICEs trigger the transfer of a specific class of mobilizable genomic islands (MGIs) through the cross-recognition of a similar *oriT*.[47] In fact, Daccord et al.[47] have recently identified a sequence similar but not identical to the origin of transfer of SXT in the genome of three *Vibrio* species (*V. cholerae* RC385, *V. vulnificus* YJ016 and *V. fluvialis* H-08942). Analysis of the surrounding DNA revealed that this sequence is not part of integrating conjugative elements, but that it is rather found in three similar genomic islands. The 299 bp *oriT* found in these MGIs shares over 63% identity with *oriT*$_{SXT}$ and contains the same five imperfect direct and inverted repeats that are thought to participate in the recognition of the *oriT* by the relaxase of SXT.

The three MGIs are integrated in the 3′ end of a gene encoding an uncharacterized stress-induced protein, *yicC*, and are flanked by 23-bp direct repeats that correspond to the core sequences of *attL* and *attR* attachment sites (Fig. 2). Downstream of *attL* is found a site-specific P4-type tyrosine recombinase (> 88% identity between the three MGI recombinases) that catalyzes their integration into the host chromosome.[47] In laboratory conditions, MGI*Vfl*Ind1 is transferable to *E. coli* and integrates into the 3′ end of *yicC*. Besides the conserved *oriT* (84.5% identity between the three MGIs) and the integrase found in the three MGIs, two of them also contain genes that could provide their host with selective advantages. MGI*Vch*USA1 contains two distinct toxin/antitoxin systems that are related to HipAB from *E. coli*, a TA system also involved in multidrug tolerance by its ability to induce the formation of dormant cells called persisters (Fig. 2).[15] MGI*Vvu*Tai1 contains a putative type I restriction-modification system which could provide the host with an increased ability to resist to bacteriophage infections.

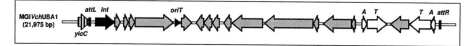

Figure 2. Structure of MGI*Vch*USA1. The ORFs are color coded as follows: integrase (Int), black; toxin (T)/antitoxin (A) system, white; genes of unknown function, gray. MGI*Vch*USA1 is found integrated in the 3′ end of *yicC* (hatched) when in *E. coli* and in the 3′ end of the orthologous gene in *V. cholerae*. The black triangle represents the origin of transfer (*oriT*).

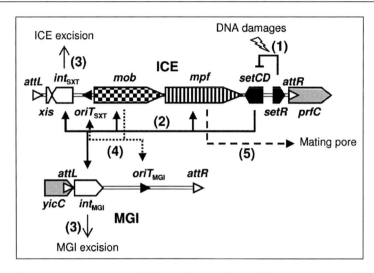

Figure 3. Schematic representation of the transfer regulation of SXT/R391 ICEs and their *trans* mobilization of MGIs. 1) DNA damaging agents induce the SOS response, alleviating SetR's repression on *setCD*. 2) SetCD activate the expression of *int*$_{SXT}$, of *int*$_{MGI}$ (directly or not) and of DNA processing (*mob*) and mating pair formation genes (*mpf*). 3) *int*$_{SXT}$ and *int*$_{MGI}$ respectively catalyze the excision of SXT and MGIs. 4) The proteins that recognize, bind and cleave the *oriT* are produced by the *mob* genes. 5) Expression of the *mpf* genes leads to the formation of the conjugative apparatus through which the nicked ICE or MGI DNA is transferred to the recipient cell. Reprinted from: Daccord A et al. Mol Microbiol 2010; 78:576-88;47 ©2010 with permission of John Wiley and Sons.

Daccord et al.[47] demonstrated that the conjugative machinery of SXT/R391 ICEs is able to specifically recognize the sequence that mimics the *oriT* of SXT found in the MGIs and promote their mobilization to a new host cell (Fig. 3). While genes necessary for the excision of the ICE (*int* and *xis*) are not required for MGI transfer, the 2 ICE-encoded DNA processing genes *traI* and *mobI* are absolutely required for the efficient transfer of the MGIs, suggesting that the proteins they encode specifically recognize the MGI-borne *oriT*.[47] The MGI-encoded integrase is also required for efficient ICE-mediated transfer of the MGI and its expression in the recipient cell is required, presumably to promote integration of the newly acquired MGI.

Interestingly, while MGIs are capable of excising from their host chromosome, their excision is dependent on the expression of the two main transcriptional activators of the SXT/R391 transfer genes, SetC and SetD. As can be expected, the induction of the SOS response also markedly stimulates both excision of the MGIs and their ICE-mediated conjugative transfer.

Mobilization of Chromosomal DNA

ICEs of the SXT/R931 family have previously been shown to mobilize chromosomal DNA in a manner similar to the Hfr transfer catalyzed by plasmid F.[46] As excision of SXT from the chromosome is not required prior to the expression of its transfer functions, the conjugative process can be initiated at the *oriT* of the integrated element, and thus lead to mobilization of the chromosomal DNA located 3′ of *prfC*.[46] Following transfer, this DNA can be integrated in the recipient's chromosome by homologous recombination. Using MGI*Vfl*Ind1 and its co-resident ICE, ICE*Vfl*Ind1, Daccord et al.[47] found that MGIs can also promote a similar type of transfer and mobilize up to 1Mb of chromosomal DNA found 5′ of *yicC* in *E. coli*. Interestingly, they also demonstrated that ICE*Vfl*IndI is able to mobilize up to 1.5 Mb of DNA, three times more than what was previously demonstrated for SXT, suggesting that the impact of ICE-mediated transfer of chromosomal DNA is a lot greater than originally thought.[46]

Conclusion

The recent advancements made in the study of the biology and evolution of SXT/R391 ICEs has uncovered that these elements participate in the plasticity of their hosts' genomes at unpredicted levels, by triggering the transfer of mobilizable genomic islands, chromosomal DNA and virulence plasmids. Moreover, these elements also have the ability to stimulate their own plasticity, by the presence of a recombination system that facilitates the formation of hybrid elements, and to influence the adaptation of their host strains by providing them with alternative metabolic pathways. The recent discovery that conjugation leads to the induction of the SOS response in *V. cholerae*[48] takes the impact of SXT/R391 ICEs on their host strains to yet another level, by inducing integron activation and co-resident bacteriophages. Such a phenomenon will facilitate the acquisition of new resistance genes by pathogenic hosts of SXT/R391 ICEs and enhance their potential to impact genome plasticity. Even more worrisome is the finding that most antibiotics trigger the SOS response in *V. cholerae*.[49] By inducing the SOS response, antibiotics are likely to facilitate the transfer of SXT/R391 ICEs to new hosts and to favor the formation of new elements by promoting inter-ICE recombination. Thus, the use of a single antibiotic could have as final consequence the dissemination of a considerable number of antibiotic resistance genes.

Along with the analysis of genomic islands, single nucleotide variations and CTX phage variants, the analysis of SXT/R391 ICEs carried by most *V. cholerae* epidemic strains becomes a useful tool for tracking the spread of this pathogen around the globe and understanding the origin of highly epidemic strains. In fact, careful analysis of SXT/R391 ICEs circulating in Indian *V. cholerae* strains has recently shown that these strains contain ICEs belonging to three different SXT/R391 subgroups, and that it is possible to correlate the molecular structure of these elements with the strain's epidemiological profile.[50] Indeed, Ceccarelli and colleagues[50] have shown that the three main *V. cholerae* variants responsible for the last epidemic events in the Indian Subcontinent are related to specific ICEs of the SXT/R391 family: *V. cholerae* O139 carries SXT or its sibling ICE*Vch*Ind4; *V. cholerae* O1 El Tor hybrid variants harbor ICE*Vch*Ind5 or siblings (ICE*Vch*Ind1, ICE*Vch*Ban5 and ICE*Vch*Ban10); and *V. cholerae* O1 El Tor Matlab variant associates with ICE*Vch*Ban9.

A remarkable example of this new approach is the very recent analysis of the *V. cholerae* strain responsible for the severe ongoing cholera outbreak in Haiti that began in October 2010. Indeed, two independent studies proved the relatedness of the Haitian strain to the main variant circulating in the Indian Subcontinent (*V. cholerae* O1 CIRS101) thanks to the ICE content of both strains, among other genetic markers.[51,52]

Acknowledgments

We thank Daniela Ceccarelli for helpful comments on the manuscript. GG is the recipient of a PhD scholarship from FQRNT. VB holds a Canada Research Chair in molecular biology, impact and evolution of bacterial mobile elements.

References

1. Burrus V, Pavlovic G, Decaris B, et al. Conjugative transposons: the tip of the iceberg. Mol Microbiol 2002; 46:601-10. PMID:12410819 doi:10.1046/j.1365-2958.2002.03191.x
2. Wozniak RA, Waldor MK. Integrative and conjugative elements: mosaic mobile genetic elements enabling dynamic lateral gene flow. Nat Rev Microbiol 2010; 8:552-63. PMID:20601965 doi:10.1038/nrmicro2382
3. Kaper JB, Morris JG Jr., Levine MM. Cholera. Clin Microbiol Rev 1995; 8:48-86. PMID:7704895
4. Waldor MK, Tschape H, Mekalanos JJ. A new type of conjugative transposon encodes resistance to sulfamethoxazole, trimethoprim, and streptomycin in Vibrio cholerae O139. J Bacteriol 1996; 178:4157-65. PMID:8763944
5. Coetzee JN, Datta N, Hedges RW. R factors from Proteus rettgeri. J Gen Microbiol 1972; 72:543-52. PMID:4564689
6. Hedges RW. R factors from Providence. J Gen Microbiol 1974; 81:171-81. PMID:4362618
7. Hedges RW. R factors from Proteus mirabilis and P. vulgaris. J Gen Microbiol 1975; 87:301-11. PMID:1095684

8. Hochhut B, Beaber JW, Woodgate R, et al. Formation of chromosomal tandem arrays of the SXT element and R391, two conjugative chromosomally integrating elements that share an attachment site. J Bacteriol 2001; 183:1124-32. PMID:11157923 doi:10.1128/JB.183.4.1124-1132.2001

9. Beaber JW, Burrus V, Hochhut B, et al. Comparison of SXT and R391, two conjugative integrating elements: definition of a genetic backbone for the mobilization of resistance determinants. Cell Mol Life Sci 2002; 59:2065-70. PMID:12568332 doi:10.1007/s000180200006

10. Burrus V, Marrero J, Waldor MK. The current ICE age: biology and evolution of SXT-related integrating conjugative elements. Plasmid 2006; 55:173-83. PMID:16530834 doi:10.1016/j.plasmid.2006.01.001

11. Hochhut B, Waldor MK. Site-specific integration of the conjugal Vibrio cholerae SXT element into prfC. Mol Microbiol 1999; 32:99-110. PMID:10216863 doi:10.1046/j.1365-2958.1999.01330.x

12. Burrus V, Waldor MK. Control of SXT integration and excision. J Bacteriol 2003; 185:5045-54. PMID:12923077 doi:10.1128/JB.185.17.5045-5054.2003

13. Beaber JW, Hochhut B, Waldor MK. SOS response promotes horizontal dissemination of antibiotic resistance genes. Nature 2004; 427:72-4. PMID:14688795 doi:10.1038/nature02241

14. Beaber JW, Waldor MK. Identification of operators and promoters that control SXT conjugative transfer. J Bacteriol 2004; 186:5945-9. PMID:15317801 doi:10.1128/JB.186.17.5945-5949.2004

15. Wozniak RA, Fouts DE, Spagnoletti M, et al. Comparative ICE genomics: insights into the evolution of the SXT/R391 family of ICEs. PLoS Genet 2009; 5:e1000786. PMID:20041216 doi:10.1371/journal.pgen.1000786

16. Kulaeva OI, Wootton JC, Levine AS, et al. Characterization of the umu-complementing operon from R391. J Bacteriol 1995; 177:2737-43. PMID:7751283

17. Ceccarelli D, Daccord A, Rene M, et al. Identification of the origin of transfer (oriT) and a new gene required for mobilization of the SXT/R391 family of ICEs. J Bacteriol 2008; 190:5328-38. PMID:18539733 doi:10.1128/JB.00150-08

18. O'Halloran JA, McGrath BM, Pembroke JT. The orf4 gene of the enterobacterial ICE, R391, encodes a novel UV-inducible recombination directionality factor, Jef, involved in excision and transfer of the ICE. FEMS Microbiol Lett 2007; 272:99-105. PMID:17504243 doi:10.1111/j.1574-6968.2007.00747.x

19. Beaber JW, Hochhut B, Waldor MK. Genomic and functional analyses of SXT, an integrating antibiotic resistance gene transfer element derived from Vibrio cholerae. J Bacteriol 2002; 184:4259-69. PMID:12107144 doi:10.1128/JB.184.15.4259-4269.2002

20. Lindsey RL, Fedorka-Cray PJ, Frye JG, et al. IncA/C plasmids are prevalent in multidrug-resistant Salmonella enterica isolates. Appl Environ Microbiol 2009; 75:1908-15. PMID:19181840 doi:10.1128/AEM.02228-08

21. Galimand M, Guiyoule A, Gerbaud G, et al. Multidrug resistance in Yersinia pestis mediated by a transferable plasmid. N Engl J Med 1997; 337:677-80. PMID:9278464 doi:10.1056/NEJM199709043371004

22. Fricke WF, Welch TJ, McDermott PF, et al. Comparative genomics of the IncA/C multidrug resistance plasmid family. J Bacteriol 2009; 191:4750-7. PMID:19482926 doi:10.1128/JB.00189-09

23. Pan JC, Ye R, Wang HQ, et al. Vibrio cholerae O139 multiple-drug resistance mediated by Yersinia pestis pIP1202-like conjugative plasmids. Antimicrob Agents Chemother 2008; 52:3829-36. PMID:18710912 doi:10.1128/AAC.00375-08

24. Welch TJ, Fricke WF, McDermott PF, et al. Multiple antimicrobial resistance in plague: an emerging public health risk. PLoS ONE 2007; 2:e309. PMID:17375195 doi:10.1371/journal.pone.0000309

25. Garriss G, Waldor MK, Burrus V. Mobile antibiotic resistance encoding elements promote their own diversity. PLoS Genet 2009; 5:e1000775. PMID:20019796 doi:10.1371/journal.pgen.1000775

26. Toleman MA, Walsh TRIS. CR elements are key players in IncA/C plasmid evolution. Antimicrob Agents Chemother 2010; 54:3534. PMID:20634542 doi:10.1128/AAC.00383-10

27. Toleman MA, Bennett PM, Walsh TRIS. CR elements: novel gene-capturing systems of the 21st century? Microbiol Mol Biol Rev 2006; 70:296-316. PMID:16760305 doi:10.1128/MMBR.00048-05

28. Hochhut B, Lotfi Y, Mazel D, et al. Molecular analysis of antibiotic resistance gene clusters in Vibrio cholerae O139 and O1 SXT constins. Antimicrob Agents Chemother 2001; 45:2991-3000. PMID:11600347 doi:10.1128/AAC.45.11.2991-3000.2001

29. Bordeleau E, Brouillette E, Robichaud N, et al. Beyond antibiotic resistance: integrating conjugative elements of the SXT/R391 family that encode novel diguanylate cyclases participate to c-di-GMP signalling in Vibrio cholerae. Environ Microbiol 2010; 12:510-23. PMID:19888998 doi:10.1111/j.1462-2920.2009.02094.x

30. Wozniak RA, Waldor MK. A toxin-antitoxin system promotes the maintenance of an integrative conjugative element. PLoS Genet 2009; 5:e1000439. PMID:19325886 doi:10.1371/journal.pgen.1000439

31. Hengge R. Principles of c-di-GMP signalling in bacteria. Nat Rev Microbiol 2009; 7:263-73. PMID:19287449 doi:10.1038/nrmicro2109

32. Römling U, Amikam D. Cyclic di-GMP as a second messenger. Curr Opin Microbiol 2006; 9:218-28. PMID:16530465 doi:10.1016/j.mib.2006.02.010

33. Beyhan S, Odell LS, Yildiz FH. Identification and characterization of cyclic diguanylate signaling systems controlling rugosity in Vibrio cholerae. J Bacteriol 2008; 190:7392-405. PMID:18790873 doi:10.1128/JB.00564-08

34. Dziewit L, Jazurek M, Drewniak L, et al. The SXT conjugative element and linear prophage N15 encode toxin-antitoxin-stabilizing systems homologous to the tad-ata module of the Paracoccus aminophilus plasmid pAMI2. J Bacteriol 2007; 189:1983-97. PMID:17158670 doi:10.1128/JB.01610-06

35. Hayes F. Toxins-antitoxins: plasmid maintenance, programmed cell death, and cell cycle arrest. Science 2003; 301:1496-9. PMID:12970556 doi:10.1126/science.1088157

36. Böltner D, MacMahon C, Pembroke JT, et al. R391: a conjugative integrating mosaic comprised of phage, plasmid, and transposon elements. J Bacteriol 2002; 184:5158-69. PMID:12193633 doi:10.1128/JB.184.18.5158-5169.2002

37. Osorio CR, Marrero J, Wozniak RA, et al. Genomic and functional analysis of ICEPdaSpa1, a fish-pathogen-derived SXT-related integrating conjugative element that can mobilize a virulence plasmid. J Bacteriol 2008; 190:3353-61. PMID:18326579 doi:10.1128/JB.00109-08

38. Burrus V, Quezada-Calvillo R, Marrero J, et al. SXT-related integrating conjugative element in New World Vibrio cholerae. Appl Environ Microbiol 2006; 72:3054-7. PMID:16598018 doi:10.1128/AEM.72.4.3054-3057.2006

39. Marrero J, Waldor MK. Interactions between inner membrane proteins in donor and recipient cells limit conjugal DNA transfer. Dev Cell 2005; 8:963-70. PMID:15935784 doi:10.1016/j.devcel.2005.05.004

40. Marrero J, Waldor MK. Determinants of entry exclusion within Eex and TraG are cytoplasmic. J Bacteriol 2007; 189:6469-73. PMID:17573467 doi:10.1128/JB.00522-07

41. Marrero J, Waldor MK. The SXT/R391 family of integrative conjugative elements is composed of two exclusion groups. J Bacteriol 2007; 189:3302-5. PMID:17307849 doi:10.1128/JB.01902-06

42. Burrus V, Waldor MK. Formation of SXT tandem arrays and SXT-R391 hybrids. J Bacteriol 2004; 186:2636-45. PMID:15090504 doi:10.1128/JB.186.9.2636-2645.2004

43. Datta S, Costantino N, Zhou X, et al. Identification and analysis of recombineering functions from Gram-negative and Gram-positive bacteria and their phages. Proc Natl Acad Sci USA 2008; 105:1626-31. PMID:18230724 doi:10.1073/pnas.0709089105

44. Little JW. An exonuclease induced by bacteriophage lambda. II. Nature of the enzymatic reaction. J Biol Chem 1967; 242:679-86. PMID:6017737

45. Stahl MM, Thomason L, Poteete AR, et al. Annealing vs. invasion in phage lambda recombination. Genetics 1997; 147:961-77. PMID:9383045

46. Hochhut B, Marrero J, Waldor MK. Mobilization of plasmids and chromosomal DNA mediated by the SXT element, a constin found in Vibrio cholerae O139. J Bacteriol 2000; 182:2043-7. PMID:10715015 doi:10.1128/JB.182.7.2043-2047.2000

47. Daccord A, Ceccarelli D, Burrus V. Integrating conjugative elements of the SXT/R391 family trigger the excision and drive the mobilization of a new class of Vibrio genomic islands. Mol Microbiol 2010; 78:576-88. PMID:20807202 doi:10.1111/j.1365-2958.2010.07364.x

48. Baharoglu Z, Bikard D, Mazel D. Conjugative DNA transfer induces the bacterial SOS response and promotes antibiotic resistance development through integron activation. PLoS Genet 2010; 6:e1001165. PMID:20975940 doi:10.1371/journal.pgen.1001165

49. Baharoglu Z, Mazel D. Vibrio cholerae triggers SOS and mutagenesis in response to a wide range of antibiotics, a route towards multi-resistance. Antimicrob Agents Chemother 2011; 55: 2438-41 DOI:. PMID:21300836 doi:10.1128/AAC.01549-10

50. Ceccarelli D, Spagnoletti M, Bacciu D, et al. ICEVchInd5 is prevalent in epidemic Vibrio cholerae O1 El Tor strains isolated in India. Int J Med Microbiol 2011; 301:318-24. PMID:21276749 doi:10.1016/j.ijmm.2010.11.005

51. Ceccarelli D, Spagnoletti M, Cappuccinelli P, et al. Origin of Vibrio cholerae in Haiti. Lancet Infect Dis 2011; 11:262. PMID:21453867 doi:10.1016/S1473-3099(11)70078-0

52. Chin CS, Sorenson J, Harris JB, et al. The origin of the Haitian cholera outbreak strain. N Engl J Med 2011; 364:33-42. PMID:21142692 doi:10.1056/NEJMoa1012928

53. Yokota T, Kuwahara S. Temperature-sensitive R plasmid obtained from naturally isolated drug-resistant Vibrio cholerae (biotype El Tor). Antimicrob Agents Chemother 1977; 11:13-20. PMID:319746

54. Iwanaga M, Toma C, Miyazato T, et al. Antibiotic resistance conferred by a class I integron and SXT constin in Vibrio cholerae O1 strains isolated in Laos. Antimicrob Agents Chemother 2004; 48:2364-9. PMID:15215082 doi:10.1128/AAC.48.7.2364-2369.2004

55. Toma C, Nakasone N, Song T, et al. Vibrio cholerae SXT element, Laos. Emerg Infect Dis 2005; 11:346-7. PMID:15759340

56. Bani S, Mastromarino PN, Ceccarelli D, et al. Molecular characterization of ICEVchVie0 and its disappearance in Vibrio cholerae O1 strains isolated in 2003 in Vietnam. FEMS Microbiol Lett 2007; 266:42-8. PMID:17233716 doi:10.1111/j.1574-6968.2006.00518.x

57. Ehara M, Nguyen BM, Nguyen DT, et al. Drug susceptibility and its genetic basis in epidemic Vibrio cholerae O1 in Vietnam. Epidemiol Infect 2004; 132:595-600. PMID:15310160 doi:10.1017/S0950268804002596

58. Korichi MN, Belhocine S, Rahal K. Inc J plasmids identified for the first time in Vibrio cholerae El Tor. Med Trop (Mars) 1997; 57:249-52. PMID:9513150

59. Taviani E, Grim CJ, Chun J, et al. Genomic analysis of a novel integrative conjugative element in Vibrio cholerae. FEBS Lett 2009; 583:3630-6. PMID:19850044 doi:10.1016/j.febslet.2009.10.041

60. Das B, Halder K, Pal P, et al. Small chromosomal integration site of classical CTX prophage in Mozambique Vibrio cholerae O1 biotype El Tor strain. Arch Microbiol 2007; 188:677-83. PMID:17618421 doi:10.1007/s00203-007-0275-0

61. Taviani E, Ceccarelli D, Lazaro N, et al. Environmental Vibrio spp., isolated in Mozambique, contain a polymorphic group of integrative conjugative elements and class 1 integrons. FEMS Microbiol Ecol 2008; 64:45-54. PMID:18318712 doi:10.1111/j.1574-6941.2008.00455.x

62. Thungapathra M, Amita, Sinha KK, et al. Occurrence of antibiotic resistance gene cassettes aac(6')-Ib, dfrA5, dfrA12, and ereA2 in class I integrons in non-O1, non-O139 Vibrio cholerae strains in India. Antimicrob Agents Chemother 2002; 46:2948-55. PMID:12183252 doi:10.1128/AAC.46.9.2948-2955.2002

63. Ahmed AM, Shinoda S, Shimamoto T. A variant type of Vibrio cholerae SXT element in a multidrug-resistant strain of Vibrio fluvialis. FEMS Microbiol Lett 2005; 242:241-7. PMID:15621444 doi:10.1016/j.femsle.2004.11.012

64. Matthew M, Hedges RW, Smith JT. Types of beta-lactamase determined by plasmids in gram-negative bacteria. J Bacteriol 1979; 138:657-62. PMID:378931

65. Harada S, Ishii Y, Saga T, et al. Chromosomally encoded blaCMY-2 located on a novel SXT/R391-related integrating conjugative element in a Proteus mirabilis clinical isolate. Antimicrob Agents Chemother 2010; 54:3545-50. PMID:20566768 doi:10.1128/AAC.00111-10

66. Pearson MM, Sebaihia M, Churcher C, et al. Complete genome sequence of uropathogenic Proteus mirabilis, a master of both adherence and motility. J Bacteriol 2008; 190:4027-37. PMID:18375554 doi:10.1128/JB.01981-07

67. Juíz-Rio S, Osorio CR, de Lorenzo V, et al. Subtractive hybridization reveals a high genetic diversity in the fish pathogen Photobacterium damselae subsp. piscicida: evidence of a SXT-like element. Microbiology 2005; 151:2659-69. PMID:16079344 doi:10.1099/mic.0.27891-0

68. Pembroke JT, Piterina AV. A novel ICE in the genome of Shewanella putrefaciens W3-18-1: comparison with the SXT/R391 ICE-like elements. FEMS Microbiol Lett 2006; 264:80-8. PMID:17020552 doi:10.1111/j.1574-6968.2006.00452.x

69. Peters SE, Hobman JL, Strike P, et al. Novel mercury resistance determinants carried by IncJ plasmids pMERPH and R391. Mol Gen Genet 1991; 228:294-9. PMID:1886614 doi:10.1007/BF00282479

CHAPTER 14

Excision and Transfer of *Bacteroides* Conjugative Integrated Elements

Abigail A. Salyers, Jeffrey F. Gardner and Nadja B. Shoemaker*

Abstract

Integrated conjugative elements (ICEs), previously called conjugative transposons (CTns), are proving to be as widespread in bacteria as conjugative plasmids, and they are making a significant contribution to bacterial evolution. CTns carry a variety of accessory genes, such as antibiotic resistance and virulence genes, but even the ones that do not carry such genes (cryptic CTns) are poised to acquire and spread accessory genes in the future. The mechanisms of excision and integration of these mobile elements have only recently begun to be understood. The most extensively studied CTn excision and integration system is the one used by a family of *Bacteroides* CTns related to CTnDOT. Members of this family of CTns are clearly very successful mobile elements since they are now found in over 70% of natural *Bacteroides* isolates, an increase from 20–30% before 1970. Information about the complex excision system CTnDOT has not only provided information about why CTnDOT type elements are so successful at being spread and maintained in the colonic ecosystem, but has also expanded the paradigm of excision that is based on earlier studies of the lysogenic phages and the Gram-positive CTn, Tn916. The regulatory cascade that regulates excision and transfer genes in response to stimulation of donor cells by tetracycline is complex and is unusual in that some of the proteins that regulate expression of transfer genes play a structural role in excision. This complex regulatory system may help to explain why CTns in this family are so stably maintained in colonic *Bacteroides* and related genera, even in the absence of selection.

Introduction

Steps in the Transfer of a Conjugative Transposon (CTn)

Conjugative transposons (CTns) are proving to be widespread in a number of different phyla of bacteria, including the Bacteroidetes, the Firmicutes and the Proteobacteria. Shared features of CTns are shown in Figure 1. The recent designation suggested for integrative and conjugative elements, "ICE," is more accurate than the older designation "CTn" because these mobile elements are very different from transposons such as Tn3 or *Tn10* that move within cells. Nonetheless, we will use the CTn designation in this chapter because this has been used until very recently in virtually all of the published accounts describing these elements.

CTns are normally integrated in the bacterial chromosome. The first step in transfer is excision from the donor cell chromosome to form a double-stranded circular intermediate. The circular intermediate is nicked at an internal transfer origin (*oriT*) and transfer of a single stranded copy of the intermediate occurs by conjugation to a recipient cell. In the recipient cell, and presumably

*Department of Microbiology, University of Illinois, Urbana, Illinois, USA.
Corresponding Author: Abigail A. Salyers—Email: abigails@uiuc.edu

Bacterial Integrative Mobile Genetic Elements, edited by Adam P. Roberts and Peter Mullany.
©2013 Landes Bioscience.

also in the donor cell, the single stranded copy is replicated to produce the double stranded circular form, which is then integrated back into the bacterial chromosome. The most extensively studied of the *Bacteroides* CTns has been CTnDOT. CTnDOT is the subject of this review because it serves to illustrate a number of features that seem to be common to CTns. It also illustrates the diversity of excision and regulatory mechanisms now being found in CTns.

CTns were first discovered as elements that carried antibiotic resistance genes,[1-5] but it is now clear that they can carry other types of genes. For example, a CTn found in *Mesorhizobium loti* carries nitrogen fixation genes.[6] Recently, some of the so-called "pathogenicity islands," integrated DNA segments that carry virulence genes and have a %GC content different from that of the rest of the chromosome, have been shown to be capable of conjugative transfer and are probably CTns.[7-9] It is important to keep in mind how much DNA a CTn can transfer in a single step. The largest CTn found so far is 500 kb in size, the size of some of the smaller bacterial chromosomes. In *Bacteroides* species, the sizes of CTns so far identified are smaller, ranging from 50 to 180 kbp,[10-12] but are still capable of transferring appreciable amounts of DNA. The smallest CTn known is the Gram positive CTn Tn*916*, which is 18 kbp in size and may represent the minimal size of an element capable of excising and transferring itself.[13] Judging from the transfer genes found on Tn*916* and other CTns, these elements seem to encode a pilus-independent form of conjugation. That is, all of the transfer genes encode inner or outer membrane proteins that form the mating apparatus. This may be the reason that CTns seem to require solid surfaces and high concentrations of bacteria for transfer.

Studies of natural isolates indicate that CTns are maintained very stably, even in the absence of selection, so in most cases the entire CTn becomes, in effect, a part of the chromosome. In the case of Bacteroidetes, a phylum that contains not only human colonic *Bacteroides* but also human oral bacteria such as *Porphyromonas* and *Prevotella* species, genome sequence analysis has revealed that carriage of multiple different CTns by natural isolates is the rule rather than the exception. In fact, we have yet to find a *Bacteroides* strain that is free of CTns. Most of these CTns are cryptic. That is, they carry no recognizable accessory genes such as antibiotic resistance or virulence genes. These cryptic CTns could, however, easily acquire accessory genes and thus become vehicles for transferring them.

Wide Range of CTn Transfer Capabilities

The ability of a CTn to introduce new traits into a recipient cell can extend beyond the process shown in Figure 1. We have shown that *Bacteroides* CTns can mobilize co-resident plasmids and integrated unlinked and unrelated DNA segments called mobilizable transposons (MTns).[12,14,15] The first MTn to be discovered was the *Bacteroides* element, NBU1, which is mobilized by CTnDOT.[16,17] Transfer of NBU1 requires two functions encoded by CTnDOT. First, CTnDOT regulatory proteins activate expression of NBU1-encoded excision genes,[17] thus leading to excision and circularization of NBU1. Second, CTnDOT provides the mating apparatus that is used to transfer a single stranded copy of NBU1 to the recipient cell..

A unexpected, recently discovered feature of CTnDOT that extends its effects on a recipient cell even further is that two regulatory proteins it carries, RteA and RteB,[18] activate or repress expression of chromosomal genes outside the CTn.[19] When CTnDOT is in a cell and the cell is exposed to tetracycline, at least 60 chromosomal genes exhibit either increased or decreased expression. Some of the upregulated genes proved to be carried on defective CTns, but other genes appear to be housekeeping genes. RteA and RteB are responsible for some of these changes in gene expression, but the participation of other CTnDOT proteins has not been ruled out. Nor is it clear whether the changes in gene expression affect the ability of the donor cell to transfer DNA. Our study of the effect of CTnDOT on chromosomal gene expression was the first of its kind. It remains to be seen whether other CTns or self-transmissible plasmids that carry regulatory genes have similar global effects. If they do, a better understanding of the phenomenon could expand considerably our view of the impact of CTns, and possibly plasmids that carry regulatory genes, on the overall physiological state of their bacterial host.

Excision of CTnDOT

Components of the Excision Reaction, the First Step in CTn Transfer

The first step in transfer of a CTn is excision to form the circular transfer intermediate (Fig. 1). The excision reaction of the lambdoid phages[20] and the Gram-positive CTn Tn*916*[21,22] requires at least three components: the integrase (Int), a noncatalytic directionality protein that causes a reversal of the integration reaction (Xis) and one or more host factors. In the case of CTnDOT, as in the case of the lambdoid phages and Tn*916,* the catalytic center of the excision system is the element's integrase. The CTnDOT integrase is encoded by a gene (*intDOT*) which is located at one end of CTnDOT. IntDOT and a host factor[23] are sufficient to catalyze integration. The CTnDOT excision system is more complex than previously described excision systems. Instead of a single directionality-determining protein like Xis, there are three proteins: Xis2c (formerly Orf2c), Xis2d (formerly Orf2d) and Exc.[24,25]

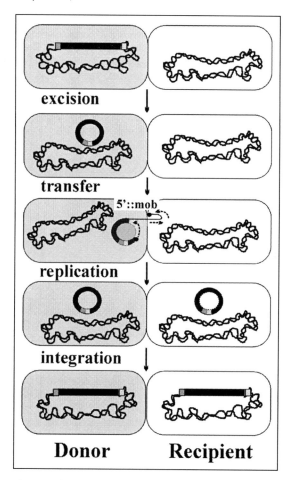

Figure 1. Steps in the transfer of a CTn (or ICE). The CTn is normally integrated into the chromosome of the donor cell. Excision produces a double stranded intermediate that is nicked at the internal transfer origin (*oriT*) by the mobilization proteins (Mob proteins). A single-stranded copy of the CTn is transferred to the recipient cell through a multiprotein complex formed by the transfer proteins (Tra proteins). In the recipient, and probably also in the donor, the single-stranded copy is made double-stranded and integrates into the chromosome.

At least one host factor is also required. The *Bacteroides* host factor that is involved in CTnDOT integration and excision is almost certainly very different from the proteobacterial integration host factor (IHF) that is involved in activities of the lambdoid phages and in many other replication and genomic rearrangement activities.

Most of the work on CTnDOT has been done with *Bacteroides thetaiotaomicron* as the host. *B. thetaiotaomicron* is one of the most abundant *Bacteroides* species in the human colon and is also an opportunistic human pathogen. There are only a few genes in the *Bacteroides thetaiotaomicron* genome sequence that have any similarity at the amino acid sequence level to IHF, and this similarity is low (30–40%). Of these candidate genes, three can be disrupted without affecting integration and excision of CTnDOT (M. Guss, unpublished data). A fourth candidate, disruption of which was lethal, now appears to be the actual host factor required for integration and excision of CTnDOT (K Ringwald, unpublished data). Evidence for this comes from the observation that the *Bacteroides* host factor (BHFa) replaces IHF in the in vitro integration assay and replaces an uncharacterized crude extract in the in vitro excision assay. The *bhf* gene is found widely in other *Bacteroides* species beside *B. thetaiotaomicron*

Of the CTnDOT excision proteins, both Xis2c and Xis2d resemble lambda Xis and Tn916 Xis in that they are small basic proteins with no apparent enzymatic activity. These proteins presumably act similarly to lambda Xis. That is, they bind DNA and interact with IntDOT to form intasomes that undergo synapsis and strand exchanges to excise CTnDOT from the chromosome. Why two Xis proteins are needed rather than one is still unclear. The function of the third excision protein, Exc, is an even greater mystery. It is a much larger protein than the Xis proteins (696 aa) and has topoisomerase activity in vitro.[24] In the CTnDOT in vitro system, Xis2c and Xis2d are essential, whereas Exc appears to be only stimulatory (C Keeton, unpublished). Also, the in vitro topoisomerase activity of Exc may be misleading because a mutant form of Exc that lacks topoisomerase activity is still active in vivo.[24] The current guess as to the role of Exc in excision is that it may stabilize the excision complex that is formed by IntDOT, Xis2c, Xis2d and BHFa. Recently, we have found that in an in vitro system based on a target that contains *attL* and *attR* on a single circular molecule, Exc aids in the excision reaction, especially if the distance between *attL* and *attR* is large (C. Keeton, unpublished data). Thus, Exc could have a DNA bending function as well as a stabilization function, although direct evidence for this activity is not yet available.

The CTnDOT Integrase IntDOT

The focus of the present chapter is on excision and transfer of CTnDOT, but it is important to start with a description of the integrase because that enzyme is the catalytic center of the excision reaction. A detailed picture of the excision and integration reactions are shown in Figure 2. Although these reactions resemble superficially the excision and integration reactions performed by the phage lambda, there are important differences. The most obvious one is that excision and integration of CTnDOT involve the formation of a short region of heterology that is then resolved in some way that has yet to be determined.[26,27]

IntDOT has been purified to near homogeneity, although no crystal structure has yet been obtained. It is a 47 kDA protein that is a member of the tyrosine recombinase super-family. This family contains a number of bacteriophage integrases, including lambda integrase, as well as recombinases such as XerC/D and Flp that are involved in DNA replication and rearrangements. The fact that IntDOT appears to be a member of this family, despite the fact that it clearly catalyzes a different type of reaction, raises the question of what differences in IntDOT compared with the other integrases in the tyrosine recombinase family account for the unusual heterology-based reaction catalyzed by IntDOT.

IntDOT was originally identified as a member of the tyrosine recombinase family because it has most of the C-terminal catalytic region amino acids that form the "signature" common to the tyrosine recombinases. Also, it catalyzes a reaction that involves the formation of Holliday Junction (HJ) intermediates, although these intermediates must be resolved by an unusual reaction due to the presence of the region of heterology.[28,29] The differences between IntDOT and the other

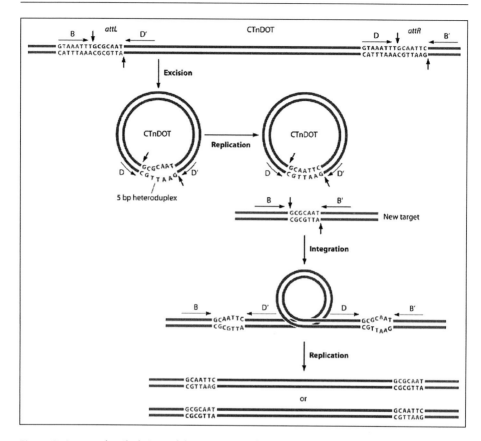

Figure 2. A more detailed view of the excision and integration reactions of a CTn, showing the region of heterology that forms during these reactions. The vertical arrows at either end of the CTn indicate the sites at which single stranded cuts are made. These sequences, which come from the chromosome and not from the CTns, have been called coupling sequences. These sequences are often flanked with imperfect repeats (B, B', D, D'). In the case of CTnDOT, the sequences at the ends do not base pair with each other, except in the case of a GC pair at the end of the coupling sequence. As a result of the bases that do not pair with a complementary base, a region of heterology is formed when the cut DNA segments are ligated. In the case of the circular form, the heterology is probably resolved during conjugative transfer but the possibility that heterology is resolved in the donor has not been ruled out. During the integration reaction, similar regions of heterology form due to lack of sequence complementarity between the coupling sequences in the circular form (*attCTn*) and the integration site (*attB*). How these regions of heterology are resolved has not yet been determined.

members of the tyrosine recombinase family have led to interest in the mechanistic bases for these differences. A detailed summary of what is known about these differences is beyond the scope of this review but can be found in a recent review.[30] A model for integration is summarized below.

Alignment of the catalytic domain of IntDOT with other tyrosine recombinases showed that it likely contained 5 of the 6 conserved catalytic residues (Arg I, Lys, His I, Arg II, His II, and Tyr; Figure 3). However the protein contained a serine residue (S 259) at the position of Arg I found in members of the tyrosine recombinase family. In order to assess the importance of the conserved residues in catalysis, we changed the five residues that we identified from the alignment and measured the activities of the proteins in the integration assay in vivo. We found that the Y381F,

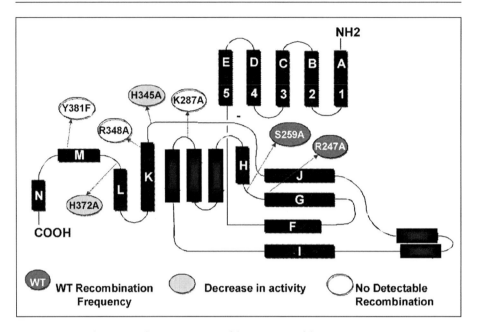

Figure 3. A two-dimensional representation of the structure of the CTnDOT integrase (IntDOT), showing mutational changes made in the C-terminal "signature" amino acids that placed CTnDOT in the tyrosine recombinase superfamily. Effects of these mutations are indicated by the key at the bottom of the figure and in the text.

R348A, and K287A proteins showed no detectible recombination and the H372A protein was 100-fold defective. These results suggest that Y381 is the catalytic tyrosine and that the R348, K287, and H372 residues are important for catalytic activity of IntDOT. The mutational analysis has so far shown that there are many structural and mechanistic similarities between IntDOT and the tyrosine recombinases, enough to justify the inclusion of IntDOT in this superfamily of recombinases. IntDOT, however, expands the range of diversity of these enzymes and raises the question of whether still other "aberrant" members of this group exist.

The current model for how the IntDOT integrase reaction works,[30] is based on observations from other tyrosine recombinases and our results. Four IntDOT monomers and an unknown number of host factor proteins assemble on *attDOT* (the joined ends of the CTn circular form) and form a nucleoprotein structure called the intasome. The catalytic domains of 2 monomers interact with the sites adjacent to the integration point, leaving the other 2 catalytic domains free to interact with *attB* (the integration site on the chromosome). The complex encounters an *attB* site and forms a synaptic complex. Two IntDOT monomers are activated and cleave the *attDOT* and *attB* sites to form 3′-phosphotyrosyl-protein linkages and free 5′-OH ends at the sites of cleavage. The complex then attempts to perform strand exchanges where the free 5′-OH groups perform nucleophilic attacks on the phosphotyrosyl bonds of the partner sites. The first pair of ligation reactions forming the Holliday junction (HJ) intermediate are successful only if complementary base pairs form at the site of ligation allowing the reaction to proceed.

In this case the reaction proceeds to the HJ stage where the 2 other IntDOT monomers become active and perform strand exchanges at the bottom strand sites that (usually) do not have identical base pairs. Thus the second reaction does not require base pairing at the sites of ligation. The changed conformation of the HJ that occurs between the strand exchanges must proceed by a mechanism that differs from other tyrosine recombinases because the overlap regions of the sites contain heterology. Thus this step is homology independent.

The other half of the time the synapsis will pair sites so that the ligation reaction cannot occur because complementary bases are not found when the strand exchanges are attempted. In this case, the reaction is reversed forming the parental substrates and the complex eventually dissociates.

Excision of CTnDOT

In the excision reaction that creates the circular transfer intermediate, IntDOT is still the catalytic center of the reaction. But now the number of players becomes larger. As already mentioned, the reaction involves at least three directionality-directing proteins that probably interact with IntDOT. These are Xis2c, Xis2d and Exc. Xis2c and Xis2d are both small basic proteins, similar to lambda Xis and the Xis of the CTn Tn916. Xis2c and Xis2d share limited amino acid sequence homology. Not surprisingly, this sequence similarity between them lies in a region that contains a helix-turn-helix motif.

As expected from these properties, both Xis2c and Xis2d are DNA binding proteins, as indicated by their ability to bind specifically in gel shift experiments to DNA at the ends of CTnDOT (*attL* and *attR*). These proteins are presumably part of a nucleoprotein complex that contains Xis2c, Xis2d, IntDOT and the host factor BHFa, but the structure of this complex remains to be determined. An in vitro system that catalyzes excision has been developed. Xis2c and Xis2d are both essential components of this in vitro reaction. The first version of this in vitro system catalyzed the joining of two plasmids, one containing *attL* and one containing *attR*. That is, although the in vitro system catalyzed excision in the sense that it caused the interaction of the joined ends of CTnDOT to form *attDOT* and *attB*, the sequences where the integration took place, it appeared more like an integration system.

There is now an in vitro system in which *attL* and *attR* are located on the same circular molecule, a configuration that more closely resembles the actual integrated element in its natural state. This turns out to have implications for the function of Exc, as will be explained shortly. The distance between *attL* and *attR* on this construct affects the efficiency of the reaction, with the longer distance between the sites being associated with a higher frequency of excision.

A third protein involved in excision, Exc, is a 79 kDa protein that has high amino acid sequence similarity to typeIA topoisomerases and purified Exc has topoisomerase activity in vitro.[24] Whereas Xis2c and Xis2d appear to bind at specific sequences, Exc does not gel shift DNA and does not seem to have any DNA sequence specificity (J. Park, unpublished). When a mutation was made in the catalytic tyrosine of Exc, which eliminated the in vitro topoisomerase activity, and the resulting construct was introduced into *Bacteroides,* the protein was still as active in excision as the wild type.[24] Thus, the activity of Exc in excision appears not to rely on alterations of DNA structure.

A clue as to a possible function of Exc came recently from studies involving the in vitro system. Exc proved not to be essential in the excision reaction but rather had a stimulatory role. This is also true in vivo because if a PCR assay for excision is used, excised product is still detected in the absence of *exc* even though the activity is lower than in its presence. In the in vitro system, the effect of Exc was more marked in the case of a substrate that had a longer stretch of DNA between *attL* and *attR*. Possible effects of Exc on binding of Xis2c and Xis2d have not yet been determined, but Exc may play a role in forming a stable excisasome.

Hfr Type Transfer of CTnDOT

Since CTnDOT is an integrated conjugative element, an obvious question is whether integrated CTnDOT could perform Hfr type transfer. This had not been determined for any of the other CTns. To do this, CTnERL was used instead of CTnDOT. CTnERL is virtually identical to CTnDOT except that it does not contain an *ermF* gene, a fact that makes it possible to use *ermF* as a marker in strain construction. In this experiment, *intERL* was disrupted to prevent the CTn from excising but not from transferring. Both ends of the element were marked with an antibiotic resistance gene and the mutant element was tested for transfer of one of these resistance genes. Mutant CTnERL did indeed transfer a portion of itself, even if it could not excise, although the frequency of this type of transfer was much lower than that of normal transfer.[31] The potential significance of this is that fragments of CTns may be transferred and, if they encounter homologous DNA segments on a CTn in the recipient, may introduce new resistance genes or even new transfer or regulatory genes into the recipient CTn.

Regulation of CTnDOT Excision

Regulation of CTn Genes

Excision of CTnDOT is stimulated by exposure of the donor cells to tetracycline. Tn*916* also exhibits tetracycline regulated transfer, although it is not clear whether both excision and transfer are stimulated by tetracycline.[22] In any event, the regulatory cascade that triggers CTnDOT is much more complex than that which controls the transfer of Tn*916*. Regulation of excision has also been seen in the case of the *Bacillus* CTn, ICEBs1, although in that case, the stimulatory trigger is DNA damage.[32] Whether the regulation of CTnDOT excision is typical of a larger group of CTns remains to be seen.

So far, there seem to be no shared regulatory themes among CTns. In fact, excision of many of them appears to be unregulated. A note of caution is appropriate when making general statements about regulation of transfer of mobile elements. Take the example of transfer of Inc.P plasmids such as R751. This transfer is commonly described as "constitutive." Yet, under anoxic conditions, R751 transfer drops to an undetectable level. Nitrate does not substitute as an energy source; oxygen seems to be required. Thus, R751 transfer is regulated by the presence of oxygen in its environment. There are probably other such examples that have not been noted because only one set of conditions has been used in mating experiments. For this reason, it is wise to amend statements about regulation of mobile element transfer by saying that transfer is or is not regulated under the conditions tested so far. Moreover, the stimulatory signal identified in laboratory studies, tetracycline in the case of CTnDOT and Tn*916*, may not be the true ecological stimulus but rather a condition that simulates the naturally occurring condition.

Production of the CTnDOT Central Regulatory Proteins, RteA, RteB and RteC

The regulatory cascade that controls CTnDOT excision starts with two genes, *rteA* and *rteB* (*r*egulation of *t*ransfer and *e*xcision), which are part of a three gene operon (*tetQ-rteA-rteB;* Figure 4A). The tetracycline resistance gene, *tetQ,* is the first gene in the operon that contains *rteA* and *rteB*. RteA and RteB are members of a two component regulatory system, with RteA being the sensor and RteB being the DNA-binding response regulator. Oddly enough, RteA seems not to be responsible for tetracycline stimulation of production of RteA and RteB. So far, the signal recognized by RteA, if any, has not been identified.

RteA is needed to activate RteB by phosphorylating it,[33] but RteA is not the sensor of tetracycline. Instead, the production of both RteA and RteB is regulated by a translational attenuation mechanism that is affected by slowing of ribosomal movement along the *tetQ* leader region.[34,35] That is, ribosomes and not RteA are the "sensors" that respond to tetracycline. It is not clear what, if anything is sensed by the putative sensor protein RteA. There is another level of regulation of RteA and RteB production: growth phase associated control of RteA and RteB. In this case, initial exposure of cells to tetracycline causes a diminished production of TetQ. But as growth continues, enough TetQ is produced to allow full translation of RteA and RteB (via translational attenuation), resulting in an increased level of excision that peaks in late exponential phase.[33]

Activated RteB binds to the promoter region of a downstream gene, *rteC,* which is in a separate transcriptional unit from *rteA* and *rteB*. RteB activates transcription of *rteC*. RteC is the regulatory protein that in turn activates expression of the excision genes, which are located in an operon downstream of *intDOT*.[36] RteC protein was initially something of a mystery because, unlike RteA and RteB, it had no homologs in the amino acid sequence databases when the primary amino acid sequence of RteC was searched. By using instead a homology-based approach, the Robetta algorithm, it was possible to predict the three dimensional structure and to search the databases for structural homologs of this predicted structure. Four homologs were found using this approach. All of them were DNA binding proteins, which had a winged helix DNA binding motif. The effects of mutations based on this predicted structure showed that the part of the winged helix motif predicted to bind DNA did in fact affect in vitro binding of RteC to the region upstream of the promoter of the excision gene operon, P_E.[37]

Excision Operon

The proteins that interact with IntDOT to catalyze excision are encoded by a four gene operon located near one end of the CTn. This operon is separated from the *intDOT* gene by a 13 kbp region that contains the *ermF* gene,[38] so it is not immediately contiguous to *intDOT* (Fig. 4A). The excision operon contains two genes that encode small DNA binding proteins (*xis2c* and *xis2d*), an open reading frame of unknown function which can be deleted without affecting excision (*orf3*), and a fourth gene, *exc*, which seems to have a stimulatory role rather than an essential role in excision. As already mentioned, Exc may act to stabilize the excision complex and/or to bend DNA so that contacts between the ends (*attL*, *attR*) can form.

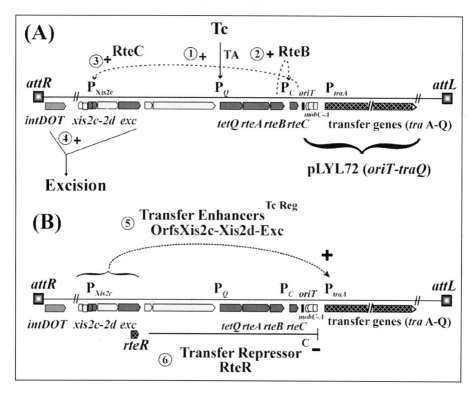

Figure 4. An overview of the complex regulatory cascade that controls the tetracycline stimulation of excision and transfer of CTnDOT. Panel (A) shows the first steps in which the central regulatory proteins RteA and RteB activate increased expression of a third regulatory gene, *rteC*. Increased production of RteA (the sensor of the two component system) and RteB (the response regulator) is mediated by translational attenuation (TA). RteC binds upstream of the promoter of the excision gene operon, P_E, and activates transcription of this operon. The bracket labeled pLYL72 indicates a region of CTnDOT that, when cloned into a nontransmissible plasmid, allows the plasmid to be transferred by conjugation. Transfer is constitutive and the frequency is intermediate between that of the fully tetracycline-induced frequency and that of the repressed frequency seen in the absence of tetracycline. Panel (B) shows the second series of regulatory steps that lead to the regulation of CTnDOT transfer. The excision proteins (Xis2c, Xis2d and Exc), which normally work with IntDOT to catalyze excision, bind to DNA upstream of the *mob* and *tra* operons (see Fig. 5) and increase the level of transcription of these operons. The effect on transfer frequency of pLYL72 is 100 to 1000 fold. Repression of transfer is mediated by a small noncoding RNA, RteR. The repressive effect of RteR on the transfer frequency of pLYL72 is 100 to 1000 fold.

Expression of the excision operon is controlled by RteC, which binds upstream of the excision promoter, P_E. The fact that expression of the excision genes is regulated may account in part for the stability of CTnDOT in the absence of selection. The stability of CTnDOT is evident from the observation that over 70% of *Bacteroides* isolates now carry a copy of CTnDOT.[39] Continuous excision and transfer of CTnDOT would be expected to take an energy toll on the cell that carries it. But, since both excision and transfer occur only when cells are stimulated with tetracycline, the genes encoded by CTnDOT seem to be silent under most conditions and thus exact less of an energy toll. So far, of the genes on CTnDOT that have been checked, only the *tetQ* operon, *intDOT* and *ermF* are expressed constitutively.

Transfer of CTnDOT

The tra *and* mob *Operons*

The operons that contain genes associated with nicking the circular form to initiate transfer (mobilization or *mob* genes) and genes encoding the proteins that form the mating bridge (transfer or *tra* genes) are located immediately downstream of *rteC*. These two operons are transcribed divergently and are separated by a 238 bp DNA segment (Fig. 5). The CTnDOT mobilization (Mob) proteins have not been characterized at all except to show that this region of DNA contains genes that allow mobilization of a plasmid into which they were cloned. The *mob* genes of CTnDOT are nearly identical to those characterized for the *Bacteroides* CTn341[40] and thus the function of the Mob proteins are presumably the same. At least some of the transfer (Tra) proteins have been localized. As expected, they are located in the inner or outer membrane.[41,42] Disruptions within the *tra* operon abolish transfer, but since the genes are part of an operon, this result is not sufficient to show which genes are essential. Disruptions in the last two genes, *traP* and *traQ*, had no effect on transfer, so these genes are not essential. Their function is still unknown.

In general, few of the Tra proteins have homologs in the databases. TraG protein seems to be the main exception to this rule. Proteins with amino acid similarity to the CTnDOT TraG have been found not only in closely related CTns such as CTnERL and CTn341[43] but also in the self-transmissible *Bacteroides* plasmid, pBF4, and in suspected cryptic CTns found in the genome sequence of *Bacteroides thetaiotaomicron*. The function of TraG is unknown but it could be is similar to that of VirB4, with which it shares some amino acid sequence similarity.

A puzzling feature of the first four *tra* genes, *traA-traD*, is that although other genes in the operon are virtual identical at the amino acid sequence level to the corresponding *tra* genes in

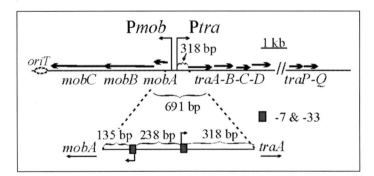

Figure 5. Divergent transcription of the operons containing the mobilization (*mob*) and transfer (*tra*) genes of CTnDOT. The promoters of these two operons are separated by 238 bp. Presumably, the excision proteins bind upstream of the two promoters (P) to upregulate expression. Both operons also have relatively long mRNA leader regions, which may have a secondary effect on regulation. This effect, if any, remains to be determined. The target of the small RNA RteR has not yet been precisely localized but appears to lie downstream of the *traA* gene.

CTnERL, The TraA – TraD proteins have only 30 to 65% amino acid sequence similarity.[41] The *traA-traD* genes of CTnDOT are definitely part of the *tra* operon and are regulated in the same way as the other *tra* genes, but *traD* is separated from *traE* by a short 100bp segment. Whatever happened during the evolution of this operon may have involved an insertion that changed the first four *tra* genes, but did so in a way that preserved the operon structure and the overall function of the mating bridge proteins, in the sense that regulation and frequency of transfer have remained unchanged.

Both the *mob* and *tra* operons are regulated at the transcriptional level in response to tetracycline stimulation. A surprising recent finding is that the expression of the *tra* operon is not regulated by RteA, RteB, RteC or a combination of these proteins. Instead transcription is stimulated by the excision proteins Xis2c, Xis2d and possibly Exc (Fig. 4B). There have been a number of reports of bifunctional proteins, but the involvement of the CTnDOT excision proteins in regulation of transfer genes may be the first example of a bifunctional complex.

Preliminary findings indicate that Xis2c and Xis2d bind DNA upstream of the promoter of the *tra* operon (P_{tra}; Park and Salyers, unpublished). No such role has yet been established for Exc. It is also not clear whether Xis2c and Xis2d are involved in the regulation of the *mob* operon by binding upstream of the promoter of the *mob* operon, although such binding might be expected if in fact the mob genes are regulated similarly to the *tra* genes.

RteR, a Small Regulatory RNA

The excision proteins are responsible for increased expression of the *tra* operon above the basal level that permits constitutive transfer of pLYL72 (a plasmid into which the transfer region was cloned), but in the absence of tetracycline, expression of the *tra* operon and transfer of CTnDOT are decreased almost to an undetectable level. Since cloning the *tra* and *mob* genes on a plasmid causes the plasmid to be transferred constitutively, the reduction of transfer gene expression and transfer frequency of CTnDOT in the absence of tetracycline indicates that there is a repressive factor that lowers the basal constitutive level of expression.

The repressive factor is encoded by a 100 bp gene that is located immediately downstream of *exc*. The repressive factor is not a peptide but a small RNA, called RteR. Although *rteR* is closely linked to *exc,* expression of *rteR* is not regulated. Instead, it is produced constitutively.[44] A model for the interaction between RteR and the excision proteins is that RteR prevents expression of essential *tra* genes when the excision proteins are not being made but that this repressive effect of RteR is overcome by upregulated expression of the excision genes. Downregulation of *tra* genes by RteR under conditions that are not conducive to CTnDOT transfer may provide part of the explanation for why CTnDOT is maintained so stably in the absence of selection.

Regulation and the Stability of CTnDOT

The complex mode of CTnDOT regulation, in which excision proteins are needed for expression of transfer genes, may be responsible for the stability and maintenance of CTnDOT. This mode of regulation ensures that excision occurs before transfer begins. As already mentioned, CTnDOT can transfer a portion of itself without excision, but this type of transfer is somewhat abortive since only a portion of the CTn is transferred and the fragment that is transferred cannot integrate in the normal way because it lacks the joined ends of the circular form. Hfr type transfer is much less efficient than normal transfer involving a circular intermediate. Thus, transfer of an intact form of CTnDOT is favored.

Excision of Other *Bacteroides* Integrated Mobile Elements

Do other *Bacteroides* mobile elements have a complex excision system like that of CTnDOT? CTnDOT and related CTns such as CTnERL and CTn341[43] are clearly the most successful CTns in the Bacteroidetes phylum. The evidence for this is that CTnDOT-like elements have been found in every *Bacteroides* species, in *Parabacteroides* spp, and in oral *Porphyromonas* and *Prevotella* spp No other CTn known, with the possible exception of the Gram-positive CTn, Tn916, has such a wide distribution. Thus, the excision, transfer and integration apparatus of CTnDOT must be highly efficient. Three resistance gene-carrying *Bacteroides* CTns besides CTn341, CTnDOT and CTnERL have been described: CTnGERM (which carries *ermG*), CTnBST (which carries *ermB*) and CTn12256 (which carries *tetQ* and *ermF*). These CTns are not nearly as widespread as the CTnDOT-related CTns: CTnBST appears to have a simpler, single protein excision system, similar to the phage lambda system. Its transfer is not regulated, at least under any of the conditions we tested. Our study of CTnBST produced an unexpected dividend. CTnBST contained a 7 kbp segment that was identical to a segment of DNA found in a Gram-positive bacterium, *Arcanobacterium pyogenes*. This is the first indication that DNA from Gram-positive bacteria can move into *Bacteroides* spp in a natural setting.

CTn12256 is a chimeric element composed of a cryptic CTn unrelated to CTnDOT (CTnBf3) with a copy of CTnDOT integrated into it. A close relative of the cryptic CTn (CTnBf3) was found previously.[10] Although the copy of CTnDOT that is integrated into this chimeric element is capable of excising and transferring itself, the CTnBf3 portion of the element clearly controls excision and transfer of the entire element. Excision and transfer are not regulated. Cryptic elements such as CTnBf3 have so far been found in every sequenced *Bacteroides* strain, so the CTnBf3 portion of CTn12256 can serve as a model for many of these elements. The excision system of this cryptic CTn appears to consist of only a single excision protein, but the copy of CTnDOT contains the same excision proteins as CTnDOT alone. The excision genes of the inserted copy of CTnDOT in CTn12256 are still active. That is, CTnDOT can transfer independently of the other part of CTn12256. When it does so, however, it makes large deletions in CTnBf3 that render CTnBf3 unable to transfer. Thus, in this case, excision of CTnDOT seems to have effects on the surrounding DNA that are more drastic than those normally seen when CTnDOT excises from a chromosomal site. This finding may indicate that there is some interaction between CTnDOT and the other part of CTn12256 during excision and transfer of the CTnDOT as it separates itself from CTnBf3.[45]

A mobilizable transposon, NBU1, which was originally discovered in our laboratory, has a more complex excision system than CTn12256 or CTnBST. As in the case of CTnDOT, transcription of the excision genes is stimulated by tetracycline. This is due to the fact that CTnDOT is controlling the excision and transfer of NBU1. NBU1 does not use the excision system of CTnDOT. Rather RteA and RteB on CTnDOT activate expression of NBU1's own excision genes. Then, the NBU1 circular form uses the Tra proteins of CTnDOT for its conjugative transfer. Although NBU1 does not produce its own transfer genes, it does provide its own mobilization gene. In contrast to CTnDOT, which has multiple mobilization genes, NBU1 has only a single gene.[46]

The excision genes of NBU1 were unrelated at the amino acid sequence level to Xis2c, Xis2d and Exc. MTns that are closely related to NBU1 have now been found in a variety of *Bacteroides* strains and some carry antibiotic resistance genes. Mobilizable transposons are now being found in other genera of bacteria, such as *Clostridium* and may prove to be much wider spread than previously appreciated.[47] Mobilizable transposons are easy to miss because not only is it necessary to have an intact element in the donor cell and the right conditions to stimulate excision and transfer, but the mobilizing element must also be present. So far, the examples of elements that excise and transfer mobilizable transposons are all CTns, but it is possible that self-mobilizing plasmids may have the capability to trigger excision and transfer of mobilizable transposons.

Conclusion

The paradigm for excision and transfer of integrated mobile elements is clearly not limited to that illustrated by lysogenic phages and CTns such as Tn916. The *Bacteroides* CTn, CTnDOT, has a much more complex excision system than those described previously. CTnDOT also has a very complex regulatory cascade that controls excision and transfer. Although CTnDOT appears to be a very successful mobile element, in the sense that it has spread widely among colonic bacteria under natural conditions, it may not be typical of CTns found in the Bacteroidetes. Other *Bacteroides* CTns such as CTnBST, the CTnBf3 portion of CTn12256 and CTnGERM appear to have simpler excision than CTnDOT and their transfer may not be regulated. It is important to note, however, that these CTns may not be a good indicators of general features of native *Bacteroides* CTns if these examples of "*Bacteroides*" CTns actually into *Bacteroides* species from other bacterial phyla, as some of their sequence properties indicate they may have. Because there is so much sequence diversity among the *Bacteroides* CTns, it is difficult to draw many conclusions not only about the ones that carry accessory genes, like CTnBST, but also about the numerous cryptic CTns that are being found in the *Bacteroides* genome sequences. We know that at least some of these cryptic CTns are active and others that are defective are nonetheless mobilizable by active CTns.[10,19] So far, however, excision and transfer genes of the cryptic CTns have not been identified and characterized. If the sequence indicators of the possible presence of a CTn can be trusted, much of the *B. thetaiotaomicron* genome consists of active and inactive CTns. This may prove to be true of other *Bacteroides* species and may be an even more widely spread phenomenon in other genera and phyla. If so, it is likely that the diversity of excision and transfer systems of integrated conjugative elements has only begun to be understood. In general, widespread distribution of CTns and MTns needs to be taken into account by those interested in evolution and genomic plasticity in bacteria.

Acknowledgments

Work described in this chapter was supported by NIH grants AI22383 (AAS) and GM28717 (JFG).

References

1. Salyers A, Shoemaker NB. Reservoirs of antibiotic resistance genes. Anim Biotechnol 2006; 17:137-46; PMID:17127525; http://dx.doi.org/10.1080/10495390600957076.
2. Shoemaker NB, Barber RD, Salyers AA. Cloning and characterization of a Bacteroides conjugal tetracycline-erythromycin resistance element by using a shuttle cosmid vector. J Bacteriol 1989; 171:1294-302; PMID:2646276.
3. Shoemaker NB, Smith MD, Guild WR. DNase-resistant transfer of chromosomal cat and tet insertions by filter mating in Pneumococcus. Plasmid 1980; 3:80-7; PMID:6278526; http://dx.doi.org/10.1016/S0147-619X(80)90036-0.
4. Sloan J, McMurry LM, Lyras D, et al. The Clostridium perfringens Tet P determinant comprises two overlapping genes: tetA(P), which mediates active tetracycline efflux, and tetB(P), which is related to the ribosomal protection family of tetracycline-resistance determinants. Mol Microbiol 1994; 11:403-15; PMID:8170402; http://dx.doi.org/10.1111/j.1365-2958.1994.tb00320.x.
5. Smith CJ, Welch RA, Macrina FL. Two independent conjugal transfer systems operating in Bacteroides fragilis V479-1. J Bacteriol 1982; 151:281-7; PMID:7085560.
6. Sullivan JT, Ronson CW. Evolution of rhizobia by acquisition of a 500-kb symbiosis island that that integrates into a phe-tRNA gene. Proc Natl Acad Sci USA 1998; 95:5145-9; PMID:9560243; http://dx.doi.org/10.1073/pnas.95.9.5145.
7. Hacker J, Kaper JB. Pathogenicity islands and the evolution of microbes. Annu Rev Microbiol 2000; 54:641-79; PMID:11018140; http://dx.doi.org/10.1146/annurev.micro.54.1.641.
8. Qiu X, Gurkar AU, Lory S. Interstrain transfer of the large pathogenicity island (PAPI-1) of Pseudomonas aeruginosa. Proc Natl Acad Sci USA 2006; 103:19830-5; PMID:17179047; http://dx.doi.org/10.1073/pnas.0606810104.
9. Harrison EM, Carter MEK, Luck S, et al. The pathogenicity islands PAPI-1 and PAPI-2 contribute individually and synergistically to virulence of Pseudomonas aeruginosa strain PA14. Infect Immun 2010; 78:1437-46; http://dx.doi.org/10.1128/IAI.00621-09; PMID:20123716.
10. Wang GR, Shoemaker NB, Jeters RT, Salyers AA. CTn12256, a chimeric Bacteroides conjugative transposon that consists of two independently active mobile elements. Plasmid 2011 in press. Plasmid 2011; http://dx.doi.org/10.1016/j.plasmid.2011.06.003; PMID:21777612.
11. Bedzyk LA, Shoemaker NB, Young KE, Salyers AA. Insertion and excision of Bacteroides conjugative chromosomal elements. J Bacteriol 1992; 174:166-72; PMID:1309516.
12. Whittle G, Shoemaker NB, Salyers AA. The role of Bacteroides conjugative transposons in the dissemination of antibiotic resistance genes. Cell Mol Life Sci 2002; 59:2044-54; PMID:12568330; http://dx.doi.org/10.1007/s000180200004.
13. Clewell DB, Jaworski DD, Flannagan SE, et al. The conjugative transposon Tn916 of Enterococcus faecalis: structural analysis and some key factors involved in movement. Dev Biol Stand 1995; 85:11-7; PMID:8586160.

14. Salyers AA. Cooper, Shoemaker NB. Lateral broad host range gene transfer in nature: how and how much? In: Syvanen M, Cato C, eds. In: Horizontal Gene transfer: Chapman and Hall, London, 1998:40-51.
15. Salyers AA, Whittle G, Shoemaker NB. Chapter 8: Conjugative and mobilizable transposons. In: Miller RV, Day MJ, eds. Microbial Evolution: Gene Establishment, Survival, and Exchange. Washington, D. C.: ASM Press, 2004:125-43.
16. Shoemaker NB, Wang GR, Salyers AA. The Bacteroides mobilizable insertion element, NBU1, integrates into the 3′ end of a Leu-tRNA gene and has an integrase that is a member of the lambda integrase family. J Bacteriol 1996; 178:3594-600; PMID:8655559.
17. Shoemaker NB, Wang GR, Stevens AM, Salyers AA. Excision, transfer, and integration of NBU1, a mobilizable site-selective insertion element. J Bacteriol 1993; 175:6578-87; PMID:8407835.
18. Stevens AM, Shoemaker NB, Li LY, Salyers AA. Tetracycline regulation of genes on Bacteroides conjugative transposons. J Bacteriol 1993; 175:6134-41; PMID:8407786.
19. Moon K, Sonnenburg J, Salyers AA. Unexpected effect of a Bacteroides conjugative transposon, CTnDOT, on chromosomal gene expression in its bacterial host. Mol Microbiol 2007; 64:1562-71; PMID:17555438; http://dx.doi.org/10.1111/j.1365-2958.2007.05756.x.
20. Pato ML. Lambda excision revisited: testing a model for synapsis of prophage ends. J Bacteriol 2001; 183:5206-8; PMID:11489876; http://dx.doi.org/10.1128/JB.183.17.5206-5208.2001.
21. Caparon MG, Scott JR. Excision and insertion of the conjugal transposon Tn916 involves a novel recombination mechanism. Cell 1989; 59:1027-34; PMID:2557157; http://dx.doi.org/10.1016/0092-8674(89)90759-9.
22. Marra D, Scott JR. Regulation of excision of the conjugative transposon Tn916. Mol Microbiol 1999; 31:609-21; PMID:10027977; http://dx.doi.org/10.1046/j.1365-2958.1999.01201.x.
23. Cheng Q, Wesslund N, Shoemaker NB, et al. Development of an in vitro integration assay for the Bacteroides conjugative transposon CTnDOT. J Bacteriol 2002; 184:4829-37; PMID:12169608; http://dx.doi.org/10.1128/JB.184.17.4829-4837.2002.
24. Sutanto Y, Shoemaker NB, Gardner JF, Salyers AA. Characterization of Exc, a protein required for the excision of the Bacteroides conjugative transposon, CTnDOT. Mol Microbiol 2002; 46:1239-46; PMID:12453211; http://dx.doi.org/10.1046/j.1365-2958.2002.03210.x.
25. Sutanto Y, DiChiara JM, Shoemaker NB, et al. Factors required in vitro for excision of the Bacteroides conjugative transposon, CTnDOT. Plasmid 2004; 52:119-30; http://dx.doi.org/10.1016/j.plasmid.2004.06.003; PMID:15336489.
26. Cheng Q, Paszkiet BJ, Shoemaker NB, et al. Integration and excision of a Bacteroides conjugative transposon, CTnDOT. J Bacteriol 2000; 182:4035-43; PMID:10869083; http://dx.doi.org/10.1128/JB.182.14.4035-4043.2000.
27. DiChiara JM, Salyers AA, Gardner JF. In vitro analysis of sequence requirements for the excision reaction of the Bacteroides conjugative transposon, CTnDOT. Mol Microbiol 2005; 56:1035-48; http://dx.doi.org/10.1111/j.1365-2958.2005.04585.x; PMID:15853888.
28. Malanowska K, Salyers AA, Gardner JF. Characterization of a conjugative transposon integrase, IntDOT. Mol Microbiol 2006; 60:1228-40; PMID:16689798; http://dx.doi.org/10.1111/j.1365-2958.2006.05164.x.
29. Malanowska K, Yoneji S, Salyers AA, Gardner JF. CTnDOT integrase performs ordered homology-dependent and homology-independent strand exchanges. Nucleic Acids Res 2007; 35:5861-73; PMID:17720706; http://dx.doi.org/10.1093/nar/gkm637.
30. Rajeev L, Malanowska K, Gardner JF. Challenging a Paradigm: the Role of DNA Homology in Tyrosine Recombinase Reactions. Microbiol Mol Biol Rev 2009; 73:300-9; PMID:19487729; http://dx.doi.org/10.1128/MMBR.00038-08.
31. Whittle G, Hamburger N, Shoemaker NB, Salyers AA. A Bacteroides conjugative transposon, CTnERL, can transfer a portion of itself by conjugation without excising from the chromosome. J Bacteriol 2006; 188:1169-74; PMID:16428422; http://dx.doi.org/10.1128/JB.188.3.1169-1174.2006.
32. Auchtung JM, Lee CA, Monson RE, et al. Regulation of a Bacillus subtilis mobile genetic element by intercellular signaling and the global DNA damage response. Proc Natl Acad Sci USA 2005; 102:12554-9; PMID:16105942; http://dx.doi.org/10.1073/pnas.0505835102.
33. Song B, Wang G-R, Shoemaker NB, Salyers AA. An unexpected effect of tetracycline concentration: growth phase-associated excision of the Bacteroides mobilizable transposon NBU1. J Bacteriol 2009; 191:1078-82; PMID:18952794; http://dx.doi.org/10.1128/JB.00637-08.
34. Wang Y, Rotman ER, Shoemaker NB, Salyers AA. Translational control of tetracycline resistance and conjugation in the Bacteroides conjugative transposon CTnDOT. J Bacteriol 2005; 187:2673-80; PMID:15805513; http://dx.doi.org/10.1128/JB.187.8.2673-2680.2005.

35. Wang Y, Shoemaker NB, Salyers AA. Regulation of a Bacteroides operon that controls excision and transfer of the conjugative transposon CTnDOT. J Bacteriol 2004; 186:2548-57; PMID:15090494; http://dx.doi.org/10.1128/JB.186.9.2548-2557.2004.

36. Moon K, Shoemaker NB, Gardner JF, Salyers AA. Regulation of excision genes of the Bacteroides conjugative transposon, CTnDOT. J Bacteriol 2005; 187:5732-41; PMID:16077120; http://dx.doi.org/10.1128/JB.187.16.5732-5741.2005.

37. Park J, Salyers AA. Characterization of the Bacteroides CTnDOT regulatory protein RteC. J Bacteriol 2011; 193:91-7; PMID:21037014; http://dx.doi.org/10.1128/JB.01015-10.

38. Whittle G, Hund BD, Shoemaker NB, Salyers AA. Characterization of the 13 kb ermF region of Bacteroides conjugative transposon, CTnDOT. Appl Environ Microbiol 2001; 67:3488-95; PMID:11472924; http://dx.doi.org/10.1128/AEM.67.8.3488-3495.2001.

39. Shoemaker NB, Vlamakis H, Hayes K, Salyers AA. Evidence for extensive resistance gene transfer among Bacteroides spp and between Bacteroides and other genera in the human colon. Appl Environ Microbiol 2001; 67:561-8 PMCID: PMC92621.

40. Peed L, Parker AC, Smith CJ. Genetic and functional analyses of the mob operon on conjugative transposon CTn341 from Bacteroides spp. J Bacteriol 2010; 192:4643-50; PMID:20639338; http://dx.doi.org/10.1128/JB.00317-10.

41. Bonheyo G, Graham D, Shoemaker NB, Salyers AA. Transfer region of a Bacteroides conjugative transposon, CTnDOT. Plasmid 2001; 45:41-51; PMID:11319931; http://dx.doi.org/10.1006/plas.2000.1495.

42. Bonheyo GT, Hund BB, Shoemaker NB, Salyers AA. Transfer region of a Bacteroides conjugative transposon contains regulatory as well as structural genes. Plasmid 2001; 46:202-9; PMID:11735369; http://dx.doi.org/10.1006/plas.2001.1545.

43. Bacic M, Parker AC, Stagg J, et al. Genetic and structural analysis of the Bacteroides conjugative transposon CTn341. J Bacteriol 2005; 187:2858-69; PMID:15805532; http://dx.doi.org/10.1128/JB.187.8.2858-2869.2005.

44. Jeters RT, Wang G-R, Moon K, et al. Tetracycline-associated transcriptional regulation of transfer genes of the Bacteroides conjugative transposon CTnDOT. J Bacteriol 2009; 191:6374-82; PMID:19700528; http://dx.doi.org/10.1128/JB.00739-09.

45. Wang GR, Shoemaker NB, Jeters RT, Salyers AA. CTn12256, a chimeric Bacteroides conjugative transposon that consists of two independently active mobile elements. Plasmid 2012 in press. Plasmid 2011; http://dx.doi.org/10.1016/j.plasmid.2011.06.003; PMID:21777612.

46. Shoemaker NB, Wang GR, Salyers AA. Multiple gene products and sequences required for excision of the mobilizable integrated Bacteroides element NBU1. J Bacteriol 2000; 182:928-36; PMID:10648516; http://dx.doi.org/10.1128/JB.182.4.928-936.2000.

47. Adams V, Lyras D, Farrow KA, Rood JI. The clostridial mobilisable transposons. Cell Mol Life Sci 2002; 59:2033-43; PMID:12568329; http://dx.doi.org/10.1007/s000180200003.

CHAPTER 15

Integrative and Conjugative Elements Encoding DDE Transposases

Violette Da Cunha,[1] Romain Guérillot,[1] Mathieu Brochet[2] and Philippe Glaser*,[1]

Abstract

Conjugative transposons were known for many years and were initially discovered for their capacity to disseminate antibiotic resistance genes. They were first named transposons even if they rely on phage-type integrases for their mobility and are now classified as Integrative and Conjugative Elements (ICE). Recently, two families of mobile genetic elements combining DDE transposition and conjugation have been described for the first time in *Streptococcus agalactiae* (TnGBSs) and in *Staphylococcus aureus* (ICE6013), respectively. These two families of ICEs are widespread in their respective host species but show a restricted host range. The conjugation imposes constraints on the transposition mechanisms as donor and recipient DNA molecules are in two different cellular compartments. It requires in particular the formation of an independent circular intermediate of transposition. Analysis of TnGBS transfer revealed that conjugation mechanisms are similar to that of integrase encoding ICEs despite unrelated excision and insertion mechanisms.

Introduction

Until recently, the mobility of all characterized Integrative and Conjugative Elements (ICEs) was shown to rely on a conserved mechanism involving an integrase (also called recombinase) and, in some cases an excisionase.[1] ICE excision occurs in a donor cell and leads to a covalently closed circular molecule. A relaxase subsequently initiates transfer by nicking duplex DNA at a specific *oriT* sequence and by transferring a single stranded DNA molecule to the recipient cell in association with the type 4 secretion system (T4SS) of the conjugation machinery. Transferred DNA is afterward converted to a double strand recircularized DNA molecule by the cellular replication machinery. Insertion of the circular ICE is finally catalyzed by the integrase in the donor and the recipient cells.[1] In ICEs, genes involved in a same function like recombination or conjugation are grouped in distinct functional modules. Each mobile element corresponds to a particular combination of such modules and combinatorial shuffling between modules allows for the assembly of highly diverse elements. ICEs share properties with plasmids and phages, and were initially named with a heterogeneous nomenclature including conjugative transposon or integrative plasmid.[2] They may use a serine recombinase, like Tn5397 from *Clostridium difficile*[3] or a tyrosine recombinase like Tn916 that is widespread among Gram-positive bacteria.[4] Despite a diverse range of module combinations, no ICE was reported to depend on a DDE type transposase for its mobility. Although transposases and integrases catalyze co-integration of two

[1]Laboratoire Evolution et Génomique Bactériennes, CNRS URA2171, Institut Pasteur, Paris, France; [2]The Wellcome Trust Sanger Institute, Cambridge, UK.
*Corresponding Author: Philippe Glaser—Email: pglaser@pasteur.fr

Bacterial Integrative Mobile Genetic Elements, edited by Adam P. Roberts and Peter Mullany.
©2013 Landes Bioscience.

DNA molecules, the mechanisms of action are intrinsically different. In contrast to site-specific recombinases, transposases do not covalently bind to DNA.

Recently, ICEs relying on a transposase for their mobility have been described in the human opportunistic pathogens *Streptococcus agalactiae* and *Staphylococcus aureus*. Two ICEs, Tn*GBS1* and Tn*GBS2* are integrated in the genome of *S. agalactiae* strain NEM316.[5] Both ICEs express a closely related DDE transposase belonging to a new class but encode two different conjugation machineries. In *S. aureus* ICE*6013* expresses an IS*30*-like transposase and carries a conjugative module similar to that of ICE*Bs1*.[6] In the present chapter we discuss the specificity of these two families of conjugative elements, the putative constraints of the conjugation process explaining the paucity of reported cases of such an association and the diversity of these elements together with their relationship to other conjugative elements.

Organization and Diversity of Tn*GBS1* and Related Transposons

Tn*GBS1* is 47 kb long and carries 49 protein encoding genes. All genes except one (*gbs0376**) are transcribed in the same orientation. It is present in three copies in *S. agalactiae* strain NEM316. Its GC content of 37.6% is rather uniform along the sequence and is slightly higher than the average value of the NEM316 genome (35.6%). The transposase gene is located at the right end of the element with the stop codon located 335 bases far from the end of the transposon (Fig. 1). Among these 49 genes, 16 encode proteins similar to proteins encoded by diverse conjugative plasmids and putative ICEs identified in various Firmicutes. In addition to the transposition module (the transposase and the two inverted repeats—IRs), different modules were identified (Fig. 1). The mobilization module was defined by similarity with the predicted relaxase and the *oriT* sequence from pHTβ and pMG2200 plasmids.[7] It includes the gene encoding the putative relaxase (*gbs0384*) preceded by four genes encoding proteins of unknown function. An *oriT* site similar to the *oriT* sequence characterized in pHTβ was identified in the promoter region of *gbs0380* (Fig. 1). This mobilization module represents a new type not previously categorized.[8] In addition, two genes encoding a putative topoisomerase I (*gbs0387*) and a protein containing a primase domain (*gbs0386*) could also be involved in the DNA processing during intercellular transfer. The conjugation module was identified by the presence of a VirD4 homolog (*gbs0396*), a distantly related VirB4 protein (*gbs0402*), and genes encoding surface proteins homologous to aggregation and exclusion proteins from Gram-positive conjugative plasmids.[9] In addition, a protein with a predicted CHAP domain (Gbs0404) is possibly implicated in the cleavage of the peptidoglycan for building type IV structures across the wall of the bacteria, as previously reported for the CsiA protein encoded by *Lactococcus lactis* MG 1363 sex factor.[10] Tn*GBS1* encodes also two proteins homologous to plasmid replication proteins: RepA from diverse Gram-positive plasmids and a pAMβ1 RepE homolog[11] (Fig. 1). The prediction of these functional modules suggests that Tn*GBS1* is conjugative and may under certain circumstances be able to replicate autonomously. Accordingly, Tn*GBS1* was shown to be transferable by conjugation as Tn*GBS1* carrying an inserted erythromycin resistance genes (*erm*) was readily transferred among *S. agalactiae* strains with a frequency of 10^{-5}–10^{-6} per donor cell (V. Da Cunha and R. Guérillot, unpublished).

DNA-array hybridizations of a collection of 92 *S. agalactiae* isolates of human and animal origins revealed the presence of this transposon or of a related transposon in 12 strains. These strains were all of animal origin but belong to distant clonal complexes (CC23, 26 and 67).[5] Until recently, there was no sequence similar to Tn*GBS1* in public sequence libraries. However, with the increasing number of genomes available, search for similar sequences revealed the presence of seven complete or partial sequences of ICEs similar to Tn*GBS1* (Table 1). Six of them were found in streptococci and a more distantly related one in *Lactobacillus salivarius*. In this last case, genes similar to Tn*GBS1* genes are located on two contigs (Table 1). These transposons share the functions predicted in Tn*GBS1* including transposition, mobilization, conjugation and replication.

---

*For Tn*GBS1* gene names correspond to the first of the three copies of this ICE in the genome of strain NEM316

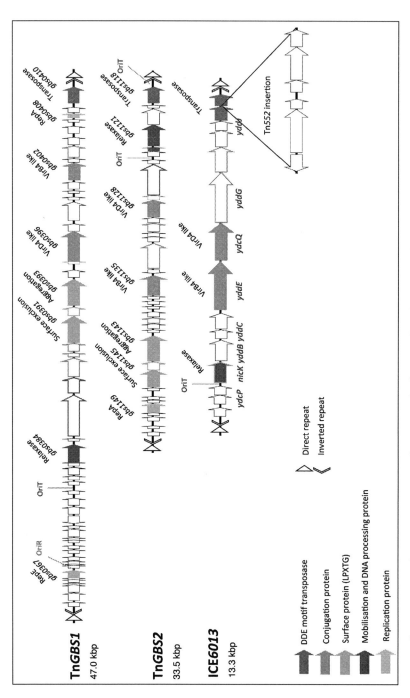

Figure 1. General organization of Tn*GBS1*, Tn*GBS2* and ICE*6013*. The functional modules are defined according to the color code shown on the figure. Colored filled arrows correspond to genes with predicted functions. Colored empty arrows indicate genes predicted to belong to a functional module without defined function. Black arrows represent genes of unknown functions. Tn*GBS1* is present in three copies in strain NEM316, *gbs* numbers are given for the first copy. For ICE*6013*, the gene names underneath the arrows correspond to the names of ICE*Bs1* orthologous genes.

They are therefore likely mobile. Additional genes encoding proteins of unknown function are conserved and may be involved in the transfer process. For example, these elements express a highly conserved ClpA homolog. ClpA is the ATPase/protein unfoldase subunit of the ATP-dependent ClpAP protease.[12] The chaperone activity of this protein may be required for DNA transfer by this atypical conjugation machinery.

Organization and Diversity of Tn*GBS2* and Related Elements

Tn*GBS2* is 33.5 kb long and is predicted to encode 36 genes. Its G+C content of 38% is, like for Tn*GBS1*, slightly higher than the *S. agalactiae* core genome G+C content. As observed in Tn*GBS1*, all genes except *gbs1134* are transcribed in the same orientation (Fig. 1). The transposase gene is located 335 bases from the right-end of the transposon. The mobilization module belongs to the MOB$_{P7}$ class[8] and encodes two proteins: a relaxase (Gbs1121) and a relaxase accessory protein (Gbs1122) as defined previously.[5] We also predict a conserved sequence similar to previously characterized *oriT* regions upstream this module[13] (R. Guérillot, unpublished data). This sequence may represent a second *oriT* region of Tn*GBS2* in addition to the one previously identified downstream the transposase gene.[5] Comparative analysis allowed identification of a conjugative module similar to that of the pheromone inducible conjugative plasmid pCF10.[5] Therefore Tn*GBS2* encodes a T4SS unrelated to the one identified in Tn*GBS1*. The conjugation module encodes all essential proteins identified for the conjugative process of pCF10 (Fig. 1 and ref. 14). In agreement with this observation, Tn*GBS2* is efficiently transferred among *S. agalactiae* strains with a frequency of 10^{-5} per donor cell.[5] Despite the unrelated T4SS used by Tn*GBS2* and Tn*GBS1*, three proteins carrying a peptidoglycan anchoring motif (LPXTG) and possibly involved in the DNA transfer are highly similar in both ICEs: Gbs1143 is 92% identical to Gbs0393 (annotated as a putative cell aggregation protein), Gbs1144 is 41% identical to Gbs0392 and Gbs1145 (annotated as a putative exclusion protein) is 48% identical to Gbs0391. This level of similarity reflects the likely exchange of these genes between different conjugation machineries and may be linked to the host specificity of these elements among streptococci. The conjugation process involves the formation of a circular intermediate catalyzed by the transposase as observed in Tn*GBS1*.[5] The abundance of the circular form of Tn*GBS2* was estimated by quantitative PCR as 10^{-4} copies per chromosome. This value is in a similar range of the conjugation frequency. Tn*GBS2* is therefore efficiently transferred by conjugation and its circularization represents the limiting step. Like in Tn*GBS1*, we identified a putative *repA* gene suggesting that it may also replicate during the conjugation process.

DNA-array analysis of a collection of strains showed that Tn*GBS2* and related transposons are common among *S. agalactiae* isolates.[5,15] For unknown reason, it was observed as particularly abundant among isolates of animal origin, where it was detected in 85% of the strains. Consistently, only NEM316 carries a Tn*GBS2* element among the eight published genomes of *S. agalactiae*, all of human origin. This puzzling observation may correspond to a counter selection for strains carrying this element among human isolates.[5] The CRISPR-cas system provides immunity against phages and mobile genetic elements.[16] The protection is mediated by short RNA (crRNA) transcribed from short sequences (spacers) captured from the invaders and inserted into the CRISPR locus. Spacers targeting Tn*GBS2* were identified in four published genomes (2603V/R, A909, CJB111, 18RS21, V. Da Cunha, Unpublished). The distribution of Tn*GBS2* among *S. agalactiae* strains may therefore be controlled by the immunity provided by the CRISPR-cas system.

Similarity searches at the time of the initial characterization of Tn*GBS2* failed to detect any similar transposon among sequenced genomes. However, with the increasing number of published genomes, we have now identified 13 Tn*GBS2*-like elements in different streptococcal genomes (Table 1). These transposons are frequent in the pyogenic, bovis and mitis groups (Table 1). The transposition, mobilization and conjugation modules are conserved, and all these elements express a RepA homolog. These Tn*GBS*-related elements are thus likely mobile apart from the ICE identified in *Streptococcus infantarius* in which several of the genes essential for mobility are pseudogenes or are missing.

Table 1. General characteristics of ICEs related to TnGBS1 and TnGBS2

Species—Strain	Type	Insertion Site[1]	Size kb (nb of genes)	Cargo Gene	Acc. Number
S. agalactiae NEM316	TnGBS1	gbs0411-gbs0741-gbs0968	47 (49)		NC_004368
S. agalactiae NEM316	TnGBS2	xpt	33 (36)		NC_004368
S. agalactiae ATCC 13813	TnGBS1	galE	partial		AEQQ01000019
S. agalactiae ATCC 13813	TnGBS2	glyQ	36 (31)		AEQQ01000097–98
S. infantarius ATCC BAA-102	TnGBS2	rpmH	64 (70)	Heavy metal resistance	NZ_ABJK02000018
S. gallolyticus UCN34	TnGBS2	rpmB	43 (45)	Sortase A	NC_013798
S. gallolyticus ATCC BAA-2069-A	TnGBS2	rpmH	36 (37)	Glucan-binding protein	NC_015215
S. gallolyticu ATCC BAA-2069-B	TnGBS2	ND[2]	30 (35)		NC_015215
S. oralis ATCC 35037	TnGBS2	ND	29 (30)		NZ_ADMV01000022
S. taxon oral sp taxon 071 str. 73H25AP	TnGBS1	rpsO	46 (44)		NZ_AEEP01000006
S. taxon oral sp taxon 071 str. 73H25AP	TnGBS2	glyQ	31 (31)		NZ_AEEP01000011
Streptococcus sp 2_1_36FAA	TnGBS1	glmS	40 (41)		NZ_GG704942
Streptococcus sp 2_1_36FAA	TnGBS2	rpsF	25 (27) partial		NZ_GG704942
S. mitis SK597	TnGBS2	pyrF	32 (32)		NZ_AEDV01000027
S. sanguinis ATCC 49296	TnGBS2	rpmB	32 (30)		NZ_AEPO01000010

continued on next page

Table 1. Continued

Species—Strain	Type	Insertion Site[1]	Size kb (nb of genes)	Cargo Gene	Acc. Number
S. sanguinis SK1058	Tn*GBS1*	*pyrG*	Partial		AFBF01000017
S. sanguinis SK49	Tn*GBS1*	*relA2*	Partial		AFFO01000015
S. anginosius F0211	Tn*GBS1*	*murF*	50(50)		AECT01000063
S. ictaluri 707–05	Tn*GBS2*	*gbs1262* orthologous gene	28 partial		AEUX01000023
S. ictaluri 707–05	Tn*GBS2*	*rimP*	37 partial		AEUX01000037 AEUX01000014
S. urinalis 2285–97	Tn*GBS1*	*rodA*	45		AEUZ01000004
L. salivarius	Tn*GBS1*	ND	32 + 73 partial	Pilus locus,	NZ_ACGT01000033 NZ_ACGT01000030

1. Name of the gene upstream which the ICE is inserted. 2. Not determined

An important question is the selective advantage these transposons may bring to their host cell. Only the ICE identified in *Streptococcus infantarius* expresses a typical cargo gene annotated as conferring resistance to heavy metals (Table 1), but this ICE is probably non-conjugative. We identified in Tn*Gallo1* from *Streptococcus gallolyticus* strain UCN34 a *srtA* paralogous gene encoding a type A sortase.[17] The sortase A catalyzes the anchoring of surface proteins carrying an LPXTG motif to the peptidoglycan. The function of this second Sortase A protein is unclear, but may contribute to the fitness of the host by contributing to protein anchorage. Interestingly, as mentioned above, the most conserved gene between Tn*GBS1* (*gbs0393*) and Tn*GBS2* (*gbs1143*) codes for an aggregation protein similar to the one found in *Enterrococcus faecalis* pCF10 plasmid and in the sex factor of *Lactococcus lactis*. This protein was shown to promote efficient transfer of these conjugative elements. Furthermore this protein is also linked to streptococcal antigen I/II polypeptide family adhesins. A member of this protein family have been shown to be involved in tooth attachment, biofilm formation, tissue invasion and immune modulation in *Streptococcus mutans*.[18] This observation suggests that this protein may also confer a selective advantage in some conditions in addition to their role in the conjugative transfer of the Tn*GBSs*. Finally, the ICE identified in *L. salivarius* carries a putative pilus operon, which may contribute to the adhesion of this bacterium or to the formation of biofilm as shown in other Gram-positive species.[19] Nevertheless, these ICEs are probably also selfish elements, imposing a limited fitness cost to the bacterial host as a result of their specificity of insertion in intergenic regions (see below).[5]

Organization of ICE*6013*

ICE*6013* is 13.3 kb long and related to the ICE*Bs1* family.[20] Its G+C content of 30.1% is slightly below the average G+C content of *S. aureus* genomes of 32.7%. A functional map is presented in Figure 1, highlighting the similarity with ICE*Bs1*. However, the integrase encoding gene is replaced by a transposase. The relaxase gene is similar to the *nicK* gene encoded by ICE*Bs1*.[20] The identification of such protein is indicative that ICE*6013* may be conjugative. ICE*6013* was identified in 6 out of the 15 strains of *S. aureus* sequenced at that time, where it was inserted at five different chromosomal *loci*. It was also shown to be widespread among *S. aureus* strains. In addition, these ICEs show a high level of polymorphism and signatures of recombination.[6] These observations strongly suggest that they are active and likely mobile. In the *S. aureus* ST239 strain where it was initially described, ICE*6013* carries a Tn*552* transposon encoding an ampicillin resistance gene. This gene was used as a marker for conjugation assays by using different AmpS EryR strains of *S. aureus* as recipient cells.[6] No transfer was detected in filter assays. However, in this ICE, Tn*552* is inserted into the putative transposase gene, explaining these negative results. Therefore, whether this ICE is fully functional remains to be demonstrated.

ICE*6013* was identified by PCR or by DNA gel blot in 19 out of 44 (43%) of clonally diverse strains of *S. aureus*. In most cases, these elements do not carry Tn*552* or genes providing an obvious selective advantage. It seems therefore that this ICE spread among *S. aureus* genomes even in the absence of any cargo gene.[6] This ICE was also characterized independently within the SCCmec cassette of strain JCSC6826.[21] A more distantly related ICE was identified in the ED98 strain of poultry origin (GenBank: CP001781). This ICE is 75% identical to ICE*6013* over its complete length, showing that these ICEs are probably ancient. In addition, similarity search at the NCBI nr nucleotide sequence database revealed a *S. aureus* plasmid (SAP077B, GenBank: GQ900429) almost identical to ICE*6013*. This plasmid corresponds to the circular intermediate of an ICE with a three nucleotide intervening sequence located between the two IRs: acaccacc_aat_ggcagtgtgt. This organization is identical to the one described for the junction of the circular form characterized for ICE*6013* expressing a full-length transposase. These three bases correspond to the three bases duplicated at the insertion site. It has been shown that ICE*Bs1* is capable of autonomous plasmid-like replication of its circular form after excision. The NicK protein, conserved in ICE*6013*, has been shown to be responsible for initiating the autonomous plasmid-like replication of ICE*Bs1*.[22] It is therefore likely that ICE*6013* may also be replicative under certain conditions as suggested for Tn*GBS* elements.

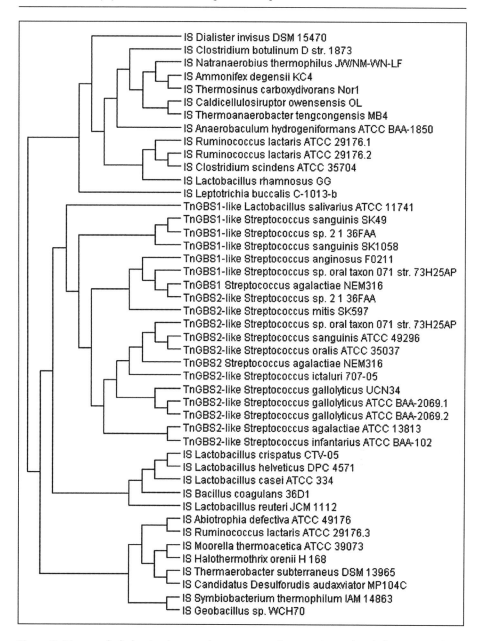

Figure 2. Unrooted phylogenetic tree of transposases from Tn*GBS*-related elements and ISs. Species and strain names are indicated. Transposase sequences were retrieved at the NCBI databank after psi-blast search using Tn*GBS2* transposase as query. Protein sequences were aligned with MUSCLE using default parameters. The tree was built using the Jalview software with the average distance method and the BLOSUM62 matrix. Only a sample of divergent IS transposases representative of the major lineages were kept in the final alignment.

The Transposition Module

Both Tn*GBS1* and Tn*GBS2* show a typical organization of transposable elements with two imperfect IRs at each extremity flanked by nine or ten base direct repeats corresponding to the duplication of the insertion site. Both elements carry at the right extremity a gene encoding a DDE transposase corresponding to a new sub-class. The transposases of both ICEs did not show significant similarity to any characterized transposase. However, motif searches at Pfam indicate that they belong to the uncharacterized protein family UPF0236 that is part of the Ribonuclease H superfamily which includes diverse families of DDE transposases. Blast searches revealed an ever-growing number of similar transposases mostly in Firmicutes. Some of these sequences are present in multiple copies in the host genome (up to 21 in *Caldicellulosiruptor owensensis* strain OL) and the analysis of the surrounding sequences revealed that they are also frequently flanked by inverted repeats and direct repeats of 7 to 10 bases.[5] These transposase genes together with the IR sequences thus represent a new family of Insertion Sequences (ISs). Phylogenetic analysis of transposases encoded by either ISs or conjugative transposons shows that Tn*GBS*-related transposases from streptococci cluster in a specific branch (R. Guérillot and V. Dacunha, unpublished). Despite the diversity of ISs in different species, each IS appears to be associated to a specific host (R. Guérillot and V. Da Cunha, unpublished).

Tn*GBS*-related transposases show a unique specificity among bacterial transposable elements. The ICEs (Table 1) and the ISs are almost systematically inserted in a defined orientation 15 or 16 bases upstream σA dependent promoters.[5] This specificity, unique among prokaryotes, is reminiscent of that of eukaryotic retroelements, like for instance Ty3 from the yeast *Saccharomyces cerevisiae*. Ty3 inserts at the transcription start site of RNA polymerase III (PolIII) dependent promoters.[23] It has been shown that this targeting involves a direct interaction between the Ty3 integrase and the transcription factor TFIIIB or TFIIIC.[24] Similarly, the insertion specificity of the Tn*GBS* subfamily of transposases may rely on an interaction with the RNA polymerase. This specific interaction could have contributed to the observed restricted host range of these transposable elements, particularly if it involves variable parts of the RNA polymerase subunits. Although this insertion specificity is probably not essential for the coupling with conjugation, it has been shown that it correspond to "safe" localizations which does not affect significantly the expression of the downstream genes, minimizing the burden of the transposition on the bacterial recipient host.[5]

The transposase of ICE*6013* is related to the well characterized IS*30* family of transposases, but shows important differences. In particular it is split in two subunits and it misses the N-terminal helix turn helix motif conserved in IS*30* transposases.[6] In strains carrying an intact transposase gene, a circular form of the ICE*6013* was detected.[6,21] This observation is in agreement with a previous study demonstrating that a circular IS*30* element acts as an intermediate for simple insertion.[25] Interestingly, it has been demonstrated that IS*30* transposase from *Escherichia coli* can replace the recombinase function in an experimental lambda phage system.[26] This experiment reproduces the natural replacement observed between ICE*6013* and ICE*Bs*1 types of ICE.

Unlike for the IS*30* transposase, the mechanism of transposition of Tn*GBS* transposases has not been characterized in depth. However, the transposition processes of Tn*GBS*-related transposases and of ICE*6013* are compatible with conjugation. In Tn*GBS1* and Tn*GBS2*, transposases catalyze the circularization of the transposon and the coupling sequence corresponding to the two IRs and the duplicated target could be amplified by PCR by using divergent primers located at both ends of the transposon.[5] A circular form of the ISs inserted into *Lactobacillus reuteri* genome was also detected by PCR and the coupling sequences corresponded to the four different insertion sites (V. Da Cunha, unpublished). Like for the integrase, this reaction produces a substrate for the conjugation machinery. However, Tn*GBS*s and IS-*L. reuteri* transposases do not restore the original insertion site of the element as opposed to integrases which catalyze a precise excision from the chromosome.. This suggests a replicative mechanism of transposition for Tn*GBS*-related transposases.[5]

Conclusion

In 2008, the association of a conjugative module to a DDE transposase in Tn*GBS1* and Tn*GBS2* may have been anecdotal. However, a similar combination was reported one year later in *S. aureus* ICE*6013* and with the increasing number of sequenced genomes, these combinations happen to be much more frequent than anticipated at this time. Tn*GBS*s contribute to the conjugative transfer of chromosomal DNA by an HFr type mechanism and have a major role in the evolution of *S. agalactiae*.[27] It has been proposed that ICE*6013* may promote similar transfers in *S. aureus*.[6] Comparison of the organization of Tn*GBS1*, Tn*GBS2* and ICE*6013* revealed two conserved features important for their conjugative transfer: the formation of a circular intermediate in the transposition process and the capacity to transiently replicate this circular form. The latter property shows the close proximity of these ICEs and conjugative plasmids. Furthermore the general transfer process of these ICEs follows the same scheme described for other ICEs. The further characterization of these mobile elements will bring new insight into the diversity of ICEs. A key issue will be to understand the regulation and the coupling between the transposition process, the conjugation and the transient replication in the dissemination of these elements.

Acknowledgments

This work was supported by the French National Research Agency (grant ANR-08-GENM-027–001) and by the Fondation pour la Recherche Médicale (FRM).

References

1. Wozniak RA, Waldor MK. Integrative and conjugative elements: mosaic mobile genetic elements enabling dynamic lateral gene flow. Nat Rev Microbiol 2010; 8:552-63. PMID:20601965 doi:10.1038/nrmicro2382
2. Burrus V, Pavlovic G, Decaris B, Guedon G. Conjugative transposons: the tip of the iceberg. Mol Microbiol 2002; 46:601-10. PMID:12410819 doi:10.1046/j.1365-2958.2002.03191.x
3. Wang H, Roberts AP, Lyras D, Rood JI, Wilks M, Mullany P. Characterization of the ends and target sites of the novel conjugative transposon Tn5397 from Clostridium difficile: excision and circularization is mediated by the large resolvase, TndX. J Bacteriol 2000; 182:3775-83. PMID:10850994 doi:10.1128/JB.182.13.3775-3783.2000
4. Poyart-Salmeron C, Trieu-Cuot P, Carlier C, Courvalin P. Molecular characterization of two proteins involved in the excision of the conjugative transposon Tn1545: homologies with other site-specific recombinases. EMBO J 1989; 8:2425-33. PMID:2551683
5. Brochet M, Da Cunha V, Couve E, Rusniok C, Trieu-Cuot P, Glaser P. Atypical association of DDE transposition with conjugation specifies a new family of mobile elements. Mol Microbiol 2009; 71:948-59. PMID:19183283 doi:10.1111/j.1365-2958.2008.06579.x
6. Smyth DS, Robinson DA. Integrative and sequence characteristics of a novel genetic element, ICE6013, in Staphylococcus aureus. J Bacteriol 2009; 191:5964-75. PMID:19648240 doi:10.1128/JB.00352-09
7. Tomita H, Ike Y. Genetic analysis of transfer-related regions of the vancomycin resistance Enterococcus conjugative plasmid pHTbeta: identification of oriT and a putative relaxase gene. J Bacteriol 2005; 187:7727-37. PMID:16267297 doi:10.1128/JB.187.22.7727-7737.2005
8. Smillie C, Garcillan-Barcia MP, Francia MV, Rocha EP, de la Cruz F. Mobility of plasmids. Microbiol Mol Biol Rev 2010; 74:434-52. PMID:20805406 doi:10.1128/MMBR.00020-10
9. Grohmann E, Muth G, Espinosa M. Conjugative plasmid transfer in gram-positive bacteria. Microbiol Mol Biol Rev 2003; 67:277-301. PMID:12794193 doi:10.1128/MMBR.67.2.277-301.2003
10. Stentz R, Wegmann U, Parker M, et al. CsiA is a bacterial cell wall synthesis inhibitor contributing to DNA translocation through the cell envelope. Mol Microbiol 2009; 72:779-94. PMID:19400771 doi:10.1111/j.1365-2958.2009.06683.x
11. Wilcks A, Smidt L, Okstad OA, Kolsto AB, Mahillon J, Andrup L. Replication mechanism and sequence analysis of the replicon of pAW63, a conjugative plasmid from Bacillus thuringiensis. J Bacteriol 1999; 181:3193-200. PMID:10322022
12. Wickner S, Gottesman S, Skowyra D, Hoskins J, McKenney K, Maurizi MR. A molecular chaperone, ClpA, functions like DnaK and DnaJ. Proc Natl Acad Sci USA 1994; 91:12218-22. PMID:7991609 doi:10.1073/pnas.91.25.12218
13. Smith MC, Thomas CD. An accessory protein is required for relaxosome formation by small staphylococcal plasmids. J Bacteriol 2004; 186:3363-73. PMID:15150221 doi:10.1128/JB.186.11.3363-3373.2004

14. Dunny GM. The peptide pheromone-inducible conjugation system of Enterococcus faecalis plasmid pCF10: cell-cell signalling, gene transfer, complexity and evolution. Philos Trans R Soc Lond B Biol Sci 2007; 362:1185-93. PMID:17360276 doi:10.1098/rstb.2007.2043

15. Brochet M, Couve E, Glaser P, Guedon G, Payot S. Integrative conjugative elements and related elements are major contributors to the genome diversity of Streptococcus agalactiae. J Bacteriol 2008; 190:6913-7. PMID:18708498 doi:10.1128/JB.00824-08

16. Horvath P, Barrangou R. CRISPR/Cas, the immune system of bacteria and archaea. Science 2010; 327:167-70. PMID:20056882 doi:10.1126/science.1179555

17. Rusniok C, Couve E, Da Cunha V, et al. Genome sequence of Streptococcus gallolyticus: insights into its adaptation to the bovine rumen and its ability to cause endocarditis. J Bacteriol 2010; 192:2266-76. PMID:20139183 doi:10.1128/JB.01659-09

18. Brady LJ, Maddocks SE, Larson MR, et al. The changing faces of Streptococcus antigen I/II polypeptide family adhesins. Mol Microbiol 2010; 77:276-86. PMID:20497507 doi:10.1111/j.1365-2958.2010.07212.x

19. Konto-Ghiorghi Y, Mairey E, Mallet A, et al. Dual role for pilus in adherence to epithelial cells and biofilm formation in Streptococcus agalactiae. PLoS Pathog 2009; 5:e1000422. PMID:19424490 doi:10.1371/journal.ppat.1000422

20. Lee CA, Grossman AD. Identification of the origin of transfer (oriT) and DNA relaxase required for conjugation of the integrative and conjugative element ICEBs1 of Bacillus subtilis. J Bacteriol 2007; 189:7254-61. PMID:17693500 doi:10.1128/JB.00932-07

21. Han X, Ito T, Takeuchi F, et al. Identification of a novel variant of staphylococcal cassette chromosome mec, type II.5, and Its truncated form by insertion of putative conjugative transposon Tn6012. Antimicrob Agents Chemother 2009; 53:2616-9. PMID:19364875 doi:10.1128/AAC.00772-08

22. Lee CA, Babic A, Grossman AD. Autonomous plasmid-like replication of a conjugative transposon. Mol Microbiol 2010; 75:268-79. PMID:19943900 doi:10.1111/j.1365-2958.2009.06985.x

23. Chalker DL, Sandmeyer SB. Ty3 integrates within the region of RNA polymerase III transcription initiation. Genes Dev 1992; 6:117-28. PMID:1309715 doi:10.1101/gad.6.1.117

24. Kirchner J, Connolly CM, Sandmeyer SB. Requirement of RNA polymerase III transcription factors for in vitro position-specific integration of a retroviruslike element. Science 1995; 267:1488-91. PMID:7878467 doi:10.1126/science.7878467

25. Kiss J, Olasz F. Formation and transposition of the covalently closed IS30 circle: the relation between tandem dimers and monomeric circles. Mol Microbiol 1999; 34:37-52. PMID:10540284 doi:10.1046/j.1365-2958.1999.01567.x

26. Kiss J, Szabo M, Olasz F. Site-specific recombination by the DDE family member mobile element IS30 transposase. Proc Natl Acad Sci USA 2003; 100:15000-5. PMID:14665688 doi:10.1073/pnas.2436518100

27. Brochet M, Rusniok C, Couve E, et al. Shaping a bacterial genome by large chromosomal replacements, the evolutionary history of Streptococcus agalactiae. Proc Natl Acad Sci USA 2008; 105:15961-6. PMID:18832470 doi:10.1073/pnas.0803654105

CHAPTER 16

The *clc* Element and Related Genomic Islands in *Proteobacteria*

Ryo Miyazaki, Marco Minoia, Nicolas Pradervand, Vladimir Sentchilo, Sandra Sulser, Friedrich Reinhard and Jan Roelof van der Meer*

Abstract

Genomic islands are large DNA segments, present in most bacterial genome, that are acquired via horizontal gene transfer and contribute to the rapid bacterial evolution and adaptation of the host cell. Here we focus on the *clc* element (or ICE*clc*), a 103-kb genomic island first discovered in *Pseudomonas knackmussii* B13, as a model of this diverse group of mobile genetic elements. ICE*clc* is normally integrated in the host bacterial chromosome but can excise and transfer to a new host by conjugation. In this chapter we review the basic features of ICE*clc*, the mechanisms of its life-style as well as evolutionary relationships with other known and unknown elements in a variety of *Proteobacteria*.

Introduction

Recent large-scale sequencing efforts have resulted in more than 1,000 completed bacterial genome sequences. Comparative genomics approaches have revealed their structural plasticity, the occurrence of rearrangements, inversions or deletions, and have highlighted the extent and regions of possible acquisition of foreign genetic material. Genome sequencing efforts also have established more clearly the widespread occurrence of a relatively new class of mobile genetic elements named "genomic islands" (GEIs).[1] More generally speaking, the term GEI refers to a large DNA segment (typically 10–500 kb) that has been acquired via horizontal gene transfer and may or may not be self-transferable. Because of their large size, GEIs carry a significant amount of genes that can offer selective advantages for the host bacterium, such as the provision of antibiotic resistance determinants, symbiosis factors, or enzymes for metabolism of aromatic compounds.[1,2]

Pseudomonas knackmussii B13 was discovered more than 30 years ago because of its propensity to metabolize chlorocatechols.[3] Metabolism of chlorocatechols is mediated via the enzymes encoded by the *clcRABDE* gene cluster. Chlorocatechols occur as intermediates in a number of different aromatic compound degradation pathways, and carrying the *clc* genes was a critical factor to utilize 3-chlorobenzoate (3-CBA) as sole source of carbon.[3,4] Since this capacity was found in a range of other bacteria,[5] the *clc* genes were supposed to have been distributed via horizontal gene transfer. Indeed, several groups demonstrated the mobility of the *clc* genes using a technique called 'molecular breeding'. This consisted of bacterial matings with *P. knackmussii* B13 as a donor and proper recipient strains, which allowed the selection of transconjugants that metabolized chlorobenzenes or chlorobiphenyl.[6-8] Initially it had been assumed that the *clc* genes were encoded on and transferred by a conjugative plasmid, which, however, was never consistently isolated.[9,10] Later in the 1990s, it was then demonstrated that the *clc* genes were located on a 103-kb DNA

*Department of Fundamental Microbiology, University of Lausanne, Lausanne, Switzerland.
Corresponding Author: Jan Roelof van der Meer—Email: janroelof.vandermeer@unil.ch

Bacterial Integrative Mobile Genetic Elements, edited by Adam P. Roberts and Peter Mullany.
©2013 Landes Bioscience.

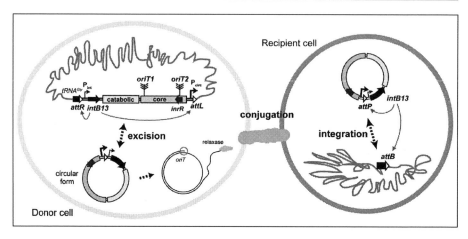

Figure 1. Schematic model of "life-style" of ICE*clc*. In the donor cell, ICE*clc* is integrated in a *tRNA^Gly* gene and the integrase gene (*intB13*) is under the weak promoter P_{int}. In stationary phase, the activation factor InrR induces the production of IntB13 that catalyzes the excision of the ICE*clc* to a circular form. Conjugation starts with DNA processing proteins forming a relaxosome at one or both of the origins of transfer (*oriT*). The single-strand DNA and relaxase complex is then supposed to be transferred into recipient cells through a type IV secretion system. In the recipient, the incoming DNA is recircularized and used as a template to reconstruct the second strand. Once in circular form, the strong constitutive promoter P_{circ} is placed in front of *intB13*. This results in a temporary overexpression of IntB13 that mediates the chromosomal integration of ICE*clc* in the integration site *attB* of a *tRNA^Gly* gene. The return to an integrated state restores the control of the weak promoter P_{int} over *intB13*. Meanwhile in the donor cell, the circularized ICE*clc* that is not transferred may be reintegrated into one of several *tRNA^Gly* genes.

fragment flanked between 2 identical repeat sequences (*attR* and *attL*) in the B13 chromosome.[11] The GEI, at that time called the *clc* element, is normally inserted in the chromosome but can excise and circularize as an extra-chromosomal molecule (Fig. 1). This excised molecule, which may have been the molecule that had been thought of previously as the plasmid, can transfer to a new recipient and integrate into the recipient's chromosome. Excision and transfer mostly occurs from stationary phase cells that have been precultured with 3-CBA as sole carbon source. Further studies then showed that excision and integration of the *clc* element specifically occurred at the 3' 18 bp of the *tRNA^Gly* gene (named the *attB* site).[12] Integration and excision are dependent on the IntB13 integrase, which is encoded on the *clc* element and belongs to the bacteriophage P4-family. Since autonomous replication was not observed the element was reclassified as an integrative and conjugative element and, therefore, renamed ICE*clc*.[13]

Genetic Structure of ICE*clc*

More recently the complete nucleotide sequence of ICE*clc* was determined (GenBank Accession number AJ617740).[14] The size of ICE*clc* is around 103 kb and is predicted to contain some 107 protein-encoding open reading frames (ORFs, Fig. 2). Megablast sequence comparisons of ICE*clc* result in a large number of consistent hits to other known genetic elements and genome regions, which suggested this family of elements consist of three characteristic parts (Fig. 3). The first part (with reference to ICE*clc*) is formed by the *intB13* integrase gene and is located directly near the *attR* end. The second part consists of a ~50 kb variable region, which in case of ICE*clc* contains the genes for chlorocatechol and 2-aminophenol (2-AP) metabolism (Fig. 2). The third part comprises the remainder ~50 kb region until *attL*, that is highly conserved but mostly composed of genes with unknown functions, apart from some that display homologies to conjugative processes (see

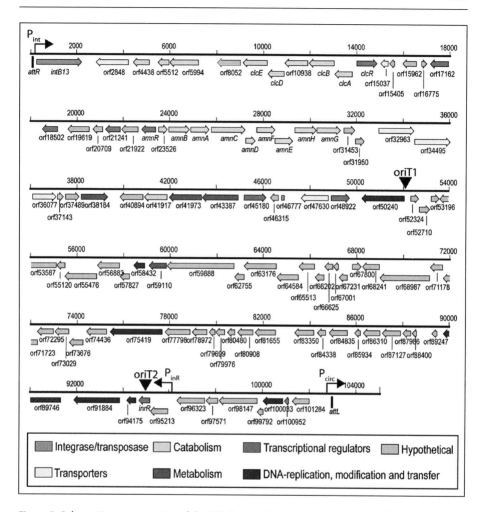

Figure 2. Schematic representation of the ICE*clc* genetic organization. Arrows indicate orientation of putative ORFs. Shades correspond to predicted functional classes. Thick vertical lines represent the border of ICE*clc*. A color version of this figure is available at www.landesbioscience.com/curie.

below) or to phages.[14] It was therefore hypothesized that this region is the 'back-bone' or 'core' region of ICE*clc* and its relatives (Fig. 3), and must encode the functions necessary for the element's life-style, i.e., excision, transfer and regulation of transfer.[14]

GEI/ICE Related to ICE*clc*

The family of elements having core regions highly similar to ICE*clc* mostly occurs in *Proteobacteria* (Table 1). The most similar element to ICE*clc* was discovered in the genome of *Burkholderia xenovorans* LB400, a strain able to degrade polychlorobiphenyls.[15] This ICE of 124 kb is basically identical to ICE*clc* (Fig. 4), the only differences being two regions which are absent in ICE*clc* and which carry the genes for *o*-halobenzoate degradation.[14] A second highly identical element to ICE*clc* was partially sequenced from a groundwater isolate of *Ralstonia*, in which a large insertion of a gene cluster for a monochlorobenzene dioxygenase had occurred.[16] Further 'catabolic' variants of ICE*clc* are found in e.g., *Acidovorax* sp strain JS42 (Fig. 5), in which the variable region contains genes for heavy metal resistance. This suggests that ICE*clc*-like elements

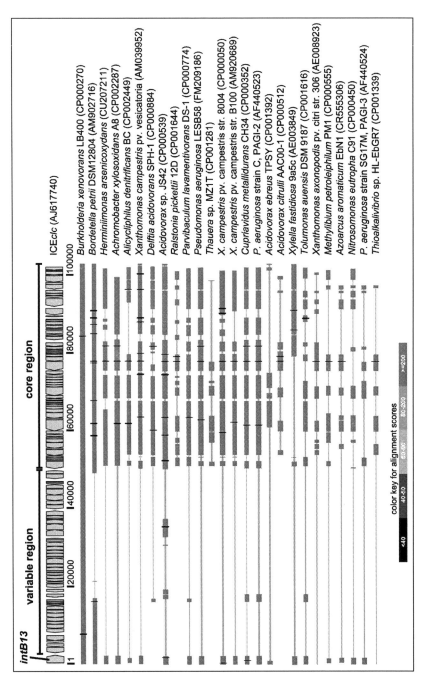

Figure 3. Megablast analysis of ICE*clc* and related members in GenBank (accessed in March 2011). Gray (color) shaded bars correspond to homologous regions with ICE*clc* (top row: gray (color) pentagons point to ORFs). Black lines indicate three characteristic regions of ICE*clc*. Small black vertical bars point to insertions in the subject compared with the ICE*clc* query sequence. Accession numbers are shown in parentheses.

Table 1. Characteristics of experimentally described GEI/ICE related to ICEclc

Name	Host	Size (kb)	Insertion Site	Accessory Features
ICEclc	Pseudomonas knackmussii	103	tRNA^{Gly}	Chlorocatechol and 2-amino-phenol degradation
GI1	Bordetella petrii	255	tRNA^{Gly}	Phthalate to protocatechuate metabolism
GI2	Bordetella petrii	143	tRNA^{Gly}	Aromatic compound metabolism
GI3	Bordetella petrii	102	tRNA^{Gly}	Chlorocatechol degradation
GI6	Bordetella petrii	159	tRNA^{Gly}	None known
ICEHin1056	Hemophilus influenzae	59	tRNA^{Leu}	Antibiotic, metal and antiseptic resistance
PAGI-2	Pseudomonas aeruginosa	105	tRNA^{Gly}	Complexation and transport of heavy metal
PAGI-3	Pseudomonas aeruginosa	103	tRNA^{Gly}	Metabolic functions and antibiotic resistance
PAPI-1	Pseudomonas aeruginosa	108	tRNA^{Lys}	Virulence factors, biofilm formation
CMGI-1	Cupriavidus metallidurans	109	tRNA^{Gly}	Heavy metal resistance
No name	Burkholderia xenovorans	124	tRNA^{Gly}	Chlorocatechol degradation

rapidly diverge in their variable gene regions but can contribute to distributing and rearranging a variety of catabolic functions within *Proteobacteria*. Interestingly, an ICE*clc* like element with *clc* genes also occurs in *Bordetella petrii* DSM12804, in which the element is named GI3. *B. petrii* harbors three other GEIs (named GI1, GI2 and GI6) with high similarities (42 to 98% nucleotide identity) to the 'core' region of ICE*clc* (Fig. 5). The GEIs of *B. petrii* seem to be 'active' and capable of excision because their free circular forms were detected, and transfer of GI3 from *B. petrii* to *B. bronchiseptica* PS2 was detected.[17] Further ICE*clc* relatives can be found in species, such as *Parvibaculum lavamentivorans, Herminiimonas arsenoxydans, Delftia acidovorans, Ralstonia pickettii, Xylella fastidiosa* or *Xanthomonas campestris* (Fig. 3).

One of the most abundant GEI families in *P. aeruginosa* is the PAGI-2/PAGI-3-type,[18] which can be found in more than 40% of clinical and environmental isolates.[19] PAGI-2 encodes proteins for the complexation and transport of heavy metals, while PAGI-3 carries metabolic functions and resistance capacities.[18] PAGI-2 and PAGI-3 are less highly related but still significantly homologous in their core region to ICE*clc* (Fig. 4). In contrast to ICE*clc* self-transfer of PAGI-2 and 3 has not been detected so far. Another genomic island in *P. aeruginosa* that shares homologous regions with ICE*clc* is PAPI-1 (Fig. 4), which encodes the cytotoxic protein ExoU and is thus classified as a pathogenicity island.[20] Excision of PAPI-1 in a population of PA14 cells and transfer of PAPI-1 between *P. aeruginosa* strains have been detected. Isolated from metallurgical sediment in Belgium, *Cupriavidus metallidurans* CH34 harbors CMGI-1, a GEI basically identical to PAGI-2 in terms of core region. CMGI-1 encoded core proteins have between 80 and 100% amino acid identities to corresponding ones in ICE*clc* (Fig. 4).[18] CMGI-1 is no longer mobile because of the interruption of its integrase gene by the transposon Tn*6049*.[21]

Finally, ICE*clc* is related to ICE*Hin*1056, an ICE carrying antibiotic resistance genes in *Hemophilus influenzae*.[22] Although the overall homology between ICE*Hin*1056 and ICE*clc* is

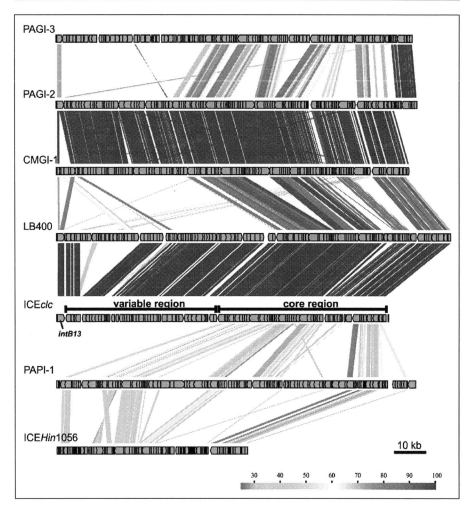

Figure 4. Comparison of a number of ICE*clc* related elements. Gray (color) pentagons point to ORFs. Gray (color) shades indicate the level of amino acids identity (%). Sequence sources are obtained from accession numbers shown in Figure 3.

relatively low and limited to some 30 kb within the core region with amino acid identities between 20 and 50% (Fig. 4), analysis of some mutants in this region of ICE*Hin*1056 showed that these genes encode part of the conjugative transfer system.[23] It is thus likely that the orthologous genes on ICE*clc* are responsible for the conjugative transfer as well. We can thus conclude that a large and widely variable family of GEI/ICE exists that bear significant similarities to ICE*clc* in terms of the core and integrase regions, which are implicated in transferring and distributing a wide variety of adaptive functions.

Integrase Mediates the Integration/Excision Reaction

As mentioned above, ICE*clc* is normally located in the host chromosome and inherited via chromosome replication and cell division. Therefore the integration and excision reactions are key steps deciding the state and possible transfer of ICE*clc*. Integration of ICE*clc* is mediated by IntB13 and occurs via site-specific recombination between the 18 bp of the 3′ end of a *tRNA^{Gly}*

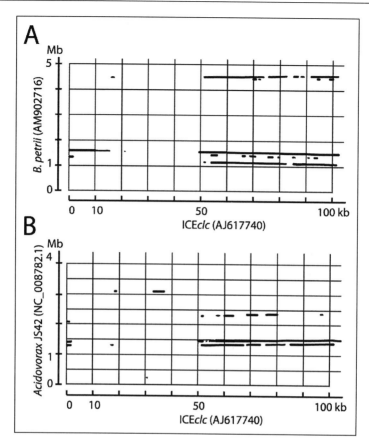

Figure 5. Genome scale comparison between ICE*clc* and *Bordetella petrii* DSM 12804 (A) or *Acidovorax* sp JS42 (B). Note the presence of multiple partial ICE*clc* core-like regions in the two genomes. Homologous regions are shown in black lines and dots. Accession numbers are shown in parentheses.

gene (*attB*) and a homologous sequence on ICE*clc* (*attP*). Generally *Proteobacteria* contain more than one *tRNA*[Gly] gene copy, which can lead to ICE*clc* inserting at different positions in the chromosome of the host.[24] On the other hand, not all 3′ 18-bp *tRNA*[Gly] gene sequences serve as efficient targets and the integration efficiency depends on the sequence homology between the 18 bp recombination sites, the presence of downstream structures or binding sites.[13,24] For instance, ICE*clc* integrated in four out of six *tRNA*[Gly] genes in *P. putida* UWC1 and three out of five in *C. necator*.[24] Although direct replication of ICE*clc* has never been observed, transconjugants in *P. putida* or *C. necator* can bear two or multiple tandem copies of ICE*clc*, probably as a result of the selection procedure for recovery of the transconjugants.[8,24,25] Even *P. knackmussii* B13 itself has 2 ICE*clc* copies.[11] The IntB13 integrase is essential for integration,[12] but strains in which the integrase gene is interrupted do not produce any excised ICE*clc* nor do they transfer the element by conjugation.[26] Thus, IntB13 integrase is also necessary for excision, a process which may be assisted by other (as yet unknown) proteins. Excised ICE*clc* can be detected by PCR or by hybridization as free circular form in which both *attL* and *attR* have recombined into *attP*.[11,27]

ICE*clc* excision and circularization lead to displacement of a strong promoter (P_{circ}) facing outwards from the left end of the element in the integrated form, directly in front of the integrase gene (Figs. 1, 6). Placement of P_{circ} in front of *intB13* during circularization and transfer may be important for the success of integration in a new cell as it may allow temporary overexpression of IntB13 from the circular form of ICE*clc* and increase the favorability of integration. Activity of the P_{circ} promoter can be detected with promoter-less hybrid *egfp*-containing integration sites.[24] Evidence has been obtained by using the PCR that excision and reintegration must also occur in growing cultures of strains containing ICE*clc*.[24] Expression of the *intB13* integrase in *E. coli* is strongly inhibitory to the cell's viability unless the gene is brought under very tight expression control, such as with the T7 promoter system.[12] In the integrated form expression of *intB13* is controlled from a promoter called P_{int}, the activity of which is limited to a few percent of cells in stationary phase (see above).[28] ICE*clc* excision is not stimulated by typical SOS response as a consequence of UV irradiation, nor in the presence of toxicants, or by heat stress.[28] No effects of cell density or quorum-sensing control on ICE*clc* activation have been discovered.[28]

Transfer System of ICE*clc*

ICE*clc* has been experimentally demonstrated to transfer from *P. knackmussii* to *Beta*- and *Gamma-proteobacteria*,[24,29] and ICE*clc*-family elements have been found in *Alpha*, *Beta* and *Gamma*-proteobacterial genomes (see below). Whereas other ICE families, such as ICE*SXT*, ICE*Ml*Sym^{R7A} or ICE*Bs1*,[2,30,31] display clearly identifiable transfer genes with similar sequence and organization to those of conjugative plasmids, ICE*clc* and its relatives show little sequence homology to known plasmid conjugative systems. Recently a first attempt to unravel parts of the ICE*clc* transfer mechanism was reported.[26] First of all, a functional screening revealed that ICE*clc* unlike conjugative plasmids and other ICEs carries two separate origins of transfer (*oriT*s), with different sequence context but containing a similar repeat motif. Conjugation experiments with eGFP-labeled ICE*clc* variants showed that both *oriT*s are used for transfer and with indistinguishable efficiencies. Frequency calculations suggested that having two *oriT*s increases ICE*clc* transfer rates by a factor of four compared with a situation of having a single *oriT*, which may have been a selective advantage for the appearance of an element with two *oriT*s. In fact, several GEIs related to ICE*clc* have conserved both *oriT* regions, underscoring their functional importance. A gene for a relaxase essential for ICE*clc* transfer was also identified (*orf50240*), which is acting on both *oriT*s but in a different manner.[26]

Weak but significant homology exists between ICE*clc* genes in the conserved core area and the proposed transfer system of ICE*Hin*1056, an ICE in *H. influenzae*.[22] Transfer deficient mutants of ICE*Hin*1056 have been produced that lacked pili formation, suggesting that ICE*Hin*1056 transfer (and that of ICE*clc*) is mediated by a type IV secretion system.[23] However, phylogenetic analysis showed those genes to be evolutionarily very distant from all previously described plasmid type IV secretion system genes.

Regulation of ICE*clc* Transfer

The first step in self-transfer of ICE*clc* requires the expression of the *intB13* integrase gene, resulting in excision of the element from the host chromosome. As explained above, *intB13* is transcribed from P_{int} in the integrated state, but—as shown by single copy P_{int}-*egfp* reporter fusions, in only a small proportion (3–5%) of individual cells in a population during stationary phase conditions of cells pregrown on 3-CBA (Fig. 6). Both the proportion of excised ICE*clc* in cultures as well as the ICE*clc* transfer rates correlated with the observed proportion of cells with induced P_{int}-*egfp* fusion in stationary phase, suggesting that indeed cells that activate P_{int} are those that excise and transfer ICE*clc*. eGFP expression was significantly higher in cultures that had been grown on 3-CBA and was basically absent in cultures grown on succinate or glucose. As mentioned above, no increase of eGFP expression from P_{int} was observed when cultures were treated with heat shock, osmotic shock, UV irradiation, or with alcohol.[28]

Bistable activation of P_{int} is a critical step in the onset of ICE*clc* transfer and it must be controlled by a mechanism generating or propagating bistability. At least one factor was found to be encoded on ICE*clc*, named InrR, that is important to maintain the 3–5% percent of integrase active cells in

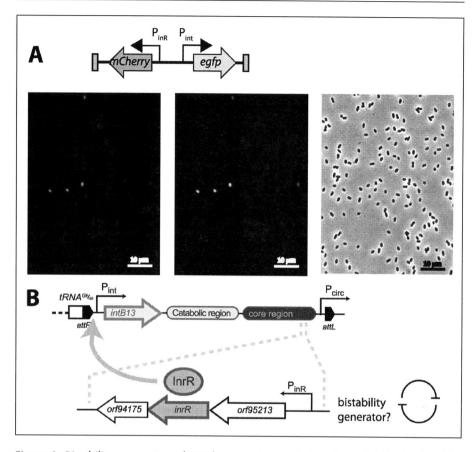

Figure 6. Bistability expression of ICEclc promoters in *P. knackmussii* B13 visualized by colocalized expression of eGFP and eCherry in stationary phase from P_{int}- and P_{inR}- promoters, respectively. A) Phase-contrast micrographs (right panel) at 1,000x magnification and corresponding epifluorescence images for eGFP (middle panel) or eCherry (left panel). B) Schematic outline of the signal transmission by InrR on the integrase promoter and the possible bistability generation at the *inrR* promoter.

stationary phase conditions.[32] Single and double deletion mutants of *inrR* in *P. knackmussii* B13 (because of two copies of ICE*clc* in B13) resulted in a decrease of ICE*clc* transfer rates as well as lower expression of eGFP from P_{int}.[33] Interestingly, transcriptional fusions of *egfp* or *eCherry* with the promoter of the *inrR* gene were also only measurably induced in a subpopulation of B13 cells during stationary phase, but deletions of *inrR* did not influence expression from P_{inR}. Moreover, time lapse imaging on strains labeled with both P_{int} and P_{inR} transcription fusions indicated that both markers are expressed in the same cell (Fig. 6) but that expression of *inrR* is preceding the expression of *intB13*.[33] This showed that bistability may in fact arise at the *inrR* operon and is then relayed by InrR to the integrase expression (Fig. 6B). Mapping the transcription start site of P_{inR} located a -10/-35 box closely resembling the consensus sequence for the stationary phase sigma factor and therefore suggesting control via RpoS. However, the mechanism of bistability generation at P_{inR} remains unknown.

By using several complementary transcriptome approaches, 15 transcripts were mapped in the ICE*clc* core region, most of which were concertedly upregulated during stationary phase on 3-CBA, but not on succinate or glucose.[34] This suggests that not only *intB13* but a complete program encoded on ICE*clc* is bistably induced. This program might consist of integrase expression, excision, formation of the transfer proficient ICE*clc* molecule and actual conjugation. The concerted timing of the program is reminiscent of a phage propagation program during the lytic phase.

Effects of ICE*clc* on the Host Cell

As far as known ICE*clc* is not an autonomous replicating element. In the integrated state, its survival and proliferation must therefore rely on the host replication and cell division machinery. This dependence suggests a host-ICE*clc* relationship that reflects both parasitism (i.e., replication, horizontal gene transfer apparatus) and commensalism (i.e., provision of catabolic pathways). Since it had been shown that acquisition of conjugative plasmids often results in initial fitness loss in the host,[35-37] the effect of ICE*clc* on host fitness was examined in more detail.[29] Hereto, ICE*clc* was transferred to a non-native host, *Pseudomonas aeruginosa* PAO1, and a single copy insertion transconjugant was examined. The selective advantage for PAO1 of having ICE*clc* is that it permits growth on 3-CBA or 2-aminophenol as sole carbon and energy source. Such selective benefits for growth carried by ICE*clc* are illustrated by bioreactor experiments in which spontaneous transconjugants of a *Ralstonia* species having acquired ICE*clc* outcompeted the original B13 donor strain during continuous supply of 3-CBA.[25] No fitness loss at a detection level of 1% difference was observed between PAO1-ICE*clc* and PAO1 during competitive growth on a non-selective substrate like succinate or acetoin.[29] Moreover, no differences in substrate utilization or stress resistance were found among a large number of tests in BIOLOG format. The only detectable significant difference was that PAO1-ICE*clc* formed thinner biofilms than PAO1 itself during growth on succinate, glucose, or acetate. Since the two strains did not significantly vary in growth rates and yields on those substrates, this implies that the difference of biofilm formation is a consequence of having ICE*clc*.

The transcriptomic profiles of the PAO1 common chromosomal genes of wild type and ICE*clc*-containing strain revealed none (in exponential phase) and < 1% (in stationary phase) significantly differentially expressed genes when grown on succinate as C- and energy-source. The differentially expressed genes in stationary phase were involved in processes with unknown ties to ICE*clc*-encoded functions, such as small molecule transport, acetoin and glycolate catabolism.[29]

Taken together, these experiments indicated that the fitness cost for the host of maintenance of ICE*clc* under non-selective conditions is less than 1%, which may be (another) one of the reasons for the wide distribution of this type of element. The mechanisms that lead to the absence of measurable fitness cost to the host cell may be a combination of ICE*clc* specific traits, such as the typical mode of site-specific integration into and restoration of the *tRNA*^{Gly} gene, the limited size of the element, a bistable expression affecting only 3–5% of the total cell population, and the possible lack of interferences between the chromosome expression network and that of ICE*clc*.

Conclusion

Several types of GEI/ICE have been discovered to date in both Gram-positive and Gram-negative bacteria. Although they have a common life-style as a large group of mobile genetic elements, in the sense of being able to integrate, excise and transfer, every type of element has unique features and mechanisms that must have evolved under the specific environmental or physiological circumstances it was exposed to.[2] From this point of view the unique traits for ICE*clc* seem, first, its bistable expression in stationary phase cells to prepare a small proportion of individuals for transfer. The advantage of doing this may lay in a reduced metabolic burden of expressing ICE*clc* in the population as a whole. Second, while excision occurs in only a small proportion of cells in a population, the dual *oriT* system may contribute to more efficient transfer ($\sim 10^{-2}$ per donor). Finally, stable maintenance of ICE*clc* is supported by a low fitness cost in new hosts and favored by using a widely conserved tRNA gene as an integration target.

Acknowledgments

Research on ICE*clc* was supported by grants from the Swiss National Science Foundation (3100A-108199, 31003A-124711). R.M. was supported by a postdoctoral fellowship from the Japan Society for the Promotion of Science for Research Abroad.

References

1. Juhas M, van der Meer JR. Gaillard M etal. Genomic islands: tools of bacterial horizontal gene transfer and evolution. FEMS Microbiol Rev 2009; 33:376-93. PMID:19178566 doi:10.1111/j.1574-6976.2008.00136.x
2. Wozniak RA, Waldor MK. Integrative and conjugative elements: mosaic mobile genetic elements enabling dynamic lateral gene flow. Nat Rev Microbiol 2010; 8:552-63. PMID:20601965 doi:10.1038/nrmicro2382
3. Dorn E, Hellwig M, Reineke W, Knackmuss H-J. Isolation and characterization of a 3-chlorobenzoate degrading Pseudomonad. Arch Microbiol 1974; 99:61-70. PMID:4852581 doi:10.1007/BF00696222
4. Frantz B, Chakrabarty AM. Organization and nucleotide sequence determination of a gene cluster involved in 3-chlorocatechol degradation. Proc Natl Acad Sci USA 1987; 84:4460-4. PMID:3299368 doi:10.1073/pnas.84.13.4460
5. Chatterjee DK, Kellogg ST, Hamada S, Chakrabarty AM. Plasmid specifying total degradation of 3-chlorobenzoate by a modified ortho pathway. J Bacteriol 1981; 146:639-46. PMID:7217013
6. Bruhn C, Bayly RC, Knackmuss H-J. The in vivo construction of 4-chloro-2-nitrophenol assimilatory bacteria. Arch Microbiol 1988; 150:171-7 doi:10.1007/BF00425158.
7. Oltmanns RH, Rast HG, Reineke W. Degradation of 1,4-dichlorobenzene by constructed and enriched strains. Appl Microbiol Biotechnol 1988; 28:609-16 doi:10.1007/BF00250421.
8. Ravatn R, Zehnder AJB, van der Meer JR. Low-frequency horizontal transfer of an element containing the chlorocatechol degradation genes from Pseudomonas sp. strain B13 to Pseudomonas putida F1 and to indigenous bacteria in laboratory-scale activated-sludge microcosms. Appl Environ Microbiol 1998; 64:2126-32. PMID:9603824
9. Chatterjee DK, Chakrabarty AM. Genetic homology between independently isolated chlorobenzoate-degradative plasmids. J Bacteriol 1983; 153:532-4. PMID:6294059
10. Weisshaar M-P, Franklin FCH, Reineke W. Molecular cloning and expression of the 3-chlorobenzoate-degrading genes from Pseudomonas sp. strain B13. J Bacteriol 1987; 169:394-402. PMID:3025183
11. Ravatn R, Studer S, Springael D, et al. Chromosomal integration, tandem amplification, and deamplification in Pseudomonas putida F1 of a 105-kilobase genetic element containing the chlorocatechol degradative genes from Pseudomonas sp. strain B13. J Bacteriol 1998; 180:4360-9. PMID:9721270
12. Ravatn R, Studer S, Zehnder AJB, et al. Int-B13, an unusual site-specific recombinase of the bacteriophage P4 integrase family, is responsible for chromosomal insertion of the 105-kilobase clc element of Pseudomonas sp. strain B13. J Bacteriol 1998; 180:5505-14. PMID:9791097
13. van der Meer JR, Ravatn R, Sentchilo VS. The clc element of Pseudomonas sp. strain B13 and other mobile degradative elements employing phage-like integrases. Arch Microbiol 2001; 175:79-85. PMID:11285744 doi:10.1007/s002030000244
14. Gaillard M, Vallaeys T, Vorholter FJ, et al. The clc element of Pseudomonas sp. strain B13, a genomic island with various catabolic properties. J Bacteriol 2006; 188:1999-2013. PMID:16484212 doi:10.1128/JB.188.5.1999-2013.2006
15. Denef VJ, Klappenbach JA, Patrauchan MA, et al. Genetic and genomic insights into the role of benzoate-catabolic pathway redundancy in Burkholderia xenovorans LB400. Appl Environ Microbiol 2006; 72:585-95. PMID:16391095 doi:10.1128/AEM.72.1.585-595.2006
16. Müller TA, Werlen C, Spain J, van der Meer JR. Evolution of a chlorobenzene degradative pathway among bacteria in a contaminated groundwater mediated by a genomic island in Ralstonia. Environ Microbiol 2003; 5:163-73. PMID:12588296 doi:10.1046/j.1462-2920.2003.00400.x
17. Lechner M, Schmitt K, Bauer S, et al. Genomic island excisions in Bordetella petrii. BMC Microbiol 2009; 9:141. PMID:19615092 doi:10.1186/1471-2180-9-141
18. Larbig KD, Christmann A, Johann A, et al. Gene islands integrated into tRNAGly genes confer genome diversity on a Pseudomonas aeruginosa clone. J Bacteriol 2002; 184:6665-80. PMID:12426355 doi:10.1128/JB.184.23.6665-6680.2002
19. Klockgether J, Würdemann D, Reva O, et al. Diversity of the abundant pKLC102/PAGI-2 family of genomic islands in Pseudomonas aeruginosa. J Bacteriol 2007; 189:2443-59. PMID:17194795 doi:10.1128/JB.01688-06

20. He J, Baldini RL, Deziel E, et al. The broad host range pathogen Pseudomonas aeruginosa strain PA14 carries two pathogenicity islands harboring plant and animal virulence genes. Proc Natl Acad Sci USA 2004; 101:2530-5. PMID:14983043 doi:10.1073/pnas.0304622101

21. Janssen PJ, Van Houdt R, Moors H, et al. The complete genome sequence of Cupriavidus metallidurans strain CH34, a master survivalist in harsh and anthropogenic environments. PLoS ONE 2010; 5:e10433. PMID:20463976 doi:10.1371/journal.pone.0010433

22. Mohd-Zain Z, Turner SL, Cerdeño-Tárraga AM, et al. Transferable antibiotic resistance elements in Haemophilus influenzae share a common evolutionary origin with a diverse family of syntenic genomic islands. J Bacteriol 2004; 186:8114-22. PMID:15547285 doi:10.1128/JB.186.23.8114-8122.2004

23. Juhas M, Crook DW, Dimopoulou ID, et al. Novel type IV secretion system involved in propagation of genomic islands. J Bacteriol 2007; 189:761-71. PMID:17122343 doi:10.1128/JB.01327-06

24. Sentchilo V, Czechowska K, Pradervand N, et al. Intracellular excision and reintegration dynamics of the ICEclc genomic island of Pseudomonas knackmussii sp. strain B13. Mol Microbiol 2009; 72:1293-306. PMID:19432799 doi:10.1111/j.1365-2958.2009.06726.x

25. Springael D, Peys K, Ryngaert A, et al. Community shifts in a seeded 3-chlorobenzoate degrading membrane biofilm reactor: indications for involvement of in situ horizontal transfer of the clc-element from inoculum to contaminant bacteria. Environ Microbiol 2002; 4:70-80. PMID:11972616 doi:10.1046/j.1462-2920.2002.00267.x

26. Miyazaki R, van der Meer JR. A dual functional origin of transfer in the ICEclc genomic island of Pseudomonas knackmussii B13. Mol Microbiol 2011; 79:743-58. PMID:21255116 doi:10.1111/j.1365-2958.2010.07484.x

27. Sentchilo V, Zehnder AJB, van der Meer JR. Characterization of two alternative promoters and a transcription regulator for integrase expression in the clc catabolic genomic island of Pseudomonas sp. strain B13. Mol Microbiol 2003; 49:93-104. PMID:12823813 doi:10.1046/j.1365-2958.2003.03548.x

28. Sentchilo V, Ravatn R, Werlen C, Zehnder AJB, van der Meer JR. Unusual integrase gene expression on the clc genomic island of Pseudomonas sp. strain B13. J Bacteriol 2003; 185:4530-8. PMID:12867462 doi:10.1128/JB.185.15.4530-4538.2003

29. Gaillard M, Pernet N, Vogne C, et al. Host and invader impact of transfer of the clc genomic island into Pseudomonas aeruginosa PAO1. Proc Natl Acad Sci USA 2008; 105:7058-63. PMID:18448680 doi:10.1073/pnas.0801269105

30. Auchtung JM, Lee CA, Monson RE, et al. Regulation of a Bacillus subtilis mobile genetic element by intercellular signaling and the global DNA damage response. Proc Natl Acad Sci USA 2005; 102:12554-9. PMID:16105942 doi:10.1073/pnas.0505835102

31. Sullivan JT, Trzebiatowski JR, Cruickshank RW, et al. Comparative sequence analysis of the symbiosis island of Mesorhizobium loti strain R7A. J Bacteriol 2002; 184:3086-95. PMID:12003951 doi:10.1128/JB.184.11.3086-3095.2002

32. van der Meer JR, Sentchilo VS. Genomic islands and evolution of catabolic pathways in bacteria. Curr Opin Biotechnol 2003; 14:248-54. PMID:12849776 doi:10.1016/S0958-1669(03)00058-2

33. Minoia M, Gaillard M, Reinhard F, et al. Stochasticity and bistability in horizontal transfer control of a genomic island in Pseudomonas. Proc Natl Acad Sci USA 2008; 105:20792-7. PMID:19098098 doi:10.1073/pnas.0806164106

34. Gaillard M, Pradervand N, Minoia M, et al. Transcriptome analysis of the mobile genome ICEclc in Pseudomonas knackmussii B13. BMC Microbiol 2010; 10:153. PMID:20504315 doi:10.1186/1471-2180-10-153

35. Dahlberg C, Chao L. Amelioration of the cost of conjugative plasmid carriage in Escherichia coli K12. Genetics 2003; 165:1641-9. PMID:14704155

36. Lee SW, Edlin G. Expression of tetracycline resistance in pBR322 derivatives reduces the reproductive fitness of plasmid-containing Escherichia coli. Gene 1985; 39:173-80. PMID:3005111 doi:10.1016/0378-1119(85)90311-7

37. Nguyen TNM, Phan QG, Duong LP, et al. Effects of carriage and expression of the Tn10 tetracycline-resistance operon on the fitness of Escherichia coli K12. Mol Biol Evol 1989; 6:213-25. PMID:2560115

CHAPTER 17

Pathogenicity Island Evolution:
A Distinct New Class of Integrative Element or a Mosaic of Other Elements?

Michael G. Napolitano and E. Fidelma Boyd*

Abstract

Pathogenicity islands are large chromosomal regions that encode multiple virulence genes along with the genes required for site specific integration and excision from the host chromosome. Among Gamma-Proteobacteria pathogens, PAIs are highly prevalent as measured by the presence of a tyrosine recombinase (TR) integrase flanking regions encoding multiple virulence factors. Within the genus *Vibrio*, PAIs are present in all species examined and usually the only conserved feature among these regions is the presence of a TR integrase, its cognate recombination directionality factor/excisionase and insertion at a tRNA locus. Phylogenetic analysis of TR integrases from PAIs and phages demonstrate that each element forms a distinct cluster separate from one another indicating different evolutionary histories.

Original Definition of a Pathogenicity Island (PAI)

As with other mobile and integrative genetic elements (MIGEs), the definition of a pathogenicity island is in a constant flux, changing due to the discovery of new elements that lay on the edge of previously defined elements in term of their overall characteristics. This is further complicated by the fact that to date there has been no identified marker that is unique and universal for a pathogenicity island as has been found for other mobile elements. For example, bacteriophages are defined by their ability to encapsulate their genetic material, plasmids by their ability to self replicate, and conjugative transposons by their ability for conjugal transfer and integration into the host chromosome (Fig. 1). In addition, the tendency of many different MIGEs to integrate into the same chromosomal loci such as tRNAs can result in two different regions at the same attachment site leading to misidentification.[1-4] Hacker, along with Kaper, and others in a series of papers beginning in 1997 laid out a set of criterion that most, if not all, pathogenicity islands (PAIs) identified at the time fell into.[5-7] These include:

- The candidate pathogenicity island consists of a large (> 10 kb) region of contiguous DNA that is present in pathogenic isolates and absent from non-pathogenic isolates.
- The pathogenicity island encodes gene products that provide a benefit to the host genome when expressed, such as virulence factors. Despite the name "Pathogenicity Island," these encoded genes do not have to be virulence genes, but can include genes for the metabolism of alternative carbon and nitrogen sources or other genes that give a bacterial cell a fitness advantage in a new niche.

*Department of Biological Sciences, University of Delaware, Newark, Delaware, USA.
Corresponding Author: E. Fidelma Boyd—Email: fboyd@udel.edu

Bacterial Integrative Mobile Genetic Elements, edited by Adam P. Roberts and Peter Mullany.
©2013 Landes Bioscience.

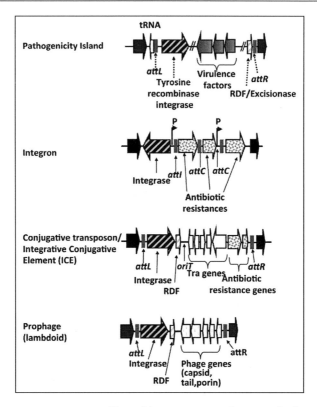

Figure 1. Schematic representation of four of the major types of MIGEs. Block arrows represent genes encoding proteins. Black arrows represent core proteins presents in all strains of a species, these flank regions acquired by a subset of strains within a species. *att* = attachment site, RDF = recombination directionality factor. Note the different type of cargo carried on each element—as a good first rule elements carry the type of gene described.

- The candidate island has a percent guanine plus cytosine (G+C) content and/or codon bias that is significantly different from the host chromosome, which is indicative of horizontal gene transfer.
- The island encodes an integrase or transposase that allows for integration into the host chromosome.

However, these features/characteristics of the conventional definition of a PAI are very broad and can also apply to a number of different MIGEs and thus many elements were named pathogenicity islands even thought they are clearly another type of MIGE (Table 1). Thus, we believe that additional features should be taken into account before a definition of a PAI is applied.

Key additional features of PAIs:

- The candidate island should contain an integrase that allows for site specific recombination at an attachment (*att*) site usually associated with a tRNA locus.
- Phylogenetic analysis of the integrase should demonstrate that it clusters with PAIs and not other MIGEs.
- The candidate island should also contain a recombinational directionality factor (RDF)/ excisionase that partners with the integrase allowing for excision of the region.
- The candidate island should not have genes that are consistent with another form of mobile element, such as a phage tail, coat proteins, or Tra/Conjugation genes. Likewise, it should not replicate when excised from the genome.

Table 1. General features of mobile and integrative genetic elements

	PAIs	Integrons	ICEs	Prophages
Site-specific integration	YES	YES	YES	YES
Integration site	tRNA	Varies	Varies	Varies
Excision	YES	YES	YES	YES
Self-transfer	NO	YES	YES	YES
Self-replication	NO	NO	NO	YES

To explain further our reasoning for narrowing the definition of pathogenicity islands, it is first necessary to examine in more detail how the definition of a PAI has evolved, starting with the first set of papers published on what became known as PAIs.

The Discovery and Elucidation of the First PAIs

As early as 1983 in the pre-genomic sequencing era, researchers working with uropathogenic *Escherichia coli* had noticed that certain strains had close linkages between virulence factors (the hemolysin and the hemagglutinin) encoded on the chromosome, and that these regions were unstable and co-excised.[8,9] Moreover, these regions did not appear to be associated with any known mobile element. Knapp and colleagues identified a 75-kb region and located flanking repeats and determined that loss of the region affected not only the hemolytic phenotype but also a reduction in serum resistance and the loss of mannose-resistant hemagglutination caused by the presence of S-type fimbriae (sfa).[9,10] Additional research on various strains of uropathogenic *E. coli* showed that this spontaneous deletion of virulence factors was not uncommon, by 1988 another region nearly 60 kb in length had been discovered in *E. coli* 20025, a "cluster of urovirulence determinants.[11]" These elements were generically termed as virulence gene clusters, and when it was shown that several of these virulence gene clusters can co-excise as a large unit, Hacker and colleagues called these regions Pathogenicity DNA islands, eventually shortened to Pathogenicity Islands.[12] At the time, the main definition of a PAI was simply that it had several virulence genes in genetic linkage that deleted in unison and lacked the signatures of other mobile elements. Quickly additional elements were discovered, including the Locus of Enterocyte Effacement (LEE) in enteropathogenic *E. coli*, the *Salmonella* Pathogenicity Islands (SPIs) in *S. enterica*, the *Yersinia* High Pathogenicity Island (HPI) in *Yersinia pestis*, and the Cag pathogenicity island in *Helicobacter pylori*.[13-19] As these elements carried large numbers of genes involved in various bacterial functions most noticeable virulence and had signatures of acquisition by horizontal gene transfer, Groisman and Ochman,[20] coining the term "evolution in quantum leaps" to highlight the large and dramatic changes in bacterial phenotype caused by the acquisition and loss of PAIs.

Pathogenicity Island Encoded Virulence Factors

PAI encoded genes are involved in a myriad of different functions required for microbial pathogenesis beginning with bacterial cell attachment to their target host cell, and in the case of intracellular pathogens, uptake, survival and replication within the host cell, with many of these steps requiring the alteration of host cell signaling pathways (Fig. 2). For example, uropathogenic *E. coli* (UPEC) pathovars, several PAIs encode Type I and Type IV pili and various adhesins required for bacterial cell attached to host cells (Table 2). In general *E. coli* PAIs can encode a range of toxins, metabolic pathways, in addition to siderophore synthesis and iron uptakes systems (Table 2). Some *E. coli* pathovars and all *Salmonella enterica* isolates contain PAIs that encode type three secretion systems (T3SSs), sometimes called injectisome that translocate effector proteins directly into the eukaryotic cell (Fig. 2). In enterohemorrhagic

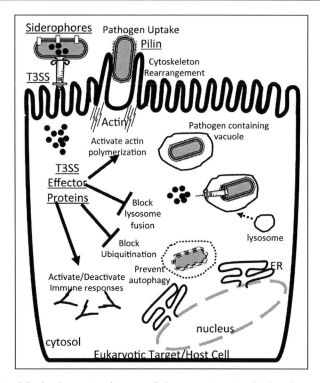

Figure 2. A simplified schematic of some of the steps in microbial pathogenesis and the virulence genes encoded by PAI, which are underlined. The function of the type 3 secretion system (T3SS) injectisome is the translocation of effectors into the host target cell cytosol. The initial attachment of the bacterial cell to the host cell requires a range of pili and adhesins, many of which are encoded on PAIs. Both T3SSs and their effectors are encoded on a range of PAIs. Within the host cell the effector proteins can have a range of effects: Effectors can cause actin microfilament formation and cytoskeleton rearrangement, which results in pedestal formation and subsequent pathogen internalization. Effectors can prevent lysosome fusion to pathogen containing vacuole (PCV), prevent autophagy, can cause activation and deactivation of the host immune responses. Once within the PCV, a T3SS can then translocate effectors into the cell cytosol to elicit a different set of responses that aid in survival and pathogen replication.

E. coli (EHEC), a T3SS is presence within the LEE island that is required for translocation of the Tir protein into the host cell that is subsequently involved in EHEC internalization (see ref. 21 for a review of this topic). In *S. enterica* two T3SSs are present, each on a PAI named Salmonella Pathogenicity Island-1 (SPI-1) and SPI-2.[13] Non-choleragenic *V. cholerae* isolates that cause gastroenteritis encode a T3SS that is absent from the toxigenic *V. cholerae* serogroup O1 and O139 isolates.[22,23] The T3SS genes in *V. cholerae* are within the VPI-2 region that encodes the genes required for sialic acid scavenging, transport and catabolism.[24] These genes are also present in toxigenic isolates within VPI-2 but the 5′ region of the island is replaced by the T3SS in pathogenic *V. cholerae* non-O1/non-O139 serogroup isolates.[24] In pathogenic isolates of *V. parahemolyticus* that cause an inflammatory diarrhea two T3SSs are present, one on each chromosome.[25] The T3SS on chromosome I named T3SS-1 is present in all isolates while the copy on chromosome II named T3SS-2 is only present in pathogenic isolates.[25] In *V. parahemolyticus* RIMD2210633, T3SS-2 is encoded within an 80 kb island that is flanked by transposase genes named *V. parahemolyticus* island-VII (VPaI-VII).[26] In *V. parahemolyticus* isolates, two distinct

Table 2. ORF content of PAIs from two pathovars of Escherichia coli

PAI	Virulence Factors Identified
CFT073 (O6:K2:H1) Uropathogenic Escherichia coli *(UPEC)*	
PAI$_{CFT073}$	3-oxoacyl-synthase I, II,- reductase, acyl carrier, MchBCD, Microcin, hemo receptor, FepA, F1C, Fimbriae (sfa), Iro, RadC
PAI-VII$_{CFT073}$	PilV, type IV pilin protein, ENDO3c
PAI-IV$_{CFT073}$	Salicylate synthase, ABC transporter, Yesiniabactin
PAI$_{CFT073}$	Thioesterase, polyketide synthases, peptide synthases, transacylase, Acetyl CoA dehydrogenase
PAI-I$_{CFT073}$	Pap fimbriae, ShiA, HemolysinCABD
PAI$_{CFT073}$	ShiA, Glucosidases, Propanol dehydrogenase, Propionate kinase, CpcA, EutJ, EutN, EutE, Maturases, SAM synthtase, RadC
PAI-II$_{CFT073}$	Pap fimbria, PgtABCP, RadC
PAI$_{CFT073}$	Helicases, SAM transferase
EDL933 (O157:H7) Enterohemorrhagic E. coli *(EHEC)*	
PAI$_{EDL933}$	Helicases; Urease, Colicin and Tellurite resistance; Enterobactin, RadC
PAI$_{EDL933}$	Helicases; Urease, Colicin and Tellurite resistance; Enterobactin, RadC
PAI$_{EDL933}$	PagC, Entertoxin, Cytoxins
PAI$_{EDL933}$	LEE region Type 3 Secretion System (T3SS)
PAI$_{EDL933}$	Resolvase, Helicases,

phylotypes of T3SS-2 are known depending on the isolate examined, T3SS-2a or T3SS-2b.[27] In *V. cholerae*, the T3SS shows closer homology to T3SS-2 than T3SS-1 from *V. parahemolyticus* and similarly, depending on the strain examined two different phylotypes are found within VPI-2 in *V. cholerae*.[24] More recently it was shown that the VPI-2 region present in *V. mimicus* pathogenic isolates also contains a T3SS-2 homolog and again depending on the strain, either T3SS-2a or T3SS-2b is present.[28-30] In addition to the diversity of T3SSs present within PAIs in pathogenic species, the number and type of proteins (named effector proteins) translocated by these systems varies between and within species (see Dean[31] for a review of effector structure and function among bacterial pathogens). In *E. coli* and *S. enterica* the best studied systems to date, up to a hundred different effector proteins have been identified (see reviews by refs. 32,33). The effector proteins can be encoded within the same region that contains its cognate T3SS or the effectors can be dispersed on the chromosome within prophages or other PAIs. Thus, the repertoire of effectors varies from strain to strain and also results in redundancy of functions encoded by these genes. The effectors proteins have a range of effects within the eukaryotic cell such as activation and deactivation of cytoskeleton rearrangement, preventing lysosome fusion with a pathogen containing vacuole, inhibiting ubiquitination, and activation and deactivation of host immune responses (Fig. 2). All these functions aid in bacterial uptake, survival, replication and intra and intercellular transfer within their target cell. For example, in the case of *S. enterica* serovar Typhimurium the effectors secreted by T3SS-1 encoded on SPI-1 are required for host cell invasion by triggering membrane ruffling and internalization within a vacuole. This *Salmonella* containing vacuole (SCV) then secretes effector proteins into the host cell cytosol to modulate intracellular survival (see Ibarra and Steele-Mortimer[34] for a review of this topic). Both SPI-2 encoded T3SS-2 and SPI-1 encoded T3SS-1 are involved in translocating proteins from the SCV into the host cell cytosol to aid in bacterial cell survival and replication.[34]

The PAI Concept is Broadened

The late 1990s into the 2000s were a boom time for the identification of not just pathogenicity islands (PAIs), but mobile genetic elements in general as a consequence of the exponential increase in whole genome sequencing. During this time period, PAIs in many *Vibrio* species were uncovered, several additional PAIs in *E. coli* as well as additional PAIs in *S. enterica*, the *M. loti* symbiosis island, and the *S. aureus* Pathogenicity islands (SaPIs) were all defined.[26,28,29,35-41] Most of these PAIs were identified by bioinformatic approaches using available whole genome sequence data.[42-44] Indeed a number of bioinformatic tools were developed to uncover islands, that is regions acquired by horizontal transfer which included phages, ICEs, integrons and PAIs, in newly sequenced bacterial genomes, (see Brinkman and colleagues[45] for a comprehensive review of this topic). Moreover, there was a shift toward a looser definition of the term PAI, with a movement away from a focus on pathogenicity toward a recognition that the island may have other, non-pathogenicity, fitness benefits (the terms genomic Island or fitness island, were coined around this time).[6,46] Presently, the term genomic/pathogenicity island is now more synonymous with the generic term mobile genetic element than having a precise specific meaning for a group of novel elements.

The first steps toward a hypothesis for the understanding of the evolution of the PAIs themselves was taken in a set of reviews proposing a degenerative model for the creation of pathogenicity islands (Fig. 3A). This hypothesis suggests that PAIs represent genetic driftwood, former phages or plasmids that have lost most of the genes once on the island involved in mobilization and transfer, and are slowly on the way to becoming a part of the genetic backbone (Fig. 3A).[7,46,47] Although the degenerative hypothesis is attractive, it has some drawbacks. PAIs cannot be degenerative prophages because all known bacterial virulence genes encoded within phages only carry a single virulence gene, rarely multiple virulence genes, and certainly never whole gene clusters required for complete pathways such as T3SSs, pilin gene cluster, siderophore synthesis etc.[48,49] The fact that PAIs use the same mechanism of site specific integration into the host as other elements may have wrongly suggested a common origin. In some cases, PAIs were misnamed and are in fact phages, ICEs or plasmids. The naming of previously described cryptic/satellite phages as PAIs has muddled and blurred the boundary between a PAI and a phage. *Staphylococcus aureus* is a Gram-positive bacterium that can induce a variety of diseases in humans and animals such as toxic shock syndrome, which results from production of toxic shock syndrome toxin-1 (TSST-1). TSST-1, a potent superantigen, is encoded by *tst*, which is found at two different sites on the chromosome and was initially believed to be part of a transposon. Subsequently, Novick's group demonstrated that in *S. aureus* strain RN4282, *tst* is found within a 15-kb genetic element which they named *S. aureus* pathogenicity island (SaPI).[37] They demonstrate that the generalized transducing phages 13 and 80φ can encapsidate and transduce the SaPI1 to recipient strains, where it integrates. SaPI1 is induced to excise and replicate specifically by 80φ {Lindsay, 1998 #788; Ruzin, 2001 #482}. We suggested that SaPI-1 has more phage-like characteristics than PAI features such as the presence of an integrase that is homologous to the staphylococcal bacteriophage φ PVL integrase, insertion into the *tyrB* gene and not a tRNA locus and has a 31% G/C content which is similar to the *S. aureus* chromosome.[37,50,51] In addition, a recent study demonstrates that SaPI gene regulation is controlled by a lambdoid Cro/CI-like repressor again suggesting a phage ancestry.[52] Since the initial characterization of SaPI-1 a large number of similar SaPIs have been identified among *S. aureus* isolates that share significant sequence identity across their genome.[53,54] SaPIs in general differ from one another by the virulence genes encoded, single toxins and superantigens similar to other phage-encoded bacterial virulence factors. Recently, Novick has proposed that these regions, as we had previously suggested, are defective/satellite phages which require a helper phage for morphogenesis.[54] To the best of our knowledge, PAIs similar to those found in *E. coli, S. enterica* and *Vibrio* species have not been identified in Gram-positive bacteria.

In other cases, the PAI is a mosaic, the result of multiple genetic elements integrating at the same chromosomal attachment site giving a Russian doll effect (Fig. 3B). For example, Schubert and colleagues identified the presence of multiple MIGEs at one insertion site in *E. coli* strain ECOR31 the High Pathogenicity Island (HPI) and an ICE. (Fig. 3B).[55] Initially, they proposed

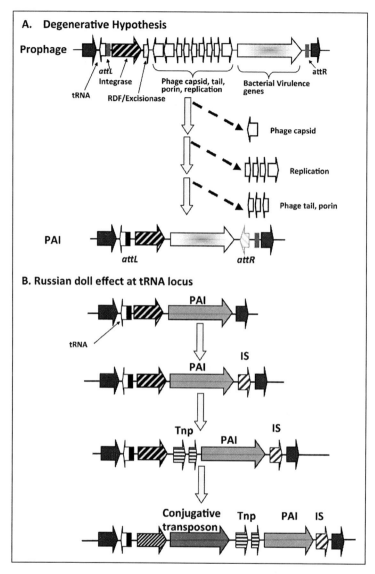

Figure 3. Hypothetical evolutionary scenarios for PAIs development. A) The degenerative hypothesis predicts that PAIs are degenerative forms of prophages or other MIGEs. Over time these MIGEs lose key feature such as phage structural and transfer genes to give rise to a PAI. However two key problems with this hypothesis are that phage never encode multiple virulence genes generally a single toxin is present, second the integrase gene found on PAIs is not related to phage integrases. B) Russian doll effect to explain why some PAIs may have features of other MIGEs. Many investigators have shown that different MIGEs can and do integrate into the host genome at the same site leading to a region with feature of two or more MIGEs.

that the High Pathogenicity Island (HPI) region, which was identified in a number of *E. coli* strains, was formed from a conjugative transposon/ ICE to explain the scattered distribution of the HPI among different strains and species. HPI was first identified on the chromosome of *Y. pestis, Y. pseudotuberculosis*, and *Y. enterocolitica* biovar 1B.[55-57] HPI contains the genes involved in yersiniabactin synthesis, a siderophore required for iron acquisition.[56] Schubert and colleagues found that in one strain of *E. coli* ECOR31 HPI is self mobilizable and they demonstrated transfer between different strains.[55] Closer examination of the HPI region in this single strain ECOR31 showed that it was twice as large as other known HPIs. In ECOR 31 the HPI contained the genes for a mating pair system, pilus assembly, and a DNA processing system resembling the ICE conjugative system,[55] in addition to previously known contents of HPI. It was proposed that HPI in ECOR31 resembles a conjugative transposon which they named ICE*Ec*1 and was the progenitor that gave rise to the *Yersinia* HPI.[55] From our work examining the phylogeny of integrases belonging to the tyrosine recombinase family, the ECOR31 HPI integrase does not cluster with integrases identified in HPIs from other *Yersinia* species or other *E. coli* isolates. We suggest that an ICE integrated into the host chromosome at the same attachment site as the HPI in ECOR 31 displacing the original HPI integrase. The HPI region in ECOR 31 can "hitchhike" along with the ICE to allow transfer between strains. It is important to note at this point that the mechanism of transfer for most PAIs has not been identified to date. The PAI regions described in *E.coli, S. enterica, Vibrio* species, *H. pylori* and *Yersinia* species do not encode self mobility genes and are probably co-mobilized with other mobile gene elements.

One would expect under a degenerative hypothesis for the origin of PAIs that the genes encoded on an island would be intermixed with genes from phages, plasmids, or ICEs. In fact, the only genes found in common across all known PAIs are the integrase and its cognate RDF.[4,58] That there are only a few exceptions that are found can be explained by the fact that many PAIs integrate into the host chromosome using site specific recombination catalyzed by the encoded integrase – without an integrase the island is an evolutionary dead end, and the few PAIs that lack integrases often "borrow" the excision machinery of other elements. The sites of attachment are highly conserved DNA sequences and thus not all regions of the chromosome are equally likely to act as integration sites. Many different MIGEs integrate at tRNA genes, which are highly conserved within and between species relative to most other genes.[1-4] For example, a recent study by Touchon and colleagues of pathogenic *E. coli* genomes showed that there are conserved chromosomal hotspots for the insertion of horizontally acquired DNA and different MIGEs are present within these hotspots depending on the strain examined.[59] Thus, the presence of multiple MIGEs at one site is possible leading to a Russian doll like effect, where several MIGEs can integrate into the same place or into each other, stacking and causing recombination events that muddle the phylogenetic history of the elements (Fig. 3B).

An Alternative Hypothesis of Pathogenicity Island Evolution

Instead of the mixed lineage suggested by the simple degeneration hypothesis, recent work in our group provides evidence that PAIs are phylogentically more related to one another than to other MIGEs.[4,58] This conclusion is based on phylogenetic analysis of the recombination module within each PAI, specifically the integrase. The recombination module consists of an integrase that catalyzes the integration event, attachment sites (*attP*), and a recombination directionality factor/excisionase that is required for excision along with the integrase. Tyrosine recombinases are the predominant integrases found among PAIs in Gamma-Proteobacteria, with a few rare exceptions involving serine recombinases. We used the *Vibrio cholerae* Vibrio Pathogenicity Island-2 (VPI-2) encoded integrase, IntV2 (VC1758 Accession NP_231394), as a probe to search the databases for homologs and identified 168 IntV2-like integrases.[4] Interrogation of the genomic context of these integrase found that all of the regions associated with these integrases fell within the PAI definition and did not encode features of other MIGEs. Phylogenetic comparison of these 168 integrases with several dozen other tyrosine recombinase (TR) integrases from phages, ICEs, or other MIGEs demonstrated that all integrases encoded on PAIs formed a distinct lineage from other MIGE integrases. Two exceptions where noted, as an integrase from phage Sf6, and P4 a satellite phage clustered in within the PAI integrases.[4] None of the PAI integrases examined clustered within the phage integrase family.

In a more recent study, our group examined the phylogenetic relationships among PAI and phage integrases encoded within a group of *E. coli* strains whose whole genome sequence was completely annotated.[58] We identified and downloaded all the protein sequences of untruncated integrases from the tyrosine recombinase family encoded within five pathogenic *E. coli* strains 536, CFT073, APEC O1, O157:H7 Sakai, and O157:H7 EDL933 in the genome database. The uropathogenic strain *E. coli* 536 genome contained seven integrases associated with seven regions that were all previously described as PAIs and each PAI encoded multiple virulence factors (PAI-I$_{536}$-PAI-VII$_{536}$) (Table 3).[38,60] The PAIs range in size from 22 to 106 kb, and their GC content from 38% to 55%; compared with the GC content of *E. coli* 536 genome of 51%. Six of the integrases are directly downstream of a tRNA locus, and within one PAI an integrase is present both at the 5′ end and at the 3′ end of the island. The UPEC strain CFT073 genome contains 12 integrases, 4 integrases within prophages. The additional eight integrases are encoded on PAIs, which range in size from 16 kb to 128 kb, with their GC content ranging from 38% to 56% (Table 3). Six of the integrases are directly adjacent to a tRNA locus, and for two PAIs, the integrases mark the beginning of the PAIs while the tRNA loci mark the end. The avian pathogenic strain APEC O1 encodes 12 integrases also, 9 are within prophages, and 3 within PAIs that integrated at tRNA loci. In strain O157:H7 Sakai, 18 integrases are present, 13 within prophages and 5 within PAIs. The PAIs range in size from 86 kb to 23 kb and all had GC content below 48% and inserted within a tRNA loci (Table 3). Nineteen integrases were identified in *E. coli* O157:H7 strain EDL933, 13 present in prophages. Six integrases were associated with PAIs and these PAIs ranged in size from 24 kb to 88kb with a GC content ranging from 41% to 48% (Table 3). Each PAI inserted at a tRNA locus. Thus, we identified a total of 68 integrases within these five strains. We classified each region associated with the these integrases as either a PAI or phage using the criteria that to be defined as a phage at least 8 of the genes within the region were phage related genes either structural, assembly, replication, recombination or lysis proteins. To be defined as a PAI, the region should have multiple virulence genes, have an anomalous GC content and not contain any phage related genes or integrative plasmid functions. All 27 PAIs identified were integrated adjacent to tRNA loci and had a percent GC content that differed from the host chromosome and contained multiple virulence genes, which was not the case for the 39 phage regions identified.[58] We reconstructed the phylogenetic history of these 68 integrases and found that the PAI-encoded integrases form a distinct lineage from phage-encoded integrases.[58] We wanted also to determine whether the phylogenetic relationships among *E. coli*-encoded PAI and phage integrases hold true for other MIGEs and PAIs from other species. Phylogenetic comparisons of these 68 integrases with 53 TR integrases from other characterized MIGEs (additional PAIs, phages, ICEs and integrons) showed that PAI-encode integrases form a distinct lineage demonstrating that PAIs in *E. coli* are not related to other MIGEs (Fig. 4).[58]

What these studies demonstrate is that PAI-encoded integrases at least those found *E. coli* and *Vibrio* species are unrelated to those from phages and thus PAIs are probably an evolutionarily distinct class of MIGEs. Additionally this data suggests that integrases can be used as a marker to only examine the evolutionary relationships among PAIs and as a mechanism to classify and name a MIGE, previously a task relegated sometimes to the gut feeling of the discovering microbiologist. These data show that by only using the integrase as a marker for a region, different MIGEs group/ cluster according to their classification phage, ICE etc (Fig. 4). Thus, this indicates that an accurate and simple way of predicting and classifying any newly identified MIGE would be to examine the phylogeny of the encoded integrase.

Pathogenicity Island Excision from the Host Chromosome

The integrase encoded within a MIGE mediates insertion into the host chromosome via site-specific recombination between an attachment site *attP*, present in the MIGE, and *attB*, present in the bacterial genome.[2,61] The integrase also mediate excision of a MIGE from the genome. PAIs can excise from their host genome and form circular non-replicative intermediates catalyzed by their encoded integrase (Fig. 5).[24,62-65] This PAI excision process has been identified and studied

Table 3. PAI features from five E. coli pathogenic strains

PAI	Size (kb)	%GC	tRNA	Integrase
Strain 536				
PAI-III$_{536}$	75.7	47.0	ThrW	Ec536_0274
PAI-VII$_{536}$	22.1	37.7	SerU	Ec536_1909
PAI-IV$_{536}$	30.7	55.3	AsnT	Ec536_1936
PAI-VI	55.3	53.0	AsnV	Ec536_1962
PAI-V	68.0	42.2	PheV	Ec536_2962
PAI-I$_{536}$	76.8	46.0	SelC	Ec536_3765
PAI-II$_{536}$	106.2	46.6	LeuX	Ec536_4521
Strain CFT073				
PAI$_{CFT073}$	113.2	48.8	SerX	EcCFT_1165
PAI-VII	22.1	37.7	SerU	EcCFT_2392
PAI-IV$_{CFT073}$	32.3	56.3	AsnT	EcCFT_2418
PAI-III$_{CFT073}$	54.0	53.2	AsnV	EcCFT_2449
PAI-I$_{CFT073}$	54.1	47.2	PheV	EcCFT_3556
PAI-II$_{CFT073}$	51.8	47.6	PheU	EcCFT_5216
PAI$_{CFT073}$	15.7	48.1	LeuX	EcCFT_5371
Strain APECO1				
PAI-IV$_{APECO1}$	31.5	55.2	Asn	EcAPEC_1051
PAI $_{APECO1}$	63.0	47.8	Phe	EcAPEC_2254
PAI-I $_{APECO1}$	38.7	46.5	Phe	EcAPEC_3534
Strain Sakai				
PAI-I$_{Sakai}$	85.9	47.9	Ser	Ecs_1299
PAI-II$_{Sakai}$	23.3	46.6	Phe	Ecs_3843
PAI-LEE	43.1	40.9	SelC	Ecs_4534
PAI-III$_{Sakai}$	44.2	47.4	LeuX	Ecs_5242/Esc5253
Strain EDL933				
PAI-I$_{EDL933}$	87.6	48.0	SerW	EcEDL_1120
PAI-II$_{EDL933}$	87.5	47.9	SerX	EcEDL_1559
PAI-III$_{EDL933}$	23.5	46.3	PheV	EcEDL_4313
PAI-LEE	43.4	40.9	SelC	EcEDL_5087
PAI-IV$_{EDL933}$	44.4	47.4	LeuX	EcEDL_5878/5890

extensively in *Escherichia coli, Vibrio* and *Yersinia* species. In *E. coli* 536, Hacker and colleagues demonstrated that PAIs PAI-I, PAI-II, PAI-III and PAI-V can excise from the chromosome by site-specific recombination and the different PAIs excised at different frequencies depending on the growth conditions.[63,65] Interestingly Hacker and colleagues did not identify an excisionase or RDF associated with the excision process even though all TR integrases to this point are known to require this.

In addition to the work done by Hacker and colleagues in *E. coli*, the pathogenicity islands of the *Vibrio* species have been studied including *V. cholerae* and *V. parahemolyticus*, several PAIs have been identified.[26,28,29,35,66,67] Within *V. cholerae* O1 serogroup isolates four PAIs are found and

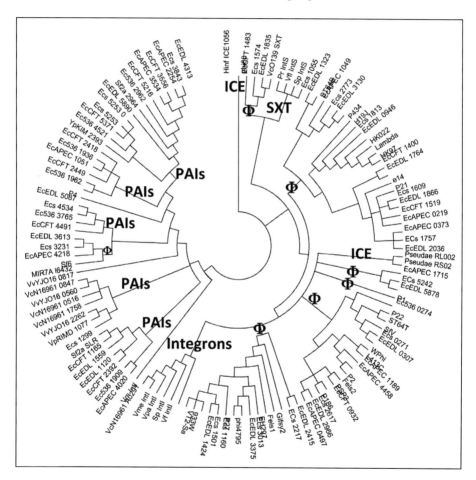

Figure 4. Evolutionary relationships of integrases from *E. coli and Vibrio* species PAIs and additional MIGEs from other strains and species in the published literature. Tags indicate strain and locus number of each integrase examined. Ec, *E. coli*; Ecs, strain O157:H7 sakai; EcEDL, strain O157:H7 EDL933; EcCFT, strain CFT073; EcAPEC, strain APEC O1; Ec536, strain 536; VcN16916, *Vibrio cholerae* N16961; VcO139, *V. cholerae* O139 MO10; VvYJO16, *V. vulnificus* YJO16; VpRIMD, *V. parahemolyticus* RIMD 2210633; YpKIM, *Yersinia pestis* KIM; MlR7A, *Mesorhizobium loti* R7A; Pr, *Providencia rettgeri*; Sp, *Shewanella putrefaciens*; Vf, *V. fluvialis*; Pseudae, *Pseudomonas aeruginosa*; Hinf, *Hemophilus influenzae*; Vm, *V. mimicus* ATCC33653; Vme, *V. metschnikovii* CIP A267. The type of MIGE present within each major lineage is indicated at internal branch points.

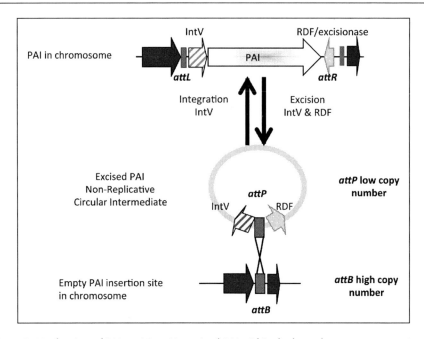

Figure 5. Mechanism of PAI excision. Linearized PAI within the host chromosome contains an integrase, its cognate RDF and site specific attachment sites attL and attR. Excision requires that action of an integrase and RDF as well as host encoded proteins to form a circularized form of the PAI. After excision of the PAI, the host chromosome contains an empty attachment site attB that can be measured by PCR.

within *V. parahemolyticus* RIMD2210633 seven PAIs are described.[26,28,35,66] *Vibrio cholerae* is the causative agent of the diarrheal disease cholera that infects millions each year. Toxigenic strains of *V. cholerae* that encode cholera toxin, the main cause of the profuse rice water diarrheal characteristic of cholera, encode two PAIs named Vibrio Pathogenicity island or TCP island and VPI-2 or Sialic acid catabolism (SAC) island (Table 4). The VPI island is a 40 kb region that encodes a TR integrase, *att* site, and the genes required for the synthesis and assembly of the type IV pilus named the Toxin coregulated pilus (TCP), the accessory colonization factor (ACF), and several virulence gene regulators ToxT and TcpPH.[35] The VPI-2 region is 57 kb in size encompassing ORFs VC1758 to VC1809 on the genome of *V. cholerae* N16961.[28] Like the VPI region, VPI-2 displays all of the characteristics of a PAI: present in pathogenic isolates, absent in non-pathogenic isolates, aberrant G+C content, encodes a TR integrase, *att* site and inserts at a tRNA-locus.[28] VPI-2 is separated into three distinct functional regions.[28] The first of these regions (in the 5′ to 3′ direction) encodes the integrase, a helicase and a type-1 restriction modification system (*hsdR, hsdM,* and *hsdS*), as well as chemotaxis proteins, the second region encodes genes for sialic acid transport (*dctPQM*), catabolism (*nanM, nanA, nanEK,* and *nagA*) and scavenging (*nanH*), and the third region encodes two RDFs, an IS element, and Mu-phage-like genes among others.[24,28] Our group has shown that the capacity to utilize sialic acid, a nine carbon keto sugar, as a carbon and energy source confers a competitive advantage to *V. cholerae* in the mucus-rich environment of the gut, where sialic acid availability is high.[68] Two additional PAIs are present only in biotype El Tor isolates named the *Vibrio* seventh pandemic island-I and VSP-II regions encode a number of hypothetical proteins but as of yet no genes involved in virulence (Table 4).[39,66] In addition, among non-choleragenic pathogenic isolates a T3SS is present within the VPI-2 region and novel island regions are present at the insertion sites of PAI from O1 serogroup isolates (Table 4).[24] VSP-I does not encode an integrase or insert at a tRNA site, instead VSP-I encodes a transposase and inserts at a glutathione reductase gene (Table 4).

Table 4. Vibrio cholerae *pathogenicity islands among different strains*

Name	tRNA	Content
V. cholerae 0395 serogroup O1 biotype classical		
VPI	tmRNA	Type IV pilus TCP, Accessory Colonization Factor, ToxT, TcpPH
VPI-2	tRNAserine	Type I RM, Sialic catabolism and transport, sialidase
V. cholerae N16961 serogroup O1 biotype El Tor		
VPI	tmRNA	Toxin Coregulated Pilus, Accessory Colonization Factor, ToxT, TcpPH
VPI-2	tRNA-serine	Type I RM, Sialic catabolism and transport, sialidase
VSP-I	Glutathione reductase	Patatin-related protein, deoycytidylate deaminase related protein
VSP-II	tRNA-met	Methyl-accepting chemotaxis proteins, RadC, Ribonuclease H, pilin IV protein
V. cholerae MO2 serogroup O139		
VPI	tmRNA	Toxin Coregulated Pilus, Accessory Colonization Factor, ToxT, TcpPH
VPI-2	tRNAserine	Type I RM, Sialic catabolism and transport, sialidase
VSP-I	Glutathione reductase	patatin-related protein, deoycytidylate deaminase related protein
VSP-II	tRNA-met	Methyl-accepting chemotaxis proteins, RadC, Ribonuclease H, pilin IV protein
V. cholerae 623–38 non-O1 serogroup		
VPI-2V	tRNAserine	T3SS-2a, Sialic catabolism and transport, sialidase
VCI-5	tRNA-met	UmuDC, HipA
V. cholerae V51 non-O1 serogroup		
VPI	tmRNA	Type IV pilus TCP, Accessory Colonization Factor, ToxT, TcpPH
VPI-2V	tRNAserine	prophage, T3SS-2b, Sialic catabolism and transport, sialidase
V. cholerae MZO-3 non-O1 serogroup		
VCI-8	tRNAmet	Chemotaxis protein, Insertion Sequence, RadC, type IV pilin

We demonstrated that VPI-2 from *V. cholerae* N16961, an O1 serogroup seventh pandemic cholera strain, can excise from its chromosomal insertion site (Fig. 6). It has also been demonstrated that the VPI (TCP) island can excise from the chromosome and does so at very low rates.[62] Also it is found that each PAI excises at different rate when grown under the same conditions.[24,62] Excision of PAIs is performed by site specific recombination mediated by the TR integrase encoded within each PAI.[24,62] A second protein, a RDF or excisionase was shown to be essential in the excision process for a number of prophages, ICEs and PAIs, along with the integrase.[69] RDF proteins are small, basic, and highly diverse, but play a role in the structural formation of the integration/ excision complex.[69] In short, PAIs encoding an integrase/RDF pair have two flanking direct repeats delineating the end of the island called *attL* and *attR* (attachment site left and right). In an excision event, the two *att* sites circle around and line up on top of one another, mediated by a tetramer of RDFs.[69] The integrase attaches to the formed complex and makes two separate cuts, the first forming a holliday junction, the second migrating down and resolving it. This produces a circular

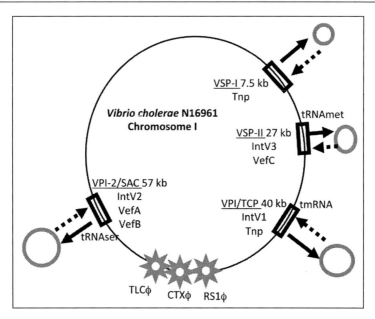

Figure 6. MIGEs identified in *V. cholerae* N16961. The genome of *V. cholerae* N16961 sero-group O1 biotype El Tor contains seven different MIGEs all encoded on chromosome I. The CTX prophage that encodes cholera toxin is flanked by two satellite prophages TLC and RS1. Four islands are present on chromosome I; VSP-I and VSP-II associated with seventh pandemic O1 serogroup biotype El Tor isolates and VPI and VPI-2 present in all O1 serogroup toxigenic isolates. Block arrows represent excision of the four island regions, which has been demonstrated for each. Broken arrows represent integration of each element, which has not been demonstrated experimentally. Associated with each PAI is the tyrosine recombinase integrase (IntV1, IntV2, and IntV3) for VPI, VPI-2 and VSP-II respectively and a transposase (Tnp) for VSP-I and and RDFs (VefA, VefB, VefC).

intermediate form of the island/phage/ICE. As a result of excision, two new *att* sites are formed, called *attB and attP* (bacterial and phage, PAI or ICE, respectively, see Fig. 1). In some cases, the *att* sites will be damaged and unable to reverse during the integrase process, this is thought to be a mechanism that forces the equilibria of the excised/integrated state toward the excision state. The exact mechanism of this process is still unclear due to a lack of good crystal structures and structural study of the problem.

Since RDFs are small in size (between 70 to 90 amino acids) and can use alternative start codons, they tend to be missed in genome annotations. PAIs from *E. coli* were initially thought not to encode RDFs, however when we examined by BLAST analysis using known RDFs as seeds the 27 PAI regions present in the five pathogenic *E. coli* genomes we identified 14 putative RDFs.[58] Similarly, using BLAST analysis with a known RDF as a seed we interrogated *V. cholerae* PAIs and identified three RDFs within two PAIs, two RDFs named VefA and VefB (for *Vibrio* excision factor) within VPI-2 and one RDF named VefC within VSP-II region (Fig. 6).[70] Phylogenetic analysis suggests that VefB is the cognate RDF from IntV2 within VPI-2, VefA is the cognate RDF for IntV1 within VPI and VefC cognate RDF for IntV3 within VSP-II (Boyd, unpublished data). We experimentally demonstrated that both VefA and VefB are RDFs by demonstrating that when each RDF is ectopically expressed in a wild-type strain VPI-2 excision occurs (Boyd, unpublished data). However, our experimental data suggests the integrase IntV2 and VefA are both essential but not VefB for efficient excision of VPI-2 from the chromosome.[24,70] These results showed that excision of PAIs is controlled by two conserved factors within PAIs, an integrase and an RDF similar to other MIGEs.

Conclusion

MIGEs and genomic islands are broad term that encompasses a diverse group of genetic elements; phages, integrative plasmids, integrons, conjugative transposons, ICEs, and PAIs. These MIGEs integrate into their host chromosome by site specific recombination, catalyzed by an integrase. We suggest that the integrase, which is the only protein common to all these MIGEs, can be used as a marker to examine the evolutionary relationship among MIGEs. Integrase phylogeny demonstrates that integrases from each of the different groups of MIGEs branch according to their predicted classification PAI, ICE etc (Fig. 4). Thus, this simple analysis can clear up problems of nomenclature and allow for the correct naming of PAIs, ICEs, conjugative transposons and integrative plasmids.[68,71,72] In the past, researchers have had a tendency to name regions that encode virulence factors that are present only in pathogenic isolates of a species as PAIs regardless of the presence of other features that distinguish them as a phage, ICE, or integrative plasmid. Phylogenetic analysis demonstrates that integrases are an accurate genetic marker to study the evolutionary relationships among PAIs and between PAIs and MIGEs, and PAIs are not amalgamation of other MIGEs but, evolutionarily, a group of their own.

Acknowledgments

I thank members of my group for their enthusiasm and hard work; Megan Carpenter, Seth L. Blumerman, Sarah Gaffney, Brandy Haines, Nityananda Chowdhury, Joseph J. Kingston, Jean Bernard Lubin, Michael G. Napolitano, Serge Ongagna, W. Brian Whitaker, and Claire Yang. Work in the Boyd group is support by a National Science Foundation CAREER grant DEB-0844409. Literature citations have been limited in may cases to review articles due to space consideration and I apologize to those authors whose primary research is not included or cited.

References

1. Reiter WD, Palm P, Yeats S. Transfer RNA genes frequently serve as integration sites for prokaryotic genetic elements. Nucleic Acids Res 1989; 17:1907-14 doi:10.1093/nar/17.5.1907. PMID:2467253
2. Williams KP. Integration sites for genetic elements in prokaryotic tRNA and tmRNA genes: sublocation preference of integrase subfamilies. Nucleic Acids Res 2002; 30:866-75 doi:10.1093/nar/30.4.866. PMID:11842097
3. Germon P, Roche D, Melo S, et al. tDNA locus polymorphism and ecto-chromosomal DNA insertion hot-spots are related to the phylogenetic group of Escherichia coli strains. Microbiology 2007; 153:826-37 doi:10.1099/mic.0.2006/001958-0. PMID:17322203
4. Boyd EF, Almagro-Moreno S, Parent MA. Genomic islands are dynamic, ancient integrative elements in bacterial evolution. Trends Microbiol 2009; 17:47-53 doi:10.1016/j.tim.2008.11.003. PMID:19162481
5. Hacker J, Blum-Oehler G, Muhldorfer I, et al. Pathogenicity islands of virulent bacteria: structure, function and impact on microbial evolution. Mol Microbiol 1997; 23:1089-97 doi:10.1046/j.1365-2958.1997.3101672.x. PMID:9106201
6. Hacker J, Carniel E. Ecological fitness, genomic islands and bacterial pathogenicity. A Darwinian view of the evolution of microbes. EMBO Rep 2001; 2:376-81. PMID:11375927
7. Hacker J, Kaper JB. Pathogenicity islands and the evolution of microbes. Annu Rev Microbiol 2000; 54:641-79 doi:10.1146/annurev.micro.54.1.641. PMID:11018140
8. Hacker J, Knapp S, Goebel W. Spontaneous deletions and flanking regions of the chromosomally inherited hemolysin determinant of an Escherichia coli O6 strain. J Bacteriol 1983; 154:1145-52. PMID:6343344
9. Knapp S, Hacker J, Jarchau T, et al. Large, unstable inserts in the chromosome affect virulence properties of uropathogenic Escherichia coli O6 strain 536. J Bacteriol 1986; 168:22-30. PMID:2875989
10. Hughes C, Hacker J, Düvel H, et al. Chromosomal deletions and rearrangements cause coordinate loss of haemolysis, fimbriation and serum resistance in a uropathogenic strain of Escherichia coli. Microb Pathog 1987; 2:227-30 doi:10.1016/0882-4010(87)90024-6. PMID:2907085
11. High NJ, Hales B, Jann K, Boulnois G. A block of urovirulence genes encoding multiple fimbriae and hemolysin in Escherichia coli O4:K12:H-. Infect Immun 1988; 56:513-7. PMID:2892797
12. Hacker J, Bender L, Ott M, et al. Deletions of chromosomal regions coding for fimbriae and hemolysins occur in vitro and in vivo in various extraintestinal Escherichia coli isolates. Microb Pathog 1990; 8:213-25 doi:10.1016/0882-4010(90)90048-U. PMID:1974320

13. Groisman EA, Ochman H. How Salmonella became a pathogen. Trends Microbiol 1997; 5:343-9 doi:10.1016/S0966-842X(97)01099-8. PMID:9294889
14. Jarvis KG, Giron JA, Jerse AE, et al. Enteropathogenic Escherichia coli contains a putative type III secretion system necessary for the export of proteins involved in attaching and effacing lesion formation. Proc Natl Acad Sci USA 1995; 92:7996-8000 doi:10.1073/pnas.92.17.7996. PMID:7644527
15. McDaniel TK, Jarvis KG, Donnenberg MS, et al. A genetic locus of enterocyte effacement conserved among diverse enterobacterial pathogens. Proc Natl Acad Sci USA 1995; 92:1664-8 doi:10.1073/pnas.92.5.1664. PMID:7878036
16. Mills DM, Bajaj V, Lee C. A 40 kb chromosomal fragment encoding Salmonella typhimurium invasion genes is absent from the corresponding region of the Escherichia coli K-12 chromosome. Mol Microbiol 1995; 15:749-59 doi:10.1111/j.1365-2958.1995.tb02382.x. PMID:7783645
17. Buchrieser C, Brosch R, Bach S, et al. The high-pathogenicity island of Yersinia pseudotuberculosis can be inserted into any of the three chromosomal asn tRNA genes. Mol Microbiol 1998; 30:965-78 doi:10.1046/j.1365-2958.1998.01124.x. PMID:9988474
18. Buchrieser C, Prentice M, Carniel E. The 102-kilobase unstable region of Yersinia pestis comprises a high-pathogenicity island linked to a pigmentation segment which undergoes internal rearrangement. J Bacteriol 1998; 180:2321-9. PMID:9573181
19. Censini S, Lange C, Xiang Z, et al. cag, a pathogenicity island of Helicobacter pylori, encodes type I-specific and disease-associated virulence factors. Proc Natl Acad Sci USA 1996; 93:14648-53 doi:10.1073/pnas.93.25.14648. PMID:8962108
20. Groisman EA, Ochman H. Pathogenicity islands: bacterial evolution in quantum leaps. Cell 1996; 87:791-4 doi:10.1016/S0092-8674(00)81985-6. PMID:8945505
21. Schmidt MA. LEEways: tales of EPEC, ATEC and EHEC. Cell Microbiol 2010; 12:1544-52 doi:10.1111/j.1462-5822.2010.01518.x. PMID:20716205
22. Dziejman M, Serruto D, Tam VC, et al. Genomic characterization of non-O1, non-O139 Vibrio cholerae reveals genes for a type III secretion system. Proc Natl Acad Sci USA 2005; 102:3465-70 doi:10.1073/pnas.0409918102. PMID:15728357
23. Chen Y, Johnson JA, Pusch GD, et al. The genome of non-O1 Vibrio cholerae NRT36S demonstrates the presence of pathogenic mechanisms that are distinct from O1 Vibrio cholerae. Infect Immun 2007; 75(5):2645-7. PMID:17283087
24. Murphy RA, Boyd E. Three pathogenicity islands of Vibrio cholerae can excise from the chromosome and form circular intermediates. J Bacteriol 2008; 190:636-47 doi:10.1128/JB.00562-07. PMID:17993521
25. Makino K, Oshima K, Kurokawa K, et al. Genome sequence of Vibrio parahaemolyticus: a pathogenic mechanism distinct from that of V. cholerae. Lancet 2003; 361:743-9 doi:10.1016/S0140-6736(03)12659-1. PMID:12620739
26. Hurley CC, Quirke A, Reen FJ, Boyd EF. Four genomic islands that mark post-1995 pandemic Vibrio parahaemolyticus isolates. BMC Genomics 2006; 7:104 doi:10.1186/1471-2164-7-104. PMID:16672049
27. Okada N, Iida T, Park KS, et al. Identification and characterization of a novel type III secretion system in trh-positive Vibrio parahaemolyticus strain TH3996 reveal genetic lineage and diversity of pathogenic machinery beyond the species level. Infect Immun. 2009 Feb;77(2):904-13. Epub 2008 Dec 15. 2009;77(2):904-913.
28. Jermyn WS, Boyd EF. Characterization of a novel Vibrio pathogenicity island (VPI-2) encoding neuraminidase (nanH) among toxigenic Vibrio cholerae isolates. Microbiology 2002; 148:3681-93. PMID:12427958
29. Jermyn WS, Boyd EF. Molecular evolution of Vibrio pathogenicity island-2 (VPI-2): mosaic structure among Vibrio cholerae and Vibrio mimicus natural isolates. Microbiology 2005; 151:311-22 doi:10.1099/mic.0.27621-0. PMID:15632448
30. Okada N, Matsuda S, Matsuyama J, et al. Presence of genes for type III secretion system 2 in Vibrio mimicus strains. BMC Microbiol 2010; 10:302. PMID:21110901
31. Dean P. Functional domains and motifs of bacterial type III effector proteins and their roles in infection. FEMS Microbiol Rev. 2011; Epub before press.
32. McGhie EJ, Brawn L, Hume P, et al. Salmonella takes control: effector-driven manipulation of the host. Curr Opin Microbiol 2009; 12:117-24 doi:10.1016/j.mib.2008.12.001. PMID:19157959
33. Dean P, Kenny B. The effector repertoire of enteropathogenic E. coli: ganging up on the host cell. Curr Opin Microbiol 2009; 12:101-9 doi:10.1016/j.mib.2008.11.006. PMID:19144561
34. Ibarra JA, Steele-Mortimer O. Salmonella–the ultimate insider. Salmonella virulence factors that modulate intracellular survival. Cell Microbiol 2009; 11:1579-86 doi:10.1111/j.1462-5822.2009.01368.x. PMID:19775254
35. Karaolis DK, Johnson JA, Bailey CC, et al. A Vibrio cholerae pathogenicity island associated with epidemic and pandemic strains. Proc Natl Acad Sci USA 1998; 95:3134-9 doi:10.1073/pnas.95.6.3134. PMID:9501228

36. Blanc-Potard AB, Solomon F, Kayser J. EA. G. The SPI-3 pathogenicity island of Salmonella enterica. J Bacteriol 1999; 181:998-1004. PMID:9922266
37. Lindsay JA, Ruzin A, Ross H, et al. The gene for toxic shock toxin is carried by a family of mobile pathogenicity islands in Staphylococcus aureus. Mol Microbiol 1998; 29:527-43 doi:10.1046/j.1365-2958.1998.00947.x. PMID:9720870
38. Dobrindt U, Blum-Oehler G, Nagy G, et al. Genetic structure and distribution of four pathogenicity islands (PAI I(536) to PAI IV(536)) of uropathogenic Escherichia coli strain 536. Infect Immun 2002; 70:6365-72 doi:10.1128/IAI.70.11.6365-6372.2002. PMID:12379716
39. O'Shea YA, Finnan S, Reen FJ, et al. The Vibrio seventh pandemic island-II is a 26.9 kb genomic island present in Vibrio cholerae El Tor and O139 serogroup isolates that shows homology to a 43.4 kb genomic island in V. vulnificus. Microbiology 2004; 150:4053-63 doi:10.1099/mic.0.27172-0. PMID:15583158
40. Quirke AM, Reen FJ, Claesson MJ, et al. Genomic island identification in Vibrio vulnificus reveals significant genome plasticity in this human pathogen. Bioinformatics 2006; 22:905-10 doi:10.1093/bioinformatics/btl015. PMID:16443635
41. Ramsay JP, Sullivan J, Stuart G, et al. Excision and transfer of the Mesorhizobium loti R7A symbiosis island requires an integrase IntS, a novel recombination directionality factor RdfS, and a putative relaxase RlxS. Mol Microbiol 2006; 62:723-34 doi:10.1111/j.1365-2958.2006.05396.x. PMID:17076666
42. Fouts DE. Phage_Finder: automated identification and classification of prophage regions in complete bacterial genome sequences. Nucleic Acids Res 2006; 34:5839-51 doi:10.1093/nar/gkl732. PMID:17062630
43. Ou HY, Chen LL, Lonnen J, et al. A novel strategy for the identification of genomic islands by comparative analysis of the contents and contexts of tRNA sites in closely related bacteria. Nucleic Acids Res 2006; 34:e3 doi:10.1093/nar/gnj005. PMID:16414954
44. Zhang R, Zhang CT. A systematic method to identify genomic islands and its applications in analyzing the genomes of Corynebacterium glutamicum and Vibrio vulnificus CMCP6 chromosome I. Bioinformatics 2004; 20:612-22 doi:10.1093/bioinformatics/btg453. PMID:15033867
45. Langille MG, Hsiao WW, Brinkman F. Detecting genomic islands using bioinformatics approaches. Nat Rev Microbiol 2010; 8:373-82 doi:10.1038/nrmicro2350. PMID:20395967
46. Dobrindt U, Hochhut B, Hentschel U, et al. Genomic islands in pathogenic and environmental microorganisms. Nat Rev Microbiol 2004; 2:414-24 doi:10.1038/nrmicro884. PMID:15100694
47. Osborn AM, Böltner D. When phage, plasmids, and transposons collide: genomic islands, and conjugative- and mobilizable-transposons as a mosaic continuum. Plasmid 2002; 48:202-12 doi:10.1016/S0147-619X(02)00117-8. PMID:12460536
48. Boyd EF, Brussow H. Common themes among bacteriophage-encoded virulence factors and diversity among the bacteriophages involved. Trends Microbiol 2002; 10:521-9 doi:10.1016/S0966-842X(02)02459-9. PMID:12419617
49. Boyd EF. Phage-encoded bacterial virulence factors: phage-pathogenicity island and phage-phage interactions. Adv Virus Res 2011; In press.
50. Ruzin A, Lindsay J, Novick RP. Molecular genetics of SaPI1–a mobile pathogenicity island in Staphylococcus aureus. Mol Microbiol 2001; 41:365-77 doi:10.1046/j.1365-2958.2001.02488.x. PMID:11489124
51. Boyd EF, Davis B. M., Hochhut, B. Bacteriophage-Bacteriophage interactions in the evolution of pathogenic bacteria. Trends Microbiol 2001; 9:137-44 doi:10.1016/S0966-842X(01)01960-6. PMID:11303502
52. Tormo-Más MA, Mir I, Shrestha A, et al. Moonlighting bacteriophage proteins derepress staphylococcal pathogenicity islands. Nature 2010; 465:779-82 doi:10.1038/nature09065. PMID:20473284
53. Fitzgerald JR, Monday SR, Foster TJ, et al. Characterization of a putative pathogenicity island from bovine Staphylococcus aureus encoding multiple superantigens. J Bacteriol 2001; 183:63-70 doi:10.1128/JB.183.1.63-70.2001. PMID:11114901
54. Novick R, Christie GE, Penadés J. The phage-related chromosomal islands of Gram-positive bacteria. Nat Rev Microbiol 2010; 8(8):541-51.
55. Schubert S, Dufke S, Sorsa J, Heesemann J. A novel integrative and conjugative element (ICE) of Escherichia coli: the putative progenitor of the Yersinia high-pathogenicity island. Mol Microbiol 2004; 51:837-48 doi:10.1046/j.1365-2958.2003.03870.x. PMID:14731283
56. Carniel E. The Yersinia high-pathogenicity island: an iron-uptake island. Microbes Infect 2001; 3:561-9 doi:10.1016/S1286-4579(01)01412-5. PMID:11418330
57. Benedek O, Schubert S. Mobility of the Yersinia High-Pathogenicity Island (HPI): transfer mechanisms of pathogenicity islands (PAIS) revisited (a review). Acta Microbiol Immunol Hung 2007; 54:89-105 doi:10.1556/AMicr.54.2007.2.1. PMID:17899790
58. Napolitano MG, Almagro-Moreno S, Boyd EF. Dichotomy in the evolution of pathogenicity island and bacteriophage encoded integrases from pathogenic Escherichia coli strains. Infect Genet Evol 2011; 11:423-36 doi:10.1016/j.meegid.2010.12.003. PMID:21147268

59. Touchon M, Hoede C, Tenaillon O, et al. Organised genome dynamics in the Escherichia coli species results in highly diverse adaptive paths. PLoS Genet 2009; 5:e1000344 doi:10.1371/journal.pgen.1000344. PMID:19165319

60. Dobrindt U, Chowdary M, Krumbholz G, Hacker J. Genome dynamics and its impact on evolution of Escherichia coli. Med Microbiol Immunol (Berl) 2010; 199:145-54 doi:10.1007/s00430-010-0161-2. PMID:20445988

61. Rajeev L, Malanowska K, Gardner JF. Challenging a paradigm: the role of DNA homology in tyrosine recombinase reactions. Microbiol Mol Biol Rev 2009; 73:300-9 doi:10.1128/MMBR.00038-08. PMID:19487729

62. Rajanna C, Wang J, Zhang D, et al. The Vibrio pathogenicity island of epidemic Vibrio cholerae forms precise extrachromosomal circular excision products. J Bacteriol 2003; 185:6893-901 doi:10.1128/JB.185.23.6893-6901.2003. PMID:14617653

63. Middendorf B, Hochhut B, Leipold K, et al. Instability of pathogenicity islands in uropathogenic Escherichia coli 536. J Bacteriol 2004; 186:3086-96 doi:10.1128/JB.186.10.3086-3096.2004. PMID:15126470

64. Lesic B, Bach S, Ghigo JM, et al. Excision of the high-pathogenicity island of Yersinia pseudotuberculosis requires the combined actions of its cognate integrase and Hef, a new recombination directionality factor. Mol Microbiol 2004; 52:1337-48 doi:10.1111/j.1365-2958.2004.04073.x. PMID:15165237

65. Hochhut B, Wilde C, Balling G, et al. Role of pathogenicity island-associated integrases in the genome plasticity of uropathogenic Escherichia coli strain 536. Mol Microbiol 2006; 61:584-95 doi:10.1111/j.1365-2958.2006.05255.x. PMID:16879640

66. Dziejman M, Balon E, Boyd D, et al. Comparative genomic analysis of Vibrio cholerae: genes that correlate with cholera endemic and pandemic disease. Proc Natl Acad Sci USA 2002; 99:1556-61 doi:10.1073/pnas.042667999. PMID:11818571

67. Boyd EF, Cohen ALV, Naughton LM, et al. Molecular analysis of the emergence of pandemic Vibrio parahaemolyticus. BMC Microbiol 2008; 8:110 doi:10.1186/1471-2180-8-110. PMID:18590559

68. Almagro-Moreno S, Boyd EF. Sialic acid catabolism confers a competitive advantage to pathogenic Vibrio cholerae in the mouse intestine. Infect Immun 2009; 77:3807-16 doi:10.1128/IAI.00279-09. PMID:19564383

69. Lewis J, Hatfull G. Control of directionality in integrase-mediated recombination: examination of recombination directionality factors (RDFs) including Xis and Cox proteins. Nucleic Acids Res. 2001 Jun 1;29(11):2205-16. 2001;29(1):2205-2216.

70. Almagro-Moreno S, Napolitano MG, Boyd EF. Excision dynamics of Vibrio pathogenicity island-2 from Vibrio cholerae: role of a recombination directionality factor VefA. BMC Microbiol 2010; 10:306. PMID:21118541

71. Böltner D, MacMahon C, Pembroke JT, Strike P, Osborn A. R391: a conjugative integrating mosaic comprised of phage, plasmid, and transposon elements. J Bacteriol 2002; 184:5158-69 doi:10.1128/JB.184.18.5158-5169.2002. PMID:12193633

72. Roberts AP, Chandler M, Courvalin P, et al. Revised nomenclature for transposable genetic elements. Plasmid 2008; 60:167-73 doi:10.1016/j.plasmid.2008.08.001. PMID:18778731

CHAPTER 18

Staphylococcal Cassette Chromosome (SCC):
A Unique Gene Transfer System in Staphylococci

Teruyo Ito, Sae Tsubakishita, Kyoko Kuwahara-Arai, Xiao Han
and Keiichi Hiramatsu*

Abstract

Staphylococcal cassette chromosome *mec* (SCC*mec*) is a class of mobile genetic element SCC that carries the methicillin resistant determinant *mecA*. It is a mobile genetic element driven by site-specific recombinase(s) designated as cassette chromosome recombinase (*ccr*). Although many structurally-distinct SCC*mec* elements have been identified in staphylococcal species to date, these elements carry two essential components, *mec* gene complex that encodes methicillin resistance determinant and *ccr* gene complex that encodes *ccr*(s) in common. Besides methicillin resistance, SCC elements carry resistance to other antibiotics and heavy metals. Multiple types of SCC elements seem to have evolved through repeated horizontal genetic transfer among various staphylococcal species.

Introduction

One year after the introduction of methicillin, a semi-synthetic penicillin, into clinical practice, the first MRSA strain was reported from an infected patient.[1] MRSA produces an additional penicillin-binding protein PBP2' (or PBP2a), which has low binding affinities for most of the penicillin as well as cephem antibiotics.[2-4] PBP2' is encoded by the chromosomally located *mecA* gene which was exogenously acquired, since the methicillin-susceptible *S. aureus* strains do not have the *mecA* gene.[5] Soon, it became apparent that *mecA* was widely disseminated among various staphylococcal species.[6-8] Nucleotide sequencing of the region surrounding *mecA* revealed that it was located on a mobile genetic element designated as SCC*mec* integrated in the *S. aureus* chromosome.[9,10] Further study on the structure of SCC*mec* elements showed that they are highly diverse in their structural organization and genetic content.[11-15] However, all SCC*mec* elements shared several characteristics in common: (1) carriage of *mecA* gene as a part of the '*mec* gene complex', (2) carriage of *ccr* gene(s) (*ccrAB* or *ccrC*) as components of the '*ccr* gene complex', (3) integration at a specific site in the staphylococcal chromosome, referred to as the integration site sequence for SCC (ISS), which serves as the target for *ccr*-mediated recombination, and (4) the presence of flanking direct repeat sequences containing the ISS (Fig. 1).[16] Herein, we review structural diversity of SCC*mec* elements focusing on the types of *mec* gene complex, *ccr* gene complex, and the carriage of resistance genes other than *mecA*. Evolution and distribution

*Department of Infection Control Science, Graduate School of Medicine, Juntendo University, Tokyo, Japan.
Corresponding Author: Teruyo Ito—Email: teruybac@juntendo.ac.jp

Bacterial Integrative Mobile Genetic Elements, edited by Adam P. Roberts and Peter Mullany.
©2013 Landes Bioscience.

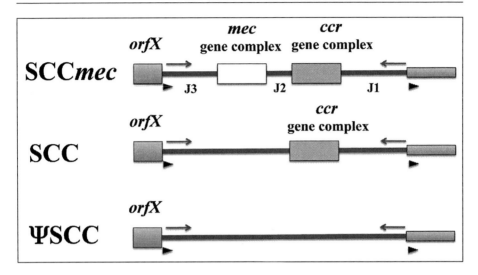

Figure 1. Schematic representation of SCC*mec*, SCC, and ΨSCC. Arrow heads indicated the location of ISS. Arrows indicate the location of inverted repeats. The J regions (Joining regions), the regions other than *mec* gene complex and *ccr* gene complex in SCC*mec* elements, are designated as J1, J2, and J3

of SCC*mec* elements is discussed based on the structural comparison of ever increasing types in SCC*mec* elements.

SCC*mec* is a SCC Specialized for the Transfer of Methicillin Resistance

Following the discovery of SCC*mec* on the MRSA chromosomes, similar elements carrying *ccr* gene complex but no *mec* gene complex were found in the chromosome of staphylococcal species. The element was found integrated at the same ISS site, and shared the same feature with SCC*mec* such as the presence of *ccr* gene complex, and characteristic direct and inverted nucleotide repeats at both extremities (Fig. 1). Therefore, we soon came to realize that SCC*mec* is one of the many classes of *ccr*-driven mobile genetic SCC elements capable of carrying various genes of different functions. Indeed, an SCC carrying a capsule gene cluster was found (designated SCC*cap1*), so was SCC*fur* carring fusidic acid resistance, and SCC*Hg* carrying a mercury resistance operon. SCC elements carrying no apparent functional genes have also been found from some staphylococcal strains. Such a SCC element is described by adding a suffix denoting the name of the strain of origin; e.g., SCC$_{476}$, an SCC identified in the chromosome of MSSA strain 476.[11,17-19]

Curiously, such chromosome regions with conserved direct and inverted repeats but no *ccr* gene complex are occasionally found integrated at the ISS. The element is considered to be precursor or a remnant of SCC, and described as ΨSCC (Fig. 1). The arginine catabolic mobile element (ACME) identified in USA300 strains, is referred to as ΨSCC. The sizes of ΨSCC elements are very diverse ranging from 105 bp of ΨSCC$_{h1435}$ identified in *S. hemolyticus* JCSC1435 to 32 kb of ACME.[20,21]

SCCs and ΨSCCs are mostly integrated at the 3′ end of *orfX*, where an ISS is located. Whole genome sequencing of staphylococcal species revealed that the *orfX* located near the replication origin, and a long stretch of region designated '*oriC* environ' is located downstream of *orfX*, where exogenously acquired genes are accumulated.[21,22] In the *oriC* environ, many SCC and ΨSCC elements are found, which can be easily distinguished from the host chromosome by their characteristic structural features: they are demarcated by flanking direct repeat sequences containing the ISS for SCC.[21,23] In addition, SCC elements carry the *ccr* gene complex.

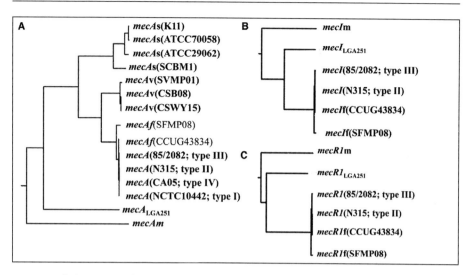

Figure 2. Phylogenetic relationships of *mecA* and its homologs (A), *mecR1* and its homologs (B), and *mecI* and its homologs (C). Phylogenetic relationships of three genes were compared using clustal W using nucleotide sequences deposited in DDBJ/EMBL/GenBank databases under following accession nos.: *S. aureus* N315 (D86923), NCTC10442 (AB033763), 85/2082 (AB037671), 85/3907(AB047089), CA05(AB063172), LGA251(FR821779.1); *S. sciuri* K11 (Y13094), ATCC70058(AB547236), ATCC79062(AB547235) ; *S. vitulinus* SVMP01(AB546780), CSB08(AM08810), CSWI15(AM08811); *Macrococcus caseolyticus* JCSC5402 (NC_011996.1). *S. fleurettii* CCUG43834(AB546266)

Identification of *mecA* Gene and Its Homologs

The *mecA* gene encodes PBP2' (or PBP2a) that has low affinity to penicillin and cephem antibiotics.[2-4] Soon after the *mecA* gene sequence was determined, it was realized that the gene had been widely disseminated among multiple staphylococcal species.[6-8,24] Recently, homologs of the *mecA* gene with similar nucleotide sequences have been identified in several staphylococcal species as well as species closely related to staphylococci.[24-29] To discriminate from the widely disseminated *mecA*, the *mecA* homologs are indicated with a strain name or the first letter of species or genus name, e.g., *mecA*LGA251 in a bovine MRSA strain LGA251 and a human isolate M10–0061, *mecAv* in *S. vitulinus*, *mecAs* in *S. sciuri*, *mecAm* in *M. caseolyticus* (the nomenclature of the *mecA* homologs will be changed to more consented ones among staphylococcal researchers, but we used previously reported ones in this chapter). The nucleotide identities of *mecA*LGA251 and *mecAm* to *mecA* are 68.7% and 61.6%, respectively. In the case of *mecAv* and *mecAs*, there are several homologs, of which identities with *mecA* are 90.7–91.0% and 79.1–80.2%, respectively (Fig. 2A). Homologs of regulatory genes, *mecR1* and *mecI*, are located upstream of *mecA*LGA251 and *mecAm*, and, downstream of the *mecA* homologs, β-lactamase gene (*blaZ*) homologs are located, constituting the gene clusters *blaZ*LGA251-*mecA*LGA251-*mecR1*LGA251-*mecI*LGA251 or *blaZm*-*mecAm*-*mecR1m*-*mecIm*, respectively. The phylogenetic distances of original *mecA*, *mecR1*, and *mecI* genes to each of the homologs are quite similar, indicating that the molecular co-evolution of the component genes had occurred in *mec* gene complex (Fig. 2B and 2C).

Structure of SCC*mec*: The *mec* Gene Complexes

The *mec* gene complex is classified into several classes based on the difference of the insertion sequence located downstream and upstream of *mecA*.[16,30] An insertion sequence, IS*431*, designated as IS*431mec*[31] is located downstream of *mecA*, and the region between IS*431* and *mecA* is nearly identical, except for the sub-region called the hyper-variable region composed of 40-bp direct repeat units (*dru*).[6,32] The nucleotide sequences of 40 bp as well as their numbers differ among MRSA strains. Therefore, differences can be used to differentiate MRSA strains; *dru* typing.[33] Upstream of *mecA*, two genes that encode regulatory proteins for *mecA*, *mecR1* (signal transducer protein) and *mecI* (repressor protein), are located.[34] Such a gene cluster, IS*431mec-mecA-mecR1-mecI* is regarded as a prototype of the *mec* gene complex and designated class A *mec* gene complex. In some cases, the *mecR1* gene is truncated by the insertion of another insertion sequence, resulting in the formation of the class B *mec* gene complex, IS*431mec-mecA-ΔmecR1-IS1272*. Class C *mec* gene complex has a structure, IS*431mec-mecA-ΔmecR1-IS431*. The class is further subdivided into two, depending on the direction of the last IS*431*: C1, the same direction; C2, the opposite direction). It is apparent that classes B and C *mec* gene complexes are the descendants of classA *mec* gene complex, with modifications of the structure by insertion of mobile genetic elements. However, the *oriC* environ is considered to suffer an extensive genetic perturbation such as IS-mediated deletion or recombination.[35] Therefore, it is not surprising that more detailed study detects extensive structural diversification of *mec* gene complex. For example, a novel class C1-like *mec* gene complex (6422 bp) in strain JCSC6945 is distinct from the class C1 *mec* gene complex (7212 bp) carried by a Swedish community-associated MRSA strain JCSC6082 in two aspects: (1) SCC*mec*JCSC6945 and SCC*mec*JCSC6082 carry different direct repeat unit (*dru*) type (14f and 10a, respectively); (2) IS*431* is inserted into the *mecR1* gene at different position (17 bp and 968 bp downstream of the start codon, respectively). Furthermore, the orientation of the class C1 *mec* gene complex of JCSC6945 is opposite to *mec* gene complexes carried by Types I-VII SCC*mec* elements.[36]

Origin of *mec* Gene Complex

Recently, a chromosomal region with 99% nucleotide identity to the genes in the *mec* gene complex, *mecA*-containing region (12 kb long), was found in *S. fleurettii* CCUG43834[37](Fig. 3). The region in *S. fleurettii* has an almost identical structure to that

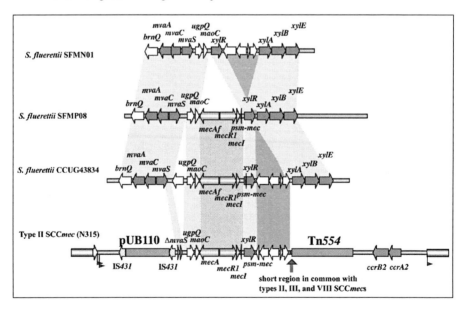

Figure 3. Structural comparisons of the region around *mecA* among Type II SCC*mec* N315, SCC*mec* in *S. pseudintermedius* KM1381 and *mecA*-containing regions of *S. fleurettii* strains. The structures of *mecA*-containing regions of *S. fleurettii* strains CCUG43834, SFMP08, and SFMN01 and SCC*mec* elements of *S. aureus* N315 and *S. pseudintermedius* KM1381 are illustrated based on the nucleotide sequences deposited in the databases; CCUG43834 (AB546266), SFMP08(AB546267), SFMN01 (AB546268), N315 (D86934) and KM1381 (AM904732).

of the class A *mec* gene complex in Type II SCC*mec*. Downstream of *mecA*, it contained an intact *mvaS*, 3-hydroxy-3-methylglutaryl CoA synthase, which is not truncated by IS*431mec*. Between the *mvaS* and *mecA* genes on the chromosome of *S. fleurettii*, the *ugpQ* and *maoC* genes, which are identical to those in Type II SCC*mec* were located. Upstream of *mecA* in *S. fleurettii* CCUG43834, *mecR1* and *mecI* genes were located. Furthermore, genes encoding a phenol-soluble modulin *mec* (*psm-mec*),[38] xylose repressor (*xylR*), and four ORFs that are identical to those of Type II SCC*mec* were identified. The *mecA* locus of *S. fleurettii* CCUG43834 carries *xylA*, *xylB*, and *xylE* genes involved in xylose utilization (xylose isomerase, xylulokinase, and the xylose transporter, respectively). In contrast the corresponding region of Type II SCC*mec* of N315 is truncated by the integration of Tn*554* (a transposon encoding macrolide and streptomycin resistance) upstream of four ORFs.

In addition, a very short stretch of DNA sequence of about 115 bp, which was not identified in the *mecA* locus of *S. fleurettii* CCUG43834, is present flanking the Tn*554*.

The *mecA* locus in *S. fleurettii* CCUG43834 was also highly homologous to the corresponding regions of Types III and VIII SCC*mec* elements in *S. aureus*[37] as well as the SCC*mec* element carried by *S. pseudintermedius* KM1381.[39] Interestingly, the short DNA region was identified in ORF12 of *S. pseudintermedius* KM1381, and the ORF was divided into two by the insertion of ΨTn*554* in Type III SCC*mec* elements. Furthermore, the *dru* type of the SCC*mec* in *S. pseudintermedius* KM1381 was identical to that of N315 (dt9a; 5a-2d-2d-4a-0–2g-3b-4e-3e), and distinct from those of *S. fleurettii* CCUG43834 (unclassified: 5a-2g-3c-IS*256*-new 40 bp-4e) and SFMP08 (new 40 bp only). These data suggest that although the class A *mec* gene complex was similar to the *mec* locus of *S. fleurettii*, other *S. fleurettii* strains that carry the *drus* similar to those identified in *S. aureus* strains and ORF12 identified in *S. pseudintermedius* KM1381 or another reservoir that supplied genes identified in *mec*-locus of *S. fleurettii* might exist. Anyhow modification should have been added presumably on the plasmids or chromosomal loci of other staphylococcal species before it was finally integrated as the component of SCC*mec*.

The *mecA* loci in *S. fleurettii* strains show considerable diversity from strain to strain. *S. fleurettii* strain SFMP08 does not carry the four ORFs between *xylR* and *xylA* genes, which are carried by *S. fleurettii* CCUG43834. *S. fleurettii* SFMN01 carried the region encoding *ugpQ*, *maoC*, and genes for xylose utilization, *xylR*, *xylA*, *xylB*, and *xylE* similar to the *mecA* locus in strain CCUG43834, but it did not carry *mecA*, *mec* regulatory genes, or *psm-mec*. This is remarkable since the *mecA* locus is considered to be vertically transmitted in staphylococcal speciation.[37] The precise mechanism for deletion or acquisition of *mecA* locus among *S. fleurettii* strains should be addressed in detail. So far there is no evidence that the deletion is mediated by a transposon or insertion sequence since there is no such mobile genetic element in the vicinity. This is in contrast to the case of *mecA*$_m$ of *Macrococcus caseolyticus*,[25] in which the *mecA*$_m$ complex is closely associated with a site-specific recombinase gene and is mobile across chromosome and plasmids.[40] How and where the *mecA*$_f$ was mobilized to form a SCC*mec* is a great question to be explored.

*ccr*s are Responsible for the Mobility of SCC*mec*s

The *ccr* genes are located in the midst of the *ccr* gene complex. Among 7–8 ORFs constituting the *ccr* gene complex, only the function of *ccr* genes has been determined. To date, three *ccr* genes, *ccrA*, *ccrB*, and *ccrC*, have been identified in staphylococcal species. Figure 4 shows phylogenetic relations of *ccr* genes in SCC*mec* elements identified in *S. aurues*. In contrast to the high similarity of the *mecA* genes widely disseminated among staphylococcal species, *ccr* genes are extremely diverse. When the *ccr* genes with less than 50% nucleotide identities are defined as distinct types of *ccr* gene, three *ccr* gene types are differentiated. In each type, *ccr* genes with more than 85% nucleotide identities are classified as a subgroup. So far, two *ccr* types, *ccrA* (1350 bp in size) and *ccrB* (1,629 bp), are further classified into subgroups, *ccrA1–7* and *ccrB1–6*, respectively. Although, the sizes of the genes belonging to the third *ccr* type, *ccrC*, range from 1,554 bp (*ccrC* in *S. aurues* strain 85/2082) to 1,683 bp (*ccrC* in *S. saprophyticus* strain ATCC10350), *ccrC* so far contains only 1 subgroup, *ccrC1*.

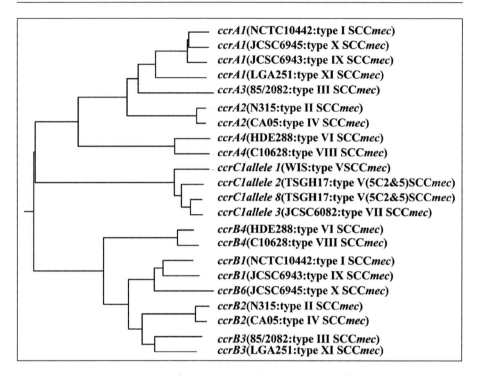

Figure 4. Phylogenetic relations of *ccr* genes identified in 11 types of SCC*mec* elements. The nucleotide sequences of 22 *ccr* genes (9 *ccrA* genes, 9 *ccrB* genes, 4 *ccrC* genes) were aligned by using the ClustalW program. In parenthesis, names of *S. aureus* strains as well as SCC*mec* types carrying *ccr* genes are indicated. Ten nucleotide sequences were used for *ccrA* and *ccrB* genes:*ccrA1ccrB1*, NCTC10442 (AB033763), JCSC6943 (AB505628):*ccrA2ccrB2*, N 315 (D86934), CA05 (AB063172); *ccrA3ccrB3*, 85/2082 (AB037671); *ccrA4ccrB4*,HDE288 (AF411935), and C10684 (FJ390057). Four nucleotide sequences were used for *ccrC* genes: WIS (AB121219), TSGH17 (AB512767), and JCSC6082 (AB373032). Nucleotide sequences of JCSC6945 (AB505630) and LGA251(FR821779.1) were used for *ccrA1ccrB6* and *ccrA1ccrB3*, respectively.

The *ccr* genes play a decisive role in the mobility of SCC*mec*. They catalyze precise excision as well as site- and orientation- specific integration of SCC*mec*. When plasmid pSR carrying *ccrA* and *ccrB* genes were introduced into an *S. aureus* strain N315, the loss of SCC*mec* occurred at a very high rate (Fig. 5). The SCC*mec* was precisely excised from the integration site of the chromosome. However, neither plasmid carrying either one of the *ccr* genes increase the rate of precise excision,[10] suggesting that both genes are required for precise excision. Upon precise excision of SCC*mec* element, a closed-circular intermediate that was formed by ligation of both termini of the element was observed similar to the case of transposon Tn*916*.[10] The closed circular intermediate could not replicate, and was lost from the cell.[10] To determine if SCC*mec* can be integrated into the chromosome from the closed circular form, miniSCC, a plasmid carrying attSCC (the sequence formed by the end to end ligation of both termini), was constructed and introduced into N315ex, a derivative of N315, from which SCC*mec* has been excised. The miniSCC was integrated into the N315ex chromosome site and orientation specific manner (Fig. 5). *ccrC*, can mediate both precise excision and site- as well as orientation specific integration. An experiment showed that a 500-bp stretch of DNA containing the sequence made by head-to-head ligation of the termini of the SCC*mec* element serves as a substrate for the *ccr*-mediated integration event.[10] Upon integration of SCC elements, the nucleotide sequence at the 3′ end of *orfX* was replaced by similar sequence

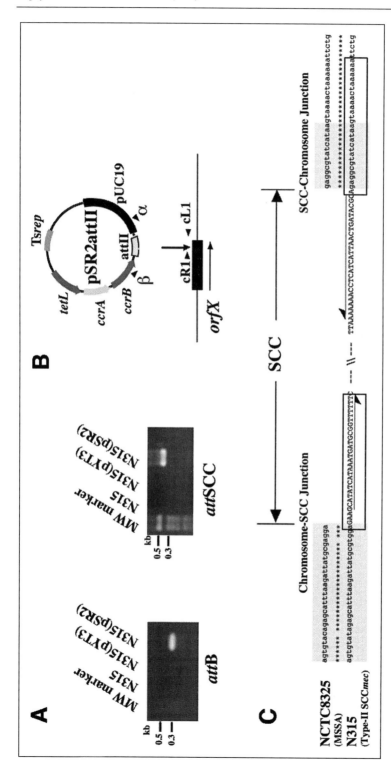

Figure 5. Precise excision and site- and orientation- specific integration of SCC*mec*. (A) PCR identification of the precise excision and closed circular formation.[10,12] (B) Integration of plasmid pSR2attII. Plasmid pSR2attII was constructed as miniSCC carrying *ccrAccrB* genes and attSCC that was located in closed-circular form generated upon excision. (C) Nucleotide sequences of the both extremities of SCC*mec*. The locations of 29 bp-oligonucleotides (aGAAGCAATATCATAAATGATGCGGTTTTT, Agaggcgtatcataagtaaaactaaaaa) containing ISS, which were proved to bind to Ccr proteins were boxed (Mizutani R, unpublished observation).

located at the terminus of the SCC*mec* element, thus direct repeats ISS1 and ISS2 containing the identical ISS are generated at both ends of the integrated copy of the SCC*mec*. Another SCC can integrate itself into the chromosome flanking the previously integrated copy of SCC by recognizing either one of the ISS1 or ISS2. In such a case, two SCC elements will be arranged in tandem. The binding sites for Ccr proteins have been investigated by constructing recombinant proteins (Mizutani R unpublished data.[41] Recombinant Ccr proteins bind to synthetic oligonucleotides containing ISS (Mizutani R unpublished data.[41]

SCC*mec* replicates as a part of the chromosome, and the methicillin resistance is stably transmitted into the daughter cells. However methicillin resistance is occasionally lost by spontaneous loss of *mecA* from the chromosome. This occurs during prolonged storage in semi-solid agar or during serial passage. Although strict quantitative data are not available, IS*431*-mediated deletion of the chromosomal region containing *mecA* seems to occur far more frequently than the *ccr*-mediated precise excision of SCC*mec*. The former type of deletion characteristically starts at either side of the IS*431mec* and the size of the deletion ranges up to 50 to 100 kb.[42,43] The *ccr*-mediated spontaneous excision of SCC*mec* seems to occur at much lower rates than the IS-mediated *mecA* deletion. There may be regulation of *ccr* gene expression that restricts frequent loss of SCC*mec* once integrated in the host cell chromosome.

SCC*mec* as a Vehicle for Multi-Drug Resistance

SCC*mec* elements so far identified in *S. aureus* are classified into 11 types based on the combination of *mec* gene complex and the *ccr* gene complex (Fig. 6).[9,11-13,15,16,27,28,36,44,45] The updated list and the classification of SCC*mec* are available at: http://www.SCC*mec*. or http://www.staphylococcus.net. Types I-VIII SCC*mec* elements were identified from MRSA strains isolated from human. Types IX, X, and XI SCC*mec* and a Type V(5&5C2) were identified from livestock-associated MRSA (LA-MRSA) strains that recently emerged and spread worldwide.

The sub-regions other than *mec* gene complex and *ccr* gene complex in SCC*mec* elements are called J (Joining) regions, which are classified into J1, J2, and J3 (Fig. 1). Many resistance determinants are identified in these regions (Table 1). The genes encoding antibiotic resistance are present at integrated copies of plasmids or transposons. The plasmid pUB110, a small-sized plasmid that encodes kanamycin and tobramycin resistance gene *aadD* and bleomycin resistance *ble*, is identified frequently in the Type II SCC*mec* element and occasionally in the Types I and IV SCC*mec* elements. The plasmid pT181, a small-sized plasmid that encodes tetracycline resistance gene *tetK*, is identified in the majority of Type III SCC*mec* element and Type V(5C2&5)c SCC*mec*. The integrated plasmids are bracketed by two copies of IS*431*. They seem to have been inserted into SCC*mec* via homologous recombination across the IS*431mec* and an IS*431* on the plasmid. Transposon Tn*554* that encodes erythromycin resistance gene *ermA* and spectinomycin resistance gene *spc* is identified in Types II and VIII SCC*mec*s, and ΨTn*554* that encodes cadmium resistance is identified in Type III SCC*mec*.

Heavy metal resistance genes are also found integrated in SCC elements either by transposition or by homologous recombination. The mercury resistance operon, which is found in SCC*Hg*, is bracketed by IS*431* and ΨTn*554* encoding cadmium resistance, indicating the contribution of either one or both of the mobile genetic elements. On the other hand, SCC*mec* elements of LA-MRSA strains carry heavy metal resistance genes, which are not associated with insertion sequences or transposons. The *cadDX* operon, which is composed of *cadD* (cadmium resistant transporter) and *cadX* (regulatory protein), is carried by two SCC*mec* elements, Type IX in JCSC6943 and Type X in JCSC6945. Both *cadD* and *cadX* genes are closely related to genes in ΨSCC*mec* of *S. hemolyticus* JCSC1435, and distantly related to *cadDX* genes identified in the chromosome or plasmids in *S. aureus*. A novel gene, *czrC* (cadmium zinc resistance C), which was shown to be responsible for cadmium and zinc resistance by Cavaco et al.,[48] is carried by Type V(5C2&5)c in strain JCSC6944. *copB* gene that may be associated with copper resistance is identified in Types IX and X SCC*mec* elements. The resistance genes for arsenate are carried by three SCC*mec* elements. Type IX SCC*mec* in JCSC6945 carry two arsenate resistance operons,

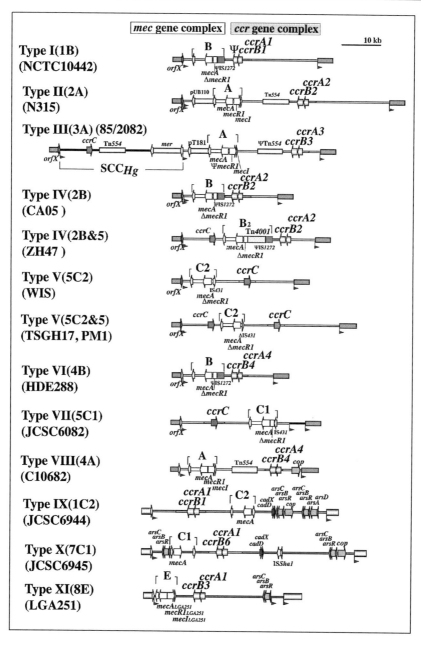

Figure 6. Schematic representation of I-XI SCC*mec* types. The structures of SCC*mec* elements of representative strains are illustrated based on the nucleotide sequences deposited in the databases; NCTC10442 (AB033763), N315 (D86934), 85/2082 (AB037671), CA05 (AB063172), ZH47 (AM292304), WIS (AB121219), TSGH17 (AB512767), PM1 (AB462393), HDE288 (AF411935), JCSC6082 (AB373032), C10684 (FJ390057), JCSC6943 (AB505628), JCSC6945 (AB505630), and LGA251(FR821779.1) Dark gray (red) arrowheads indicate the integration site sequences of SCC that comprise direct repeats. A color version of this figure is available online at www.landesbioscience.com/curie.

Table 1. Resistance determinant in SCCmec elements identified in S. aureus

	Antibiotic Resistance Genes Other Than *mecA*						Heavy Metal Resistance Genes			
	aadD and *ble* in pUB110	*tetK* in pT181	*ermA* and *spc* inTn554	*cadD* in ΨTn554	*fur*	*merA* in Mercury Resistance Operon	*copB*	*czrC*	*arsC* in Arsenate Resistance Operon	References
SCCmec Elements										
Type I	(+)									18
Type II	+		+							10,11
Type III		+	(+)	+						18
Type IV	(+)									19
Type V										9
Type V (5C2&5)										27,51
Type VI					+					45
Type VII										44
Type VIII			+				+			24
Type V (5C2&5)		+						+		46
Type IX							+		+	46
Type X							+		+	46
Type XI					+				+	47
SCC Elements										
SCCHg			+			+				18
SCCfar										18

(+) signify that the ratio of the carriage is minor.

arsR (regulatory protein*)-arsB*(a membrane-binding protein)-*arsC*(arsenate reductase), and *arsD*(regulatory protein*)-arsA*(ATPase subunit)-*arsR-arsB-arsC*. Type IX and XI SCC*mec*s carry the operon composed of the three genes; *arsR-arsB-arsC*.

The SCC*mec* elements of LA-MRSA strains so far analyzed carried at least one gene associated with heavy metal detoxification. In contrast, only one SCC*mec* element (Type VIII) identified in the human isolate carried a gene of the kind, i.e, *copB* gene encoding copper resistance. It seems that the SCC*mec* elements of LA-MRSA strains reflect the kinds of drugs used for livestock farming. Heavy metals have been widely used as growth promoters of livestocks.[49]

Horizontal Transfer and Molecular Evolution of SCC*mec*

Recent identification of *mecA* locus in *S. fleurettii* chromosome, and finding of novel *mecA* homologs, *mecA*$_{LGA251}$ in LA-MRSA and *mecAm* in macrococci, reinforced the importance of active intra-, and inter-species transfer of *mecA* gene homologs and SCC*mec* among staphylococci and related bacterial species. Nucleotide sequence comparison provided circumstantial evidence for horizontal transfer of SCC*mec* and its components. Inter-species transfer is indicated, because the same SCC*mec* element is found carried by different staphylococcal species, e.g., Type IV SCC*mec* in *S. aureus* and *S. epidermidis* strains,[50] and Type V(5C2&5) SCC*mec* shared by *S. aureus* and *S. pseudointermedius* strains.[51] In addition, there are some reports suggesting the transfer of SCC*mec* element from *S. epidermidis* to *S. aureus*[52] and *S. hemolyticus* to *S. aureus*.[53] Intra-species transfer is also likely, since the same SCC*mec* element is found in MRSA strains of different sequence types (ST) using Multilocus Sequence Typing (MLST):[47] e.g., Type IVa SCC*mec* is carried by ST1 MRSA (USA400 clone) and ST8 MRSA (USA300 clone); and TypeV(5C2&5) SCC*mec* is carried by MRSA strains of ST59 and ST91.[20,54-57]

The mechanism, through which SCC*mec* elements were transferred, is not clear. Sjostorm reported successful transformation experiment of methicillin resistance.[58] Transfer of methicillin resistance by phage transduction has also been reported.[59,60] The transduction experiment has been confirmed by Vaudaux et al.[61] We also confirmed their experiment: Type I SCC*mec* could be transferred successfully from MRSA COL to MSSA RN4220 (T. Ito unpublished data). However, transduction did not transfer a large-sized (> 45 kb) SCC*mec* that exceeded the size of phage genome. Therefore, it is not likely that such SCC*mec* elements as Type-II and Type V(5C2&5) were acquired by single transduction events. Other possibilities need to be explored. It may be worthwhile thinking about a possible contribution of conjugative plasmids as a carrier. There were some reports that methicillin resistance was carried by plasmids.[46,62,63] Trees et al. reported Tn*4291* encoding methicillin resistance was located on the plasmid.[64] Although these reports have been not paid attention for a long time, the possibility is not entirely excluded since a *mecA* gene homolog, *mecAm* is carried by *Macrococcus* plasmids.[25] If transduction did serve for the horizontal transfer for *mecA* gene, it must have established only a small-sized SCC*mec* element on the *oriC* environ of MSSA chromosome. Subsequent multiple recombination events should have shaped the large SCC*mec* elements such as Type-II and Type V(5C2&5) SCC*mec*.

Shaping a New SCC*mec*: Recombination and Fusion of SCC Elements

Combination of *mec* and *ccr* gene complexes defines a type of SCC*mec*. Therefore recombination between the *mec* and *ccr* gene complexes diversifies SCC*mec*. For example, practically identical class B *mec* gene complex is found in Types I, IV, and VI SCC*mec* elements (Fig. 6). The class C2 *mec* gene complex is carried by several Type V SCC*mec* elements as well as by ΨSCC*mec*1435 identified in *S. hemolyticus*. So far five *mec* gene complexes and eight *ccr* gene complex are combined to generate 11 SCC*mec* types (Fig. 6). It is likely that new types of SCC*mec* with novel combinations of *mec* and *ccr* gene complexes will be found in the future. Not only the *mec* and *ccr* gene complexes, the other parts, sub-regions, of SCC*mec* are distributed across different types of SCC*mec* elements. For example some distinct types of SCC*mec* elements share highly homologous J1 region: those of Type I.2 SCC*mec* in strain PL72, Type II.5 SCC*mec* in JCSC6833, and Type IVj SCC*mec* in JCSC6670; and those of Type II.2 SCC*mec* in JCSC3063 and Type IVi SCC*mec* in JCSC6668.[65,66]

It is also remarkable that the part of J1 region containing *arsRBC* operon in Type IX SC*Cmec* is highly homologous to that identified in some SCC elements in *S. hemolyticus*.

Another way to generate novel SC*Cmec* is through fusion of the two SCC elements; e.g., the fusion between an SC*Cmec* and an SCC or a ΨSCC element integrated side-by-side with the SC*Cmec*. A composite SC*Cmec* is produced by the fusion event. Type IV(2B&5) SC*Cmec* or Type V(5C2&5) SC*Cmec*, are the examples (Fig. 6). These big SC*Cmec* elements are considered as composite SC*Cmec* elements. The elements carry two sets of *ccr* genes while there is no direct and inverted repeats in between the two *ccr* gene complexes. It is speculated that an SCC element driven by *ccrC* and Type V SC*Cmec* were integrated side-by-side then the direct and inverted repeats that demarcate the two elements were deleted to generate the composite SC*Cmec*. The Type V(5C2&5) SC*Cmec* of strain JCSC6944 (43.381 bp) also seems to be a composite SC*Cmec* (Fig. 6). It is made up with an SCC carrying *ccrC1* allele2 (10 kb) and a Type V SC*Cmec* carrying *ccrC1* allele8 and class C2 *mec* gene complex (17 kb). It is very similar to Type V (5C2&5) SC*Cmec* found in Taiwanese PVL-positive ST59 MRSA strains, TSGH17 and PM1, except for their J1 region.

The biological effect of the fusion is unclear. It is possible that the loss of the repeat sequences containing ISS would protect the entire oriC environ from frequent *ccr*-mediated recombination and loss of the useful exogenous genes. It is curious in this regard that homologous recombination between the two copies of *ccrC* genes occurred resulting in the loss of *mecA* gene from Type V (5C2&5) SC*Cmec*.[67] Clearly, such a recombination further diversifies the structure of SCC elements in the oriC environ generating novel SCC or SC*Cmec* elements.

Conclusion

SC*Cmec* is a mobile genetic element SCC that carries the methicillin-resistance gene *mecA*. SCC serves as a vehicle not only for *mecA* but also for any genes useful for the survival of staphylococcal strains. SCC elements are found as genomic islands integrated near the oriC of staphylococcal chromosomes. SCC is a vehicle specifically generated for the staphylococcal and related bacterial species. It serves as an efficient tool for the staphylococcal strains to acquire useful genes of exogenous origin. The versatile nature of *S. aureus* as a hospital pathogen is greatly dependent on the SCC gene transfer system. We have only started tracing the traffic of SCC among staphylococcal species. Understanding the mechanism of the SCC gene transfer system would greatly contribute to the prevention and control of infections caused by *S. aureus*.

References

1. Jevons MP. "Celbenin"-resistant staphylococci. BMJ 1961; 124:124-5; http://dx.doi.org/10.1136/bmj.1.5219.124-a.
2. Hartman BJ, Tomasz A. Low-affinity penicillin-binding protein associated with beta-lactam resistance in Staphylococcus aureus. J Bacteriol 1984; 158:513-6; PMID:6563036.
3. Reynolds PE, Brown DF. Penicillin-binding proteins of beta-lactam-resistant strains of Staphylococcus aureus. Effect of growth conditions. FEBS Lett 1985; 192:28-32; PMID:3850810; http://dx.doi.org/10.1016/0014-5793(85)80036-3.
4. Utsui Y, Yokota T. Role of an altered penicillin-binding protein in methicillin- and cephem-resistant Staphylococcus aureus. Antimicrob Agents Chemother 1985; 28:397-403; PMID:3878127.
5. Song MD, Wachi M, Doi M, et al. Evolution of an inducible penicillin-target protein in methicillin-resistant Staphylococcus aureus by gene fusion. FEBS Lett 1987; 221:167-71; PMID:3305073; http://dx.doi.org/10.1016/0014-5793(87)80373-3.
6. Ryffel C, Tesch W, Birch-Machin I, et al. Sequence comparison of mecA genes isolated from methicillin-resistant Staphylococcus aureus and Staphylococcus epidermidis. Gene 1990; 94:137-8; PMID:2227446; http://dx.doi.org/10.1016/0378-1119(90)90481-6.
7. Suzuki E, Hiramatsu K, Yokota T. Survey of methicillin-resistant clinical strains of coagulase-negative staphylococci of mecA gene distribution. Antimicrob Agents Chemother 1992; 36:429-34; PMID:1605606.
8. Ubukata K, Nonoguchi R, Song MD, et al. Homology of mecA gene in methicillin-resistant Staphylococcus haemolyticus and Staphylococcus simulans to that of Staphylococcus aureus. Antimicrob Agents Chemother 1990; 34:170-2; PMID:1691614.

9. Ito T, Katayama Y, Hiramatsu K. Cloning and nucleotide sequence determination of the entire mec DNA of pre-methicillin-resistant Staphylococcus aureus N315. Antimicrob Agents Chemother 1999; 43:1449-58; PMID:10348769.

10. Katayama Y, Ito T, Hiramatsu K. A new class of genetic element, staphylococcal cassette chromosome mec, encodes methicillin resistance in Staphylococcus aureus. Antimicrob Agents Chemother 2000; 44:1549-55; PMID:10817707; http://dx.doi.org/10.1128/AAC.44.6.1549-1555.2000.

11. Ito T, Katayama Y, Asada K, et al. Structural comparison of three Types of staphylococcal cassette chromosome mec integrated in the chromosome in methicillin-resistant Staphylococcus aureus. Antimicrob Agents Chemother 2001; 45:1323-36; PMID:11302791; http://dx.doi.org/10.1128/AAC.45.5.1323-1336.2001.

12. Ito T, Ma XX, Takeuchi F, et al. Novel Type V staphylococcal cassette chromosome mec driven by a novel cassette chromosome recombinase, ccrC. Antimicrob Agents Chemother 2004; 48:2637-51; PMID:15215121; http://dx.doi.org/10.1128/AAC.48.7.2637-2651.2004.

13. Ma XX, Ito T, Tiensasitorn C, et al. Novel type of staphylococcal cassette chromosome mec identified in community-acquired methicillin-resistant Staphylococcus aureus strains. Antimicrob Agents Chemother 2002; 46:1147-52; PMID:11897611; http://dx.doi.org/10.1128/AAC.46.4.1147-1152.2002.

14. Oliveira DC, Tomasz A, Lencastre H. The evolution of pandemic clones of methicillin-resistant Staphylococcus aureus: identification of two ancestral genetic backgrounds and the associated mec elements. Microb Drug Resist 2001; 7:349-61; PMID:11822775; http://dx.doi.org/10.1089/10766290152773365.

15. Zhang K, McClure JA, Elsayed S, Conly JM. Novel staphylococcal cassette chromosome mec type, tentatively designated type VIII, harboring class A mec and type 4 ccr gene complexes in a Canadian epidemic strain of methicillin-resistant Staphylococcus aureus. Antimicrob Agents Chemother 2009; 53:531-40; PMID:19064897; http://dx.doi.org/10.1128/AAC.01118-08.

16. Classification of staphylococcal cassette chromosome mec (SCCmec): guidelines for reporting novel SCCmec elements. Antimicrob Agents Chemother 2009; 53:4961-7; PMID:19721075; http://dx.doi.org/10.1128/AAC.00579-09.

17. Chongtrakool P, Ito T, Ma XX, et al. SCCmec typing of MRSA strains isolated in eleven Asian countries - a proposal for a new nomenclature for SCCmec elements-. Antimicrob Agents Chemother 2006; 50:1001-12; PMID:16495263; http://dx.doi.org/10.1128/AAC.50.3.1001-1012.2006.

18. Holden MT, Feil EJ, Lindsay JA, et al. Complete genomes of two clinical Staphylococcus aureus strains: evidence for the rapid evolution of virulence and drug resistance. Proc Natl Acad Sci USA 2004; 101:9786-91; PMID:15213324; http://dx.doi.org/10.1073/pnas.0402521101.

19. Luong TT, Ouyang S, Bush K, Lee CY. Type 1 capsule genes of Staphylococcus aureus are carried in a staphylococcal cassette chromosome genetic element. J Bacteriol 2002; 184:3623-9; PMID:12057957; http://dx.doi.org/10.1128/JB.184.13.3623-3629.2002.

20. Diep BA, Gill SR, Chang RF, et al. Complete genome sequence of USA300, an epidemic clone of community-acquired meticillin-resistant Staphylococcus aureus. Lancet 2006; 367:731-9; PMID:16517273; http://dx.doi.org/10.1016/S0140-6736(06)68231-7.

21. Takeuchi F, Watanabe S, Baba T, et al. Whole-genome sequencing of staphylococcus haemolyticus uncovers the extreme plasticity of its genome and the evolution of human-colonizing staphylococcal species. J Bacteriol 2005; 187:7292-308; PMID:16237012; http://dx.doi.org/10.1128/JB.187.21.7292-7308.2005.

22. Kuroda M, Ohta T, Uchiyama I, et al. Whole genome sequencing of meticillin-resistant Staphylococcus aureus. Lancet 2001; 357:1225-40; PMID:11418146; http://dx.doi.org/10.1016/S0140-6736(00)04403-2.

23. Baba T, Takeuchi F, Kuroda M, et al eds. The Genome of Staphylococcus aureus. London: Eliis Harwood, 2004.

24. Wu S, Piscitelli C, de Lencastre H, Tomasz A. Tracking the evolutionary origin of the methicillin resistance gene: cloning and sequencing of a homologue of mecA from a methicillin susceptible strain of Staphylococcus sciuri. Microb Drug Resist 1996; 2:435-41; PMID:9158816; http://dx.doi.org/10.1089/mdr.1996.2.435.

25. Baba T, Kuwahara-Arai K, Uchiyama I, et al. Complete genome sequence of Macrococcus caseolyticus strain JCSC5402, reflecting the ancestral genome of the human-pathogenic staphylococci. J Bacteriol 2009; 191:1180-90; PMID:19074389; http://dx.doi.org/10.1128/JB.01058-08.

26. Couto I, de Lencastre H, Severina E, et al. Ubiquitous presence of a mecA homologue in natural isolates of Staphylococcus sciuri. Microb Drug Resist 1996; 2:377-91; PMID:9158808; http://dx.doi.org/10.1089/mdr.1996.2.377.

27. García-Álvarez L, Holden MT, Lindsay H, et al. Meticillin-resistant Staphylococcus aureus with a novel mecA homologue in human and bovine populations in the UK and Denmark: a descriptive study. Lancet Infect Dis 2011; 11:595-603; PMID:21641281; http://dx.doi.org/10.1016/S1473-3099(11)70126-8.

28. Shore AC, Deasy EC, Slickers P, et al. Detection of Staphylococcal Cassette Chromosome mec Type XI Carrying Highly Divergent mecA, mecI, mecR1, blaZ, and ccr Genes in Human Clinical Isolates of Clonal Complex 130 Methicillin-Resistant Staphylococcus aureus. Antimicrob Agents Chemother 2011; 55:3765-73; PMID:21636525; http://dx.doi.org/10.1128/AAC.00187-11.

29. Wu SW, de Lencastre H, Tomasz A. Recruitment of the mecA gene homologue of Staphylococcus sciuri into a resistance determinant and expression of the resistant phenotype in Staphylococcus aureus. J Bacteriol 2001; 183:2417-24; PMID:11274099; http://dx.doi.org/10.1128/JB.183.8.2417-2424.2001.

30. Katayama Y, Ito T, Hiramatsu K. Genetic organization of the chromosome region surrounding mecA in clinical Staphylococcal strains: role of IS431-mediated mecI deletion in expression of resistance in mecA-carrying, low-level methicillin- resistant Staphylococcus haemolyticus. Antimicrob Agents Chemother 2001; 45:1955-63; PMID:11408208; http://dx.doi.org/10.1128/AAC.45.7.1955-1963.2001.

31. Barberis-Maino L, Berger-Bachi B, Weber H, et al. IS431, a staphylococcal insertion sequence-like element related to IS26 from Proteus vulgaris. Gene 1987; 59:107-13; PMID:2830163; http://dx.doi.org/10.1016/0378-1119(87)90271-X.

32. Ryffel C, Bucher R, Kayser FH, Burger-Bachi B. The Staphylococcus aureus mec determinant comprises an unusual cluster of direct repeats and codes for gene product similar to the Escherichia coli sn-glycerophosphoryl diester phosphodiesterase. J Bacteriol 1991; 173:7416-22; PMID:1718947.

33. Goering RV, Morrison D, Al-Doori Z, et al. Usefulness of mec-associated direct repeat unit (dru) typing in the epidemiological analysis of highly clonal methicillin-resistant Staphylococcus aureus in Scotland. Clin Microbiol Infect 2008; 14:964-9; PMID:18828855; http://dx.doi.org/10.1111/j.1469-0691.2008.02073.x.

34. Suzuki E, Kuwahara-Arai K, Richardson JF, Hiramatsu K. Distribution of mec regulator genes in methicillin-resistant Staphylococcus clinical strains. Antimicrob Agents Chemother 1993; 37:1219-26; PMID:8328773.

35. Watanabe S, Ito T, Morimoto Y, et al. Precise excision and self-integration of a composite transposon as a model for spontaneous large-scale chromosome inversion/deletion of the Staphylococcus haemolyticus clinical strain JCSC1435. J Bacteriol 2007; 189:2921-5; PMID:17237177; http://dx.doi.org/10.1128/JB.01485-06.

36. Li S, Skov RL, Han X, et al. Novel types of staphylococcal cassette chromosome mec elements identified in clonal complex 398 methicillin-resistant Staphylococcus aureus strains. Antimicrob Agents Chemother 2011; 55:3046-50; PMID:21422209; http://dx.doi.org/10.1128/AAC.01475-10.

37. Tsubakishita S, Kuwahara-Arai K, Sasaki T, Hiramatsu K. Origin and molecular evolution of the determinant of methicillin resistance in staphylococci. Antimicrob Agents Chemother 2010; 54:4352-9; PMID:20679504; http://dx.doi.org/10.1128/AAC.00356-10.

38. Queck SY, Khan BA, Wang R, et al. Mobile genetic element-encoded cytolysin connects virulence to methicillin resistance in MRSA. PLoS Pathog 2009; 5:e1000533; PMID:19649313; http://dx.doi.org/10.1371/journal.ppat.1000533.

39. Descloux S, Rossano A, Perreten V. Characterization of new staphylococcal cassette chromosome mec (SCCmec) and topoisomerase genes in fluoroquinolone- and methicillin-resistant Staphylococcus pseudintermedius. J Clin Microbiol 2008; 46:1818-23; PMID:18305127; http://dx.doi.org/10.1128/JCM.02255-07.

40. Tsubakishita S, Kuwahara-Arai K, Baba T, Hiramatsu K. Staphylococcal cassette chromosome mec-like element in Macrococcus caseolyticus. Antimicrob Agents Chemother 2010; 54:1469-75; PMID:20086147; http://dx.doi.org/10.1128/AAC.00575-09.

41. Wang L, Archer GL. Roles of CcrA and CcrB in excision and integration of staphylococcal cassette chromosome mec, a Staphylococcus aureus genomic island. J Bacteriol 2010; 192:3204-12; PMID:20382769; http://dx.doi.org/10.1128/JB.01520-09.

42. Cohen S, Sweeney HM. Modulation of protein A formation in Staphylococcus aureus by genetic determinants for methicillin resistance. J Bacteriol 1979; 140:1028-35; PMID:533763.

43. Wada K, Wada Y, Doi H, et al. Codon usage tabulated from the GenBank genetic sequence data. Nucleic Acids Res 1991; 19:1981-6; PMID:2041796.

44. Berglund C, Ito T, Ikeda M, et al. Novel type of staphylococcal cassette chromosome mec in a methicillin-resistant Staphylococcus aureus strain isolated in Sweden. Antimicrob Agents Chemother 2008; 52:3512-6; PMID:18676883; http://dx.doi.org/10.1128/AAC.00087-08.

45. Oliveira DC, Milheirico C, de Lencastre H. Redefining a structural variant of staphylococcal cassette chromosome mec, SCCmec type VI. Antimicrob Agents Chemother 2006; 50:3457-9; PMID:17005831; http://dx.doi.org/10.1128/AAC.00629-06.

46. Lacey RW. Genetic control in methicillin-resistant strains of Staphylococcus aureus. J Med Microbiol 1972; 5:497-508; PMID:4486061; http://dx.doi.org/10.1099/00222615-5-4-497.

47. Enright MC, Day NP, Davies CE, et al. Multilocus sequence typing for characterization of methicillin-resistant and methicillin-susceptible clones of Staphylococcus aureus. J Clin Microbiol 2000; 38:1008-15; PMID:10698988.

48. Cavaco LM, Hasman H, Stegger M, et al. Cloning and occurrence of czrC, a gene conferring cadmium and zinc resistance in methicillin-resistant Staphylococcus aureus CC398 isolates. Antimicrob Agents Chemother 2010; 54:3605-8; PMID:20585119; http://dx.doi.org/10.1128/AAC.00058-10.

49. Hasman HFS, Rensing C, eds. Resistance to metals used in agricultural production. Washington DC: ASM, 2006.

50. Wisplinghoff H, Rosato AE, Enright MC, et al. Related clones containing SCCmec type IV predominate among clinically significant Staphylococcus epidermidis isolates. Antimicrob Agents Chemother 2003; 47:3574-9; PMID:14576120; http://dx.doi.org/10.1128/AAC.47.11.3574-3579.2003.

51. Black CC, Solyman SM, Eberlein LC, et al. Identification of a predominant multilocus sequence type, pulsed-field gel electrophoresis cluster, and novel staphylococcal chromosomal cassette in clinical isolates of mecA-containing, methicillin-resistant Staphylococcus pseudintermedius. Vet Microbiol 2009; 139:333-8; PMID:19604657; http://dx.doi.org/10.1016/j.vetmic.2009.06.029.

52. Bloemendaal AL, Brouwer EC, Fluit AC. Methicillin resistance transfer from Staphylocccus epidermidis to methicillin-susceptible Staphylococcus aureus in a patient during antibiotic therapy. PLoS ONE 2010; 5:e11841; PMID:20686601; http://dx.doi.org/10.1371/journal.pone.0011841.

53. Berglund C, Soderquist B. The origin of a methicillin-resistant Staphylococcus aureus isolate at a neonatal ward in Sweden-possible horizontal transfer of a staphylococcal cassette chromosome mec between methicillin-resistant Staphylococcus haemolyticus and Staphylococcus aureus. Clin Microbiol Infect 2008; 14:1048-56; PMID:19040477; http://dx.doi.org/10.1111/j.1469-0691.2008.02090.x.

54. Baba T, Takeuchi F, Kuroda M, et al. Genome and virulence determinants of high virulence community-acquired MRSA. Lancet 2002; 359:1819-27; PMID:12044378; http://dx.doi.org/10.1016/S0140-6736(02)08713-5.

55. Boyle-Vavra S, Ereshefsky B, Wang CC, Daum RS. Successful multiresistant community-associated methicillin-resistant Staphylococcus aureus lineage from Taipei, Taiwan, that carries either the novel Staphylococcal chromosome cassette mec (SCCmec) type VT or SCCmec type IV. J Clin Microbiol 2005; 43:4719-30; PMID:16145133; http://dx.doi.org/10.1128/JCM.43.9.4719-4730.2005.

56. Hisata K, Ito T, Matsunaga N, et al. Dissemination of multiple MRSA clones among community-associated methicillin-resistant Staphylococcus aureus infections from Japanese children with impetigo. J Infect Chemother 2011; 17:609-21; PMID:21327935; http://dx.doi.org/10.1007/s10156-011-0223-4.

57. Takano T, Higuchi W, Otsuka T, et al. Novel characteristics of community-acquired methicillin-resistant Staphylococcus aureus strains belonging to multilocus sequence type 59 in Taiwan. Antimicrob Agents Chemother 2008; 52:837-45; PMID:18086843; http://dx.doi.org/10.1128/AAC.01001-07.

58. Sjöström JE, Lofdahl S, Philipson L. Transformation reveals a chromosomal locus of the gene(s) for methicillin resistance in Staphylococcus aureus. J Bacteriol 1975; 123:905-15; PMID:125746.

59. Cohen S, Sweeney HM. Transduction of methicillin resistance in Staphylococcus aureus dependent on an unusual specificity of the recipient strain. J Bacteriol 1970; 104:1158-67; PMID:16559089.

60. Stewart GC, Rosenblum ED. Transduction of methicillin resistance in Staphylococcus aureus:Recipient effectiveness and Beta-lactamase production. Antimicrob Agents Chemother 1980; 18(3):424-32; PMID:6448580; PMID:6448580

61. Vaudaux PE, Monzillo V, Francois P, et al. Introduction of the mec element (methicillin resistance) into Staphylococcus aureus alters in vitro functional activities of fibrinogen and fibronectin adhesins. Antimicrob Agents Chemother 1998; 42:564-70; PMID:9517933.

62. Annear DI, Grubb WB. Methicillin-sensitive variants in ageing broth cultures of methicillin-resistant Staphylococcus aureus. Pathology 1976; 8:69-72; PMID:972766; http://dx.doi.org/10.3109/00313027609094426.

63. Dornbusch K, Hallander HO. Transduction of penicillinase production and methicillin resistance-enterotoxin B production in strains of Staphylococcus aureus. J Gen Microbiol 1973; 76:1-11; PMID:4489832.

64. Trees DL, Iandolo JJ. Identification of a Staphylococcus aureus transposon (Tn4291) that carries the methicillin resistance gene(s). J Bacteriol 1988; 170:149-54; PMID:2826391.

65. Berglund C, Ito T, Ma XX, et al. Genetic diversity of methicillin-resistant Staphylococcus aureus carrying type IV SCCmec in Orebro County and the western region of Sweden. J Antimicrob Chemother 2009; 63:32-41; PMID:19001453; http://dx.doi.org/10.1093/jac/dkn435.

66. Han X, Ito T, Takeuchi F, et al. Identification of a novel variant of staphylococcal cassette chromosome mec, type II.5, and Its truncated form by insertion of putative conjugative transposon Tn6012. Antimicrob Agents Chemother 2009; 53:2616-9; PMID:19364875; http://dx.doi.org/10.1128/AAC.00772-08.

67. Chlebowicz MA, Nganou K, Kozytska S, et al. Recombination between ccrC genes in a type V (5C2&5) staphylococcal cassette chromosome mec (SCCmec) of Staphylococcus aureus ST398 leads to conversion from methicillin resistance to methicillin susceptibility in vivo. Antimicrob Agents Chemother 2010; 54:783-91; PMID:19995931; http://dx.doi.org/10.1128/AAC.00696-09.

INDEX